ELEMENTARY
GENERAL THERMODYNAMICS

Perpetual Waterfall, by Maurits Escher. Original painting in the private collection of Cornelius Roosevelt; photograph reproduced by courtesy of Mr. Roosevelt.

MARTIN V. SUSSMAN
Tufts University

ELEMENTARY
GENERAL THERMODYNAMICS

ADDISON-WESLEY PUBLISHING COMPANY

Reading, Massachusetts · Menlo Park, California · London · Don Mills, Ontario

This book is in the

ADDISON-WESLEY SERIES IN CHEMICAL ENGINEERING

Harold E. Hoelscher, Consulting Editor

To J., A., E., and D.

PREFACE

The teaching of thermodynamics has been changing in recent years in my university and in many others throughout the United States. The most important changes include the following:

The subject, formerly studied in the junior and senior years, is now taught more frequently to students in their sophomore and even freshman years, as part of core science curricula.

There is a movement away from specialized presentations. Heat–power approaches to thermodynamics are disappearing. Chemical engineering texts are meeting competition from broader, more basic texts. The *universal* rather than the special applicability of thermodynamics is being stressed.

The concepts of statistical and irreversible thermodynamics are recognized as being useful for enriching classical concepts, and practical for determining properties and investigating coupled flow phenomena. These concepts are therefore being made part of introductory courses.

These changes create pedagogic problems. A subtle and complex subject is being presented in greater richness to younger, intellectually less-sophisticated students. The rationale for early presentation is sound, because of the universality of the subject's applications and the need to develop student deductive-logical skills as soon as possible. Thermodynamics can do the latter most effectively.

This book is designed with the above trends and pedagogic problem in mind. It presents a broad introduction to thermodynamic thought and methodology, and applications to many branches of engineering and science. It is unusual in a number of respects: in the simplicity and generality of its approach; in the smooth melding of narrative and mathematics; and in the stress on the conjectural–deductive nature of thermodynamics. Within the framework of a simple approach, the book achieves reasonable depth and sophistication, and treats generalized Gibbs equations, Maxwell relations, chemical potential, partition functions, reciprocal relations, etc.

The book is student-oriented, and tends to be a self-teaching text. Many exercises are strategically located throughout the text to enable the student to gage his comprehension. The exercises include derivations of basic equations, at points where the student has sufficient background to do the derivations himself, so that he may make these equations his own, rather than force them into more or less temporary residence in his memory. The stress on conjectural foundations and deductive structure heightens esthetic appreciation of thermodynamics. Students have enjoyed using this text, even in its earlier, mimeographed form.

The book has a cohesive structure. It is divided into eleven sections, which follow each other in what I trust is a logical sequence. The relevance of any one section to those preceding it is always explicitly stated.

Section 1 stresses the deductive nature of thermodynamics and the idea that thermodynamics relations are either abbreviated statements or necessary consequences of a few basic premises. The consequences of the first premise (energy conservation) are examined at length in Section 2; included among these consequences is thermochemistry. In Section 3 the second premise (second law) is examined, and the Carnot efficiency and entropy function are shown to be among its necessary consequences. The entropy–dependent properties (the work functions) are also introduced here. Practical power and refrigeration cycles are next concisely described in Section 4, and compared with the Carnot cycle. A treatment of statistical thermodynamics then appears in Section 5, in which the student learns the mechanical significance of entropy, so that much of the mystery surrounding that function is dispersed. The student also receives an operational—although elementary—understanding of statistical mechanics and learns how the thermodynamic properties are related to the partition function.

Having introduced the major fundamental and derived thermodynamic properties, the text then develops the necessary relationships among them in Section 6, with stress on methods of finding relationships for all properties in terms of P, V, T, and other measurable properties. As a tool for developing these relationships, the Maxwell relations are presented in concise Jacobean form: $[T, S] = [P, V]$.

Section 7 turns to the P, V, T relations (equations of state of pure substances) themselves. This section is supplemented by a short section (Section 8) on fugacity and activity. Sections 7 and 8 show how to use the equations of state to evaluate the property interrelationships developed in Section 6.

Section 9 examines the thermodynamic consequences of mixing and variation of composition, and shows these to include the possibility of obtaining work at constant P, V, and T. This leads smoothly to the concepts of chemical potential, and then to the Gibbs–Duhem equation and the phase rule. The thermodynamics of chemical reaction processes and chemical equilibrium are then presented in Section 10 as straightforward extensions of the thermodynamics of mixing.

The last section, Section 11, opens the door to newer and future thermodynamic developments by showing how thermodynamics may be extended to irreversible phenomena.

Both I and my colleagues have used mimeographed editions of this book for about three years. The text material has been presented to a variety of students: sophomore mechanical-, chemical-, civil-, and electrical-engineering students; junior chemical-engineering students; summer-school students of varied backgrounds; and summer institute participants who were high school and college chemistry teachers. The material on statistical thermodynamics formed the basis of a lecture at the June 1967 ASEE meeting and also the basis of a lecture series at the Indian Institutes of Technology, at New Delhi and Kharagpur in 1968 and 1970.

Experience with the book indicates that, for a one-semester general introductory course, the first four sections—supplemented by material drawn from Sections 5, 6, and 7 in accordance with the instructor's wishes and/or class interest—constitute a very satisfactory course package. The entire text can be covered at a comfortable pace in a two-semester course. I strongly recommend that the student do all the exercises in the text.

M.V.S.

Medford, Massachusetts
January 1972

ACKNOWLEDGMENTS

I wish to acknowledge with gratitude the assistance and suggestions of colleagues and friends in the departments of chemical and mechanical engineering of Tufts University, who used the book in mimeographed form, in particular Professors K. N. Astill, G. D. Botsaris, Ashley Campbell, and R. L. Harrington. I also wish to thank Dr. G. Hatsopoulos, who gave me generous portions of his time during formative periods of the book; Professor O. Levenspiel, who read and commented on an early version of this work, Professor R. S. Davis, who critically examined a late version; Mrs. Ellen Davis and Mrs. Alice Buttrick, for their patience and skill in typing and retyping; the many students on whom the material was honed; and A. Cerullo, Y. A. Liu, and A. Martisauskas, for assistance in preparing the manuscript.

A book of this kind is a filtration, a restructuring, and a reordering—with new nuances, embellishments, and accents—of beautiful, but largely borrowed, ideas. My borrowings have been eclectic, to an extent that may pass for research. All errors and misconceptions I claim as wholly my own.

CONTENTS

Symbols Used in This Book xvii

Section 1   Introduction

1   Introduction. 2
2   Temperature and the zeroth law 3

Section 2   Energy Conservation: the First Law of Thermodynamics

1   Energy forms 8
2   Accumulated energy 8
3   Transitory energy forms 11
4   Graphical representation of work 17
5   Reversible work processes 18
6   Efficiency 19
7   Reversible expansion and compression 19
8   Equivalence of all energy forms 21
9   Equivalence of heat and work energy 21
10   Equivalence of mass and energy 23
11   System 23
12   Surroundings 25
13   First law or the conservation of energy 27
14   The sign convention 28
15   Accumulated internal energy and the thermodynamic state . . . 29
16   U: a state property 30
17   Enthalpy 33
18   General procedure for analyzing problems 35
19   Specific heat 37
20   Phase change 41
21   Ideal gas 42
22   Reversible work: ideal-gas systems 44
23   Exact and inexact differentials; thermodynamic surfaces 50
24   Exact and inexact differentials: some mathematical characteristics . 53
25   Cyclical process 56
26   Steam: a nonideal gas 56
27   Application of the first law to irreversible processes: the free expansion
     process 61
28   The infinitesimal free expansion process 62
29   Flow processes: open systems. 63
30   Steady-flow processes 65

31   Frictionless, reversible steady-flow processes 66
32   Engineering applications 67
33   Nuclear energy 75
34   Thermochemistry 76
35   Standard enthalpies of formation 79
36   Standard enthalpy of combustion 86
37   Standard enthalpy change for any chemical reaction process at 25°C,
      1 atm. 91
38   Enthalpy of solution 95
39   Enthalpy of atomization 95
40   Enthalpy change of a chemical process at any temperature . . . 96
41   Adiabatic chemical processes: flame temperatures 100

Section 3   Entropy and the Second Law of Thermodynamics

 1   Work from heat 114
 2   The second law of thermodynamics (and spontaneous change) . . 114
 3   Heat reservoirs 115
 4   Reversibility 115
 5   Characteristics of reversible and irreversible processes 116
 6   Converting heat into work: noncyclical process 117
 7   Converting heat into work: Carnot cycle 118
 8   Thermal efficiency of a heat engine 121
 9   Refrigeration cycle 121
10   Maximum thermal efficiency 123
11   Temperature and the efficiency of a reversible engine 125
12   The thermodynamic temperature scale 126
13   Entropy 128
14   Entropy and integrating factors 131
15   Computing changes in entropy 132
16   Temperature-entropy diagram: Carnot cycle 136
17   Free expansion: entropy increase for an irreversible process . . . 137
18   Other irreversible processes 138
19   Entropy change during chemical reactions 140
20   Entropy of phase change 141
21   Trouton's law 141
22   Third law of thermodynamics: absolute entropy 141
23   Absolute entropy and molecular structure 142
24   Work functions 142
25   Gibbs free energy 143
26   dG for any system 144
27   Applications of free-energy equations 145
28   Effect of total pressure on vapor pressure: the Poynting effect . . 149
29   The Helmholtz function 150
30   Equilibrium and spontaneous change 151
31   Work functions and equilibrium 154
32   Equilibrium criteria for systems that do work other than expansion
      work 156

33 Carnot engine and the measurement of available work energy . . 157
34 Supplementary remarks on reversible work, energy and entropy change, lost work, and the Clausius inequality 158

Section 4 Power and Refrigeration Cycles

1 Gasoline engines: the Otto cycle 168
2 The Diesel engine cycle 170
3 Steam engines: the Rankine cycle 171
4 Gas turbine: the Joule cycle 174
5 Refrigeration cycles 176
6 Properties of refrigerants 176
7 Vapor-compression cycle 177
8 Air conditioners and heat pumps 178
9 Absorption refrigeration 179
10 Steam jet refrigeration 181
11 Air liquefaction 181

Section 5 Statistical Thermodynamics

1 Connecting the microscopic and macroscopic 188
2 Allowed energies 188
3 Multiparticle quantum states 191
4 Additional hypothesis 195
5 Maximizing uncertainty 195
6 Partition functions 199
7 Changing β and ψ 200
8 The nature of β 200
9 Negative temperatures 204
10 Probability and partition function for single-particle quantum states . 204
11 N-particle partition function from single-particle partition functions . 205
12 Reversible work 207
13 Work and heat 209
14 Entropy 210
15 The meaning of entropy 211
16 The Clausius inequality 211
17 Increase in entropy when heat is added to a system 212
18 Maxwell's distribution 213
19 Other distributions: open systems 215
20 The nature of α 217
21 Bose-Einstein and Fermi-Dirac distribution 219
22 Other approaches to statistical thermodynamics 220
23 Statement of basic problems 220
24 Ensemble of states 221
25 Ergodic hypothesis 222
26 Equal *a priori* probabilities 222
27 True ensemble average 223

28 Most likely condition 223
29 Mathematical necessity 225
30 Information-theory approach 226

Section 6 Relations Among Thermodynamic Properties

1 Combined first and second law 236
2 The state principle 237
3 Relations among thermodynamic properties 238
4 Differential property relationships 238
5 The Maxwell relations 239
6 The Clapeyron equation 242
7 Jacobians 244
8 Specific heats and expansion coefficients 245
9 Thermodynamic properties as functions of measurable properties . 246
10 Evaluation of changes in thermodynamic properties 249

Section 7 Equations of State

1 Deviation from ideal-gas behavior; compressibility factor . . . 256
2 Real gases 257
3 The van der Waals equation 258
4 Other two-constant equations of state 259
5 Benedict-Webb-Rubin equation 260
6 Virial equations of state 261
7 The generalized equation of state 261
8 Improved generalized correlation 263
9 Thermodynamic properties from equations of state 268
10 Effect of different equations of state on property computation . . 269
11 Departure functions 274
12 Generalized equation of state: departure functions 277
13 The need for experimental data 282
14 Presentation of thermodynamic data 283
15 Thermodynamic properties of liquids and solids 283

Section 8 Fugacity and Activity

1 Fugacity 294
2 Fugacities computed from the generalized equation of state . . . 295
3 Fugacity of liquids and solids 298
4 The effect of pressure on fugacity of liquids and solids 299
5 Activity 301
6 Activity of liquids and solids 302
7 General dependence of fugacity and activity on pressure and
   temperature 303

Section 9   Thermodynamics of Mixing and Composition Change

   1   Ideal solutions 308
   2   Ideal solutions of ideal gases; Dalton's and Amagat's laws . . . 309
   3   Ideal solutions of liquids; Raoult's law 312
   4   Entropy of mixing ideal solutions: statistical approach 314
   5   Entropy of mixing ideal solutions: macroscopic approach . . . 316
   6   Entropy, free energy, and Helmholtz function of mixing ideal gases   317
   7   Free energy of mixing: ideal liquid solutions 319
   8   Composition change and chemical work 322
   9   Minimum work of separation 323
  10   Ideal solutions of real materials 324
  11   Compressibility and fugacity of real gas solutions: pseudocritical
      properties 324
  12   Ideal solution of real gases 325
  13   Solutions of real liquids; Henry's law 326
  14   Ideal solutions of real liquids 327
  15   K factors; composition of vapor and liquid phases in equilibrium . 327
  16   Real liquid solutions: activity coefficients 328
  17   Partial molar properties 329
  18   Partial molar volume 332
  19   Partial molar enthalpies of a nonideal binary solution 334
  20   Graphical computation of partial molar properties 335
  21   Energy-conservation equation for variable work systems . . . 336
  22   Differential relations in chemical work systems 337
  23   Gibbs-Duhem equation 339
  24   Equilibrium in multiphase systems: constancy of μ_i 345
  25   Phase rule 346

Section 10   Chemical Equilibrium

   1   The chemical-equilibrium state 356
   2   ΔG of equilibrium chemical-reaction process 357
   3   ΔG of a chemical-reaction process at standard conditions 359
   4   Graphical representation of the ΔG_R^0 computation path 361
   5   The equilibrium constant 362
   6   Tabulations of $\Delta G_{formation}^0$ 363
   7   Additivity of standard free energies: calculation of $\Delta G_{reaction}^0$ . . 363
   8   Absolute entropy and standard free energy of reaction 367
   9   Composition of chemical equilibrium mixtures 371
  10   Effect of pressure on $\Delta G_{reaction}$ and the equilibrium constant K . . 373
  11   Effect of temperature on the equilibrium constant 376
  12   Equilibrium mixtures of real gases 380
  13   Homogeneous reactions in liquid or solid phases 382
  14   Heterogeneous reactions 384
  15   Simultaneous reactions 385
  16   Electrochemical processes 389

xvi   Contents

17 Standard electrode voltages 390
18 Fuel cells 392

Section 11   Irreversible Thermodynamics

1 The steady state 400
2 Coupled flows 401
3 Supplementary postulates 401
4 Applications: the thermocouple 405
5 Peltier effect 408
6 Thomson effect 409
7 Thermal transpiration 411
8 Other transport processes 413

Appendix I 416

Appendix II 418

Appendix III 428

Index 431

Answers to Selected Problems 436

SYMBOLS USED IN THIS BOOK*

Symbol			Equation in which symbol appears
a	=	activity	8–19
a	=	constant in specific heat equation, in van der Waals equation, and in other equations of state	2–20, 7–2, 6, 6a, 7, 11
$\mathbb{A}$	=	constant in van Laar equation	9–77
A	=	cross-sectional area of a fluid channel	2–43
A	=	Helmholtz free energy or arbeit function	3–34
$\mathscr{A}$	=	surface area	2–9
A_0	=	constant in BWR equation	7–11
b	=	constant in specific heat equation, in van der Waals equation, and in other equations of state	2–20, 7–2, 6, 6a, 7, 11
$\mathbf{B}$	=	second virial coefficient	7–12
$\mathbb{B}$	=	constant in van Laar equation	9–77
B_0	=	constant in BWR equation	7–11
c	=	constant in specific heat equation and in BWR equation	2–20, 7–11
c	=	velocity of light	2–10
C	=	specific heat	2–17
$\mathbf{C}$	=	third virial coefficient	7–12
C	=	number of independent components in a system	9–86
C_0	=	constant in BWR equation	7–11
D	=	diffusivity constant	11–1c
E	=	total accumulated energy	2–3
E_p	=	accumulated (potential) energy	2–2
E_k	=	accumulated (kinetic) energy	2–1
$\mathscr{E}_l$	=	energy of a multiparticle quantum state l	5–4
$\langle \mathscr{E} \rangle$	=	expected value of the energy of a multiparticle system	5–8
f	=	fugacity	8–3
F	=	force	2–4
F	=	work-energy lost through friction in a flowing fluid	2–50
F_i	=	generalized force	3–28

---

* Some special notations confined to a limited section of the book and defined in the places where they are used are not listed.

$\mathbf{F}$	= number of degrees of freedom	9–86
$\mathscr{F}$	= Faraday's constant	10–41
$\mathbb{F}_i$	= force producing the flux of i	11–1
g	= acceleration of gravity	2–2
G	= Gibbs free energy or free energy	3–22
h	= Planck's constant	5–1
H	= enthalpy	2–16
i	= quantum number	5–1
I	= current flow in amperes	11–1b
I	= moment of inertia	5–2
J	= rotational quantum number	5–2
J_i	= flux of i	11–1
k	= Boltzmann constant	5–5
$\mathbf{K}$	= K-factor; vaporization equilibrium constant	9–29
K	= equilibrium constant at 298° K	10–11
K_p	= equilibrium constant expressed in terms of reactant and product equilibrium partial pressures	10–12
K'	= Henry's law constant	9–29
K_T	= equilibrium constant at temperature $T°$K	10–30
l	= quantum number for vibrational energy	5–2
L	= length	2–6
L_{ii}	= direct or conjugate coefficient relating $\mathbb{F}_i$ to J_i	11–9
L_{ij}	= coupling coefficient relating $\mathbb{F}_j$ to J_i	11–2
m	= mass	2–1
M	= a general thermodynamic property	6–20
n	= number of moles	2–21
N	= number of particles in a system	5–50
$\overline{N}$	= Avogadro's constant, 6.02×10^{23} particles; number of particles in a mole of material	5–36
N	= a general thermodynamic property	6–20
$\widehat{p}_i$	= partial pressure of component i in a gas mixture	9–4
P	= pressure	2–7
P_L	= vapor pressure of liquid material L	3–33
$\mathbb{P}$	= total pressure of a gas mixture	3–33
P_l	= pressure exerted by a system when in multiparticle quantum state l	5–52
$\mathscr{P}_i$	= probability of quantum state i	5–34
$\mathscr{P}_l$	= probability of multiparticle quantum state l	5–4
$\mathscr{P}_i$	= probability of single-particle quantum state i	5–42
q	= electric charge	2–8
Q	= heat	2–12
R	= universal gas constant	2–22
$\mathbb{R}$	= electrical resistance	11–1b
S	= entropy	3–13

t	= time	11–1a
T	= temperature	2–17
u	= velocity	2–1
U	= accumulated internal energy	2–3
V	= volume	2–7
$\mathbf{V}$	= voltage	2–8
w	= the number of quantum states of equal likelihood	5–5a
W	= work	2–4
W_s	= shaft work	2–42
x	= elevation above a reference level	2–2
x_i	= mole fraction of component i in a solution	9–9
X_i	= generalized displacement	3–28
y_i	= mole fraction of component i in a gas mixture	9–5
z	= compressibility factor	7–13
$\mathbf{z}$	= valence or valence change	10–41
$\mathfrak{z}$	= partition function for single-particle states	5–44
Z	= a general thermodynamic property	6–20
Z	= partition function for multiparticle states	5–16
psia	= pounds$_f$ per square inch, absolute pressure	Ex. 2–6
psig	= pounds$_f$ per square inch, gage pressure	Ex. 2–18

GREEK SYMBOLS

α	= Lagrangian multiplier = μ/RT	5–77
α	= constant in BWR equation	7–11
β	= Lagrangian multiplier = $1/kT$	5–9
β_T	= coefficient of thermal expansion	6–28
γ	= ratio of specific heats C_P/C_V	2–36
γ_i	= activity coefficient of component i	9–35
γ	= constant in BWR equation	7–11
Γ	= surface tension	2–9
ϵ	= energy of single-particle state	5–1
η	= efficiency	2–9b
κ_T	= coefficient of isothermal compressibility	6–29
λ	= Lagrangian multiplier	5–43
μ	= chemical potential; partial molar free energy	5–84
ν	= vibrational frequency	5–2
π	= pi, 3.141592	5–3
Π_i	= product of a series	5–91
ρ	= density	Ex. 6–5
σ	= Thomson effect coefficient	11–29
$\sum_i$	= summation of a series	
τ	= tension	6–18a

ϕ	= number of phases	9–86
ϕ	= function symbol	9–84
Ω	= Lagrangian multiplier	5–77

SUBSCRIPTS

c	= critical conditions	7–15
e	= at equilibrium conditions	10–5
F	= of formation	Fig. 2–32
i	= of ith component	9–15
i	= of ith state; usually a single-particle state	5–41
l	= of lth state; a multiparticle state;	5–4, 5–7;
	also a vibrational energy state	5–2
$\mathbf{N}e$	= of product, at equilibrium conditions	10–34
P	= at constant pressure	2–18
r	= reduced conditions	7–14
$\mathbf{R}e$	= of reactant, at equilibrium conditions	10–34
rev	= reversible	3–9
R	= of reaction	2–54a
T	= at constant temperature	2–30
T	= Thomson heat	11–6
vap	= of vaporization	3–20
V	= at constant volume	2–19

SUPERSCRIPTS

^{0}	= standard state value (at STP) of a property	2–54c, 10–5
$^{-}$	= specific or molar property	2–26
^0T	= value of a property at standard pressure and *non-* standard temperature T	10–23
†	= of coupling	11–19
*	= property of ideal gas	7–33
$^\frown$	= partial molar property	9–29
ex	= excess function	9–48

SADI CARNOT, 1796–1832

Section 1
INTRODUCTION

1–1
Introduction

Thermodynamics concerns the transformation of energy from one form to another and the interaction of energy with matter. It is a protracted deductive exercise, based on certain few but very strict rules, of which the two most important and explicit may be considered as negative restrictions on the behavior of energy. These rules bear the austere titles: "The First Law of Thermodynamics" and "The Second Law of Thermodynamics."

The First Law can be stated as:

Energy can neither be created nor destroyed.

The Second Law can be stated as:

Heat cannot flow spontaneously from a cold body to a hot body.

Both "Laws" are in agreement with all human experience and deserve to be called "Laws of Nature." This is very important to thermodynamics as a useful science, but it is irrelevant to thermodynamics as thermodynamics, that is, as a logically consistent system. What is most relevant to the latter is that the two laws are always *assumed to be true.* In other words, these two laws are the *axioms* or *premises* forming the foundation of thermodynamics, and thermodynamics is in turn the body of conclusions that answer the question, "What logically necessary consequences must follow if we assume that the First Law and the Second Law are true?"

A capable logician may start building thermodynamics at points other than the above two,* and construct the entire logical structure from his point of departure, in a manner that is basically like the development of Euclidean geometry.

The importance of thermodynamics lies in the fact that its purely deductive conclusions about the nature of the interactions of energy and matter seem to coincide precisely with real energy and matter interactions in all chemical and physical phenomena in which such interactions can be tested. Thermodynamics, therefore, is a powerful means for both describing and predicting the course of real phenomena. (The ability of thermodynamics to deal with real phenomena presents strong evidence that the "real universe" is as logical and precise as the "thermodynamic universe.")

Practical uses of thermodynamics are unlimited. It enables us to compute the maximum efficiency of engines. It establishes ideality for energy conversions and goals toward which we can strive when making such conversions. It tells us the amount of heat or work needed to carry out physical and chemical processes. It makes possible the estimation of the temperature of flames and the heat effects of

---

* Alternative views of the fundamental postulates of thermodynamics may be found in Carathéodory, C., "Grundlagen der Thermodynamik," *Math. Ann.* **67**, 355, 1909; Hatsopoulos, G., and J. Keenan, *Principles of Generalized Thermodynamics*, Wiley, New York, 1965.

chemical reactions. It enables us to calculate the conditions of chemical and phase equilibrium. It establishes relationships between properties of materials. It indicates the direction of spontaneous change; and there are some who maintain that it even determines the inevitable direction of time.†

Classically, thermodynamics deals with chunks of matter which are very large compared with molecules, and with those properties of chunks of matter which we can sense without needing to theorize about the microscopic* or molecular nature of matter. Typical of such properties are temperature, volume, and pressure. These are properties of matter *en masse*, and are therefore called MACROSCOPIC properties.

The bridge between the conjectural unseen atomic world and the macroscopic properties of matter is constructed out of the concepts and techniques of STATISTICAL THERMODYNAMICS, which reconciles atomic theories of matter with the things we can measure with our crude instruments or senses. These concepts are essential in spectroscopy, low-temperature physics, semiconductor, laser, and catalysis studies, and in other new and important fields. They enrich classical thermodynamic concepts. They permit precise determination of certain thermodynamic properties and the derivation of important relationships called Equations of State.

Traditionally, thermodynamics has been able to deal with systems only when they were at rest, or at equilibrium. But thermodynamics is not a closed subject, and methods are being developed to apply it to nonequilibrium systems which conduct heat, matter, or electricity at a steady rate. This relatively new application is called IRREVERSIBLE THERMODYNAMICS. We encounter it when we deal with thermoelectric and thermionic generators and diffusional separation systems. It may provide unifying insights into the nature of chemical reaction rates (chemical kinetics).

1–2
Temperature and the zeroth law

The concept of "degree of hotness," or temperature, is fundamental to thermodynamic considerations. When we measure temperature we quantify the property which our senses respond to as degree of "hotness."

Devices for measuring temperature have been developed over many centuries. As early as 1592, Galileo developed a "thermoscope" which was arbitrarily calibrated and could be used to indicate relative "hotness." G. D. Fahrenheit is generally credited with the development of the forebear of the modern thermometer which made possible the reproducible estimation of temperature.

Implicit in the use of a thermometer is a primitive concept which has come to

---

† Eddington, A. S., *The Nature of the Physical World*, Macmillan, New York, 1928; Grunbaum, A., "Time and Entropy," *American Scientist*, **43**, 550–572, October 1955.
* "Microscopic" is used here in a special sense, i.e., in the sense of molecular. Submicroscopic is a more appropriate term, but it is not used by the thermodynamics experts.

be known as the ZEROTH LAW OF THERMODYNAMICS. The law resembles the fundamental mathematical principle that "things equal to the same thing are equal to each other." The Zeroth Law states that if system A is in thermal equilibrium with system B and system B is in thermal equilibrium with system C, then system A is in thermal equilibrium with system C. *Thermal equilibrium* implies that if system A and system B are allowed to interact only with each other, by placing them on opposite sides of a rigid wall that transmits only heat, it will be found that neither system will change its properties (pressure, volume, temperature, composition) with time as a result of the thermal connection made between the two systems (Fig. 1–1).

A thermometer serves the function of "system B" of the above statement. If a thermometer, in thermal equilibrium with system A, is brought into thermal contact with system C and does not change its properties (i.e., if the height of the mercury column does not change), then it can be said that A and C have the same degree of "hotness" or temperature. Temperature can therefore be considered as that property having equal magnitude in systems which are in thermal equilibrium.

The most frequently used units of temperature and the conversion factors which related them are given in Table 1–1.

If the properties of A and B are time invariant,

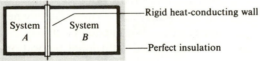

and the properties of B and C are time invariant,

then the properties of A and C are also time invariant.

Temperature is the property whose magnitude is the same in all three systems.

Fig. 1–1. The zeroth law of thermodynamics.

Exercise 1–1. A convenient way to convert °C to °F mentally is to double the °C, subtract 10% of the result, and add 32°F. Try it for 100°C. What is the corresponding way to convert °F to °C?

Table 1–1

Units of Temperature and their Conversion Factors

Unit	Symbol	Freezing point of water	Boiling point of water	Triple point of water
Degree centigrade (or degree Celsius)	°C	0.0°C	100°C	0.01°C
Absolute degree centigrade (or degree Kelvin)	°K	+273.15°K	+373.15°K	273.16°K
Degree Fahrenheit	°F	32.0°F	212.0°F	
Absolute degree Fahrenheit (or degree Rankine)	°R	491.67°R	671.67°R	

A temperature *interval* of 100°C is equivalent to an *interval* of 180°F. Therefore $1.8°F/1.0°C = 1$.

$$°F = °C \, (1.8°F/°C) + 32$$
$$°K = °C + 273.15; \quad 0°K = -273.15°C$$
$$°R = °F + 459.67; \quad 0°R = -459.67°F$$

The reference temperature to which all temperature scales are fixed is the temperature of the *triple point* of water, which is the temperature at which pure ice, water, and water vapor are in equilibrium in the absence of air. This temperature is easily and precisely reproduced, and has been arbitrarily assigned the value 273.16°K or 0.01°C.

Sections 3 and 4 contain further discussions on temperature.

Summary of Concepts

Thermodynamics: a deductive system describing the interactions of matter and energy.

Premises of Thermodynamics:

1. Energy is always conserved.
2. Heat flows down a temperature gradient.

Macroscopic Properties: properties of pieces of matter whose dimensions are large compared with molecular dimensions.

Zeroth Law: two systems which are each in thermal equilibrium with a third system are in thermal equilibrium with each other.

Temperature: the property having equal magnitude in systems in thermal equilibrium.

Units of Temperature: °F, °C, °R, °K.

JAMES PRESCOTT JOULE, 1818–1889

Section 2
ENERGY CONSERVATION:
THE FIRST LAW OF THERMODYNAMICS

2-1
Energy forms

Most of us have an intuitive feeling about energy. We realize that it is used in doing work, in running around the block, or in sawing firewood; that it is required to warm our soup, light an incandescent bulb, or charge a battery; that it is involved in stopping or starting a freight train, in compressing gas, stretching a rubber band, or lifting a weight; that it is stored in a rotating flywheel; used in large quantities to lift a rocket into outer space, and released in immense quantities during an atomic bomb explosion.

As a starting point for our considerations, we have a qualitative realization that energy, like romantic Zeus, assumes many diverse forms. We need also to realize that all forms are measured in units whose dimensions are equivalent to force multiplied by length. (Some commonly used units of energy are shown in Table 2-1 on page 21.)

It is convenient to classify the many forms of energy into two principal categories: ACCUMULATED ENERGY and TRANSITORY ENERGY.

2-2
Accumulated energy

A chunk of matter, an object, a machine, a piece of the universe can *store energy*. We shall call this stored form of energy ACCUMULATED ENERGY. Accumulated energy is the energy held within a system,* and associated with the system's matter. The amount of accumulated energy is reflected in such properties as the system's mass, composition, temperature, pressure, stress, phase, velocity, or position in a gravitational or potential field. As the amount of accumulated energy changes, the values of these properties change. [The Greek symbol Δ (delta) preceding a property symbol is used to indicate a change in the value of that property, in particular a difference between a final and an initial property value.] A curious and important characteristic of classical thermodynamics is that it deals with the *changes* in the amount of accumulated energy in a system and not with the system's absolute or total accumulated energy.

Accumulated energy may be further classified into accumulated EXTERNAL ENERGY and accumulated INTERNAL ENERGY forms. The external forms are usually independent of temperature and are described by parameters measured with respect to a reference or coordinate frame outside the system. KINETIC and POTENTIAL energies are forms of accumulated *external* energy.

a) *Kinetic Energy* $\equiv \Delta E_k = \dfrac{m}{2}(u_2^2 - u_1^2)$ (2-1)

---

* We shall use the word *system* to refer to the matter, object, machine, system, or piece of the universe that occupies our attention. All systems have *boundaries* which separate them, at least conceptually, from their surroundings. The concept of system is discussed further in Section 2-11.

This is the external energy that a body of mass m accumulates when its velocity changes from u_1 to u_2. It is the energy that must be removed from the freight train to stop it and the energy possessed by a rifle bullet about to strike a target.

Example 2–1. What is the change in kinetic energy accumulation in a 22-caliber bullet (0.0635 oz_m) having a velocity of 1000 ft/sec relative to its energy at rest?

Answer: Since there are 16 ounces in a pound, we have

$$0.0635 \ oz_m = \frac{0.0635}{16} \ lb_m = 3.97 \times 10^{-3} \ lb_m.$$

Therefore:

$$\Delta E_k = \Delta \text{ (kinetic energy)} = \frac{1}{2} \times \frac{3.97 \times 10^{-3} \ lb_m}{32.2 \ \frac{ft}{sec^2} \cdot \frac{lb_m}{lb_f}} \times (1000)^2 \ \frac{ft^2}{sec^2} * = 61.7 \ \text{ft-lb}_f.$$

---

b) Potential Energy $\equiv \Delta E_p = mg(X_2 - X_1)$ (2–2)

This is the accumulated external energy that a body of mass m has by virtue of its elevation, X_2, above a ground level, X_1, in a gravitational field whose acceleration is constant and equal to g. The term "potential energy" may also be applied to the

---

* The fundamental *mass* unit in the English system of engineering units is the SLUG, defined as

$$1 \ \text{slug} \equiv 1 \ \frac{lb_{\text{force}}}{ft/sec^2}.$$

However, strong custom favors the use of an alternative mass unit, called the POUND-MASS, which is a mass of such magnitude that it exerts a one-pound force in a gravitational field having an acceleration of 32.2 ft/sec². If this definition is to agree with Newton's law,

$$F = ma,$$

a conversion factor must be employed which equalizes the units on both sides of the equation. The factor is given the symbol g_c and has the value

$$32.2 \ \frac{ft}{sec^2} \cdot \frac{lb_m}{lb_f} .$$

Therefore, for use with pound-mass and pound-force units, Newton's law may be written

$$F = \frac{ma}{g_c},$$

where F is in units of pounds-force (lb_f), m is in units of pounds-mass (lb_m), and a is in ft/sec². The factor must be used whenever pounds-force or an energy is computed from an equation involving pounds-mass.

An analogous situation arises in the metric unit system when a GRAM-FORCE (defined as the force exerted by a one-gram mass when the gravitational acceleration is 981 cm/sec²) is used in equations of energy or motion. The unit conversion factor is then

$$981 \ \frac{cm}{sec^2} \cdot \frac{gm_m}{gm_f} .$$

energy that a body has by virtue of its position in *any* potential field. A potential field is a space in which a stationary body's energy is dependent on its position. Gravitational fields are obviously potential fields. Electrostatic, magnetic, and elastic fields may also be treated as potential fields.

Example 2–2. How much does the accumulated gravitational potential energy in a 10-kg iron ball change when it is carried to the top of the Tower of Pisa (59 m)?

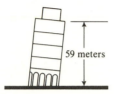

59 meters

Answer: We use Eq. (2–2) as follows:

$$\Delta E_p = \Delta \text{ (potential energy)} = 10{,}000 \text{ gm}_m \times 981 \frac{\text{cm}}{\text{sec}^2}$$

$$\times 5900 \text{ cm} = mg\,\Delta X = 5.79 \times 10^{10} \frac{\text{gm}_m\text{cm}^2}{\text{sec}^2} = 5.79 \times 10^{10} \text{ dyne–cm}$$

$$= 5790 \text{ joules} \quad \text{(see Table 2–1).}$$

c) Accumulated Internal Energy $\equiv U$

A system stores energy within and between its constituent molecules. We call this deeply stored energy accumulated *internal* energy, and use the symbol ΔU to represent a change in its magnitude. Any change in molecular velocity or vibration rate, in the bonds and forces between molecules, or in the number and kind of molecules changes the internal energy accumulation. The macroscopic indicators of internal energy change include:

Temperature change. A hot object has more accumulated internal energy than a cold one.

Phase change. Steam at 100°C has more accumulated internal energy than liquid water at the same temperature.

Composition change. Methane (CH_4) and O_2 at room temperature have more accumulated internal energy than their room-temperature combustion products, CO_2 and H_2O.

Mass change. A system that gains (or loses) mass gains (or loses) the accumulated internal energy within that mass.

Other properties also function as indicators of change in accumulated internal energy, we shall discuss these in later sections. The internal form of accumulated energy will concern us most frequently.

d) Total Accumulated Energy $\equiv E$

The total energy accumulated in a system is the *sum* of all the different accumulated energy forms the system contains. A change in total accumulated energy is represented by the symbol ΔE (without a subscript):

$$\Delta E = \Delta U + \Delta E_k + \Delta E_p. \qquad (2\text{–}3)$$

The Greek letter delta (Δ) associated with the symbol for each kind of accumulated energy, again means change or difference. It indicates that our concern is with the *change* in the amount of accumulated energy and not with the absolute amount of accumulated energy.

2–3
Transitory energy forms

Certain important energy forms *cannot be stored as such* within a system. They exist only *in transit* between the system and whatever interacts with the system. These forms are called TRANSITORY energy forms. Transitory energy is "pure" energy, *unassociated with matter*. By contrast, accumulated energy exists only in the presence of matter. Transitory energy is the energy entering or leaving a system without accompanying matter. It appears in *two forms*, HEAT and WORK.* *Heat and work are the only energy forms that a system can give to or take from its surroundings without transferring matter.*

a) Heat

This is the kind of energy that moves between systems in thermal contact whenever their temperatures differ.

Heat, as defined above, is fundamentally different from the popular conception of "heat." According to the definition, heat is energy *moving* between systems whose temperatures differ. Heat cannot be stored in a system. A system, no matter how hot, *contains no heat*. It contains accumulated internal energy. Heat changes to some form of accumulated energy as soon as it crosses the boundaries of a system, much as pedestrians change into passengers as soon as they enter a bus. It is imperative that you recognize the special narrow connotation given to the term heat throughout this text.

Heat is energy *in transit* between systems at different temperatures. *Once it enters a system it becomes some other kind of energy. Heat is designated by the symbol Q.*

---

* It is sometimes convenient to include a third form, *radiation*. This relieves us of the need to decide whether a particular radiation is WORK (coherent), or HEAT (blackbody), or a mixture of the two. The need arises when we deal with laser and other very special radiation sources and receivers. At this point in our learning process, we shall consider radiation as a form of transitory energy and thermal radiation as equivalent to heat energy.

Example 2–3. A kettle of cold water is placed on a hot stove. Eventually the water begins to boil.

Question: What kind of energy has the water received from the stove?

 Answer: Heat energy.

Question: Is there more heat in the boiling water than in the cold water?

 Answer: The energy entered as heat. Once in the water, it became part of the water's accumulated (internal) energy. The hot water has more accumulated internal energy than the cold water. There is *no* "heat" in either boiling water or cold water. No system contains "heat." It contains "accumulated energy."

b) Work

This is the energy that enters or leaves a system when a force at the system's boundary acts through a displacement. The effect of a work–energy interchange can always be reproduced by or made equivalent to the raising or lowering of a weight outside the system. A work effect requires some form of *linkage* between the system and the receiver or dispenser of work energy in the system's surroundings. The linkage may be a movable shaft, an electric cable, or, very frequently, the system's own *flexible* boundary, which is capable of moving so as to displace the surroundings.

Work energy transfer is evaluated by the equation

$$dW = F\,dX \qquad\qquad\qquad (2\text{–}4)$$

or

$$W = \int F\,dX, \qquad\qquad\qquad (2\text{–}5)$$

where

$W = \text{work}$ and $F = \text{force (or a generalized force)};$

$dX = \text{distance through which the force acts (or a generalized displacement)}.$

Examples of generalized forces are: F, force in a mechanical system; P, pressure in a system such as a gas; τ, tension in a stretched wire or rubber band; $\mathbf{V}$, voltage across a condenser or electric cell. Forces which do work are always boundary or surface forces, and never body forces. Thus a freely falling mass is not doing work or having work done on it, since it is acted on only by body forces. It is interchanging accumulated gravitational potential energy for kinetic energy.

Examples of generalized displacements corresponding to the above generalized forces are dX, the force displacement distance; dV, the volume change of a gas; $-dL$, the length change of a stretched rubber band or wire; $-dq$, the charge transfer in a condenser or cell. The sign of a displacement for a particular system is chosen so as to make the product $F \, dX$ positive $(+)$ when the system does work and negative $(-)$ when work is done on the system. For example, a stretched rubber band attached to a suspended weight raises and does work on the weight by *reducing* its length. Therefore the work done by or on a rubber-band system is

$$dW_{\text{rubber band}} = -\tau \, dL. \tag{2–6}$$

Hence dW is positive when the band does work and shrinks, and negative when the band stretches and receives work. On the other hand, a gas in a cylinder does work by *increasing* its volume. Therefore

$$dW_{\text{gas}} = +P \, dV. \tag{2–7}$$

Work, like heat, *cannot be stored as such* and *changes into an accumulated energy form* on entering a system.

Example 2–4. A mass of m grams is raised a distance of x meters, at a slow and steady pace, by a lifting machine. How much work is done on the mass?

Answer: Work $= W = \displaystyle\int_0^x F \, dX = mgx.$

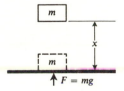

Note that the work done on the mass is equal to the gravitational potential energy accumulated by the mass (Example 2–2); also that the force exerted is assumed to be constant and equal to the weight (mg) of the mass. This implies that there is no friction and that the mass moves upward without accelerating. Such a work effect is called a *reversible work effect* (see Section 2–5).

Exercise 2–1. Assuming that the above mass is 10 tons (mass), compare the work needed to raise it 1 ft on earth $(g = 32.2 \text{ ft/sec}^2)$ and on the moon $(g = 5.47 \text{ ft/sec}^2)$.

Example 2–5. A spring has a spring constant of 2 lb/in. What is the work done *by* a machine in compressing the spring 4 in. from its no-load position? The compression occurs at a steady rate and without acceleration or friction, i.e., the work effect is being carried out reversibly (as in our previous example).

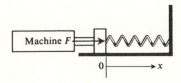

Answer: The force exerted on the spring is equal to the spring constant multiplied by the displacement, or $2x$ pounds. Therefore

$$W_{\text{machine}} = \int_0^4 F \, dX = \int_0^4 2X \, dX = [X^2]_0^4$$

$$= 16 \text{ in-lb}_f = 1.33 \text{ ft-lb}_f.$$

The work done *on* the spring is simply $(-)1.33$ ft-lb$_f$.

Example 2–6.   Steam from a large steam pipe at 50 psia pressure enters the cylinder of a simple steam engine having a piston which sweeps through a volume of 30 ft^3 during its stroke. The expansion occurs at constant pressure. How much work can the steam do?

Answer: We know that the work can be expressed as

$$W = \int F \, dX. \qquad (2\text{–}5)$$

But

$$F = \text{pressure} \times \text{area} = PA.$$

and

$$X = \text{volume displaced by piston/piston area}.$$

Therefore

$$W = \int (PA) \frac{dV}{A} = \int P \, dV,$$

which is the integrated form of Eq. (2–7). Since P is a constant (50 psia),

$$W = P \int_{V_1}^{V_2} dV = P \, \Delta V$$

$$= 50 \frac{\text{lb}_f}{\text{in}^2} \cdot 30 \text{ ft}^3 \cdot \frac{144 \text{ in}^2}{\text{ft}^2} = 216{,}000 \text{ ft-lb}_f.$$

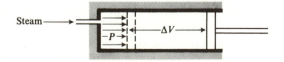

Implicit in the use of a constant value of P in integrating Eq. (2–7) is the assumption that the piston moves without acceleration.

Many other kinds of work are possible. For example, a condenser or battery does electrical work on the circuit connected to it:

$$W = -\int \mathbf{V} \, dq. \qquad (2\text{–}8)$$

A soap film, as it contracts, does work:

$$W = -\int \Gamma \, d\mathscr{A}, \tag{2–9}$$

where Γ is the surface tension and $d\mathscr{A}$ is the change in surface area.

Example 2–7. What is the work required to charge a 100-μF (microfarad) capacitor at 10 volts?

Answer: We have

$$W_{\text{elect}} = -\int \mathbf{V} \, dq.$$

But $q = C\mathbf{V}$, where C is the capacitance in farads and q is the charge in coulombs. Therefore $dq = C \, d\mathbf{V}$ and

$$W_{\text{elect}} = -\int_0^{\mathbf{V}} C\mathbf{V} \, d\mathbf{V} = -\tfrac{1}{2} C\mathbf{V}^2$$

$$= -\tfrac{1}{2}(100 \times 10^{-6}) \text{ farad} \times \left(\frac{\text{coulomb/volt}}{\text{farad}}\right)$$

$$\times (10 \text{ volts})^2 \times \frac{\text{joules}}{\text{volt-coulomb}}$$

$$= -5 \times 10^{-3} \text{ joules}.$$

The minus sign indicates that work is done on the condenser.

Example 2–8. What is the least amount of work energy that must be done on 1 liter of water to convert it into a uniform spray having droplets which are 5 μ in diameter (at 20°C)?

Answer: We have Γ (20°C) = 72.75 dynes/cm.* Let N be the number of droplets. Then

$$1000 \text{ cm}^3 = N \times \frac{\pi}{6}(5 \times 10^{-4})^3 \text{ cm}^3,$$

$$N = 1.531 \times 10^{13} \text{ droplets}.$$

$$\text{Total surface area} = 1.531 \times 10^{13} \times \pi(5 \times 10^{-4})^2 \text{ cm}^2$$

$$= 1.202 \times 10^7 \text{ cm}^2.$$

---

* See *Handbook of Physics and Chemistry* **45**, page F18.

The surface area of the original liter of water may be neglected in comparison with the total droplet area, and Γ is constant. Therefore

Work done on system to create surface $= -\Gamma \, \Delta \mathscr{A}$

$$= 72.75 \times 1.202 \times 10^7 \text{ dyne-cm}$$

$$= -87.4 \times 10^7 \text{ ergs}$$

$$= -87.4 \text{ joules} = -64.5 \text{ ft-lb}_f.$$

(From Table 2–1, 1.356 joules $=$ ft-lb$_f$.)

Work is the energy interchange between a system and its surroundings, which occurs when a generalized force acts through a corresponding generalized displacement at or across the system's boundaries. Energy effects that occur completely inside or outside these boundaries are *never* considered as work. Work energy is recognizable by the fact that, should one so desire, it can be completely and continuously expended in raising weights or compressing springs. For example, a system consisting of an electric storage battery is used to energize an electrical device in its surroundings. The energy that the battery gives to its surroundings is work because it can be imagined to run through an efficient electric motor that uses it to raise weights.

Generalized forces are usually INTENSIVE PROPERTIES of a system; that is, their value is *independent* of·the size or *extent* of the system. Generalized displacements are usually EXTENSIVE PROPERTIES of a system; that is, they *depend* on the *extent* or size of a system. Hence the usual work equations (Eqs. 2–6 through 2–9) have the form

$$W = \int (\text{intensive property}) \times d(\text{extensive property}). \qquad (2\text{–}9\text{a})$$

It is essential to realize that these integrals can be evaluated only if the intensive and extensive properties can be expressed as continuous functions of each other; or if the value of one property is known and finite at every value that the other property takes on during the course of the work process.

Time-independent functional relations between corresponding intensive and extensive properties of a system (for example, relations between pressure and volume) exist only for states in which the system's properties are uniform and time invariant. Such states are called *equilibrium states*. Hence the work integrals (Eqs. 2–6 through 2–9) can be solved only when the work process occurs via a chain of equilibrium states. Such work processes were used in Examples 2–4 through 2–8. They are called *reversible, quasi-static processes*. During such processes the properties of the system are always at, or infinitesimally close to, their equilibrium values.

2–4
Graphical representation of work

The reversible work integral can be represented as an *area* on a plane having the integral properties as coordinates. For example, reversible expansion work, $\int_1^2 P\,dV$, equals the area under the reversible process curve connecting states 1 and 2 on a PV plane, with P as ordinate and V as abscissa. The area depends on the shape of the curve connecting states 1 and 2, which in turn is determined by the functional relation between P and V as the system moves between 1 and 2.

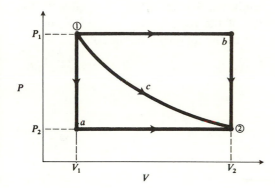

Fig. 2–1

Figure 2–1 shows three of the *infinite* variety of *reversible* process paths that might be traversed in moving from a given initial state (1) to a given final state (2). In the case shown, state (1) has a high pressure (P_1) and low volume (V_1) and state (2) has a lower pressure (P_2), and larger volume (V_2). If the process occurs reversibly, a relation exists between P and V for every point on the process curve connecting the two states. (In fact, the curve means that the relation exists. If the relation did not exist, neither would the curve.) We can evaluate the work by using the relation to express P in terms of V or dV in terms of dP, and then integrating between the initial and the final volume or between the initial and the final pressure.

For the process which moves along path 1–*a*–2: $V = V_1$ for all values of P between P_1 and $P_a(=P_2)$(process 1–*a*); and $P = P_2$ for all values of V between V_1 and V_2 (process *a*–2). The work of the process is $\int_1^a P\,dV + \int_a^2 P\,dV$. Or:

$$W_{1-a-2} = (0)_{1-a} + P_2[V_2 - V_a] = P_2(V_2 - V_1).$$

Exercise 2–2

i) Show that in Fig. 2–1 $W_{1-b-2} = P_1(V_2 - V_1)$ and that $W_{1-c-2} = K \ln (P_1/P_2)$, given that the function relating P to V along 1–*c*–2 is $PV = K$, where K is a constant.

ii) Draw a path between 1 and 2 which differs from those shown in Fig. 2–1 and evaluate $\int P\,dV$ analytically (not graphically) for your path.

2-5
Reversible work processes

The concept of a reversible work process occupies a key position in thermodynamics. It was introduced by a French general and amateur scientist named Lazare Carnot, who happened to be Napoleon's minister of war and the father of Sadi Carnot, a young man whose fundamental contributions to thermodynamics will be discussed at length in Section 3.

In an article published in 1803, the elder Carnot set down the necessary conditions for maximally efficient transfer of mechanical energy. He concluded that maximum efficiency occurred under conditions that he called "reversible." His conclusions were based on consideration of the rudimentary machine shown in Fig. 2–2, and on reasoning similar to that in the following paragraphs.

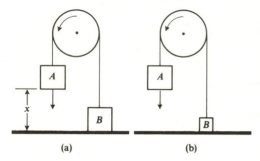

(a)   (b)

Fig. 2–2. (a) Reversible process: Mass$_A$ − mass$_B$ = differential quantity.
(b) Irreversible process: Mass$_A$ − mass$_B$ = finite quantity.

A REVERSIBLE WORK PROCESS is a process that can be undone or reversed by an infinitesimal reversal in the magnitude of the generalized forces (force, pressure, voltage, etc.) which actuate the process. For example, two masses, A and B, may be attached to each other by a weightless string which passes over a *frictionless*, massless pulley, as in Fig. 2–2a. If mass A is equal to mass B and the system is at rest, it will tend to stay at rest. If an infinitesimal weight is added to A, then A will descend very slowly and raise weight B, thereby doing work on B. The process can be reversed by moving the infinitesimal weight from A to B, at which time B will do work on A. The process is reversible and also quasi-static, because the system is always infinitesimally removed from an equilibrium condition.

If mass A is *finitely heavier* than mass B, then releasing A will permit A to do work on B (Fig. 2–2b). Mass B will be raised rapidly, and A will have performed $m_B g \, \Delta x_B$ units of work. It is obvious, however, that A, in raising the lesser mass B, has not performed all the work that it is capable of doing, since it can raise a body whose weight is only infinitesimally smaller than its own. Thus A does *maximum* work in the absence of friction, and when the unbalance actuating the work process is infinitesimal. Similarly, the *minimum* work required to raise B is work performed on B without friction and with infinitesimal unbalances actuating the work process. Such processes are called REVERSIBLE WORK PROCESSES. Work processes which proceed in the presence of friction, or under the actuation of finite unbalance, are called *irreversible*.

Exercise 2-3. Compare the maximum work performable by mass A (Fig. 2-2a) in dropping a distance x with the minimum work needed to raise A the same distance.

Compare two processes, one reversible and the other irreversible, for raising A to height x, with regard to rates, heat effects, noise, acceleration, oscillation, etc.

2-6
Efficiency

The efficiency of a process is the ratio of the energy outputs to the energy inputs. The work energy input into the Fig. 2-2 processes equals the weight of A times the distance (x_A) that A falls. The work energy output equals the weight of B times the distance B rises. The efficiency η (the Greek letter eta) is therefore

$$\eta = \frac{m_B g x_B}{m_A g x_A} = \frac{m_B}{m_A}. \tag{2-9b}$$

It should be apparent that efficiency approaches 100% as the mass of B approaches the mass of A; that is, as the process approaches reversibility:

$$\eta \to 100\% \text{ as } m_B \to m_A.$$

2-7
Reversible expansion and compression

We may apply considerations of reversibility to the expansion of a gas in a frictionless cylinder and piston (Fig. 2-3a). Figure 2-3b represents the *isothermal* (constant temperature) expansion of a gas in the engine of Fig. 2-3a from state 1 to state 2 on a pressure-volume diagram. We now ask: Under what conditions does the expansion process deliver maximum work?

As the pressure of the gas in the cylinder decreases, the volume increases and the piston moves outward. The force on the piston decreases as the pressure decreases according to the relationship,

$$\text{Force} = \text{gas pressure} \times \text{piston area.}$$

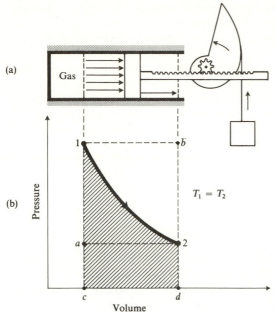

Fig. 2–3. (a) Reversible isothermal expansion. (b) A pressure–volume diagram for the gas in (a).

It follows from the conclusions drawn from the weight-and-pulley machine (Fig. 2–2) that the maximum work this force can do will be obtained when the surroundings oppose the outward movement of the piston with a force that is always only differentially smaller than that exerted by the expanding gas. The opposing force must be large at the beginning of the expansion and decrease as the piston moves outward. The mass, gear, and cam assembly actuated by the piston in Fig. 2–3a provides such an opposing force because the torque exerted by the mass decreases as the piston moves to the right.

If the external opposing force is constant and equivalent to the final gas pressure P_2, then the gas in the cylinder will expand very rapidly and irreversibly to its final volume V_2. The work done on the surroundings equals the area enclosed by *a–2–d–c* in Fig. 2–3b, which is the work done by a constant pressure expansion from the initial to the final volume. This is considerably less than the work of a *reversible* isothermal expansion (area within 1–2–d–c).

Exactly analogous considerations determine the minimum work needed to compress the gas isothermally from (2) back to (1). The *minimum work* is expended when the isothermal compression is carried out *reversibly*, that is, when the compressing force is always only infinitesimally greater than the isothermal equilibrium pressure within the cylinder. The *reversible compression work* is the area enclosed by 2–1–c–d and has the *same magnitude as the reversible expansion work, but an opposite sign.* Area 1–b–d–c represents the work expended by the surroundings on the system when a force equivalent to the final pressure P_1 is applied to the piston throughout the entire compression process.

2–8
Equivalence of all energy forms

All energy forms are equivalent and can be expressed or measured in similar units. Conversion or equivalence factors must be used in computation whenever all the energy measures are not expressed in the same units. Table 2–1 shows equivalence or conversion factors for frequently used units of energy.

Table 2–1

Energy Units and Conversion Factors (Dimensions = force × length)

Energy unit	Abs. joule	Calories	Btu	Foot-pound$_f$
		Factors		
1 calorie*	= 4.184	1	0.00397	3.086
1 (abs.) joule	= 1	0.239	0.948×10^{-3}	0.738
1 Btu	= 1,055	252	1	778.2
1 kilowatt-hour	= 3.6×10^6	860,421	3,412.1	2.655×10^6
1 foot-pound$_f$	= 1.356	0.324	0.001285	1
1 liter-atmosphere	= 101.3	24.22	0.0960	74.73

1 joule = 1 watt-sec = 10^7 ergs = 10^7 (g-cm^2)/sec^2 = 10^7 dyne-cm
        = 1 newton-meter = 0.239 calorie = 1 volt-coulomb

$$\text{Power units} \left(\text{Dimensions} = \frac{\text{energy}}{\text{time}} = \frac{\text{force} \times \text{length}}{\text{time}} \right)$$

1.0 horsepower = 550 ft-lb$_f$/sec = 0.745 kilowatt

Mass units	Length units
1.0 pound$_{mass}$ = 453.59 grams	1.0 inch = 2.54 centimeters

2–9
Equivalence of heat and work energy

The fact that heat and work are equivalent energy forms was not generally accepted until the middle of the last century. Prior to that time, thermal effects were explained by the Caloric Theory, which postulated that all matter contained an elastic,

---

* The calorie used in this table is the thermochemical calorie, defined as equivalent to 4.184 absolute joules. An International Steam Table calorie (I.T. calorie) is also found in the thermodynamic literature. It is slightly larger than the thermochemical calorie; that is, 1 I.T. calorie = 1.000669 calorie.

weightless, indestructible and "subtile" fluid called "caloric." Hot objects were thought to contain more "caloric" than cold objects. When objects at different temperatures were brought into thermal contact, "caloric" would flow from one into the other without loss, until their temperatures matched.

The caloric theory was discredited through the efforts of many men, but principally through the works of Count Rumford of Bavaria (born Benjamin Thompson in Woburn, Massachusetts, in 1753) and the English scientist (and brewer) James Prescott Joule. Thompson (or Rumford) used a horse-driven cannon-boring machine to demonstrate that unlimited quantities of heat could be generated by rubbing a blunt boring tool inside a cavity in a metal block. The heat produced depended on the work expended on the boring tool, and hence heat could not be a fluid substance pervading either the metal block or the boring tool.*

Between 1843 and 1848, Joule performed a series of definitive quantitative experiments which determined the equivalence factor relating heat to work energy. The experiments involved the expenditure of mechanical work in a container of water while measuring the water's rise in temperature. His most precise and famous experiments used the apparatus shown in Fig. 2–4, in which falling weights

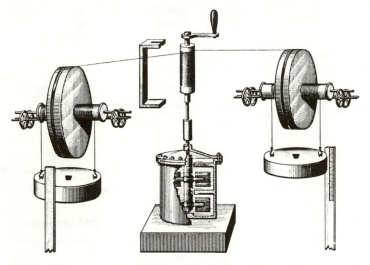

Fig. 2–4. Apparatus used by Joule to perform his experiments on the relationship between heat and work energy. (From "Scientific Papers of J. P. Joule," *Philo. Trans. Royal Society*, 1884, page 298)

---

* Rumford was a royalist who moved to England during the American Revolution. From simple beginnings as a colonial schoolteacher, he rose to become minister of war to the King of Bavaria, from whom he received his title. He married Lavoisier's widow. In the published descriptions of his "philosophical researches" on heat, he mentions that his experiments were performed on scrap metal and did not reduce armament production. He was a very practical man. (*Complete Works of Count Rumford*, Am. Acad. Arts and Sciences, Boston, 1870–1875.)

Joule was a gentleman scientist devoted to scientific study in his own private laboratory. His income came from family brewery holdings.

and a system of pulleys drove a set of paddle wheels immersed in a tank full of water. He concluded that the expenditure of 773 foot-pounds$_f$ of work produced the same temperature rise in a pound of water as did a unit of heat energy (one British Thermal Unit). Considering the crudity of Joule's instrumentation, we must marvel at the accuracy of his result, which is within 0.7% of the value obtained by the best modern instrumentation. The equivalence factor used in English engineering units today is:

$$778.2 \text{ ft-lb}_f = 1 \text{ British thermal unit (Btu).}$$

2–10
Equivalence of mass and energy

Albert Einstein, in 1905, set forth the famous equation stating the equivalence of mass, m, and energy:

$$E = mc^2, \tag{2–10}$$

where c is the velocity of light in a vacuum, 2.997×10^{10} cm/sec. Hence 1 gram (rest mass) of matter is equivalent to 9×10^{20} ergs of energy.

Exercise 2–4. Express the work done by the steam in Example 2–6, in units of Btu, calories, joules, kilowatt-hours, and grams rest-mass.

2–11
System

The word "system," as used in the preceding pages, deserves further elaboration. By system we mean whatever object, machine, material, combination of objects, or part of the universe we focus our attention on. A system is distinguished from the

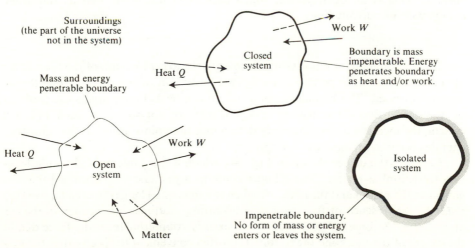

Fig. 2–5. Three kinds of system.

rest of the universe by a complete *boundary*. If *matter can cross* this boundary, the system is said to be *open*. If *matter cannot cross* the boundary, the system is said to be *closed*. If neither matter nor energy can cross the boundary, then the system is said to be *isolated* (Fig. 2–5).

Open systems may exchange transitory and accumulated energy with their surroundings because they can transfer matter from or to the surroundings as well as exchange energy by heat and work interactions.

On the other hand, a closed system can exchange *only heat and/or work* with its surroundings. It has *no other means* of gaining or losing energy.

An isolated system, by definition, cannot interact with its surroundings in any way.

Example 2–9. Designate each of the following systems as open, closed, or isolated.

i) The system is the gas contained inside the cylinder of the machine in Fig. 2–3.

Answer: Closed system. If the gas does not leak out of the cylinder, the system can receive only heat and work energies. However, if the piston leaks, and the system is defined as the *contents* of the cylinder, then we have an *open* system which exchanges (loses) matter to the surroundings. When it loses matter, the system also loses the energy accumulated in that matter.

ii) The system is as above, but the cylinder and piston are covered with perfect insulation and the piston is locked so that it cannot move.

Answer: Since we assume a rigid, insulated, and immobile container, the contents constitute an isolated system.

iii) The system is the gas contained at any instant between the inlet and outlet of a turbine (Fig. 2–22).

Answer: This is an open system which exchanges matter as well as transitory energy with its surroundings.

In thermodynamics it is essential to *specify* the *exact* location of the *boundaries* of the system so as to permit unambiguous limitation of the extent of matter we wish to consider. The boundaries of a system can be as important as the contents of a system. Because transitory energy forms are not stored as such within a system, the location of the boundary determines the points at which the transitory energies become accumulated energy.

Consider the device shown in Fig. 2–6, which is a cylinder sealed with a leak-proof, frictionless piston. The cylinder contains a gas, and is so arranged that the gas raises the piston and the mass, m, when it expands. The gas is called a *working fluid*. In the absence of friction and accelerating motions, the pressure of the gas always exerts a force on the piston which exactly counterbalances the force due to the weight of the mass and the piston. It is often convenient to specify that the sys-

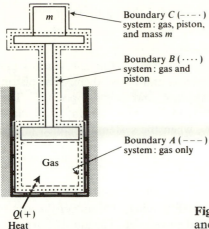

Boundary C $(-\cdot-\cdot)$
system: gas, piston,
and mass m

Boundary B $(\cdots)$
system: gas and
piston

Boundary A $(---)$
system: gas only

$Q(+)$
Heat

Fig. 2–6. The location of the boundary is arbitrary, and determines the extent of the system.

tem consists of the gas (working fluid) only. The *boundary* of the system (A in Fig. 2–6) is then an imaginary, infinitesimally thin, flexible, elastic sack which takes on the shape of the inside surface of the piston–cylinder device. The system is *closed*. The heat that enters the system is the heat entering the gas. If the gas expands, the boundary expands and the system does work. This work is expended in raising the piston, in pushing back the atmosphere,* and in elevating the mass m.

Since the *boundaries* of the system are *arbitrarily established*, we may consider, as an alternative system, the gas and piston assembly (boundary B). Now the work done *by the system* is that used to raise the mass m and push back the atmosphere. System B's work will be less than that of system A by an amount equal to the energy accumulated in the raised piston ($m_{piston}\, g\, \Delta x$).

Should we decide to place the boundary so as to enclose the mass m (boundary C), then the system consists of gas, piston, and mass m, and the work done *by the system* is only the work of pushing back the atmosphere. The increased elevation of the piston and mass is now counted as a contribution to the system's accumulated potential energy.

2–12
Surroundings

The boundary of the system serves to divide our field of attention between *the system*, which is everything *within* the boundary, and its *surroundings*, which is everything *outside* the boundary. The system, plus *all* its surroundings, encompasses the entire *universe*.

---

* Whenever a system expands, work is expended in pushing back the atmosphere. Anyone who has ever tried to open a vacuum-sealed jar of pickles may appreciate the work required for this task. (Also see Example 2–10 and Exercise 2–5.)

The following example illustrates how the choice of different system boundaries affects the work done by a system.

Example 2–10. A quantity of heat is added to the machine of Fig. 2–6, causing the gas to expand quasi-statically and to raise the piston and mass 10 cm. The piston has an area of 2.0 in². It is frictionless and leakproof, and weighs 1 pound$_m$. The mass m is 2.0 pounds$_m$. Atmospheric pressure is 14.7 pounds$_f$ per in². Calculate the work done *by the system* in units of ft-lb$_f$, when the system is defined (a) by boundary A; (b) by boundary B.

Answer

a) System A is closed, and consists of the gas only. It delivers work to its surroundings by elevating a 3-lb mass 10 cm, and displacing a volume of the atmosphere equal to the piston area times its vertical movement. The amount of work done by the system on the surroundings, as measured by the effects in the surroundings, is:

$$W = mg\,\Delta x + P_{\text{atm}}\,\Delta V$$

$$= 3\ \text{lb}_m \times 32.2\,\frac{\text{ft}}{\text{sec}^2} \times \frac{\text{sec}^2 \cdot \text{lb}_f}{32.2\ \text{ft-lb}_m} \times \frac{10.0\ \text{cm}}{30.5\ \text{cm/ft}}$$

$$+ 14.7\,\frac{\text{lb}_f}{\text{in}^2} \times 2.0\ \text{in}^2 \times \frac{10.0\ \text{cm}}{30.5\ \text{cm/ft}}$$

$$= (0.984 + 9.632)\ \text{ft-lb}_f = 10.62\ \text{ft-lb}_f.$$

b) System B is closed, and includes the gas and piston. Its surroundings consist of the mass and the atmosphere to which it delivers an amount of work:

$$W = \left(2 \times \frac{10}{30.5} + 9.63\right)\ \text{ft-lb}_f = 10.29\ \text{ft-lb}_f.$$

Note that this particular machine does more work in displacing the atmosphere than in raising the mass.

    We may calculate W in two ways: by considering the effects in the surroundings, or by considering the effects inside the system. The calculation shown evaluates W from the effects in the surroundings. Calculation of W from effects in the system is required in Exercise 2–5 (iii).

Exercise 2–5

i) Calculate the work done by the system defined by boundary C in Example 2–10.

ii) Calculate the gas pressure in the cylinder when: (1) the atmospheric pressure is 14.7 pounds per square inch absolute (psia); (2) the surroundings' absolute pressure is zero.

iii) Calculate the work done by system A in Example 2–10 from the changes in the system's properties (i.e., cylinder pressure and volume change). Compare with the result of Example 2–10(a). Repeat the calculation for the system conditions that prevail when the air pressure in the surroundings is zero. Note that $(P \Delta V)_{\text{cylinder gas}}$ is *not* the work done by systems B or C.

iv) Suggest a design change in the machine in Fig. 2–6 that would decrease the work of atmospheric displacement without changing the useful (mass-raising) work. Would this change affect the cylinder pressure?

2–13
First law or the conservation of energy

The First Law of Thermodynamics is the premise that *Energy can neither be created nor destroyed.* Another way of stating it is: *Energy is a conservative property.* Let us assume the law to be true and explore its consequences.

The law implies that a simple accounting procedure can keep track of energy moving into and out of a system. Energy may enter a system and/or leave a system. If the energy output is different from the input, then there must be a change in the amount of energy *accumulated* or stored within the system.* The change in *accumulation* can be *positive* or *negative*, depending on whether the input is greater or smaller than the output (Fig. 2–7).

$$\text{Input} = \text{output} + \text{change in accumulation.} \qquad (2-11)$$

The bookkeeping for this accounting procedure is complicated by the fact that energy takes on many different forms, which we must learn to classify as work energy, heat energy, or accumulated energy.

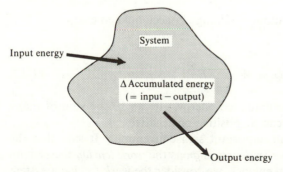

Fig. 2–7. The law of conservation of energy: Input energy = output energy + change in accumulated energy.

---

* One might say that accumulated or stored energy is an invention we are forced to make as a consequence of accepting the conservation law and the existence of transitory energy forms.

Consider, for example, the cylinder and piston device of Fig. 2–6. If heat energy enters the system (now defined as the gas), the gas in the cylinder will expand, raising the piston and elevating mass m. If the net work energy produced (the transitory energy leaving) is not equivalent to the net amount of heat introduced, then, acting on our premise, we shall seek and expect to find the difference in energy somewhere within the system. For the device under consideration, we shall find that the *temperature of the gas* in the cylinder *has risen*, which is an indication that the *gas has increased its accumulated internal energy.*

The change in internal energy accumulated equals the difference in the transitory energy input and output expressed in common units.

If the system is defined by boundary B in Fig. 2–6, then the output of work energy is the sum of the work done on the mass and atmosphere, which is less than the work done by system A. The energy accumulated in system B is, however, more than that accumulated in system A, and consists of the internal energy increase of the gas *plus* the potential energy increase of the piston.

Exercise 2–6. Using the data of Example 2–10, find the energy accumulated in the system when the system is defined by boundary A in Fig. 2–6; boundary B; boundary C. Assume that the heat added to the system is 0.1 Btu.

2–14
The sign convention

Traditional thermodynamic considerations involved steam engines and similar power devices which *perform work* when *supplied* with *heat*. It is therefore customary to write the First Law in a form which applies directly to such engines and energy transformations. Thus:

Net heat input = net work output + change in accumulated energy

or in symbols:

$$Q = W + \Delta E \qquad (2\text{--}12)$$

where ΔE is the change in total accumulated energy. It can be the sum of many forms of accumulated energy. It can be positive or negative.

Equation (2–12) establishes an important sign convention. It says that the *heat entering* a system has the *same arithmetic sign* as the *work leaving* the system. Out of long association with steam engines, we consider the *work leaving* a system as *positive* (+); therefore the *heat entering* a system is also *positive* (+). It follows that work done on a system represents a negative (−) work output; whereas heat removed from a system represents a negative (−) heat input.

The validity of the equation is not affected by the scale of operations. It applies equally well to huge transfers of energy or to differentially small ones. For the latter case, the First Law (or the equation of the conservation of energy) may be written as:

$$dQ = dW + dE.$$ (2–13)

The symbols Q, W, and E, when they appear in one equation, all refer to the *same system*.

2–15
Accumulated internal energy and the thermodynamic state

The change in accumulated energy, ΔE, that may result from a system's energetic interaction with its surroundings can be external or internal, or both. A change in external energy accumulation is seen as changes in elevation or velocity. A change in internal energy accumulation is indicated by changes in the properties of the system; for example, temperature, mass, pressure, density, tension, composition, etc. These are called *thermodynamic properties*, and the *set* of their values in a given system determines the THERMODYNAMIC STATE of that system. A change in a system's accumulated internal energy is therefore said to be indicated by changes in the system's thermodynamic state.*

The thermodynamic state of a system is the set of its thermodynamic property values. These properties are not all independent. A small number of properties usually suffices to specify all the others in the set or the thermodynamic state. The number required to specify the thermodynamic state, which is the number of independently variable properties, is *one* plus *the number of kinds of independent-reversible work* the system can perform. The foregoing statement is called the STATE PRINCIPLE. It is a consequence of the Conservation Law and the fact that dU is an exact differential. (Further discussion of the State Principle appears in Section 6.)

A system that has only *one* reversible work mode—for example, expansion work—therefore has its state specified by two properties. This means that fixing two properties suffices to fix all others. As an example which you already know, the volume or density of one mole of a gas is fixed when its temperature and pressure are specified. A system with only one reversible work mode generally has a fixed mass, composition, and phase.

It is often convenient to refer to the *intensive thermodynamic state* of a system, by which we mean that we are considering the properties of a unit mass of the system. The thermodynamic properties are then called *specific* or *molar* properties, depending on whether they refer to a pound (or gram) or a pound-mole (or gram-mole) of matter.

---

* Our discussion is confined to nonrelativistic systems (i.e., systems in which velocity and gravitational effects do not alter the mass or temperature of the system), in which accumulated external energy and internal energy are independent of each other.

2–16
U: A state property

It follows from the above paragraphs that in a closed system which has constant composition and phase, U depends only on the thermodynamic state of the system (that is, the pressure and temperature, or any two other thermodynamic properties), and each state has only one internal energy value. That this is a consequence of the Conservation Law may be demonstrated as follows:

Consider an experiment similar to Joule's experiment, carried out in a rigid-walled, closed container. The experiment consists of doing work on a pure liquid (the system) stored in the container, by means of paddles attached to a rotating shaft which extends through the container walls. Because there is no flow of heat into or out of the system and because the system changes neither its potential nor its kinetic energy, $Q = 0$, $\Delta E_k = \Delta E_p = 0$. The Conservation Law is therefore written as:

$$\overset{0}{\cancel{Q}} = W + \Delta U + \overset{0}{\cancel{\Delta E_k}} + \overset{0}{\cancel{\Delta E_p}}.$$

Therefore

$$-W = \Delta U.$$

The system is initially in state 1, and after absorbing work finds itself in some final state 2, different from the initial state because the final temperature differs from the initial temperature.

The work done on the system (negative work) is all accumulated within the system. The accumulation takes the form of an increase in internal energy, ΔU.

Imagine that a second operation is now performed on the contents of the rigid vessel. A cooling coil is immersed in the liquid and an amount of heat, $-Q$,* equivalent to the amount of work initially absorbed, is removed from the system. No work is involved in this second process, and therefore the Conservation-Law statement describing this process is:

$$Q = \overset{0}{\cancel{W}} + \Delta U \quad \text{or} \quad Q = \Delta U.$$

Because heat is removed from the system (negative heat effect), ΔU now represents a decrease in the system's accumulated internal energy. The internal energy loss of the system, as a result of the second process, must equal the internal energy gain of the first process, since the heat removed in the second process is equivalent to the work absorbed by the first process. If the initial and final states of the first process have internal energies of U_1 and U_2, respectively, a system is returned to U_1 by the second process.

But what is the state of the system, i.e., the values of the final pressure and temperature, at the end of the second process? As a result of the initial work process, the state of the system changed so that the system was at a temperature and pressure represented by T_2, P_2. How do the final values of T and P compare to the

---

* Why $(-)$?

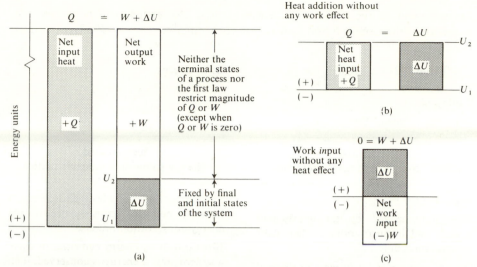

Fig. 2–8. The first law of thermodynamics. Net heat input = net work output + change in accumulated internal energy (if ΔE_K and $\Delta E_P = 0$).

initial values of T_1 and P_1 after the second cooling process? If the final T is greater than T_1, then the system may be placed in contact with any other system at T_1 and heat (energy that flows between two thermally connected systems as a result of a temperature difference between them) will flow out of the system until it cools to T_1. Similarly, if the final T is less than T_1, the system will be able to absorb heat from a source at T_1. In the former case, we shall have gotten more energy out of the system than was put in, and in the latter case we would have returned to our initial state, but somehow or other part of the universe's energy would have disappeared. Since neither of these alternatives can occur, the system, when it returns to U_1, also returns to T_1, and (for similar reasons) to P_1. It therefore returns to the initial state.

Therefore internal energy is a function of (depends on) the state of the system, and is *independent* of the process or path by which the state was produced or the past history of the system. Any process which takes a system from an initial state 1 to a given final state 2 always produces the same internal energy change:

$$\Delta U = U_2 - U_1, \tag{2–14}$$

irrespective of the nature of the process. Because of this characteristic, internal energy is called a STATE PROPERTY. It follows that the temperature, pressure, and volume are also path-independent state properties.

A path-independent property which is closer to everyday experience is elevation. The change in elevation when one travels from the base of a mountain to its peak depends only on the elevation of the peak and the elevation of the base, and is completely independent of the path that the climber chooses to ascend the mountain.

By contrast, neither heat, Q, nor work, W, are state properties. Their magnitudes depend on the path taken between states. Nevertheless, a necessary consequence of the Conservation Law is that their difference, $Q - W$ or ΔU, is independent of path (Fig. 2–8).

Table 2–2 summarizes the principal forms of energy.

Table 2–2

Principal Forms of Energy

Transitory Energy crossing a system's boundary	Accumulated Energy stored in a system's matter
Heat $\equiv Q$ Energy exchanged by an uninsulated system and its surroundings when their temperatures differ. Energy exchanged by systems in thermal contact when their temperatures differ. Work $\equiv W$ Energy exchanged between a system and its surroundings as a result of a (generalized) force acting at the system's boundary, through a (generalized) displacement, whose effect may be made equivalent to raising or lowering a weight in the surroundings. *Reversible forms:*	The form of energy whose existence is a. logically necessary consequence of the energy-conservation premise. If net transitory energy can enter or leave a system, and if energy is conserved, then it must be true that systems can accumulate energy. Internal energy $\equiv U$ Energy whose accumulation in a system is determined by the thermodynamic state of the system. A *change* in internal energy accumulation changes the state of a system, the values of its T, P, V, composition, etc. (U reflects the number, species, bonding, spacing, and motion of the molecules that make up the matter of a system.)
Mechanical $\quad W = \int F\,dX$ Expansion $\quad\;\; W = \int P\,dV$ Electrical $\quad\;\;\; W = -\int \mathbf{V}\,dq$ Surface $\quad\quad W = -\int \Gamma\,d\mathscr{A}$ Acceleration $\; W = -\int mu\,du$ Elevation $\quad\; W = -\int mg\,dx$	External energy $\equiv E_{\text{ext}}$ Energy whose accumulation in the system is independent of temperature or composition and dependent on position or velocity relative to reference coordinates.
Q and W appear only as a system goes through a process, i.e., as the system changes or produces a change. Once the process is completed, the system contains neither Q nor W. Q and W are the only kinds of energy a closed system can exchange with its surroundings.	Forms: Kinetic energy $\qquad\qquad$ Potential energy Total accumulated energy $\equiv E$ $\Delta E \equiv \Delta U + \sum \Delta E_{\text{ext}}$ (During nuclear processes, $\Delta E_{\text{mass}} = c^2\,\Delta m$ must also be considered part of a system's accumulated energy.)

Exercise 2–7

i) Which part of Fig. 2–8 (a, b, or c) best represents the energy interactions of Joule's experiment (Fig. 2–4)?

ii) Draw a diagram similar to Fig. 2–8 for a system which increases its accumulated internal energy from U_1 to U_2 as a result of a heat input and a work input.

Exercise 2–8. For centuries men have tried to make machines that will deliver work without an external energy supply. Such machines are called *perpetual motion machines*. Examples of such devices are shown in Fig. 2–9. The U.S. Patent Office refuses to consider patents on such machines. What do you think is the reason for this policy? What do you think of the machines in Fig. 2–9 and in the frontispiece? Will they work? Why?

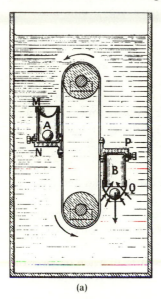

(a)

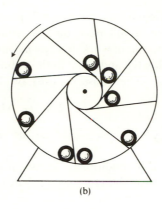

(b)

Fig. 2–9. Two examples of perpetual-motion machines. (a) Buoyant force on right appears to exceed that on left. Therefore machine runs indefinitely in direction shown by curved arrows(?). (b) The overbalancing wheel. Loose masses inside large wheel roll toward the wheel axis on the right and toward the wheel edge on the left, producing an apparent counter-clockwise moment. It is reported that during the seventeenth century a model of this wheel 14 feet in diameter was built in the Tower of London for the King of England. No information is available on the fate of the builder. (Chambers Encyclopedia 1858)

Exercise 2–9. The 10-kg iron cannonball of Example 2–2 has fallen 59 m from the top of the Tower of Pisa into an insulated vat containing 1000 kilograms of wine (Valpolicelli 1563). Neglecting air friction and assuming negligible loss of fluid from the splash, calculate W, Q, ΔU, and ΔE for the system consisting of: (a) The ball only, (b) the wine only, and (c) the wine and ball taken together.

2–17
Enthalpy

In the preceding section, we developed the idea that internal energy is a property of the thermodynamic state of the system, as are temperature, pressure, density, or volume. In the course of exploring thermodynamics, we shall find other thermodynamic (energy-related) properties which are also *uniquely determined* when the state of the system is specified.

ENTHALPY is such a state property. It is a synthetic combination of the internal energy and the expansion work exchanged with the surroundings. Its *raison d'être* is convenience.

Consider a system which is a liquid housed in a container open to the atmosphere. When heat, Q, is added to the system from some convenient heat source, the Conservation Law states that (when there are *no* external energy accumulations):

$$Q = W + \Delta U. \tag{2–15}$$

If the liquid in the system behaves like most liquids, it will expand as its temperature rises, and in so doing it will push up the atmosphere. (See Example 2–7 and Exercise 2–5.) The amount of work done *by the fluid* on the atmosphere is reversible (if the heat is added slowly), and equal to the atmospheric pressure times the volume expansion of the fluid. The Conservation Law (Eq. 2–15) therefore becomes:

$$Q = P \, \Delta V + \Delta U,$$

and the heat absorbed has been used to lift the atmosphere and increase the system's internal energy.

Heat additions under conditions of constant pressure such as the foregoing are most commonplace. It is therefore convenient to have a handle for carrying the sum of the reversible expansion work and the change of internal energy.

Just such a handle was provided by Rudolf Clausius in 1850 when he gave the name *enthalpy* to the package of properties $U + PV$. Enthalpy is symbolized by H, and is defined as:

$$\boxed{H \equiv U + PV} \tag{2–16}$$

(the three-line symbol $\equiv$ means "is defined as" or "is identical to").

Now the differential of enthalpy is:

$$dH = dU + P \, dV + V \, dP.$$

But when P is constant:

$$dH_P = dU + P \, dV \tag{2–16a}$$

Therefore when a system which does expansion work only is *heated at a constant pressure*,

$$dQ = dU + dW = dU + P \, dV = dH_P.$$

The heat entering the system is equal to the internal energy accumulation plus the work done against the atmosphere, which in turn equals the gain in enthalpy of the system. Since P, U, and V are state properties, it follows that H, the enthalpy, is also a state property.

If a system is heated at constant volume, $W = 0$ and

$$Q_V = \Delta U.$$

It therefore follows that ΔU is a measure of the *heat effect* occurring in a *constant-volume* process, whereas ΔH is the corresponding measure for a *constant-pressure* process.

Exercise 2–10. What are the dimensions of enthalpy?

2–18
General procedure for analyzing problems

The concepts we have discussed are applicable to an immense variety of problems, both complex and simple. Applying these concepts is made easier if we use a general procedure for analyzing problems. Simpler problems may not require all the analysis steps.

1. Read the problem carefully, at least twice.
2. Draw a simple schematic or block diagram of the apparatus in its initial and final states.
3. Identify the system and locate its boundaries unambiguously in these diagrams.
4. Does heat enter the system or leave it? If so, indicate by an arrow labeled Q, which points in the appropriate direction, and which crosses the system's boundaries. If there are no heat interactions, note this as $Q = 0$ on the diagram. Processes in which the heat effect is zero are called *adiabatic processes*.
5. Does work enter the system or leave it? (This requires some means of coupling the system to the surroundings, such as a flexible boundary, a shaft, an electrical conductor, etc.) If so, indicate by an arrow labeled W which crosses the system's boundaries in the appropriate direction. If there are no work interactions (rigid system without shaft or other coupling), this should be noted on the diagram as $W = 0$.
6. Does matter enter or leave the system? (Is the system open or closed?) If the system is open, indicate the flow of matter by appropriate arrows.
7. Identify the process and note which (if any) properties (particularly P, V, T, H, or U) remain constant. If possible, draw the process on an appropriate property plane, such as PV or PT.
8. Write the equation for the conservation of energy (2–12) and assign values to as many terms as the statement of the problem permits. The unassigned terms will then be calculable directly; or supplementary relations, equations, or information will be required.

The above analysis may be made easier by preparing a table such as the following. The information in the table applies to Example 2–11.

System	Material	State	P	V	T	U	H	Q	W	Remarks
Closed	Ideal gas	Initial (1)	16.2 psia					0.1 Btu	10.62 ft-lb	Process: isobaric, (constant pressure). W, as in Example 2–10(a)
		Final (2)	16.2							

Example 2–11. Calculate the change in enthalpy and internal energy of the gas in the machine in Fig. 2–6 under the conditions given in Example 2–10. The heat added to the gas is 0.10 Btu.

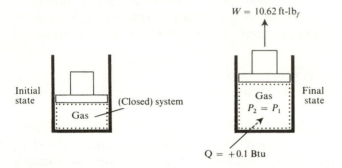

Answer: The system is the *gas*. Pressure remains constant. We know that $Q = +0.10$ Btu (given data), and we know from Example 2–10 that $W = 10.62$ ft-lb$_f$. Therefore:

$$\Delta U = Q - W = 0.10 \text{ Btu} \times \frac{778 \text{ ft-lb}_f}{\text{Btu}} - 10.62 \text{ ft-lb}_f = 67.18 \text{ ft-lb}_f.$$

At constant pressure, we have

$$\Delta H = \Delta U + P\,\Delta V,$$

where P is the pressure of the gas inside the cylinder, which must be equal to atmospheric pressure plus the pressure resulting from the force (3 lb$_f$) exerted by the piston and mass, or 16.2 psia:

$$P\,\Delta V = 16.2 \frac{\text{lb}}{\text{in}^2} \times 2.0 \text{ in}^2 \times \frac{10}{30.5} \text{ ft} = 10.62 \text{ ft-lb}_f,$$

$$\Delta H = 67.18 + 10.62 = 77.8 \text{ ft-lb}_f = 0.10 \text{ Btu} = Q.$$

In other words: *the heat added at constant pressure equals the enthalpy change* of the system (when the system does expansion work only).

In this problem, as in all our problems, we shall be computing the *change* (Δ) in H and U, rather than their absolute values. This is because the premise of thermodynamics deals with, and therefore permits conclusions about, the changes in energy content of a system, and not the absolute energy of a system.

Exercise 2–11. Will the quantity of heat energy entering the system of Example 2–11 be changed if the piston is locked in place (keeping the gas volume constant) and the internal energy accumulation is the same as in Example 2–11?

Exercise 2–12. Draw a diagram similar to Fig. 2–8 for the addition of heat, at constant pressure, to a system capable of doing expansion work only. Place the following labels on the appropriate parts of the diagram: Q input; ΔU, ΔH, $P \Delta V$ Work, $H_{initial}$, H_{final}, $U_{initial}$, U_{final}.

Exercise 2–13. A system in its initial state consists of 1 mol of a chemical A and 1 mol of a chemical B, separated by a thin partition of negligible mass, all at room temperature and atmospheric pressure. When the partition is removed, the chemicals react to form 1 mol of compound C. The reaction is exothermic (i.e., it gives off heat). Thirty thousand calories of heat energy leave the system by the time all the reactants have been converted to C and the system is again back at room temperature and pressure. It is also found that the system in the final state occupies a slightly smaller volume than it did in the initial state.

How do the enthalpy and internal energy of C compare to those of A and B?

How would the results of this experiment differ if the entire process were carried out at constant volume (in a rigid closed container)?

2–19
Specific heat

The quantity

$$C \equiv \frac{d\bar{Q}}{dT} \qquad (2\text{--}17)$$

is called the SPECIFIC HEAT or *heat capacity*. It is a measure of the heat energy that must enter a unit mass of a system to produce a unit increase in temperature.

The measure was introduced by Joseph Black (1728–1799, who was, successively, ᴾrofessor of anatomy, medicine, and chemistry at Glasgow University) to distinguish heat (then thought to be "caloric fluid") from degree of hotness or temperature. Black found that a mass of hot iron immersed in a container of cold water raised the temperature of the water more than an equally hot and massive piece of lead similarly treated. He therefore postulated that iron had more "heat capacity" than lead, and so on, for many different materials. Since we no longer think of "heat" as residing in matter, we shall call the quantity defined by Eq. (2–17) *specific heat*. The caloric-theory term "heat capacity" is still widely used.

The amount of heat required to produce a unit change in temperature in a given material depends on the conditions at which the heating occurs. A gas heated at constant volume gets hotter than one heated at constant pressure (with equal amounts of heat energy). Therefore to specify C, one must specify the conditions under which it is measured.

The most commonly used specific heats are C_P (specific heat at constant pressure)

$$C_P \equiv \left(\frac{d\bar{Q}}{dT}\right)_P = \left(\frac{\partial \bar{H}}{\partial T}\right)_P \qquad (2\text{-}18)$$

and C_V (specific heat at constant volume)

$$C_V \equiv \left(\frac{d\bar{Q}}{dT}\right)_{\bar{V}} = \left(\frac{\partial \bar{U}}{\partial T}\right)_V. \qquad (2\text{-}19)^*$$

Exercise 2-14. Using the definitions of C_P and C_V, prove that

$$C_P = \left(\frac{\partial \bar{H}}{\partial T}\right)_P \qquad \text{and} \qquad C_V = \left(\frac{\partial \bar{U}}{\partial T}\right)_V.$$

It follows from Eqs. (2-18) and (2-19) that for a constant-pressure process

$$\Delta \bar{H}_P = \int_{T_1}^{T_2} C_P \, dT = \bar{Q}_P \qquad (2\text{-}18a)$$

and for a constant-volume process

$$\Delta \bar{U}_V = \int_{T_1}^{T_2} C_V \, dT = \bar{Q}_V. \qquad (2\text{-}19a)$$

Specific heat varies with temperature. The variation is frequently given by an equation of the form

$$C_P = a + bT + cT^2, \qquad (2\text{-}20)$$

where a, b, and c are constants and T is absolute temperature. Values of the constants for some common gases are shown in Appendix I. Because specific heat changes slowly with temperature (b and c are small numbers), it may be considered constant over moderate temperature ranges. Table 2-3 lists specific heats for common substances near room temperature. Specific heat may be expressed in

---

* A *bar* above a property (for example, $\bar{H}$ or $\bar{U}$) signifies that the property is *specific*, that is, that it is being measured per unit mass of the system. The unit mass may be any consistent mass unit (g, lb_m, g-mol, lb-mol, etc.). Specific properties are called *molar* properties when expressed per unit mole of a system. Note that one can make any extensive property intensive (independent of the size of the system) by making it specific.

mass units as Btu/lb·°F or cal/g·°C, or in molar units as Btu/lb-mol·°F and cal/g-mol·°C.

Exercise 2–15. Evaluate the conversion factor for changing a specific heat given in units of Btu/lb$_m$-mol·°F into units of cal/g-mol·°K.

Table 2–3

Specific Heats

Substance	C_P Cal/g-mol·°K or Btu/lb-mol·°R
Water (liquid)	18.0
Air	7.15
O_2	7.40
N_2	7.09
CO	7.14
CO_2	10.46
CH_4 (methane)	10.26
H_2	6.98
H_2O	8.41
Iron	4.13 + 0.0064T °K
Lead (solid)	5.77 + 0.00202T°K

(Gas C_P are mean values between 25° and 400°C.) Computed from data by K. A. Kobe *et al.*, "Thermochemistry for the Petrochemical Industry," *Petrol. Ref.*, January 1949 through November 1954 (App. I).

Exercise 2–16. Suggest a connection between the relative temperature expansion characteristics of gases, liquids, and solids and the fact that C_P and C_V differ appreciably for a gas, whereas they are almost equal for most liquids and solids. In your explanation, use the facts that

$$C_P = \left(\frac{\partial \overline{H}}{\partial T}\right)_P \quad \text{whereas} \quad C_V = \left(\frac{\partial \overline{U}}{\partial T}\right)_V.$$

Example 2–12. What is the change (measured in calories) in enthalpy and internal energy of a pound of water heated from 20°C to 30°C in an open container? The coefficient of thermal expansion of liquid water is 0.000207/°C. What is C_V in this temperature range?

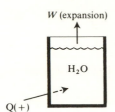

Answer: From Table 2–3,

$$C_P \text{ (H}_2\text{O liquid)} = 18.0 \frac{\text{cal}}{\text{g-mol} \cdot {}^\circ\text{K}} = 1.0 \frac{\text{cal}}{\text{g} \cdot {}^\circ\text{K}} .$$

The system is 1 lb$_m$ of water $= 453.6$ g$_m$ of water.
The pressure remains constant.
The system does expansion work. Therefore:

$$Q = W + \Delta U = P\,\Delta V + \Delta U = \Delta H_P. \tag{a}$$

From Eq. (2–18a),

$$\Delta H_P = m \int d\overline{H} = m \int_{T_1}^{T_2} C_P\, dT = mC_P\,\Delta T = 453.6 \text{ g} \times \frac{1 \text{ cal}}{\text{g} \cdot {}^\circ\text{C}} \times 10{}^\circ\text{C},$$

$$\Delta H = 4536 \text{ cal} \qquad \text{(Answer)}.$$

Now from Eq. (a) above, $\Delta U = \Delta H_P - P\,\Delta V$.
Because the container is open:

$$P = 1 \text{ atm} = 1.013 \times 10^6 \text{ dynes/cm}^2,$$

$$\Delta V = V_{\text{initial}}\,(0.000207/{}^\circ\text{C})\,\Delta T = 453.6 \text{ cm}^3(0.000207)\,10 = 0.939 \text{ cm}^3.$$

And therefore we can write

$$P\,\Delta V = 1.013 \times 10^6 \times 0.939 \text{ dyne-cm} = 0.09525 \times 10^7 \text{ dyne-cm}$$

$$= 0.2273 \text{ cal} \qquad \text{(negligible)}$$

or

$$\Delta U \approx \Delta H = 4536 \text{ calories} \qquad \text{(Answer)}.$$

We know that

$$C_V = \frac{1}{m} \left(\frac{\partial U}{\partial T} \right)_V$$

and, since V is practically constant,

$$C_V \doteq \frac{4536 \text{ cal}}{453.6 \text{ g} \times 10{}^\circ\text{C}} = 1.0 \frac{\text{cal}}{\text{g} \cdot {}^\circ\text{C}} \qquad \text{(Answer)}.$$

**2–20
Phase change**

A material interchanges heat energy with its surroundings when it undergoes a change of phase at constant pressure; that is, when it changes from a liquid to a gas, or from a solid to a liquid, or from one form of solid to another (for example, graphitic carbon changing to diamond). During a phase change at constant pressure, while two phases are present, the heat capacity of a material becomes infinite because the material's temperature does not increase with the addition of heat. Thus water at 212°F and 1 atm pressure does not gain in temperature when heat is added to it. It *boils*, generating steam, which is also at 212°F. If we continue to add heat after all the water has been converted to steam, then the temperature will again begin to rise. To keep our energy books balanced, we must say that the *vapor* has a *higher internal energy* than the *liquid*, when both are at the *same* temperature. The difference is called the *internal energy of the phase transformation*, $\Delta U_{\text{phase change}}$. The transformation may involve vaporization, fusion, condensation, sublimation, or any other phase change.

If the phase transformation occurs at constant pressure P (for example, in an open vessel), then the heat effect accompanying the change, Q_P, equals the enthalpy of the phase change, $\Delta H_{\text{phase change}}$, which differs from $\Delta U_{\text{phase change}}$ by $P \Delta V_{\text{phase change}}$. For example, water increases its specific volume by 26.78 ft^3 per pound when it changes into steam at 1 atm and 212°F. As a result, $\Delta H_{\text{vaporization}}$ is larger than $\Delta U_{\text{vaporization}}$ because part of the heat energy added to the system is used to do work on the atmosphere.

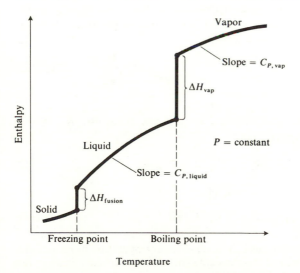

Fig. 2–10. Variation of enthalpy with temperature for a pure substance which changes from a solid to a liquid to a gas. Pressure is constant throughout.

Figure 2–10 shows how the enthalpy of a pure material, at constant pressure, varies with temperature as it passes from solid to liquid to gaseous form.

Exercise 2–17. At 212°F and 1 atm, water has the following properties:

$$\Delta \bar{H}_{vap} = 970.33 \text{ Btu/lb}_m, \qquad \Delta \bar{V}_{vap} = 26.78 \text{ ft}^3/\text{lb}_m.$$

Calculate $\Delta \bar{U}_{vap}$.

Exercise 2–18. Water expands on freezing. Is $\Delta \bar{U}_{fusion}$ $(= \bar{U}_{liquid} - \bar{U}_{solid})$ larger or smaller than $\Delta \bar{H}_{fusion}$?

2–21
Ideal gas

The young thermodynamicist's favorite working substance is a well-behaved material called the IDEAL GAS. The ideal gas is recognized by the fact that it follows the famous IDEAL GAS LAW:

$$\boxed{PV = nRT} \tag{2–21}$$

where n is the number of gram-mols or pound-mols of gas present, T is the *absolute* temperature (°R or °K) and R = UNIVERSAL GAS CONSTANT. The numerical value of R depends on the units in which it is expressed, as shown in Table 2–4. At low pressures, the behavior of most stable gases approximates that of an ideal gas. We then say that these gases are in—or are approaching—the *ideal gas state*. We may also write Eq. (2–21) in terms of molar volume:

$$P\bar{V} = RT. \tag{2–21a}$$

Other characteristics of the ideal gas are:

a) The molar specific heats are proportional to the gas constant, R. For a MONATOMIC GAS (similar to helium, He),

$$C_P = \tfrac{5}{2}R, \qquad C_V = \tfrac{3}{2}R. \tag{2–22}$$

For a DIATOMIC GAS (similar to hydrogen, H_2),*

$$C_P = \tfrac{7}{2}R, \qquad C_V = \tfrac{5}{2}R. \tag{2–23}$$

---

* The specific heats of an ideal diatomic gas can increase with temperature if the diatomic molecular model is allowed to stretch as the molecule rotates. The specific heats of real gases at very low pressures—i.e., in ideal gas states— do in fact increase with temperature. The above values correspond to theoretical statistical mechanical values for "fully excited" rigid rotors. They approximate the specific heats of real gases in the ideal state (at low pressures) when temperatures are near room temperature.

For either MONATOMIC or DIATOMIC GASES, it follows that:

$$C_P - C_V = R. \tag{2–24}$$

b) The enthalpy and internal energy of ideal gases depend on the *temperature only*, and are independent of pressure. ΔH and ΔU are both zero for constant-temperature processes. (This is true only for an ideal gas. For all other materials enthalpy *depends on pressure as well as on temperature*.)

$$\left. \begin{aligned} \Delta H_{\text{ideal gas}} &= n \int_{T_1}^{T_2} C_P \, dT \\ \Delta U_{\text{ideal gas}} &= n \int_{T_1}^{T_3} C_V \, dT \end{aligned} \right\} \begin{aligned} &\text{Pressure has no effect on the} \\ &\text{enthalpy or internal energy} \\ &\text{of an ideal gas.} \end{aligned} \tag{2–24a}$$

Table 2–4

The ideal-gas-law constant: R

Units	Value
$\dfrac{\text{Calories}}{\text{g-mol} \cdot {}^\circ\text{K}}$	1.987
$\dfrac{\text{Btu}}{\text{lb-mol} \cdot {}^\circ\text{R}}$	1.987
$\dfrac{\text{cm}^3 \text{ atm}}{\text{g-mol} \cdot {}^\circ\text{K}}$	82.06
$\dfrac{\text{ft}^3 \text{ lb}_f}{\text{in}^2 \text{ lb-mol} \cdot {}^\circ\text{R}}$	10.731

When an ideal gas is compressed isothermally (at constant temperature), *neither its enthalpy nor its internal energy changes*.

At the molecular level, the ideal gas is pictured as consisting of hard, elastic, inert, moving particles which are completely unconscious of each other's existence, except for the moment when they meet briefly in perfectly elastic collision. All the characteristics (Eqs. 2–21, 2–22, and 2–23) of the ideal gas may be shown to be necessary consequences of this molecular picture (see Section 5). Gases lose their ideality when individual molecules become conscious of the presence of neighboring molecules; that is, when they interact prior to collision.

Exercise 2–19. Calculate the density in lb_m/ft^3 of N_2, O_2, CO_2, and air (MW = 29) at 70°F and 1 atm, assuming ideal-gas behavior for each. Calculate the volume occupied by 1 lb-mol of each of these gases.

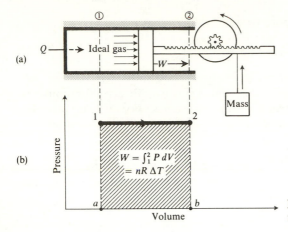

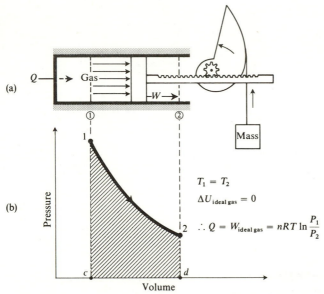

Fig. 2–11. Expansion of an ideal gas under conditions of constant pressure.

Fig. 2–12. Expansion of an ideal gas under conditions of constant temperature (isothermal conditions).

2–22
Reversible work: ideal-gas systems

A gas delivers or receives reversible work by expanding or contracting. If the gas is housed in a frictionless piston-cylinder machine (Figs. 2–11 and 2–12), then an infinitesimal change in the volume of the gas (or an infinitesimal movement of the piston) will involve an amount of work:

$$dW = P \, dV. \tag{2–7}$$

A finite change in the volume of the gas will produce an amount of reversible work:

$$W = \int_1^2 P \, dV \tag{2-25}$$

where the limits (1) and (2) refer to the initial and final states of the expansion (or compression) process.

Equation (2–25) can be integrated only if P is known at every value of V. If the gas is ideal, Eq. (2–21) provides a means for expressing P in terms of V. However, Eq. (2–21) applies to a gas whose properties, P and T, are uniform throughout its volume V. Therefore the equation's use is restricted to work processes which produce no accelerations, and no pressure or temperature nonuniformities within the volume V. The process can therefore proceed only at an infinitesimal rate and with an infinitesimal driving force. If the gas is expanding, then the force exerted on the face of the piston by the gas is only infinitesimally greater than the reaction of the piston. Conversely, if the gas is being compressed, the compressing force on the piston is only infinitesimally greater than the force resulting from the pressure of the gas on the piston. As previously stated, a process which is frictionless and occurs under the conditions described is called a *reversible process*.

For reversible processes, Eq. (2–21) permits elimination of either P or dV from the expression for work, Eq. (2–25):

$$P = \frac{nRT}{V} = \frac{RT}{\overline{V}} \tag{2-26}$$

or

$$d\overline{V} = -\frac{RT}{P^2} \, dP + \frac{R}{P} \, dT, \tag{2-26a}$$

where $\overline{V}$ is volume per mole. When we substitute Eq. (2–26) or (2–26a) in Eq. (2–25), we obtain

$$W = \int RT \, \frac{d\overline{V}}{\overline{V}}$$

or

$$W = \int \left(R \, dT - RT \, \frac{dP}{P} \right).$$

Let us now examine the work effects produced by an ideal-gas expansion process under three different constraints: (a) constant pressure, (b) constant temperature, and (c) adiabatic ($Q = 0$) conditions.

a) *Constant-pressure expansion-work process* (Fig. 2–11): An ideal gas expands from an initial state (1) to a final state (2) as heat is added to it. The pressure of the gas remains constant throughout the process.

The Conservation Law, in differential form, is

$$dQ = dW + dU. \tag{2–26b}$$

When we apply Eq. (2–7), this becomes

$$dQ = P\,dV + dU, \tag{2–27}$$

which says that the heat added to the system is in part converted to expansion work and in part used to increase the accumulation of internal energy of the gas. (The temperature of the gas rises.)

We can calculate the magnitude of the reversible work effect if we know the change in volume:

$$W_P = \int_1^2 P\,dV = P \int_1^2 dV = P\,\Delta V, \tag{2–28}$$

or if we know the change in temperature of the ideal gas:

$$W_P = P\,\Delta V = nR\,\Delta T. \quad \text{(Ideal gas only)} \tag{2–29}$$

Equation (2–28) is valid for reversible isobaric expansion work performed by any substance, and is not limited to ideal gases. Because we used Eq. (2–21) in obtaining the right-hand side of Eq. (2–29), Eq. (2–29) applies only to ideal gases.

W is represented by area 1–2–b–a in Fig. 2–11.

b) *The constant-temperature process* (Fig. 2–12): An ideal gas expands reversibly from initial state (1) to final state (2). The temperature remains *constant* throughout the process. The volume of the gas increases as the pressure of the gas decreases. The cylinder is in contact with a constant-temperature bath which adds heat to the system as the system does expansion work.

The Conservation Law requires that:

$$dQ_T = dW + dU, \qquad dQ_T = dW + 0.$$

Since temperature is constant, $dU_{\text{ideal gas}} = 0$ (Eq. 2–24a), and all the heat entering the system leaves as work. The work, which is shown by area 1–2–d–c in Fig. 2–12, is

$$W_T = \int_1^2 P\,dV.$$

Expressing P as in Eq. (2–26) or dV as in Eq. (2–26a), we obtain

$$W_T = nRT \int_1^2 \frac{dV}{V} = nRT \ln\left(\frac{V_2}{V_1}\right) \tag{2–30}$$

or

$$W_T = -nRT \int_1^2 \frac{dP}{P} = nRT \ln\left(\frac{P_1}{P_2}\right). \tag{2–31}$$

c) *The adiabatic reversible process* (Fig. 2–13): An ideal gas expands reversibly from initial state (1) to final state (2) in a cylinder which is completely *insulated*, so that there is no heat exchange with the surroundings; that is, $Q = 0$. Processes wherein $Q = 0$ are called *adiabatic*. We shall now find that the volume of the gas increases while both the pressure and the temperature of the gas decrease.

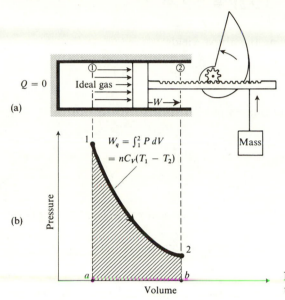

Fig. 2–13. Expansion of an ideal gas under adiabatic conditions.

The Conservation Law is written as

$$dQ = dW + dU = \underline{0} \quad \text{or} \quad dW_Q = -dU,$$

which indicates that *the work effect is produced at the expense of the system's internal energy.* Using Eq. (2–24a), we obtain

$$W_Q = -\Delta U = -nC_V(T_2 - T_1). \tag{2–32}$$

The temperature of the gas *drops* during an adiabatic reversible expansion.

We can use Eq. (2–32) to establish an important relation between P and V for an ideal gas in an adiabatic expansion. Let us write Eq. (2–32) in differential form:

$$dW = P \, dV = -nC_V \, dT, \tag{2–33}$$

then differentiate Eq. (2–21),

$$P \, dV + V \, dP = nR \, dT, \tag{2–34}$$

and use Eq. (2–33) to replace dT in Eq. (2–34),

$$P \, dV + V \, dP = -nR \frac{P \, dV}{nC_V}$$

or

$$\left(1 + \frac{R}{C_V}\right) P \, dV = -V \, dP.$$

Recalling Eq. (2–24) and rearranging the above yields

$$\frac{C_P}{C_V} \cdot \frac{dV}{V} = -\frac{dP}{P}.$$

Integrating between states (1) and (2) gives us

$$\ln \left(\frac{V_2}{V_1}\right)^{C_P/C_V} = \ln \left(\frac{P_1}{P_2}\right)$$

or

$$\boxed{P_1 V_1^{C_P/C_V} = P_2 V_2^{C_P/C_V}.}$$ (Ideal gas only)   (2–35)

Note that Eq. (2–22) and (2–23) fix the value of C_P/C_V for an ideal gas.

Example 2–13. Using the data on heat capacity given in Table 2–3, find the change in enthalpy and internal energy (in calories) of a pound of O_2 heated from 20°C to 30°C at a constant pressure of 1 atm (in a frictionless cylinder-piston machine). What is the expansion coefficient of O_2 at these conditions? Calculate C_V from ΔU.

Answer: From Table 2–3, we find that

$$C_P(O_2 \text{ gas}) = 7.40 \text{ cal/g-mol} \cdot °K = 0.231 \text{ cal/g} \cdot °K.$$

System: 1 lb_m of O_2 (treated as an ideal gas).
Pressure: constant at 1 atm.
Work is that of expansion.

Using Eq. (2–16a), we obtain

$$Q = W + \Delta U$$
$$= P \, \Delta V + \Delta U = \Delta H_P,$$ (a)

and then, using Eq. (2–24a):

$$\Delta H_P = m C_P \, \Delta T = 453.6 \text{ g} \times 0.231 \text{ cal/g} \cdot °K \times 10°K$$
$$\Delta H_P = 1048 \text{ cal.}   \text{(Answer)}$$

Now from Eq. (a) above, we have

$$\Delta U = \Delta H_P - P\,\Delta V.$$

But, for an ideal gas,

$$P\,\Delta V = nR\,\Delta T = \frac{453.6}{32} \times 1.987 \times 10 = 281.5 \text{ cal.}$$

Therefore

$$\Delta U = 1048 - 281.5 = 766.5 \text{ cal,}$$

and

$$C_V = \frac{766.5}{453.6 \times 10} = 0.1689 \text{ cal/g} \cdot {}^\circ\text{K.} \qquad \text{(Answer)}$$

The coefficient of volume expansion (β_T) has units of $\text{cm}^3/\text{cm}^3 \cdot {}^\circ\text{K}$ or ${}^\circ\text{K}^{-1}$. For an ideal gas,

$$\beta_T = \frac{1}{V}\left(\frac{\partial V}{\partial T}\right)_P = \frac{nR}{PV} = \frac{1}{T}.$$

Therefore, at the mean temperature of 25°C (298°K),

$$\beta_T = \frac{1}{298^\circ\text{K}} = 0.00335^\circ\text{K}^{-1}. \qquad \text{(Answer)}$$

Exercise 2–20. Are the values of C_P and C_V used in Example 2–13 in agreement with Eq. (2–24)? How do the values in Table 2–3 compare with Eq. (2–23)?

Exercise 2–21

a) A pound-mole of ideal diatomic gas, initially at 100°F and 10 atm, expands reversibly and adiabatically to twice its original volume. What is its final pressure? How much work is done by the expanding gas?

b) Combine Eq. (2–35) with Eq. (2–21) and derive the relation:

$$\boxed{\frac{T_2}{T_1} = \left(\frac{P_2}{P_1}\right)^{(\gamma-1)/\gamma}} \qquad (2\text{--}36)$$

and

$$\boxed{\frac{T_2}{T_1} = \left(\frac{V_1}{V_2}\right)^{(\gamma-1)}} \qquad (2\text{--}37)$$

where $\gamma = C_P/C_V$.

2–23
Exact and inexact differentials; thermodynamic surfaces

We have stated that temperature, pressure, accumulated internal energy, enthalpy, and volume are properties which depend only on the state of the system, and that the state is specified by two properties if the system has a fixed mass, phase, and composition, and can engage in expansion work only. Consequently a three-dimensional plot of (molar) volume, pressure, and temperature for one mole of pure gas yields a gap-free smooth *surface* ($P\overline{V}T$ surface) which never doubles or closes on itself.*

Figure 2–14 shows a part of the $P\overline{V}T$ surface of an ideal gas. In the figure every set of P and T values has one, and only one, corresponding value of $\overline{V}$. Every set of T and $\overline{V}$ values corresponds to one, and only one, value of P.

Thermodynamic surfaces can be drawn using any three properties of a pure gas system as coordinates. As a further example, Fig. 2–15 shows the accumulated internal energy, pressure, and temperature surface ($\overline{U}PT$ surface) for one mole of ideal gas. This surface is simply a flat plane running through the P-axis and having a slope measured from the PT plane equal to C_V. For a real gas, the internal energy surface would be curved. (Why?)

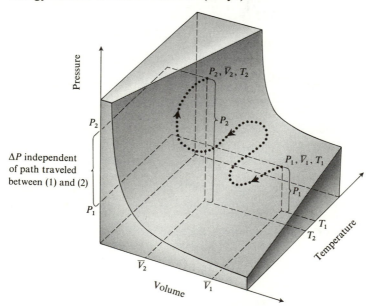

Fig. 2–14. Part of the $P\overline{V}T$ surface of an ideal gas.

* At certain conditions of temperature and pressure, a gas begins to condense to either a liquid or solid. The $P\overline{V}T$ surface demonstrates sudden changes in slope (discontinuities) when these new phases appear. The surface is smooth in regions where only one phase is present.

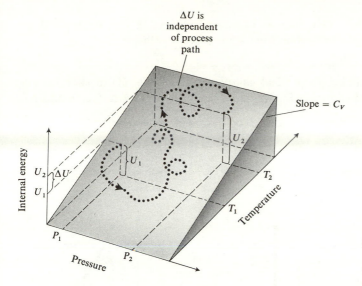

Fig. 2–15. Part of the $\overline{U}PT$ surface of an ideal gas.

Figures 2–14 and 2–15 are pictorial representations of the concept that the state of a pure single-work-mode system is fixed when any *two* of its properties are fixed (Section 2–16). The figures also serve to illustrate the fact that the change in the value of a *property* resulting from a process that takes a system from an initial state (1) to a final state (2) depends only on the coordinates of state (1) and (2) and is *independent of the process*. Thus a system may start at state (1) and wander through a great many intermediate states all over the surface. The changes in its properties depend only on initial and final states (1) and (2) and are not affected by the intermediate wanderings. Consequently we can calculate the change in properties of the system by choosing *any* convenient process or set of processes which will move the system between the desired initial and final states. By a convenient process, we mean one in which the system's properties are *calculable*, and this usually means a reversible process. (This fact will be particularly useful in Section 3, when we consider the property called entropy.)

We express the dependence of thermodynamic properties on the state of the system in mathematical language by saying that thermodynamic properties form EXACT DIFFERENTIALS. By contrast, the work W and the heat Q form INEXACT DIFFERENTIALS. Q and W are not properties of the system, but of the *process* (the functional paths) *that connects two states*. Thus, when there are no external energy effects, the first law of thermodynamics, Eq. (2–15), contains three terms, only one of which (ΔU) is a *property* of the system. When the system moves through a process, the accumulated internal energy change is fixed by the initial and final condition of the process, but no restrictions are placed on the magnitudes of the heat effect Q or the work effect W, other than that their difference must equal the change in internal energy for the process. This holds true for any and all processes, reversible or irreversible. The idea is illustrated in Fig. (2–8). Example 2–14 illustrates the dependence of heat and work on process paths.

Example 2–14. A pound-mole of ideal gas passes from state (1) to state (2) via the reversible processes shown in Fig. (2–16). Determine Q, W, ΔH, ΔU, and ΔT

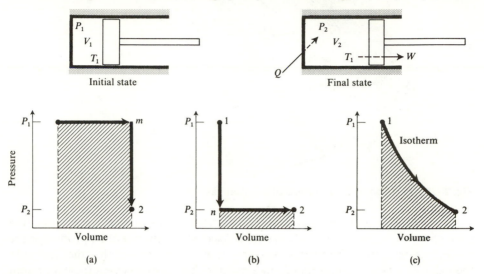

Fig. 2–16. Reversible work processes which can take an ideal gas from state (1) (initial state) to state (2) (final state).

for each process, given that $P_1 = 30$ psia, $V_1 = 200$ ft^3, $P_2 = 15$ psia, and $V_2 = 400$ ft^3.

Answer: ΔH and ΔU are property differences and depend on the terminal states only. Since the terminal states are always the same, the property changes for processes (a), (b), and (c) must all be equal:

$$\Delta H_a = \Delta H_b = \Delta H_c = 1 C_P \, \Delta T \quad \text{(Eq. 2–24a)}$$

and

$$\Delta U_a = \Delta U_b = \Delta U_c = 1 C_V \, \Delta T \quad \text{(Eq. 2–24a)}.$$

But Fig. 2–16(c) indicates that states (1) and (2) are on the same isotherm, i.e., the final temperature equals the initial temperature. Therefore, because the system is an ideal gas,

$$\Delta T = \Delta U = \Delta H = 0.$$

Applying the First Law of Thermodynamics, we find that $Q = W$.

Now let us examine the work done in each process. In (a), the path is $1 \rightarrow m \rightarrow 2$. Process 1 to m occurs at constant pressure P_1 and process m to 2 occurs at constant volume. Therefore

$$W_a = \int_1^m P\,dV + \int_m^2 P\,dV = Q_a.$$

And thus:

$$W_a = Q_a = P_1 \int_1^m dV + 0 = P_1(V_m - V_1).$$

But $V_m = V_2$, and so

$$W_a = Q_a = P_1(V_2 - V_1) = 30 \times 200 \times 144 = 864,000\ \text{ft-lb}_f.$$

In process (b), the path is $1 \to n \to 2$. Process n to 2 occurs at constant pressure P_2 and process 1 to n occurs at constant volume. Therefore

$$W_b = \int_1^n P\,dV + \int_n^2 P\,dV = Q_b.$$

And thus:

$$W_b = Q_b = 0 + P_2(V_2 - V_1) = 15 \times 200 \times 144 = 432,000\ \text{ft-lb}_f.$$

In process (c), the path is $1 \to 2$. This process is one of isothermal expansion, and therefore

$$W_c = \int P\,dV = Q_c.$$

Differentiating the ideal gas law at constant T, we obtain

$$dV = -nRT\,\frac{dP}{P^2}.$$

Therefore

$$W_c = Q_c = -nRT \int_1^2 \frac{dP}{P} = nRT\,\ln\frac{P_1}{P_2} = P_1 V_1 \ln 2 = 598,000\ \text{ft-lb}_f.$$

2-24
Exact and inexact differentials: some mathematical characteristics

An *exact differential* is a function which is the derivative of another function. For example,

$$x\,dy + y\,dx$$

is an exact differential because it is equal to the derivative of the function xy. That is,

$$d(xy) = x\,dy + y\,dx.$$

An interesting property of exact differentials is that the value of their integrals depends only on the integration limits and not on the integration path. This means that

$$\int_1^2 (y\,dx + x\,dy) = x_2 y_2 - x_1 y_1, \tag{a}$$

irrespective of the equation or equations used to relate y to x. To carry out the integration of the left side of Eq. (a), we must use a continuous function or a series of continuous functions, $y = f(x)$, which relates y to x over any path connecting x_1, y_1 to x_2, y_2. The value of the integral will be the same for all paths that start at a particular x_1, y_1, and end at a particular x_2, y_2. That is, the integral of an exact differential depends only on the integration limits, or the final and initial state of the function.

As has been stated, all thermodynamic properties form exact differentials, and their integrals therefore depend only on final and initial thermodynamic states.

An *inexact differential* is a differential which cannot be formed by differentiation. For example,

$$x\,dy - y\,dx$$

is an inexact differential because there is no function ϕ such that

$$d\phi = x\,dy - y\,dx. \tag{b}$$

This does not mean that we cannot find

$$\int_1^2 (x\,dy - y\,dx).$$

We can find it by assigning functional relations, $y = f(x)$, that make possible the integration between the limits. However, and this is the important idea, the value of the integral *depends on the function path* between 1 and 2, unlike the integral of the exact differential, which is the same for all function paths between 1 and 2. Thus dQ and dW are inexact differentials and path dependent, whereas

$$dU = dQ - dW \tag{2–26b}$$

is exact and independent of path. (When we evaluate dU from the above equation, both dQ and dW must follow the same path.)

Exercise 2–22. Divide the right-hand side of Eq. (b) by x^2 and again try to find the integral function. Here $1/x^2$ is called an *integrating factor* because it converts the inexact differential, $x\,dy - y\,dx$, into an exact differential.

Example 2–15

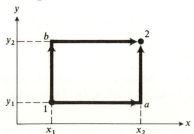

Let us integrate the function

$$\int_1^2 (x\,dy - y\,dx)$$

via path 1–a–2 and via path 1–b–2.

Answer: For path 1–*a*–2: The relation between x and y from 1 to a is

$$y = \text{constant} = y_1,$$

and from a to 2,

$$x = \text{constant} = x_2.$$

Thus we may write

$$\int_{(1-a-2)} (x\,dy - y\,dx) = \int_1^a (x\,dy - y\,dx) + \int_a^2 (x\,dy - y\,dx)$$

$$= 0 - \int_1^a y_1\,dx + \int_a^2 x_2\,dy - 0$$

$$= -y_1(x_a - x_1) + x_2(y_2 - y_a).$$

But $x_a = x_2$ and $y_a = y_1$. Therefore

$$\int_1^a + \int_a^2 = y_1 x_1 - y_1 x_2 + x_2 y_2 - x_2 y_1$$

$$= y_1 x_1 - 2y_1 x_2 + y_2 x_2. \qquad \text{(Answer)} \qquad \text{(c)}$$

For path 1–*b*–2: The relation between x and y from 1 to b is $x = x_1$. From b to 2, $y = y_2$. Therefore

$$\int_{(1-b-2)} (x\,dy - y\,dx) = \int_1^b (x\,dy - y\,dx) + \int_b^2 (x\,dy - y\,dx)$$

$$= x_1(y_2 - y_1) + 0 + 0 - y_2(x_2 - x_1)$$

$$= x_1 y_2 - x_1 y_1 - y_2 x_2 + x_1 y_2$$

$$= -y_1 x_1 + 2x_1 y_2 - y_2 x_2. \qquad \text{(Answer)} \qquad \text{(d)}$$

The value of the integral over path 1–*b*–2 differs from that over path 1–*a*–2. It therefore also follows that the *cyclic* integral (that starts and ends at state 1) is not zero for an inexact differential.

Exercise 2–23. Integrate the functions $(P\,dV + V\,dP)$ and $(P\,dV - V\,dP)/V^2$ over integration paths on a *PV* plane similar to paths 1–*a*–2 and 1–*b*–2 in the previous example. Are these exact or inexact differentials? Compare the value of the integrals for each path. Also re-examine Exercise 2–22.

2–25
Cyclical process

A *cyclical* process is one whose initial and final states are *identical*. A system may wander through any number of different states. If it returns to its initial pressure, volume, and composition, it has completed a cyclical process (Fig. 2–17).

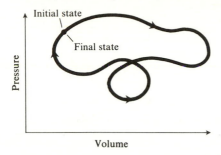

Fig. 2–17. A cyclical process.

Since the initial and final states are the same, it follows that the initial and final properties are also equal. As a result:

$$Q = W + \cancel{\Delta U}^{\,0}$$

or

$$\boxed{Q_{cycle} = W_{cycle}.} \tag{2–38}$$

This may be written as

$$\oint (dQ - dW) = 0 = \oint dU, \tag{2–38a}$$

where $\oint$ means the integral around a closed path.

Equation (2–39) is the *distinguishing characteristic* of a thermodynamic property or exact differential:

$$\boxed{\oint d \ (\text{ANY THERMODYNAMIC PROPERTY}) = 0.} \tag{2–39}$$

2–26
Steam: A nonideal gas

Steam is a fluid that is commonly used as the working substance in energy-conversion devices. Figure 2–18 shows the relation between its pressure, specific volume, and temperature. Clearly this relation is highly complex, and is not described by the ideal gas law except in the very low-pressure, high-temperature region. Steam is therefore a NONIDEAL GAS or vapor, and we *cannot* use thermodynamic expressions based on ideal-gas behavior, such as Eqs. (2–29) through (2–37), to estimate its energy interchanges or changes in its thermodynamic properties.*

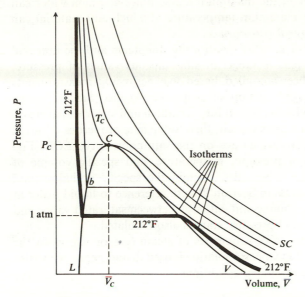

Fig. 2–18. Relationship between pressure and specific volume of liquid and gaseous water at various temperatures.

The complexity of the behavior of steam may be illustrated by considering the constant-temperature line at 212°F (heavy line in Fig. 2–18). At pressures above 1 atm, the line is in the liquid phase (liquid water) and is nearly vertical because liquid water is nearly incompressible. The specific volume of liquid water hardly changes with pressure.

Following the 212°F isotherm down to 1-atm pressure, the isotherm intersects the SATURATED-LIQUID line (LC), which is the locus of pressure and temperature (or specific volume) conditions at which liquid water can exist in equilibrium with steam (or water vapor), or at which liquid begins to boil. The vapor formed from the liquid has a specific volume, considerably greater than that of the liquid, given by the intersection of the 212°F isotherm (or the 1-atm constant-pressure line) with the SATURATED-VAPOR line (VC). The latter line is the locus of water vapor (steam) states that can coexist with liquid water at the same temperature and pressure.

At pressures below 1 atm, 212°F water is a *superheated* vapor. The 212°F

---

* A major objective of this section is to persuade you *not* to use the ideal-gas law to solve all thermodynamic problems. Previous sections emphasized ideal-gas behavior because it yields elementary thermodynamic relations. But these relations apply only at a limiting condition, i.e., the ideal-gas state. The behavior of a real fluid is always more complex. Further discussion is given in Sections 7 and 8.

isotherm drops off to the right as pressure decreases, and begins to take on the hyperbolic shape (PV = constant) of an ideal gas isotherm. Isotherm SC represents the P–V relationship of steam in the *supercritical* temperature range. This is the temperature range above 705.4°F, the maximum temperature at which water can exist in the liquid phase. The maximum temperature at which any material can exist as a liquid is called its *critical temperature*.

Because of the complexity of the $P\overline{V}T$ relationship for steam and the practical need for such information, extensive tables of simultaneous values of pressure, volume, and temperature have been compiled, based on experimental measurements of these properties. An abridged version of such tables (called *Steam Tables*) appears in Appendix II. The table is divided into saturated and superheated steam sections. The saturated-steam sections present thermodynamic properties of liquid and gaseous water in those states which are on the saturation curve LCV (Fig. 2–18). In addition to values of pressure, temperature, and specific volume of saturated liquid and vapor, the tables also contain the specific enthalpy (and entropy, a property we shall discuss in Section 3) with reference to liquid water at 32°F. The difference in specific volume and enthalpy between saturated vapor and liquid in the same temperature and pressure state is also tabulated.

The superheated-steam tables give properties of steam (gas or vapor) in the region to the right of curve CV. Linear extrapolation is used to find steam properties at conditions lying between the tabulated conditions.

a) *Work processes: constant-pressure expansion*

If a fixed quantity of steam is heated at constant pressure, it expands, and can deliver reversible work to its surroundings. (Steam replaces the ideal gas in the machine in Fig. 2–11.) The reversible work of a constant-pressure expansion process is, as before,

$$W_P = P\,\Delta V, \tag{2–28}$$

which is easily evaluated if the pressure and change in volume are given. However, if the pressure *and the temperature* change are given, then we must use the steam table to find the change in specific volume, and then substitute this value in Eq. (2–28). Note that we cannot use Eq. (2–29) because it applies to an ideal gas only.

Example 2–16. One-half pound of water (liquid) is contained in the cylinder in Fig. 2–11 at a constant pressure of 60 psia and an initial temperature of 100°F. The water is heated to 400°F. Determine the change in total volume, the reversible work, the heat added, and the change in enthalpy of the water.

Answer: The process path is superimposed on the saturation curve in Fig. 2–19. Liquid water begins to boil at point (*a*) and at point (*b*) liquid has disappeared, and only vapor is present in the cylinder. From (*b*) to (2), the vapor is heated from its saturation temperature at 60 psia to 400°F.

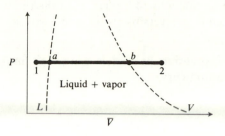

Fig. 2–19

In state (1): $P_1 = 60$ psia, $T_1 = 100°F$, and, from Steam Table 1, we find that

$$\overline{V}_{1\ \text{liq}} = 0.0161\ \text{ft}^3/\text{lb}.$$

(This is the $\overline{V}$ of liquid water at 100°F and 0.949 psia. In view of the incompressibility of liquid water, we assume it to be the same as the $\overline{V}$ of liquid water at 60 psia.) Again from Steam Table 1,

$$\overline{H}_1 = \overline{H} \text{ of saturated liquid at } 100°F = 67.97\ \text{Btu/lb}.$$

In state (2): $P_2 = 60$ psia, $T_2 = 400°F$, and, from Steam Table 3,

$$\overline{V}_2 = 8.353\ \text{ft}^3/\text{lb} \quad \text{and} \quad \overline{H}_2 = 1234.0\ \text{Btu/lb};$$

$$\Delta \overline{V} = 8.353 - 0.016 = 8.337\ \text{ft}^3/\text{lb}.$$

Because the cylinder contains only $\frac{1}{2}$ pound of water, the total volume change is:

$$\text{Total } \Delta V = 0.5(8.337) = 4.169\ \text{ft}^3. \quad \text{(Answer)}$$

For the reversible work:

$$W_P = 60\ \frac{\text{lb}_f}{\text{in}^2} \times 144\ \frac{\text{in}^2}{\text{ft}^2} \times 4.169\ \text{ft}^3 \times 778\ \frac{\text{Btu}}{\text{ft-lb}_f} = 46.4\ \text{Btu}. \quad \text{(Answer)}$$

The equation for the heat added and the enthalpy change is:

$$Q_P = \Delta U + P\,\Delta V = \Delta H.$$

But

$$\Delta H = 0.5(\overline{H}_2 - \overline{H}_1) = 0.5(1234.0 - 67.97)$$

$$= 583.0\ \text{Btu} = Q_P. \quad \text{(Answer)}$$

Exercise 2–24. Find the saturation temperature and the ΔU of the steam in the previous example. Use the relation $\Delta U = \Delta H - \Delta(PV)$, which is a consequence of Eq. (2–16.)

---

b) *Isothermal Expansion: Steam*

Example 2–17. Heat is added to a pound of liquid water contained in a reversible expansion engine, at 200 psia and 381.8°F. The liquid is converted to steam at

381.8°F and the steam is then expanded isothermally to 14.7 psia. Find the total volume and enthalpy change and the amount of reversible work and heat exchanged.

Answer: The isothermal vaporization and expansion process is shown as line 1–2–3 in Fig. 2–20, superimposed on a saturation curve.

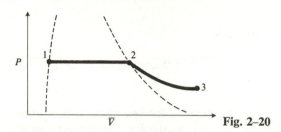

Fig. 2–20

$$P_1 = 200 \text{ psia}; \ T_1 = 381.8°F; \ \bar{V}_1 = 0.0184 \text{ ft}^3/\text{lb}; \ \bar{H}_1 = 355.4 \text{ Btu/lb};$$

$$P_2 = 200 \text{ psia}; \ T_2 = T_1; \ \bar{V}_2 = 2.288 \text{ ft}^3/\text{lb}; \ \bar{H}_2 = 1198.7 \text{ Btu/lb};$$

$$P_3 = 14.7 \text{ psia}; \ T_3 = T_1; \ \bar{V}_3 = 33.92 \text{ ft}^3/\text{lb}; \ \bar{H}_3 = 1230.3 \text{ Btu/lb}.$$

State (1) and state (2) properties are from Steam Table 2. State (3) properties are from Superheated Steam Table 3, by interpolation between 300 and 400°F. Therefore the change in properties as one moves from state (1) to state (3) is:

$$\Delta V = \bar{V}_3 - \bar{V}_1 = 33.90 \text{ ft}^3, \qquad \Delta H = \bar{H}_3 - \bar{H}_1 = 874.9 \text{ Btu} \qquad \text{(Answers)}$$

We may evaluate the reversible work by adding the work of process 1–2 to that of process 2–3. Process 1–2, since it is isobaric (constant pressure) as well as isothermal, is evaluated by Eq. (2–28). Process 2–3 requires the integration

$$W_{2-3} = \int_2^3 P \, d\bar{V}.$$

Because the relation between P and V is not available in equation form, we have to perform the integration graphically by plotting $\bar{V}$ at 381.8°F (obtained by interpolation between saturation temperatures and 400°F) versus pressure, and measuring the area between the curve and the $\bar{V}$ axis between 200 and 14.7 psia. (The graphical integration may be avoided by using tabulated entropy values, as we shall see in Section 3.) Thus $W_{1-3} = 236$ Btu.

Once we know W, then we can write $Q = \Delta U + W$, and

$$\Delta U = \Delta H - \Delta(PV) = (\bar{H}_3 - \bar{H}_1) - (P_3\bar{V}_3 - P_1\bar{V}_1),$$

which enables us to compute that $Q = 1019$ Btu.

Exercise 2–25. Evaluate W_{1-2} and W_{2-3} for the isothermal reversible expansion of one pound of saturated liquid water at 381.9°F and 200 psia to steam at 381.8°F and 14.7 psia Also find Q_{1-2} and Q_{2-3}.

Evaluate W_{2-3} by integrating graphically on a plot of P versus $\overline{V}$ at 381.8°F. Prepare the plot by finding values of $\overline{V}$ at $P = 200, 160, 120, 80, 40,$ and 14.7 psia, by linear interpolation of steam-table data.

What are the thermal efficiencies of the $1 \rightarrow 2$ and $2 \rightarrow 3$ processes; that is, the ratios $W_{\text{out}}/Q_{\text{in}}$?

2–27
Application of the First Law to irreversible processes: the free expansion process

The first law of thermodynamics applies to all processes, whether they are reversible or irreversible; that is, energy is conserved in all possible processes. Consider the irreversible process shown in Fig. 2–21. In its initial state an insulated, rigid con-

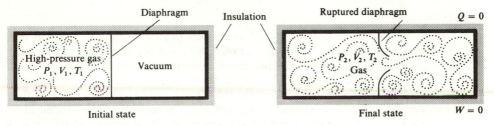

Fig. 2–21. A free expansion process.

tainer is divided into two compartments by a thin diaphragm. The compartment on the left contains gas at high pressure and room temperature. The compartment on the right is evacuated. The diaphragm separating the two compartments ruptures suddenly and gas rushes into the evacuated side. What does the First Law allow us to conclude about the final state of the system (the gas in the container)?

$$Q = W + \Delta U.$$

Since the process is adiabatic ($Q = 0$) and does no work ($W = 0$) on its surroundings,

$$\Delta U = 0 = U_{\text{final}} - U_{\text{initial}}.$$

Now if the gas is *ideal*, ΔU depends only on the temperature change:

$$\Delta U = nC_V(T_{\text{final}} - T_{\text{initial}}) = 0 \quad \text{and} \quad T_{\text{final}} = T_{\text{initial}}.$$

The process described is called a FREE EXPANSION. We come to the important conclusions that:

$$\Delta U_{\text{free expansion}} = 0 \text{ (for all gases)},$$

$$\Delta T_{\text{free expansion}} = 0 \text{ (for ideal gases } only).$$

The gas could have been carried from its initial to its final state by one or more reversible processes. For example, if the gas is ideal and is expanded *isothermally* from V_{initial} to V_{final} (in this case the gas container would have to be fitted with a frictionless piston and a heat-conducting wall):

$$Q = W + \Delta U, \qquad Q = W = nRT \ln \frac{V_{\text{final}}}{V_{\text{initial}}}.$$

ΔU is the same for both the reversible and irreversible processes, since the initial and final states are the same in both processes. However, the reversible process requires an absorption of heat equal to the work performed by the expansion.

Exercise 2–26

a) Steam at P_1, $\overline{V}_1$, T_1, and $\overline{U}_1$ loses half its pressure in a free expansion. Indicate how to find $\overline{V}_2$, P_2, T_2, and $\overline{U}_2$. Choose a set of state (1) values and do the problem.

b) Compute ΔU and the heat and work effects accompanying the following reversible processes, which move one pound-mole of the ideal gas discussed in the preceding paragraph from P_1 (= 30 psia) and V_1 (= 200 ft³) to P_2 (= 15 psia) and V_2 (= 400 ft³). The gas is first expanded reversibly and adiabatically to V_2 and then heated at V_2 until the temperature returns to the initial temperature. Repeat the computation for a free expansion between states (1) and (2).

2–28
The infinitesimal free expansion process

An infinitesimal free expansion process (from V to $V + dV$) can be used to emphasize the difference between dW and $P\,dV$.

A system does work $dW = P\,dV$, only when the system's pressure P gives rise to a force on the surroundings and the surroundings are displaced an amount depending on dV. During an infinitesimal free expansion process there is no force exerted on the surroundings and no displacement of the surroundings; therefore $dW = 0$ even though the term $P\,dV > 0$. In general $dW = P\,dV$ only when the surroundings are resisting the pressure in the system with an equivalent pressure infinitesimally smaller than P_{system}. That is,

$$dW = P\,dV \text{ for reversible processes only.} \qquad (2\text{--}7)$$

The general implication of Eq. (2–7) is that *work done by a system is a function of properties of the system only when the work is done reversibly.*

2–29
Flow processes: open systems

We shall now apply the energy conservation law to open systems. An *open system* is one whose boundaries will admit and/or discharge matter from and to the surroundings. Devices through which matter is flowing can be thought of as open systems. Important engineering examples include pipelines, pumps, turbines, steam engines, continuous chemical reactors, rockets, jet engines, heat exchangers, and so forth.

Figure 2–22 represents a generalized open system or flow process. The system is the matter enclosed by the double-lined sack which marks the boundaries of the system. Streams of matter enter and leave the boundaries through suitable conduits. A rotatable shaft penetrates the boundaries, providing means by which work can be interchanged between the system and its surroundings.

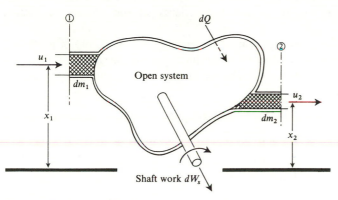

Fig. 2–22. A generalized open system or flow process.

Consider the system for a differential period of time dt during which an amount of matter dm_1 enters and dm_2 leaves, dQ units of heat enter, and dW_s units of work leave through the shaft. We call the work exchange via a moving shaft, W_s, SHAFT WORK. The shaft makes it possible for the system to exchange work energy with the surroundings in a manner other than via expansion or contraction. This work energy must appear in our energy balance as a separate term.

The system's boundaries must receive careful scrutiny. The boundary at the inlet and outlet is flexible, enclosing the matter dm_1 entering the system, and stretching over the material dm_2 that leaves the system in the time interval dt. As dm_1 is *pushed into* the system, the system is *compressed* by an amount equal to the volume of dm_1. If the pressure at the inlet (1) is P_1 and the volume of each pound of entering matter is $\overline{V}_1$ then the admission of dm_1 corresponds to a differential compression work *input* of $P_1(\overline{V}_1 dm_1)$. Similarly, at the exit (2) the *expansion* (or expulsion) of dm_2 into the surroundings corresponds to a work output of $P_2(\overline{V}_2 dm_2)$. The work required to move matter into and out of an open system is called *flow work*, and is in addition to the *shaft work*, W_s, the system may be exchanging.

In the time interval dt, the system in Fig. 2–22 receives energy as heat, dQ, and as compression (or flow) work, $P_1\bar{V}_1 \, dm_1$. Because it is open and receiving matter dm_1, the system also gains the *accumulated internal, potential,* and *kinetic energy* carried by dm_1, or

$$\bar{U}_1 \, dm_1 + gx_1 \, dm_1 + \frac{u_1^2}{2} \, dm_1.$$

Simultaneously the system gives off energy as shaft work, dW_s, expansion (flow) work, $P_2\bar{V}_2 \, dm_2$, and the accumulated energy of the exiting matter,

$$\left[\bar{U}_2 + gx_2 + \frac{u_2^2}{2}\right] dm_2.$$

Any difference between the energy entering the system and the energy leaving it must result in a change in the accumulated energy of the system's contents, dE_{system}.

Because analyzing the energetics of flow systems involves a variety of work and accumulated energies, it is convenient to write the law of conservation of energy in the form suggested by Eq. (2–11) and shown in Eqs. (2–40) and (2–41). In the time interval dt:

	Energy inputs	$-$	Energy outputs	$= \Delta$ accumulated energy
				(2–40)

	Energy inputs	Energy outputs		
Heat	dQ			
Shaft work		dW_s		
Flow work	$P_1\bar{V}_1 \, dm_1$	$P_2\bar{V}_2 \, dm_2$		
Internal energies	$\bar{U}_1 \, dm_1$	$\bar{U}_2 \, dm_2$	$= dE_{\text{system}}$	(2–41)
External energies	$\left(gx_1 + \dfrac{u_1^2}{2}\right) dm_1$	$\left(gx_2 + \dfrac{u_2^2}{2}\right) dm_2$		

If we substitute $\bar{H}$ for $(P\bar{V} + \bar{U})$, Eq. (2–41) rearranges to:

$$dQ - dW_s = (\bar{H}_2 \, dm_2 - \bar{H}_1 \, dm_1)$$
$$+ \left[\left(gx_2 + \frac{u_2^2}{2}\right) dm_2 - \left(gx_1 + \frac{u_1^2}{2}\right) dm_1\right]$$
$$+ dE_{\text{system}}. \qquad (2\text{–}42)$$

Equation (2–41) is a general equation for a flow system. It allows for changes in kinetic energy $[(\Delta u^2/2 \, dm)]$ such as those changes that occur in a jet engine whose operation depends on the immense acceleration of gases between inlet and exhaust. It also allows for changes in gravitational potential energy $(xg \, dm)$ such as may occur in a pipeline conducting a fluid from an elevated tank to a nozzle at ground level. If additional energy effects occur in a flow process, they can be added to the input and output columns of Eq. (2–41).* For many engineering applications, the general Eq. (2–41) can be considerably simplified, as we shall see.

2–30
Steady-flow processes

Many engineering flow systems are designed to operate for extended periods of time without change in their internal conditions. For example, a turbine may run for months with the same inlet and discharge steam pressure and temperature. A pump discharges the same mass of fluid that enters it and usually operates at constant speed, temperature, and pressure. *Steady-flow* conditions are typical of continuously operating engineering systems.† Such conditions imply that neither matter nor energy accumulates between inlet (1) and outlet (2) in Fig. 2–22. Therefore

$$dE_{\text{system}} = 0,$$
$$dm_1 = dm_2 = dm,$$

and

$$(2\text{--}42\text{a})$$

$$\frac{A_1 u_1}{\overline{V}_1} = \frac{A_2 u_2}{\overline{V}_2}. \qquad (2\text{--}43)$$

Equation (2–43) is simply another way of stating that matter enters the system at the same mass rate that it leaves. (A_1 and A_2 are the cross-sectional areas of the in and out conduits; u is the average velocity in the cross section.) Equation (2–43) is called a *continuity equation*.

Exercise 2–27. A fire hose with an internal diameter of 3 inches is connected to a nozzle that tapers to a discharge orifice which has an internal diameter of 1 inch. Compare the velocity of the discharge water to the velocity of the water in the body of the hose. What happens to the discharge velocity if 2% (by weight) of the water leaks out of the hose at the nozzle coupling?

At steady-flow conditions, Eq. (2–42) becomes:

$$dQ - dW_s = (\Delta \overline{H})\,dm + (\Delta x)g\,dm + \left(\frac{\Delta u^2}{2}\right) dm, \qquad (2\text{--}44)$$

---

* In Eq. (2–42) one must use the conversion factors for changing pound-mass units to pound-force units, and ft-lb$_f$ to Btu, in order to make all units in the equation the same. These conversion factors are not shown as part of the equation because it is assumed that the student knows that, in order to evaluate any equation, one must make its units homogeneous by using suitable conversion factors.
† Cyclical processes whose cycle time is short compared with the time of observation may be treated as steady-flow processes. Thus a reciprocating steam engine may be subjected to the same overall analysis as a steam turbine.

where $\Delta\overline{H}$, Δx, and Δu^2 refer to the difference between outlet (2) and inlet (1) conditions. Since $\Delta\overline{H}$, Δx, and Δu^2 are constant for steady-flow conditions, Eq. (2–44) integrates, over the finite time interval required to carry a *unit mass* of material ($m = 1$) through the system, to:

$$\overline{Q} - \overline{W}_s = \Delta\overline{H} + g\,\Delta x + \frac{\Delta u^2}{2}, \tag{2–45}$$

where $\overline{Q}$ and $\overline{W}_s$ are the heat entering and the shaft work *leaving the system per unit mass of material flowing.* When sections (1) and (2) are only a differential distance apart:

$$d\overline{Q} - d\overline{W}_s = d\overline{H} + g\,dx + u\,du. \tag{2–45a}$$

Each term in the equation must be expressed in the same units.

2–31
Frictionless reversible steady-flow processes

When fluid passes through a flow process without friction and so that it is always infinitesimally close to an equilibrium state, the flow process is reversible. During a *reversible* process, the heat energy entering or leaving the system affects only the expansion work and the internal energy. That is,

$$d\overline{Q}_{rev} = d\overline{U} + P\,d\overline{V}. \tag{2–27}$$

If we substitute Eq. (2–27) into Eq. (2–45a) and expand $d\overline{H}$ in terms of $\overline{U}$ and $P\overline{V}$, then

$$-d\overline{W}_{s,\,rev} = \overline{V}\,dP + g\,dx + u\,du. \tag{2–45b}$$

On integrating Eq. (2–45b), we obtain

$$-\overline{W}_{s,\,rev} = \int_1^2 \overline{V}\,dP + g\,\Delta x + \frac{\Delta u^2}{2}. \tag{2–46}$$

If there are no changes in the accumulations of kinetic and potential energy, the shaft work in a reversible flow process equals

$$+\overline{W}_{s,\,rev} = -\int \overline{V}\,dP. \tag{2–46a}$$

Note that $\int \overline{V}\,dP$ is not the same as expansion work $\int P\,d\overline{V}$. We can find the relation between reversible shaft work and reversible expansion work from elementary calculus:

$$d(P\overline{V}) = P\,d\overline{V} + \overline{V}\,dP.$$

Therefore

$$\int P\,d\overline{V} = \int d(P\overline{V}) - \int \overline{V}\,dP \tag{2–46b}$$

Reversible expansion work = flow work + reversible shaft work.

Exercise 2–28. Draw a reversible expansion process on a PV plane between an initial state P_1, V_1 and a final state P_2, V_2, where $P_1 > P_2$ and $V_1 < V_2$. Label the areas on the drawing equal to $\int_1^2 P\, dV$, $\int_1^2 V\, dP$, and $\int_1^2 d(PV)$. Under what conditions is $\int_1^2 d(PV) = 0$?

Integration of Eq. (2–46a) requires a functional relationship between P and $\overline{V}$. When the fluid is incompressible, $\overline{V}$ is constant and the integral becomes $\overline{V} \int dP$ or $\overline{V} \Delta P$, as shown in the example below.

2–32
Engineering applications

In the examples that follow, we apply the energy conservation equations for flow processes to typical engineering devices.

Example 2–18. FRICTIONLESS FLOW OF AN INCOMPRESSIBLE FLUID THROUGH A PIPELINE

An incompressible fluid of low viscosity (water) is pumped at a constant rate of mass flow (50 lb_m per minute) through a pipeline which has an internal diameter of 2 in. and which discharges to the atmosphere through a nozzle which has a diameter of 1 in., as shown in Fig. 2–23. Frictional losses are negligible; $x_2 =$

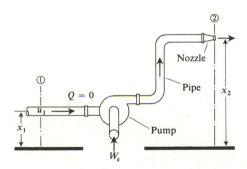

Fig. 2–23. Flow of an incompressible fluid through a pipeline.

100 ft, $x_1 = 0$ ft. Determine the minimum amount of work, $\overline{W}_s$, that must be supplied at the pump shaft, given that $P_1 = 5$ psig.

Answer: The system is the fluid within the pipeline and pump, between sections (1) and (2) of Fig. 2–23. Because frictional losses are negligible, we may assume reversible operation. Equation (2–46) applies, and because the liquid is incompressible, $\overline{V}_2 = \overline{V}_1 = \overline{V}$. This equation may be integrated to

$$-\overline{W}_s = \overline{V}(P_2 - P_1) + (x_2 - x_1)g + \frac{u_2^2 - u_1^2}{2}. \qquad (2\text{–}47)$$

The basis of our computation is 1 lb_m of water. We know that

$$P_2 = 1 \text{ atm} = 14.7 \text{ psia.}$$

Therefore

$$-\overline{W}_s = \overline{V}(P_2 - P_1) + \frac{(1)g}{g_c}(x_2 - x_1) + \frac{1}{2}\frac{(1)}{g_c}(u_2^2 - u_1^2).$$

From the data given in the problem, and from Steam Table 1, we can compute

$$\overline{V}(P_2 - P_1) = 0.0160 \frac{\text{ft}^3}{\text{lb}_m}(14.7 - 19.7)\frac{\text{lb}_f}{\text{in}^2} \times 144\frac{\text{in}^2}{\text{ft}^2} = -11.53\frac{\text{ft-lb}_f}{\text{lb}_m}.$$

We also know that

$$G \equiv \text{mass flow rate} = \frac{50 \text{ lb}_m}{\text{min}} = u_1 A_1/\overline{V} = u_2 A_2/\overline{V}.$$

Therefore

$$u_1 = \frac{50 \text{ lb}_m/\text{min}}{(62.4 \text{ lb}_m/\text{ft}^3)(2/12)^2(\pi/4)\text{ft}^2} = 36.76\frac{\text{ft}}{\text{min}} = 0.613\frac{\text{ft}}{\text{sec}}.$$

And since $u_1 A_1 = u_2 A_2$, then

$$u_2 = \left(\frac{A_1}{A_2}\right)u_1 = \frac{(2)^2(\pi/4)}{(1)^2(\pi/4)}(u_1) = 4u_1 = 2.45\frac{\text{ft}}{\text{sec}}.$$

And thus, using Eq. (2–47), we obtain

$$-\overline{W}_s = -11.53 + \frac{32.2 \text{ ft/sec}^2}{32.2\frac{\text{lb}_m \text{ ft/sec}^2}{\text{lb}_f}}(100 \text{ ft} - 0)$$

$$+ \frac{1}{2}\frac{1}{32.2\frac{\text{lb}_m \text{ ft/sec}^2}{\text{lb}_f}}(2.45^2 - 0.613^2)\frac{\text{ft}^2}{\text{sec}^2} = 89.34 \text{ ft-lb}_f/\text{lb}_m.$$

(Answer)

Exercise 2–29. Water at 70°F flows through a constricted horizontal section of pipe called a Venturi meter, such as that shown in Fig. 2–24. The cross-sectional area of the

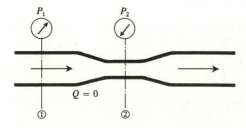

Fig. 2–24. A Venturi meter.

pipe at the constriction (2) is half that in the main section of the pipe (1). Pressure gages are placed at (1) and (2). Draw a graph relating $(P_1 - P_2)$ to the water velocity at (1), expressed in ft/min. Note that W_s and x are both zero in this instant; also that, to make the units homogeneous, the conversion factor, g_c, must be divided into the KE term. If you wish to calculate the pounds of water flowing through the line each minute, what additional information do you need? Neglect fluid friction.

Exercise 2–30. Find the minimum pressure required at the nozzle inlet (upstream) to maintain a discharge velocity of 90 ft/sec at the discharge orifice of the nozzle of the fire hose in Exercise 2–27. Discharge pressure is 1 atm and temperature is 70°F.

Example 2–19. ADIABATIC REVERSIBLE TURBINE. Figure 2–25 is a schematic representation of an expansion turbine. High-pressure, high-temperature gas or

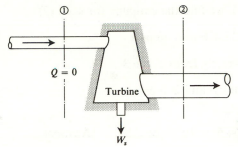

Fig. 2–25. An adiabatic, reversible expansion turbine.

vapor enters the turbine at (1), and expands through the turbine's nozzles and blades, causing the turbine rotor and shaft to rotate, and exits at (2) at a low pressure and temperature. Changes in inlet and exhaust velocities are negligible, as are elevation changes. The turbine operates at reversible steady-flow conditions. Leakage of heat into or out of it is made negligible by insulation (adiabatic). Friction is also negligible. Express the shaft work in terms of inlet and outlet properties.

Answer: The system is the fluid within the turbine between (1) and (2). From conditions stated initially, we know that

$$\bar{Q} = 0 = \Delta x = \Delta u^2.$$

Equation (2–45) takes the form

$$+\bar{W}_s = -\Delta \bar{H} = \bar{H}_1 - \bar{H}_2, \tag{2–48}$$

which means that the shaft work delivered is equal to the enthalpy loss (per unit mass of working fluid). Since $\Delta \bar{H}$ depends only on states (1) and (2), we can calculate the work done by this turbine from knowledge of the inlet and outlet temperatures and pressures. If the working fluid is an ideal gas, $\bar{H}$ will depend only on temperature and $\Delta \bar{H} = C_P \Delta T$. For steam (which is not an ideal gas), we can find the value of $\bar{H}_1$ and $\bar{H}_2$ from tables of thermodynamic properties of steam (Appendix II), if T and P at inlet and outlet are known.

To compare computations of W_s using air and steam as working fluids, let the following conditions hold:

Inlet conditions	Final conditions
$P_1 = 200$ psia	$P_2 = 40$ psia
$T_1 = 700°F$	

For steam: The discharge temperature T_2 is 335.4°F. This is not an arbitrarily chosen temperature; the discharge temperature is a function of the inlet and final conditions and the reversible adiabatic nature of the process. It is found by entropy considerations to be discussed in Section 3.

By interpolating between $\bar{H}$ values at 300°F and 400°F at 40 psia, in the Superheated Steam Table 3 (Appendix II), we find the enthalpy for state (2):

$$\bar{H}_2 = 1204.6 \text{ Btu/lb}, \qquad \text{at } 335.4°F \text{ and } 40 \text{ psia.}$$

Then we read the enthalpy for state (1) directly from Table 3:

$$\bar{H}_1 = 1372.5 \text{ Btu/lb}, \qquad \text{at } 700°F \text{ and } 200 \text{ psia.}$$

Therefore

$$\bar{W}_{s,\text{steam}} = -\Delta\bar{H} = 1372.5 - 1204.6 = 167.9 \text{ Btu/lb}. \qquad \text{(Answer)}$$

For air: we find T_2 by employing Eq. (2–36) because the expansion is reversible and adiabatic, and air may be treated as an ideal gas:

$$\frac{T_2}{T_1} = \left(\frac{P_2}{P_1}\right)^{0.4/1.4}.$$

To calculate T_2, we proceed as follows:

$$T_2 = 1160 \left(\frac{40}{200}\right)^{0.286} = 1160(0.631) = 732°R = 272°F.$$

Therefore

$$\bar{W}_s = -\Delta\bar{H} = -C_P\,\Delta T \qquad (C_P = 7.15; \text{ Table } 2\text{–}3)$$

$$= +7.15 \frac{\text{Btu}}{\text{lb-mol}°R} \times \frac{\text{lb-mol}}{29 \text{ lb air}} \times (1160\text{–}732)°R,$$

or

$$W_s = 105.4 \text{ Btu/lb}. \qquad \text{(Answer)}$$

Example 2–20. JET ENGINE. Figure 2–26 represents an idealized jet engine in steady horizontal flight through still air at velocity u_1. The engine takes in 1 pound (mass) of air at velocity u_1, temperature T_1, and enthalpy $\bar{H}_1$, and discharges it at

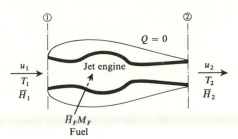

Fig. 2–26. A jet engine.

u_2, T_2, and $\bar{H}_2$. M_F is the number of pounds of fuel used, which is small compared with the number of pounds of air used. $\bar{H}_F$ is the enthalpy of the fuel per pound, and equals the energy released when a pound of fuel is burned at constant pressure.

Relate the thrust of the engine, per pound of air entering, to the enthalpies of the inlet and exit gases. Neglect losses due to friction.

Answer: We can best attack the problem by using Eq. (2–40) directly:

Energy inputs − energy outputs = Δ accumulated energy of system.

Since the system (gas content of the engine) is in steady flight, it does not accumulate energy. In addition, the air, while it is inside the engine, exchanges negligible heat with the surroundings. Hence $dU_{system} = 0 = dQ$. Also there is no shaft coupling the system with the surroundings. Therefore $W_s = 0$. In addition, Δx is zero because the flight is horizontal.

Equation (2–40) becomes (per pound of entering air):

$$\underbrace{\text{Energy inputs}}\quad - \text{ Energy outputs} = 0$$

	Energy inputs	− Energy outputs = 0
Enthalpy	$\bar{H}_1 + \bar{H}_F M_F$	$\bar{H}_2(1 + M_F)$
Kinetic energy	$\dfrac{u_1^2}{2}$	$\dfrac{u_2^2}{2}(1 + M_F)$

or

$$\bar{H}_1 - \bar{H}_2 + \bar{H}_F M_F = (u_2^2 - u_1^2)/2, \quad \text{if} \quad M_F \ll 1.$$

Given data on the enthalpy, we can calculate the discharge velocity u_2 from the above equation. We can then compute the thrust generated by this engine from the equation:

$$\text{Thrust} \left(\frac{\text{lb}_f}{\text{lb}_m \text{ of air/sec}} \right) = (u_2 - u_1) \left(\frac{\text{ft/sec}}{32.2 \dfrac{\text{ft}}{\text{sec}^2} \cdot \dfrac{\text{lb}_m}{\text{lb}_f}} \right).$$

The law of conservation of energy has enabled us to make statements about the operation of a jet engine which are independent of the mechanical details of the engine.

Example 2–21. COMPRESSIBLE ADIABATIC FLOW. Let us consider the adiabatic steady flow of a compressible fluid in a horizontal pipeline of uniform cross section, and let us assume that there are no pumps or turbines in the pipeline so that there is no shaft work. For these conditions, Q, W_s, and Δx all equal zero. Therefore Eq. (2–45) becomes:

$$\Delta \bar{H} = -\frac{\Delta u^2}{2}. \tag{2–49}$$

Dry air (an ideal gas) flows steadily through an insulated pipeline. The velocity of the air is 100 ft/sec at a point at which the temperature is 100°F and the pressure is 30 psia. What is the velocity and the specific volume of the air at a point further downstream from the point at which the air temperature is found to be 95°F?

Answer: We can use Eq. (2–24a) to find $\Delta \bar{H}$ and then use Eq. (2–49) to find the new velocity. The computation will also require the conversion factor g_c. For 1 pound of air,

$$C_P = \frac{6.95\ \text{Btu}}{29\ \text{lb}_m \cdot \text{°F}} = 0.24\ \frac{\text{Btu}}{\text{lb}_m \cdot \text{°F}},$$

$$\Delta \bar{H} = C_P\,\Delta T = 0.24\,(-5) = -1.2\ \frac{\text{Btu}}{\text{lb}_m},$$

$$-\Delta u^2 = \frac{-2 \times 1.2\ \text{Btu}}{\text{lb}_m} \times \frac{\text{ft-lb}_f}{1.285 \times 10^{-3}\ \text{Btu}} \times \frac{32.2\ \text{ft-lb}_m}{\text{sec}^2 \cdot \text{lb}_f},$$

or

$$\Delta u^2 = u_2^2 - u_1^2 = 60{,}000\ \text{ft}^2/\text{sec}^2,$$

$$u_2^2 = 60{,}000 + 10{,}000 = 70{,}000\ \text{ft}^2/\text{sec}^2,$$

$$u_2 = 264.6\ \text{ft/sec}. \quad \text{(Answer)}$$

The velocity more than doubled to produce a relatively small thermal effect. We can find the specific volume of the air at the downstream point by combining the continuity equation (2–43) with the ideal-gas equation (2–21):

$$\frac{A_1 u_1}{\bar{V}_1} = \frac{A_2 u_2}{\bar{V}_2}. \tag{2–43}$$

But $A_1 = A_2$, and therefore

$$\overline{V}_2 = \frac{u_2}{u_1}\,\overline{V}_1 = \frac{u_2}{u_1} \times \frac{RT_1}{P_1}$$

$$= \frac{264.6}{100} \times 10.731 \,\frac{\text{ft}^3\text{lb}_f}{\text{in}^2\text{lb}_m\text{-mol}^\circ\text{R}} \times \frac{560^\circ\text{R}}{30 \,\text{lb}_f/\text{in}^2}\,,$$

or

$$\overline{V}_2 = 530 \,\frac{\text{ft}^3}{\text{lb-mol}} = 18.26 \,\frac{\text{ft}^3}{\text{lb}_m}\,. \qquad \text{(Answer)}$$

Example 2–22. PIPELINE FLOW WITH FLUID FRICTION. The flow of real fluids is always accompanied by viscous friction, which dissipates energy. We may introduce the energy dissipated by fluid friction into Eq. (2–45b) by adding, to the left-hand side of the equation, the term $-d\text{F}$, which is the work energy dissipated by friction. In effect, $d\text{F}$ is the difference between reversible shaft work and the work actually delivered. Equation (2–45b) becomes:

$$-d\text{F} - d\overline{W}_s = \overline{V}\,dP + g\,dx + u\,du. \qquad (2\text{–}50)$$

The dissipation of energy due to friction is dependent on the system's geometry, fluid viscosity, and velocity. The magnitude of this dissipation is determined by the methods of fluid mechanics. Integration of $\overline{V}\,dP$ may present problems.

Example 2–23. ADIABATIC THROTTLING PROCESS. Gas at high pressure passes through an insulated throttling valve to a region of low pressure (Fig. 2–27). A

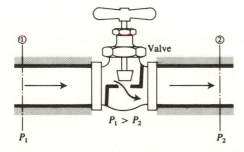

Fig. 2–27. A throttling valve.

throttling valve is simply a partially opened valve which irreversibly reduces the pressure of the fluid passing through it. A small orifice, a porous plug, or any flow-resistance device could be used in this example. The throttling valve has the advantage of offering a controlled and variable resistance to flow. The process is irreversible, adiabatic, and involves no shaft work. Also, differences in velocity upstream and downstream from the valve are usually negligible. Therefore:

$$\bar{Q} = \bar{W}_S = \Delta x = \Delta u^2 = 0.$$

As a consequence, Eq. (2–45) reduces to $\Delta \bar{H} = 0$, or

$$\bar{H}_1 = \bar{H}_2 \qquad \text{(for adiabatic throttling).} \qquad (2\text{–}51)$$

Compare the result of this process with the result of a free-expansion process (Fig. 2–21).

Exercise 2–31. The pressure of an ideal diatomic gas flowing in an insulated pipeline is reduced from 50 psia to 25 psia when it passes through a throttling valve. The initial temperature is 400°F, and kinetic and potential energy effects are negligible. What is the temperature change? Repeat the problem for steam.

Example 2–24. NON-STEADY FLOW PROCESS. Some non-steady flow processes can be treated by a closed rather than an open system analysis. For example, consider the process in Fig. 2–28, in which an empty cylinder is being filled from a large constant high-pressure supply of ideal gas. The heat exchanged with the atmosphere is negligible. Compute the final temperature of the gas in the cylinder when the pressure in the cylinder reaches supply pressure, P_1.

Answer: We can attack the problem by defining the system as a *closed system* whose boundary encloses all the gas in the supply source that will eventually find its way into the cylinder (Fig. 2–28). We then write the Conservation Law:

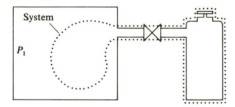

Fig. 2–28. Filling an empty gas cylinder.

$$Q = W + \Delta E, \qquad (a)$$

and note that $Q = 0$ and ΔE consists of ΔU only because the gas that is initially in the supply source and finally in the filled tank has no net velocity. Hence equation (a) becomes

$$W = -\Delta U. \qquad (b)$$

For this system, W is the work done on the system (i.e., the gas that will enter the cylinder) by the *constant-pressure* surroundings (i.e., the gas remaining in the supply source). Therefore

$$W = -P_1 V_1 = -nRT_1, \qquad (c)$$

where V_1 is the volume of gas—at supply conditions (P_1, T_1)—that will eventually flow into the cylinder.

When we substitute equation (c) into equation (b) and employ Eqs. (2–24a) and (2–24), we find that:

$$nRT_1 = nC_V(T_2 - T_1) \quad \text{and} \quad \frac{T_2}{T_1} = \frac{C_P}{C_V}.$$

Our analysis indicates that the contents of the cylinder will become quite hot.

2–33
Nuclear energy

The interconvertibility of mass and energy postulated by the theory of relativity,

$$\text{Energy} = c^2(\text{mass}), \tag{2–10}$$

and demonstrated by atomic bombs and nuclear power plants, does not constitute a violation of the Conservation Law. It simply indicates that we should look upon mass (or rest mass) as a highly condensed form of *accumulated* energy.* The factor c is the velocity of light in a vacuum, 3×10^{10} cm/sec. We may better appreciate the significance of the equation by rewriting it as:

$$\Delta E_{\text{mass}} = c^2 \, \Delta m. \tag{2–52}$$

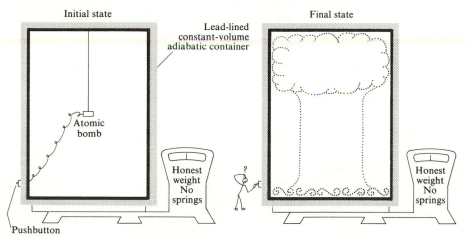

Fig. 2–29. A hypothetical experiment. Does the scale register the destruction of mass?

Consider the difficult experiment illustrated in Fig. 2–29.† Suppose that an

---

* Some theories of cosmology, such as the theory of Continuous Creation, imply that the Energy Conservation Law may be a reflection of the *provincial* nature of our observations, and were we able to make observations over astronomically large spans of space and time, we might find that energy is not conservative.
† Suggested by Dr. G. Hatsopoulos.

atom bomb were to be exploded in a huge rigid-walled, lead-lined adiabatic container, sitting on a sensitive balance (rest-mass meter). What would the change in accumulated energy and the loss of rest mass of the container be as a result of the internal explosion?

Since the process is adiabatic and does no work, $dQ = dW = 0$. Therefore, from Eqs. (2–12) and (2–52), we have

$$\Delta E = 0 = \Delta m. \tag{2–52a}$$

Is there any way for us (the observers on the outside) to know whether or not the bomb has exploded? Equation (2–52a) says that the balance reading won't change, and implies that getting information out of a system may require that there be an exchange of energy with the system.

2–34
Thermochemistry

A further application of the First Law of Thermodynamics makes it possible for us to compute the energy interchanges that accompany a process involving chemical reactions.

Consider a general process in which chemicals A and B react (at 298°K) to produce products C and D in the manner described by the equation:

$$aA + bB \rightarrow cC + dD,$$

where a, b, c, and d are the number of moles of each reaction component.

Let the reaction occur in a rigid closed vessel, as in Fig. 2–30. Initially the

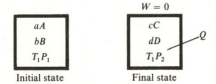

Fig. 2–30. Reaction at *constant volume* and constant temperature:
$$Q_V = \Delta U_{\text{reaction}} = U_{\text{products}} - U_{\text{reactants}}, \qquad aA + bB \rightarrow cC + dD.$$

vessel contains reactants A and B at room temperature. At the end of the process, the vessel is again at room temperature and holds only C and D. A certain quantity of heat, Q, has either left or entered the system (the contents of the vessel) as a result of the process. If heat has *left* the system and gone off into the surroundings, the reaction is called EXOTHERMIC. If the system has absorbed heat from its surroundings, it is called an ENDOTHERMIC reaction.

The Conservation Law must always be obeyed:

$$Q = W + \Delta U. \tag{2–15}$$

The vessel is closed and rigid, and has no shaft. Therefore W is zero and

$$Q_V = \Delta U_{\text{reaction}}, \qquad (2\text{–}53)$$

which is to say that the heat effect accompanying the constant-volume chemical-reaction process must have come from a change in the system's accumulated internal energy ($\Delta U_{\text{reaction}}$). If, for example, the reaction is exothermic, the change in the accumulated internal energy is negative. We are therefore compelled to conclude that the internal energies of the products C and D must be lower than those of the reactants A and B by an amount equal to the heat given off during reaction, *even though all are at the same temperature.*

If the reaction is endothermic, then the internal energies of C and D must be higher than those of A and B. In other words, to be consistent with the premise of conservation of energy, we must conclude that the amount of accumulated internal energy in a system depends on the molecular species that constitute the system (i.e., the system's composition), as well as the temperature, pressure, and phase.

If the above process is repeated in an open vessel exposed to atmospheric pressure (Fig. 2–31), the system, in addition to exchanging heat energy, may also

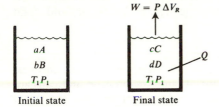

Fig. 2–31. Reaction at *constant pressure* and constant temperature:
$\quad Q_P = \Delta U_R + P\Delta V_R = \Delta H_{\text{reaction}} = H_{\text{products}} - H_{\text{reactants}}, \; aA + bB \rightarrow cC + dD.$

exchange work energy with its surroundings, by expansion or contraction, depending on whether the products occupy more or less volume than the reactants. As always, the Conservation Law holds:

$$Q = W + \Delta U.$$

The work, W, is that of constant-pressure expansion or contraction:

$$W = \int P \, dV = P \, \Delta V_{\text{reaction}} = P(V_{\text{products}} - V_{\text{reactants}}).$$

Therefore the heat effect for a constant-pressure reaction process Q_P is:

$$Q_P = P \, \Delta V_R + \Delta U_R$$

$$= P(V_{\text{products}} - V_{\text{reactants}}) + (U_{\text{products}} - U_{\text{reactants}}), \qquad (2\text{–}53a)$$

which is the difference in enthalpy between the products and the reactants, or the

ENTHALPY OF REACTION:

$$
\begin{aligned}
Q_P &= H_{\text{products}} - H_{\text{reactants}} \\
&= H_{\text{final}} - H_{\text{initial}} \\
&= \Delta H_{\text{reaction}}.
\end{aligned}
\qquad (2\text{–}54)
$$

In Eq. (2–53a), ΔU_R is the change in accumulated energy during a constant-pressure reaction process. It is approximately equal to ΔU_R in Eq. (2–53) for the constant-volume reaction process (same reaction). Substituting Eq. (2–53) into Eq. (2–53a), we obtain

$$
Q_P = P \, \Delta V_R + Q_V. \qquad (2\text{–}54\text{a})^*
$$

Hence the heat effect at constant pressure differs from the heat effect at constant volume by the amount of *work* the system interchanges with its surroundings. An *exothermic* Q_P is greater than Q_V (in an absolute sense) if work energy enters the system (the system contracts and the entering work is negative), and smaller if work energy leaves the system, that is, if the system expands. The work term is important in gaseous reactions, and is usually negligible for liquid and solid reactions.

Exercise 2–32. Compare Q_V, Q_P, ΔH_R, and ΔU_R for ideal-gas, endothermic chemical reaction processes that take place under conditions of constant pressure and constant temperature, and that occur (a) without a change in volume, (b) with an increase in volume, and (c) with a decrease in volume.

Example 2–25. A chemist places 1 g-mol of carbon monoxide and $\frac{1}{2}$ g-mol of oxygen in an ideal piston-cylinder machine at a temperature of 25°C and a pressure of 1 atm. The piston is fitted with a locking device which keeps it stationary. He then ignites the mixture by a tiny spark, and the gases react to form one g-mol of carbon dioxide:

$$
CO + \tfrac{1}{2}O_2 \rightarrow CO_2.
$$

After the initial explosion (which is contained by the piston-cylinder machine), the chemist releases the piston lock and allows the contents of the cylinder to expand or contract to a pressure of 1 atm. At the same time, he cools the outside of the cylinder to bring the temperature back to 25°C. By careful measurement, he finds that 67,636 cal of heat have left the system (the contents of the cylinder)

---

* Generally the pressure changes during a constant-volume reaction process. The accumulated energy in any substance (except ideal gases) depends on both temperature and pressure, but is far less sensitive to pressure than to temperature. Therefore ΔU_R for a constant-pressure process only approximates ΔU_R for a constant-volume process. Strictly speaking, Eq. (2–54a) is precise only for ideal-gas reactions.

between the start of the reaction process and the time the product, CO_2, is cooled back to 25°C, and the pressure is brought back to 1 atm. At this time he also observes that the volume of the contents of the cylinder has decreased, even though no material has leaked out past the piston. Compute the following: (a) The change in volume, (b) $\Delta H_{reaction}$, (c) Q_P, (d) $\Delta U_{reaction}$, (e) the heat that would have been given off if the piston had been kept in its initial locked position and the cylinder had been cooled back to 25°C, and (f) the final pressure in the cylinder under condition (e).

Answer

a) *The change in volume.* The reaction equation shows that the process starts with $1\frac{1}{2}$ g-mol of gas ($1CO$ plus $\frac{1}{2}O_2$) and ends with 1 mol of CO_2. P and T are the *same* in the initial and final states. Therefore

$$V_1 = n_1 \frac{RT_1}{P_1}, \qquad V_2 = n_2 \frac{RT_2}{P_2},$$

$$\Delta V_{reaction} = V_2 - V_1 = \Delta n \frac{RT}{P}$$

$$= -\tfrac{1}{2} \text{ g-mol} \times \frac{82.06 \text{ cm}^3 \cdot \text{atm}}{\text{g-mol} \cdot {}^\circ K} \times \frac{298{}^\circ K}{1 \text{ atm}} = -11{,}207 \text{ cm}^3$$

b) $\Delta H_{reaction} = \Delta U + P\,\Delta V = Q_P = -67{,}636$ cal

c) $Q_P = -67{,}636$ cal (given in statement of problem)

d) $\Delta U_R = \Delta H_R - P\,\Delta V_R = \Delta H_R - \Delta nRT$ (from Eq. 2–54a)

$$= -67{,}636 - \left(-\tfrac{1}{2} \text{ g-mol} \times \frac{1.987 \text{ cal}}{\text{g-mol}{}^\circ K} \times 298{}^\circ K \right) = -67{,}339 \text{ cal}$$

e) $Q_V = \Delta U_V = -67{,}339$ cal

f) $P_{final} = n_2 \dfrac{RT}{V_1} = \dfrac{n_2 RTP_1}{n_1 RT} = \tfrac{2}{3}$ atm

2–35
Standard enthalpies of formation

Experiments tell us that heat and work effects are produced by most chemical reactions, without the intercession of external sources of heat or work energy. This leads to the conclusion that, if the Energy Conservation Law holds, *the amount of accumulated energy in different compounds must differ,* or the enthalpies and internal energies of reactants must differ from those of their products.

Table 2–5 Standard Enthalpies of Formation and Solution

Reference Conditions: 25°C (298.16°K), 1 atm pressure, gaseous
substances in ideal state.

ΔH_F^0 = standard enthalpy of formation, kcal per g-mole
ΔH_s^0 = standard integral enthalpy of solution, kcal per g-mole

Multiply values by 1000 to obtain g-cal per g-mole, or kcal per kg-mole.
Multiply values by 1800 to obtain Btu per lb-mole.

Source: *Selected Values of Chemical Thermodynamic Properties*, Tech. note
270–3 and 270–4, Jan. 1968 and May 1969, D. D. Wagman *et al.*, National
Bureau of Standards.

Abbreviations

c = crystalline state	∞ = infinite dilution	
l = liquid state	ppt = precipitated solid	
g = gaseous state	$amorph$ = amorphous state	
dil = in dilute aqueous solution		

Compound	Formula	State	ΔH_F^0, enthalpy of formation	Moles of water	ΔH_s^0, enthalpy of solution
Acetic acid	CH₃COOH	l	−116.4	∞	−0.343
Aluminum chloride	AlCl₃	c	−166.2	600	−79.3
Aluminum hydroxide	Al(OH)₃	amorph	−304.2		
Aluminum oxide	Al₂O₃	Corundum, c	−399.09		
Aluminum silicate	Al₂SiO₅	Sillimanite, c	−648.9		
Aluminum sulfate	Al₂(SO₄)₃	c	−820.98	∞	−76.12
Ammonia	NH₃	g	−11.04	∞	−8.28
Ammonia	NH₃	l	−16.06	∞	−3.26
Ammonium carbonate	(NH₄)₂CO₃	dil	−225.11		
Ammonium bicarbonate	(NH₄)HCO₃	c	−203.7	∞	+6.78
Ammonium chloride	NH₄Cl	c	−75.38	∞	+3.62
Ammonium hydroxide	NH₄OH	*in* 1 H₂O	−87.64		
Ammonium nitrate	NH₄NO₃	c	−87.27	∞	+6.16
Ammonium oxalate	(NH₄)₂C₂O₄	c	−268.72	2100	+8.12
Ammonium sulfate	(NH₄)₂SO₄	c	−281.86	∞	+1.48
Ammonium acid sulfate	(NH₄)HSO₄	c	−244.83	800	−0.76
Antimony trioxide	Sb₂O₃	c	−168.4	∞	+1.9
Antimony pentoxide	Sb₂O₅	c	−234.4	∞	+8.0
Antimony sulfide	Sb₂S₃	c	−43.5		
Arsenic acid	H₃AsO₄	c	−215.2	∞	+0.4
Arsenic trioxide	As₂O₃	monoclinic, c	−156.4	∞	+6.7
Arsenic pentoxide	As₂O₅	c	−218.6	∞	−6.0
Arsine	AsH₃	g	41.0		
Barium acetate	Ba(C₂H₃O₂)₂	c	−355.1	400	−6.4
Barium carbonate	BaCO₃	c	−291.3		
Barium chlorate	Ba(ClO₃)₂	c	−181.7	∞	+6.1
Barium chloride	BaCl₂	c	−205.56	∞	−3.16
Barium chloride	BaCl₂·2H₂O	c	−349.35	∞	+4.00
Barium hydroxide	Ba(OH)₂	c	−226.2	∞	−12.38
Barium oxide	BaO	c	−133.4	∞	−36.9
Barium peroxide	BaO₂	c	−150.5		
Barium silicate	BaSiO₃	c	−359.5		
Barium sulfate	BaSO₄	c	−350.2	∞	+4.63
Barium sulfide	BaS	c	−106.0	∞	−12.4
Bismuth oxide	Bi₂O₃	c	−137.9		
Boric acid	H₃BO₃	c	−260.2	∞	+5.0
Boron oxide	B₂O₃	c	−302.0	∞	−3.45
Bromine chloride	BrCl	g	+3.51		

Table 2–5 (*continued*)

Compound	Formula	State	ΔH_F^0, enthalpy of formation	Moles of water	ΔH_S^0, enthalpy of solution
Cadmium chloride	$CdCl_2$	c	−93.0	∞	−4.39
Cadmium oxide	CdO	c	−60.86		
Cadmium sulfate	$CdSO_4$	c	−221.36	∞	−12.84
Cadmium sulfide	CdS	c	−34.5		
Calcium acetate	$Ca(C_2H_3O_2)_2$	c	−355.0	∞	−7.5
Calcium aluminate	$CaO \cdot Al_2O_3$	glass	−545		
Calcium aluminate	$2CaO \cdot Al_2O_3$	glass	−695		
Calcium aluminate	$3CaO \cdot Al_2O_3$	glass	−848		
Calcium aluminum silicate	$3CaO \cdot Al_2O_3 \cdot 2SiO_2$	c	−1303		−7.5
Calcium aluminum silicate	$CaO \cdot Al_2O_3 \cdot 6SiO_2$	c	−1828		
Calcium carbide	CaC_2	c	−15.0		
Calcium carbonate	$CaCO_3$	Calcite, c	−288.45		
Calcium chloride	$CaCl_2$	c	−190.0	∞	−19.82
Calcium chloride	$CaCl_2 \cdot 6H_2O$	c	−623.15	∞	+3.43
Calcium fluoride	CaF_2	c	−290.3		
Calcium hydroxide	$Ca(OH)_2$	c	−235.80	∞	−3.88
Calcium nitrate	$Ca(NO_3)_2$	c	−224.0	∞	−4.51
Calcium oxalate	$CaC_2O_4 \cdot H_2O$	c	−399.1		
Calcium oxide	CaO	c	−151.9	∞	−19.46
Calcium phosphate	$Ca_3(PO_4)_2$	c, α	−986.2		
Calcium silicate	$CaSiO_3$	(Wallastonite, c)	−378.6		
Calcium silicate	Ca_2SiO_4	c, β	−538.0		
Calcium sulfate	$CaSO_4$	Anhydrite, c	−342.42	∞	−4.25
Calcium sulfide	CaS	c	−115.3	∞	−4.5
Carbon graphite	C	c	0		
Diamond	C	c	+0.4532		
Amorphous (in coke)	C	amorph	+2.6		
Carbon monoxide	CO	g	−26.4157		
Carbon dioxide	CO_2	g	−94.0518	∞	−4.64
Carbon disulfide	CS_2	g	+27.55		
Carbon disulfide	CS_2	l	+21.0		
Carbon tetrachloride	CCl_4	g	−25.50		
Carbon tetrachloride	CCl_4	l	−33.34		
Chloric acid	$HClO_3$	dil	−23.50		
Chromium chloride	$CrCl_3$	c	−134.6		
Chromium chloride	$CrCl_2$	c	−94.56	∞	−18.64
Chromium oxide	Cr_2O_3	c	−269.7		
Chromium trioxide	CrO_3	c	−138.4	80	−2.5
Cobalt oxide	CoO	c	−57.2		
Cobalt oxide	Co_3O_4	c	−210		
Cobalt chloride	$CoCl_2$	c	−77.8		
Cobalt sulfide	CoS	ppt	−21.4		
Copper acetate	$Cu(C_2H_3O_2)_2$	c	−213.2	∞	−2.7
Copper carbonate	$CuCO_3$	c	−142.2		
Copper chloride	$CuCl_2$	c	−52.3	in aq. HCl	−6.3
Copper chloride	CuCl	c	−32.5		
Copper nitrate	$Cu(NO_3)_2$	c	−73.4	800	−10.4
Copper oxide	CuO	c	−37.1		
Copper oxide	Cu_2O	c	−39.84		
Copper sulfate	$CuSO_4$	c	−184.00	∞	−17.51
Copper sulfide	CuS	c	−11.6		
Copper sulfide	Cu_2S	c	−19.0		
Cyanogen	C_2N_2	g	+73.60		
Hydrobromic acid	HBr	g	−8.66	∞	−20.24
Hydrochloric acid	HCl	g	−22.063	∞	−17.960
Hydrocyanic acid	HCN	g	+31.2	∞	−6.0
Hydrofluoric acid	HF	g	−64.2	∞	−14.46

Table 2–5 (*continued*)

Compound	Formula	State	ΔH_F^0, enthalpy of formation	Moles of water	ΔH_s^0, enthalpy of solution
Hydriodic acid	HI	g	+6.20	∞	−19.57
Hydrogen oxide	H_2O	g	−57.7979		
Hydrogen oxide	H_2O	l	−68.3174		
Hydrogen oxide (heavy water)	D_2O	g	−59.5628		
Hydrogen oxide (heavy water)	D_2O	l	−70.4133		
Hydrogen peroxide	H_2O_2	l	−44.84	∞	−0.84
Hydrogen sulfide	H_2S	g	−4.815	∞	−4.58
Iron acetate	$Fe(C_2H_3O_2)_2$	in 1800 H_2O	−353.8		
Iron carbide	Fe_3C	c	+5.0		
Iron carbonate	$FeCO_3$	c	−178.70		
Iron chloride	$FeCl_2$	c	−81.5	∞	−19.5
Iron chloride	$FeCl_3$	c	−96.8	∞	−31.1
Iron hydroxide	$Fe(OH)_2$	c	−135.8		
Iron hydroxide	$Fe(OH)_3$	c	−197.0		
Iron nitride	Fe_4N	c	−2.55		
Iron oxide	FeO	c	−64.3		
Iron oxide	$Fe_{0.95}O$	Wustite, c	−63.7		
Iron oxide	Fe_3O_4	c	−267.0		
Iron oxide	Fe_2O_3	c	−196.5		
Iron silicate	$FeO·SiO_2$	c	−276		
Iron silicate	$2FeO·SiO_2$	c	−343.7		
Iron sulfate	$Fe_2(SO_4)_3$	in 3000 H_2O	−653.62		
Iron sulfate	$FeSO_4$	c	−220.5	200	−15.5
Iron sulfide	FeS	c	−22.72		
Iron sulfide	FeS_2	Pyrites, c	−42.52		
Lead acetate	$Pb(C_2H_3O_2)_2$	c	−230.5	∞	−2.1
Lead carbonate	$PbCO_3$	c	−167.3		
Lead chloride	$PbCl_2$	c	−85.85	∞	+6.20
Lead nitrate	$PbNO_3$	c	−107.35	∞	+9.00
Lead oxide (yellow)	PbO	c	−52.07		
Lead peroxide	PbO_2	c	−66.12		
Lead suboxide	Pb_2O	c	−51.2		
Lead sesquioxide	Pb_3O_4	c	−175.6		
Lead sulfate	$PbSO_4$	c	−219.50		
Lead sulfide	PbS	c	−22.54		
Lithium chloride	$LiCl$	c	−97.70	∞	−8.877
Lithium hydroxide	$LiOH$	c	−116.45	∞	−5.061
Magnesium carbonate	$MgCO_3$	c	−266		
Magnesium chloride	$MgCl_2$	c	−153.40	∞	−37.06
Magnesium hydroxide	$Mg(OH)_2$	c	−221.00		
Magnesium oxide	MgO	c	−143.84		
Magnesium silicate	$MgSiO_3$	c	−357.9		
Magnesium sulfate	$MgSO_4$	c	−305.5	∞	−21.81
Manganese carbonate	$MnCO_3$	c	−213.9		
Manganese carbide	Mn_3C	c	−1		
Manganese chloride	$MnCl_2$	c	−115.3	400	−16.7
Manganese oxide	MnO	c	−92.0		
Manganese oxide	Mn_3O_4	c	−331.4		
Manganese oxide	Mn_2O_3	c	−232.1		
Manganese dioxide	MnO_2	c	−124.5		
Manganese dioxide	MnO_2	$amorph$	−117.0		
Manganese silicate	$MnO·SiO_2$	c	−302.5		
Manganese silicate	$MnO·SiO_2$	$glass$	−294.0		
Manganese sulfate	$MnSO_4$	c	−254.24	∞	−14.96
Manganese sulfide	MnS	c	−48.8		

Table 2–5 (*continued*)

Compound	Formula	State	ΔH_F^0, enthalpy of formation	Moles of water	ΔH_s^0, enthalpy of solution
Mercury acetate	Hg(C₂H₃O₂)₂	c	−199.4	∞	+4.2
Mercury bromide	HgBr₂	c	−40.5	∞	+3.4
Mercury chloride	HgCl₂	c	−55.0	∞	+3.2
Mercury chloride	Hg₂Cl₂	c	−63.32		
Mercury nitrate	Hg(NO₃)₂	dil	−58.0		
Mercury nitrate	Hg₂(NO₃)₂·2H₂O	c	−206.9		
Mercury oxide	HgO	red, c	−21.68		
Mercury oxide	HgO	yellow, c	−21.56		
Mercury oxide	Hg₂O	c	−21.8		
Mercury sulfate	HgSO₄	c	−168.3		
Mercury sulfate	Hg₂SO₄	c	−177.34		
Mercury sulfide	HgS	Cinnabar, c	−13.90		
Mercury thiocyanate	Hg(CNS)₂	c	+48.0		
Molybdenum oxide	MoO₂	c	−130		
Molybdenum oxide	MoO₃	c	−180.33	∞	−7.77
Molybdenum sulfide	MoS₂	c	−55.5		
Nickel chloride	NiCl₂	c	−75.5	10,000	−19.63
Nickel cyanide	Ni(CN)₂	c	+27.1		
Nickel hydroxide	Ni(OH)₃	c	−162.1		
Nickel hydroxide	Ni(OH)₂	c	−128.6		
Nickel oxide	NiO	c	−58.4		
Nickel sulfide	NiS	c	−17.5		
Nickel sulfate	NiSO₄	c	−213.0	∞	−19.2
Nitrogen oxide	NO	g	+21.600		
Nitrogen oxide	N₂O	g	+19.49		
Nitrogen oxide	NO₂	g	+8.091		
Nitrogen pentoxide	N₂O₅	g	+3.6	∞	−34.03
Nitrogen pentoxide	N₂O₅	c	−10.0	∞	−20.43
Nitrogen tetroxide	N₂O₄	g	+2.309		
Nitrogen trioxide	N₂O₃	g	+20.0		
Nitric acid	HNO₃	l	−41.404	∞	−7.968
Oxalic acid	H₂C₂O₄·2H₂O	c	−340.9	2100	+8.70
Oxalic acid	H₂C₂O₄	c	−197.6	2100	+2.03
Perchloric acid	HClO₄	l	−11.1	∞	−20.31
Phosphoric acid (meta)	HPO₃	c	−228.2	∞	−6.7
Phosphoric acid (ortho)	H₃PO₄	c	−306.2	3000	−3.2
Phosphoric acid (pyro)	H₄P₂O₇	c	−538.0	∞	−7.9
Phosphorous acid (hypo)	H₃PO₂	l	−143.2	∞	−2.4
Phosphorous acid (ortho)	H₃PO₃	l	−229.1	∞	−3.1
Phosphorus trichloride	PCl₃	g	−73.22		
Phosphorus pentoxide	P₂O₅	c	−360.0		
Platinum chloride	PtCl₄	c	−62.9	∞	−19.5
Platinum chloride	PtCl	c	−17.7		
Potassium acetate	KC₂H₃O₂	c	−173.2	∞	−3.68
Potassium carbonate	K₂CO₃	c	−273.93	1000	−7.63
Potassium chlorate	KClO₃	c	−93.50	∞	+9.96
Potassium chloride	KCl	c	−104.175	∞	+4.115
Potassium chromate	K₂CrO₄	c	−330.49	∞	+4.49
Potassium cyanide	KCN	c	−26.90	∞	+2.80
Potassium dichromate	K₂Cr₂O₇	c	−485.90	2000	+17.20
Potassium fluoride	KF	c	−134.46	∞	−4.24
Potassium nitrate	KNO₃	c	−117.76	∞	+8.35
Potassium oxide	K₂O	c	−86.4	∞	−75.28
Potassium sulfate	K₂SO₄	c	−342.66	∞	+5.68
Potassium sulfide	K₂S	c	−100	400	−9.9
Potassium sulfite	K₂SO₃	c	−266.9	∞	−2.2
Potassium thiosulfate	K₂S₂O₃	aq	−274		

Table 2–5 (*continued*)

Compound	Formula	State	ΔH_F^0, enthalpy of formation	Moles of water	ΔH_s^0, enthalpy of solution
Potassium hydroxide	KOH	c	−101.78	∞	−13.22
Potassium nitrate	KNO₃	c	−117.76	∞	+8.348
Potassium permanganate	KMnO₄	c	−194.4	4000	+10.5
Selenium oxide	SeO₂	c	−55.0	∞	+0.93
Silicon carbide	SiC	c	−26.7		
Silicon tetrachloride	SiCl₄	l	−153.0		
Silicon tetrachloride	SiCl₄	g	−145.7		
Silicon dioxide	SiO₂	Quartz, c	−205.4		
Silver bromide	AgBr	c	−23.78		
Silver chloride	AgCl	c	−30.362		
Silver nitrate	AgNO₃	c	−29.43	∞	+5.37
Silver sulfate	Ag₂SO₄	c	−170.50	∞	+4.50
Silver sulfide	Ag₂S	α, c	−7.60		
Sodium acetate	NaC₂H₃O₂	c	−169.8	∞	−4.322
Sodium arsenate	Na₃AsO₄	c	−365	500	−16.5
Sodium tetraborate	Na₂B₄O₇	c	−777.7	900	−10.2
Sodium borate	Na₂B₄O₇·10H₂O	c	−1497.2	900	+26.1
Sodium bromide	NaBr	c	−86.030	∞	−0.15
Sodium carbonate	Na₂CO₃	c	−270.3	400	−5.6
Sodium carbonate	Na₂CO₃·10H₂O	c	−975.6	400	+16.5
Sodium bicarbonate	NaHCO₃	c	−226.5	∞	+4.0
Sodium chlorate	NaClO₃	c	−85.73	∞	+4.95
Sodium chloride	NaCl	c	−98.232	∞	+0.930
Sodium cyanide	NaCN	c	−21.46	200	+0.26
Sodium fluoride	NaF	c	−136.0	∞	+0.06
Sodium hydroxide	NaOH	c	−101.99	∞	−10.246
Sodium iodide	NaI	c	−68.84	∞	−1.81
Sodium nitrate	NaNO₃	c	−101.54	∞	−5.111
Sodium oxalate	NaC₂O₄	c	−314.3	600	+3.8
Sodium oxide	Na₂O	c	−99.4	∞	−56.8
Sodium triphosphate	Na₃PO₄	c	−460	1000	−13.9
Sodium diphosphate	Na₂HPO₄	c	−417.4	1000	−6.04
Sodium monophosphate	NaH₂PO₄	in 300 H₂O	−367.7		
Sodium phosphite	Na₂HPO₃	c	−338	800	−9.5
Sodium selenate	Na₂SeO₄	c	−258	∞	−1.7
Sodium selenide	Na₂Se	c	−63.0	∞	−19.9
Sodium sulfate	Na₂SO₄	c	−330.90	∞	−0.56
Sodium sulfate	Na₂SO₄·10H₂O	c	−1033.48	∞	+18.85
Sodium bisulfate	NaHSO₄	c	−269.2	200	−1.4
Sodium sulfide	Na₂S	c	−89.2	800	−15.16
Sodium sulfide	Na₂S·4½H₂O	c	−416.9	800	+5.11
Sodium sulfite	Na₂SO₃	c	−260.6	∞	−3.2
Sodium bisulfite	NaHSO₃	dil	−206.6		
Sodium silicate	Na₂SiO₃	glass	−360		
Sodium silicofluoride	Na₂SiF₆	c	−677	600	+5.8
Sulfur dioxide	SO₂	g	−70.96	10,000	9.90
Sulfur trioxide	SO₃	g	−94.45	∞	−54.13
Sulfuric acid	H₂SO₄	l	−193.91	∞	−22.99
Tellurium oxide	TeO₂	c	−77.69	∞	+1.31
Tin chloride	SnCl₄	l	−130.3	aq HCl	−29.9
Tin chloride	SnCl₂	c	−83.6	aq HCl	−0.4
Tin oxide	SnO₂	c	−138.8		
Tin oxide	SnO	c	−68.4		
Titanium oxide	TiO₂	amorph	−207		
Titanium oxide	TiO₂	Rutile, c	−218.0		
Tungsten oxide	WO₂	c	−136.3		
Vanadium oxide	V₂O₅	c	−373		

Table 2–5 (*continued*)

Compound	Formula	State	ΔH_F^0, enthalpy of formation	Moles of water	ΔH_S^0, enthalpy of solution
Zinc acetate	Zn(C₂H₃O₂)₂	c	−258.1	800	−9.8
Zinc bromide	ZnBr₂	c	−78.17	∞	−16.06
Zinc carbonate	ZnCO₃	c	−194.2		
Zinc chloride	ZnCl₂	c	−99.40	∞	−17.08
Zinc hydroxide	Zn(OH)₂	c	−153.5		
Zinc iodide	ZnI₂	c	−49.98	∞	−13.19
Zinc oxide	ZnO	c	−83.17		
Zinc sulfate	ZnSO₄	c	−233.88	∞	−19.45
Zinc sulfide	ZnS	c	−48.5		
Zirconium oxide	ZrO₂	c	−258.2		

As a consequence, one can determine the difference in enthalpy between a chemical compound and its constituent elements by measuring the heat effect. accompanying the formation of the compound from its elements, at a specified pressure. This heat effect depends only on the terminal states of the reaction process, because enthalpy is a state property. Therefore a given reaction always yields the same heat effect under the same terminal conditions. When the temperature and pressure of the elemental reactants and the product compound are all 25°C (298°K) and 1 atm, the difference in enthalpy between compound and elements is called the STANDARD ENTHALPY OF FORMATION of the *compound* and has the symbol ΔH_F^0. *The enthalpies of the elements in their normal states of aggregation at 25°C and 1 atm are assumed to be zero.* For example, if 1 mol of HCl gas is produced from its elements at 25°C and 1 atm pressure, then 22,063 calories of heat leave the system, provided that the product gas is cooled back to 25°C:

$$0.5\ H_{2\ (g)} + 0.5\ Cl_{2\ (g)} \rightarrow HCl_{(g)}.$$

[We shall use a subscript (g) to refer to a material which is in a gaseous state, a subscript (c) or (a) to refer to one that is in a crystalline or amorphous state, and a subscript (*l*) to refer to one that is in a liquid state.]

The heat effect (Eq. 2–54) is:

$$Q_P = (\text{enthalpy of products} - \text{enthalpy of reactants}) \text{ at } 25°C$$

$$= \Delta H_F^0\ (\text{HCl}) - \left(0.5\ \Delta H_F^0\ (H_2) + 0.5\ \Delta H_F^0\ (Cl_2)\right)$$

$$= -22{,}063\ \frac{\text{cal}}{\text{g-mol HCl}}.$$

The enthalpies of formation of diatomic hydrogen and chlorine are *assigned* zero values and the *entire* heat effect is *debited* to the enthalpy of formation of the *compound*. Thus Table 2–5, Standard Enthalpies of Formation,* lists HCl$_{(g)}$ as −22.063 kcal/g-mol.

---

* Such tables are commonly called *tables of heat of formation*. The name grows out of the dated practice of using the term *heat content* as a synonym for enthalpy.

When liquid water is formed from its elements, the total heat released by the process—which starts with hydrogen and oxygen at 25°C and ends with liquid water at 25°C—equals $-68,317$ cal per mole of water formed. If we take the enthalpies of hydrogen and oxygen (considered as the normally occurring diatomic molecules) to be zero, it follows from Eq. (2–54) that the enthalpy of liquid water at 25°C and 1 atm with respect to its elements (at 25°C and 1 atm) must be $-68,317$ cal per g-mol. This is the value that appears in Table 2–5 as the standard enthalpy of formation of liquid H_2O.

The enthalpies in Table 2–5 are presented with respect to elements in *normal states of aggregation* at 25°C and 1 atm pressure. Some solid elements can exist in various crystalline and amorphous forms at this standard state. The table, therefore, must also designate which solid form is being used as the reference basis. For example, the standard enthalpies of formation of carbon compounds are given with reference to carbon in the form of *graphite*, and not with reference to the diamond or amorphous forms of carbon.

Exercise 2–33. Use data from Table 2–5 to find the heat effect that accompanies the conversion of graphite to diamond at standard conditions of pressure and temperature.

2–36
Standard enthalpy of combustion

The heat effect that results from the complete combustion, at constant pressure, of 1 mol of a chemical compound with molecular oxygen reflects the difference in enthalpy between the compound and its combustion products. This enthalpy difference is called the enthalpy of combustion (or the heat of combustion).

When the combustion process starts with a reactant and oxygen at 25°C and 1 atm and ends with the complete combustion products cooled back down to 25°C and 1 atm, the accompanying change in enthalpy is called the STANDARD ENTHALPY OF COMBUSTION, ΔH_c^0. Table 2–6 lists values for common organic compounds. These values are given with reference to specified combustion products. For example, the standard enthalpy of combustion of propane is listed as $-530,605$ cal per g-mol. This is the change in enthalpy at 25°C and 1 atm for the reaction process:

$$C_3H_{8\,(g)} + 5O_{2\,(g)} \rightarrow 3CO_{2\,(g)} + 4H_2O_{(l)}.$$

Note that all the carbon has been converted to CO_2 and all the hydrogen to *liquid* water, both at 25°C and 1 atm. The process need not be carried out at a constant temperature of 25°C. All that is required to obtain the change in standard enthalpy

Table 2–6 Standard Enthalpies of Combustion

Reference conditions: 25° C (298.16° K), 1 atm pressure, gaseous
substances in ideal state

ΔH_c^0 = standard enthalpy of combustion, kcal per g-mole

Multiply values by 1000 to obtain g-cal per g-mole, or kcal per kg-mole.
Multiply values by 1800 to obtain Btu per lb-mole.

Abbreviations

s = solid   l = liquid   g = gaseous

Hydrocarbons

Final Products: $CO_2(g)$, $H_2O(l)$

Compound	Formula	State	$-\Delta H_c^0$
Carbon (graphite)	C	s	94.0518
Carbon monoxide	CO	g	67.6361
Hydrogen	H_2	g	68.3174
Methane	CH_4	g	212.798
Ethyne (acetylene)	C_2H_2	g	310.615
Ethene (ethylene)	C_2H_4	g	337.234
Ethane	C_2H_6	g	372.820
Propyne (allylene, methylacetylene)	C_3H_4	g	463.109
Propene (propylene)	C_3H_6	g	491.987
Propane	C_3H_8	g	530.605
1,2-Butadiene	C_4H_6	g	620.71
2-Methylpropene (isobutylene, isobutene)	C_4H_8	g	646.134
2-Methylpropane (isobutane)	C_4H_{10}	g	686.342
n-Butane	C_4H_{10}	g	687.982
1-Pentene (amylene)	C_5H_{10}	g	806.85
Cyclopentane	C_5H_{10}	l	786.54
2,2-Dimethylpropane (neopentane)	C_5H_{12}	g	840.49
2-Methylbutane (isopentane)	C_5H_{12}	g	843.24
n-Pentane	C_5H_{12}	g	845.16
Benzene	C_6H_6	g	789.08
Benzene	C_6H_6	l	780.98
1-Hexene (hexylene)	C_6H_{12}	g	964.26
Cyclohexane	C_6H_{12}	l	936.88
n-Hexane	C_6H_{14}	l	995.01
Methylbenzene (toluene)	C_7H_8	g	943.58
Methylbenzene (toluene)	C_7H_8	l	934.50
Cycloheptane	C_7H_{14}	l	1086.9
n-Heptane	C_7H_{16}	l	1151.27
1,2-Dimethylbenzene (o-xylene)	C_8H_{10}	g	1098.54
1,2-Dimethylbenzene (o-xylene)	C_8H_{10}	l	1088.16
1,3-Dimethylbenzene (m-xylene)	C_8H_{10}	g	1098.12
1,3-Dimethylbenzene (m-xylene)	C_8H_{10}	l	1087.92
1,4-Dimethylbenzene (p-xylene)	C_8H_{10}	g	1098.29
1,4-Dimethylbenzene (p-xylene)	C_8H_{10}	l	1088.16
n-Octane	C_8H_{18}	l	1307.53
1,3,5-Trimethylbenzene (mesitylene)	C_9H_{12}	l	1241.19
Naphthalene	$C_{10}H_8$	s	1231.6
n-Decane	$C_{10}H_{22}$	l	1620.06
Diphenyl	$C_{12}H_{10}$	s	1493.5
Anthracene	$C_{14}H_{10}$	s	1695
Phenanthrene	$C_{14}H_{10}$	s	1693
n-Hexadecane	$C_{16}H_{34}$	l	2557.64

(*continued*)

Table 2–6 (*continued*)

Alcohols

Final Products: $CO_2(g)$, $H_2O(l)$

Compound	Formula	State	$-\Delta H_c^0$
Methyl alcohol	CH_4O	g	182.59
Methyl alcohol	CH_4O	l	173.65
Ethyl alcohol	C_2H_6O	g	336.82
Ethyl alcohol	C_2H_6O	l	326.70
Ethylene glycol	$C_2H_6O_2$	l	284.48
Allyl alcohol	C_3H_6O	l	442.3
n-Propyl alcohol	C_3H_8O	g	494.26
n-Propyl alcohol	C_3H_8O	l	483.56
Isopropyl alcohol	C_3H_8O	g	493.02
Isopropyl alcohol	C_3H_8O	l	481.11
Glycerol	$C_3H_8O_3$	l	396.27
n-Butyl alcohol	$C_4H_{10}O$	g	649.98
n-Butyl alcohol	$C_4H_{10}O$	l	638.18
Amyl alcohol	$C_5H_{12}O$	l	786.7
Methyl-diethyl carbinol	$C_6H_{14}O$	l	926.9

Acids

Final Products: $CO_2(g)$, $H_2O(l)$

Compound	Formula	State	$-\Delta H_c^0$
Formic (monomolecular)	CH_2O_2	g	75.70
Formic	CH_2O_2	l	64.57
Oxalic	$C_2H_2O_4$	s	58.82
Acetic	$C_2H_4O_2$	g	219.82
Acetic	$C_2H_4O_2$	l	208.34
Acetic anhydride	$C_4H_6O_3$	g	432.34
Acetic anhydride	$C_4H_6O_3$	l	426.00
Glycolic	$C_2H_4O_3$	s	166.54
Propionic	$C_3H_6O_2$	g	378.36
Propionic	$C_3H_6O_2$	l	365.41
Lactic	$C_3H_6O_3$	s	325.8
d-Tartaric	$C_4H_6O_6$	s	274.9
n-Butyric	$C_4H_8O_2$	l	520
Citric (anhydrous)	$C_6H_8O_7$	s	474.3
Benzoic	$C_7H_6O_2$	s	771.5
o-Phthalic	$C_8H_6O_4$	s	770.8
Phthalic anhydride	$C_8H_4O_3$	s	781.4
o-Toluic	$C_8H_8O_2$	s	928.6
Palmitic	$C_{16}H_{32}O_2$	s	2379
Stearolic	$C_{18}H_{32}O_2$	s	2628
Elaidic	$C_{18}H_{34}O_2$	s	2663
Oleic	$C_{18}H_{34}O_2$	l	2668
Stearic	$C_{18}H_{36}O_2$	s	2697

Carbohydrates, Cellulose, Starch, etc.

Final Products: $CO_2(g)$, $H_2O(l)$

Compound	Formula	State	$-\Delta H_c^0$
d-Glucose (dextrose)	$C_6H_{12}O_6$	s	673
l-Fructose	$C_6H_{12}O_6$	s	675
Lactose (anhydrous)	$C_{12}H_{22}O_{11}$	s	1350.1
Sucrose	$C_{12}H_{22}O_{11}$	s	1348.9

g-cal per gram

Starch	4177
Dextrin	4108
Cellulose	4179
Cellulose acetate	4495

Table 2-6 (*continued*)

Other CHO Compounds

Final Products: $CO_2(g)$, $H_2O(l)$

Compound	Formula	State	$-\Delta H_c^0$
Formaldehyde	CH_2O	g	134.67
Acetaldehyde	C_2H_4O	g	284.98
Acetone	C_3H_6O	g	435.32
Acetone	C_3H_6O	l	427.79
Methyl acetate	$C_3H_6O_2$	g	397.5
Ethyl acetate	$C_4H_8O_2$	g	547.46
Ethyl acetate	$C_4H_8O_2$	l	538.76
Diethyl ether	$C_4H_{10}O$	l	652.59
Diethyl ketone	$C_5H_{10}O$	l	738.05
Phenol	C_6H_6O	g	747.55
Phenol	C_6H_6O	l	731.46
Pyrogallol	$C_6H_6O_3$	s	639
Amyl acetate	$C_7H_{14}O_2$	l	1040
Camphor	$C_{10}H_{16}O$	s	1411

Nitrogen Compounds

Final Products: $CO_2(g)$, $N_2(g)$, $H_2O(l)$

Compound	Formula	State	$-\Delta H_c^0$
Urea	CH_4N_2O	s	151.05
Cyanogen	C_2N_2	g	261.70
Trimethylamine	C_3H_9N	l	578.4
Pyridine	C_5H_5N	l	660
Trinitrobenzene (1,3,5)	$C_6H_3N_3O_6$	s	664.0
Trinitrophenol (2,4,6)	$C_6H_3N_3O_7$	s	620.0
o-Dinitrobenzene	$C_6H_4N_2O_4$	s	703.2
Nitrobenzene	$C_6H_5NO_2$	l	739
o-Nitrophenol	$C_6H_5NO_3$	s	689
o-Nitroaniline	$C_6H_6N_2O_2$	s	766
Aniline	C_6H_7N	l	812
Trinitrotoluene (2,4,6)	$C_7H_5N_3O_6$	s	821
Nicotine	$C_{10}H_{14}N_2$	l	1428

Halogen Compounds

Final Products: $CO_2(g)$, $H_2O(l)$, dil.sol. of HCl

Compound	Formula	State	$-\Delta H_c^0$
Carbon tetrachloride	CCl_4	g	92.01
Carbon tetrachloride	CCl_4	l	84.17
Chloroform	$CHCl_3$	g	121.8
Chloroform	$CHCl_3$	l	114.3
Methyl chloride	CH_3Cl	g	182.81
Chloracetic acid	$C_2H_3ClO_2$	s	172.24
Ethylene dichloride	$C_2H_4Cl_2$	l	296.77
Ethyl chloride	C_2H_5Cl	g	339.66

Sulfur Compounds

Final Products: $CO_2(g)$, $SO_2(g)$, $H_2O(l)$

Compound	Formula	State	$-\Delta H_c^0$
Carbonyl sulfide	COS	g	132.21
Carbon disulfide	CS_2	g	263.52
Carbon disulfide	CS_2	l	256.97
Methyl mercaptan	CH_4S	g	298.68
Dimethyl sulfide	C_2H_6S	g	457.12
Dimethyl sulfide	C_2H_6S	l	450.42
Ethyl mercaptan	C_2H_6S	l	448.0

References:
  1. Selected Values of Physical and Thermodynamic Properties of Hydrocarbons and Related Compounds, *Am. Petroleum Inst. Research Proj.* **44**, edited by F. D. Rossini, Carnegie Institute of Technology (1952).
  2. *International Critical Tables*, vol. V (1929). The values taken from this source were converted to a reference temperature of 25° C.
  3. John H. Perry, *Chemical Engineers Handbook,* 4th Ed., McGraw-Hill (1963).

is that the reactions start at 25°C and 1 atm and the reaction products be cooled down to the standard conditions of 25°C and 1 atm. In other words, the change in enthalpy is independent of the path and depends only on the initial and final states. For hydrocarbons the final state for standard enthalpy of combustion tables is usually liquid H_2O and gaseous CO_2 at 25°C and 1 atm.

Many compounds cannot be made directly from their elements. One can determine enthalpies of formation for these compounds indirectly by a suitable combination of data on the enthalpy of combustion for the compound in question and the enthalpies of formation of the combustion products.

Example 2–26. Compute the standard enthalpy of formation of C_3H_8 (propane) from data on the standard enthalpy of combustion and standard enthalpy of formation.

Answer: You can compute the ΔH_F^0 for C_3H_8 by combining the reversed combustion reaction of C_3H_8 with the formation reactions of the reaction products (Fig. 2–32):

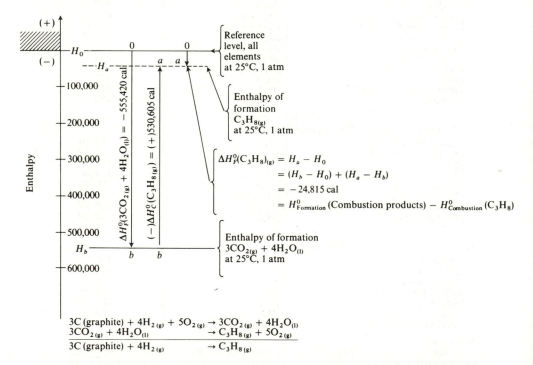

Fig. 2–32. The computation path $(0 \rightarrow b \rightarrow a)$ for finding the enthalpy of formation of propane from data on enthalpy of combustion.

$$3CO_{2 (g)} + 4H_2O_{(l)} \rightarrow C_3H_{8 (g)} + 5O_{2 (g)}; \qquad (-)\,\Delta H_C^0 = +530{,}605 \text{ cal}$$
$$\text{(Table 2–6),}$$

$$3C_{(graphite)} + 3O_{2 (g)} \rightarrow 3CO_{2 (g)}; \qquad 3\Delta H_F^0 = -282{,}152$$
$$\text{(Table 2–5),}$$

$$4H_{2 (g)} + 2O_{2 (g)} \rightarrow 4H_2O_{(l)}; \qquad 4\Delta H_F^0 = -273{,}268$$
$$\text{(Table 2–5),}$$

$$3C_{(graphite)} + 4H_{2 (g)} \rightarrow C_3H_{8 (g)}; \qquad \Delta H_F^0 = -24{,}815.$$
$$\text{(Answer)}$$

Note the change of sign in ΔH_F^0 when the direction of a reaction is reversed.

Similar calculations make it possible for one to determine indirectly the standard enthalpy of formation of many of the compounds listed in Table 2–5.

2–37
Standard enthalpy change for any chemical reaction process at 25°C, 1 atm

Since enthalpy is a property, we can evaluate the change in enthalpy in going from the intial to the final state of *any* chemical process by any calculable path that connects the initial and final states. When the terminal states of the process consist of compounds, for example,

$$CaCO_3 \rightarrow CaO + CO_2 \qquad \text{(25°C and 1 atm),}$$

the most convenient calculation path is one that (a) starts with the reactants and moves to their elements, and then (b) moves from the product elements to the products (Fig. 2–33). The change in enthalpy for (a) is minus the standard enthalpy of formation of the reactants. The change in enthalpy for (b) is the standard enthalpy of formation of the products. Hence the change in standard enthalpy and the heat effect of any chemical reaction that *begins and ends at 25°C and 1 atm* is:

$$\Delta H_R^0 = \sum \Delta H_{F \text{ (products)}}^0 - \sum \Delta H_{F \text{ (reactants)}}^0, \qquad (2\text{–}54c)$$

where the $\sum$'s refer to all the substances in the balanced chemical equation. Data on standard enthalpy of combustion may be used in an analogous fashion when standard enthalpies of formation are not available.

The computation is shown schematically in Fig. 2–33, and in the examples that follow. Note that the direction of the process path is always shown by the arrow in the balanced chemical equation, and this direction determines the sign of the enthalpy change.

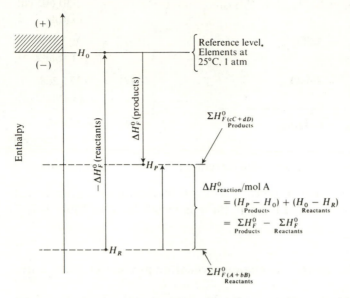

Fig. 2–33. The computation path ($H_R \rightarrow H_0 \rightarrow H_P$) for finding the change in enthalpy ($\Delta H^0_{reaction}$) which takes place during a chemical reaction process at 25°C and 1 atm: $aA + bB \rightarrow cC + dD$.

Example 2–27. STANDARD ENTHALPY OF REACTION FROM STANDARD ENTHALPIES OF FORMATION. Calculate the standard enthalpy of reaction for the conversion of the calcite form of calcium carbonate to lime.

$$CaCO_{3\,(calcite)} \rightarrow CaO_{(c)} + CO_{2\,(g)}, \qquad \text{at 25°C and 1 atm.}$$

The computation makes use of Eq. (2–54c):

$$\Delta H^0_R = \sum \Delta H^0_{F\,(products)} - \sum \Delta H^0_{F\,(reactants)} \qquad (2\text{–}54c)$$

$$= \Delta H^0_F(CaO_{(c)}) + \Delta H^0_F(CO_{2\,(g)}) - \Delta H^0_F(CaCO_{3\,(calcite)})$$

$$= -151,900 - 94,052 - (-283,450)$$

or

$$\Delta H^0_R = +42,498 \text{ cal/g-mol } CaCO_3.$$

The reaction is endothermic, and will require at least this much heat if the process is carried out at standard pressure and temperature.

Example 2–28. STANDARD ENTHALPY OF REACTION FROM STANDARD ENTHALPIES OF COMBUSTION. Calculate the standard enthalpy of reaction for the conversion of ethyl alcohol to diethyl ether.

Answer: At 25°C, we have

$$2C_2H_5OH_{(l)} \rightarrow C_2H_5OC_2H_{5\,(l)} + H_2O_{(l)}.$$

From Table 2–6 (note that the ΔH_C^0 in Table 2–6 are represented with respect to $H_2O_{(l)}$), we find the enthalpies of combustion of ethyl alcohol and ether and combine them as shown here:

$$2C_2H_5OH_{(l)} + 6O_{2\,(g)} \rightarrow 4CO_{2\,(g)} + 6H_2O_{(l)};$$

$$\Delta H_C^0 = 2(-326,700) = -653,400 \text{ cal/2 g-mol},$$

$$4CO_{2\,(g)} + 5H_2O_{(l)} \rightarrow C_2H_5OC_2H_{5\,(l)} + 6O_{2\,(g)};$$

$$-\Delta H_C^0 = 652,590 \text{ cal/g-mol}.$$

When we add the two equations, we obtain the desired reaction and its ΔH:

$$2C_2H_5OH_{(l)} \rightarrow C_2H_5OC_2H_{5\,(l)} + H_2O_{(l)};$$

$$\Delta H_R^0 = -810 \text{ cal/g-mol of ether}.$$

Example 2–29. Calculate H_R^0 for the nitration of benzene with 100% HNO_3.

Answer: $C_6H_{6\,(l)} + HNO_{3\,(l)} \rightarrow C_6H_5NO_{2\,(l)} + H_2O_{(l)};$ 25°C.

The ΔH_C^0 data are available for the organic compounds, and ΔH_F^0 for the acid. The former enthalpies are presented with respect to $CO_{2\,(g)}$, liquid H_2O, and $N_{2\,(g)}$ at 25°C and 1 atm, whereas the latter enthalpy is given with respect to the elements H_2, O_2, and N_2. Before we can combine these data, we must express all of them with respect to the same reference state. Either we can convert the ΔH_C^0 data to ΔH_F^0 data, as we did previously (Example 2–26), or we can use the ΔH_F^0 for HNO_3 to find the enthalpy of HNO_3 with respect to $H_2O_{(l)}$, $O_{2\,(g)}$, and $N_{2\,(g)}$, which are the reference materials for ΔH_C^0. Let us do the latter.

We can obtain ΔH_F^0 for $H_2O_{(l)}$ and $HNO_{3\,(l)}$ from Table 2–5 and combine them as shown below so as to produce the needed ΔH_C^0:

$$HNO_{3\,(l)} \rightarrow \tfrac{1}{2}H_{2\,(g)} + \tfrac{1}{2}N_{2\,(g)} + \tfrac{3}{2}O_{2\,(g)};$$

$$-\Delta H_F^0 = +41,404 \text{ cal/g-mol } HNO_3.$$

$$\tfrac{1}{2}H_{2\,(g)} + \tfrac{1}{4}O_{2\,(g)} \rightarrow \tfrac{1}{2}H_2O_{(l)};$$

$$\Delta H_F^0 = -34,159,$$

$$HNO_{3\,(l)} \rightarrow \tfrac{1}{2}H_2O_{(l)} + \tfrac{1}{2}N_{2\,(g)} + \tfrac{5}{4}O_{2\,(g)};$$

$$-\Delta H_C^0 = +7,245 \text{ cal/g-mol } HNO_3.$$

We now combine the above enthalpy with the standard enthalpies of combustion taken from Table 2–6:

$$C_6H_6{}_{(l)} + 7\tfrac{1}{2}O_2{}_{(g)} \rightarrow 6CO_2{}_{(g)} + 3H_2O_{(l)};$$

$$\Delta H_C^0 = -780,980,$$

$$HNO_3{}_{(l)} \rightarrow \tfrac{1}{2}H_2O_{(l)} + \tfrac{1}{2}N_2{}_{(g)} + \tfrac{5}{4}O_2{}_{(g)};$$

$$\Delta H_C^0 = +7,245,$$

$$6CO_2{}_{(g)} + \tfrac{5}{2}H_2O_{(l)} + \tfrac{1}{2}N_2{}_{(g)} \rightarrow C_6H_5NO_2{}_{(l)} + \tfrac{25}{4}O_2{}_{(g)};$$

$$\Delta H_C^0 = +739,000.$$

All of which can be summed to yield the equation for benzene nitration:

$$C_6H_6{}_{(l)} + HNO_3{}_{(l)} \rightarrow C_6H_5NO_2{}_{(l)} + H_2O_{(l)};$$

$$\Delta H_R^0 = -\frac{34,730 \text{ cal}}{\text{g-mol } C_6H_6}.$$

Exercise 2–34. Compute ΔH_F^0 for C_6H_6 and $C_6H_6NO_2$ from values of ΔH_C^0, and use these computed enthalpies of formation to find the ΔH_R^0 for the nitration of benzene, via the reaction shown in Example 2–29.

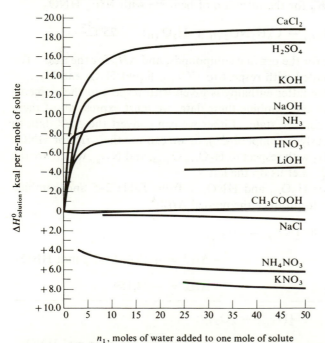

n_1, moles of water added to one mole of solute

Fig. 2–34. Integral enthalpy of solution at 25°C. (Data from *Selected Values of Chemical Thermodynamic Properties*, Nat. Bur. Standards Circ. No. 500, plus supplements, 1952)

2–38
Enthalpy of solution

Thermal effects accompany the dissolution of most chemicals, and hence the accumulated internal energy of the solution generally differs from that of the pure solvent and solute.

We can conveniently measure enthalpies of solutions by adding incremental amounts of solvent to a known quantity of solute and measuring the heat released or absorbed by the process. The data obtained are frequently presented as in Fig. 2–34, and are called the INTEGRAL ENTHALPY (OR HEAT) OF SOLUTION. The *reference state* for the figure corresponds to *pure solvent and solute* at 25°C. The solution-forming process corresponds to the equation:

1 mol solute A + n mols solvent $B \rightarrow$

$\qquad\qquad\qquad (1 + n)$ mols solution of A in B;   25°C.

The integral enthalpy of solution is usually expressed as calories or Btu per mole of solute, at a specified number of moles of solvent.

Example 2–30. Calculate the heat effect accompanying the preparation of a 10-mol % solution of sulfuric acid from pure water and 100% H_2SO_4 at 25°C and 1 atm.

Answer: At constant pressure, the heat effect would be

$$Q_P = \Delta H^0_{solution}$$

A 10-mol % solution corresponds to 9 mols of H_2O (solvent) for every 1 mol of H_2SO_4 (solute). From Fig. 2–34, we see that, for $n = 9$,

$$\Delta H^0_{solution} = -15,250 \text{ cal/g-mol } H_2SO_4.$$

The total heat effect accompanying the dissolution of 1 mol of solute in a large quantity of solvent is referred to as the *standard integral enthalpy of solution*. It corresponds to the value of $\Delta H^0_{solution}$ at infinite dilution, or at a concentration at which the curves of Fig. 2–34 become horizontal. Table 2–5 (right-hand column) gives selected values of standard enthalpies of solution.

2–39
Enthalpy of atomization

The disruption of elemental gas molecules into their atomic components is a process that usually involves the absorption of large amounts of energy. Hence the enthalpy of atoms with respect to the molecules that are their more usual state of agglomeration is very high. Table 2–7 gives such information. The enthalpies ($\Delta H^0_{\text{atomization}}$) correspond to the reaction process,

$$\frac{1}{n} G_{n\,(\text{g})} \rightarrow G_{(\text{g})},$$

where n is usually 2 and the temperature is 25°C.

Table 2–7

Enthalpy of Formation of Atoms

$\frac{1}{2}O_{2(\text{g})} = O_{(\text{g})},$	$\Delta H^0_{25} = +59{,}159$ cal/g-atom
$\frac{1}{2}H_{2(\text{g})} = H_{(\text{g})},$	$\Delta H^0_{25} = +52{,}089$ cal/g-atom
$\frac{1}{2}N_{2(\text{g})} = N_{(\text{g})},$	$\Delta H^0_{25} = +85{,}566$ cal/g-atom
$\frac{1}{2}F_{2(\text{g})} = F_{(\text{g})},$	$\Delta H^0_{25} = +32{,}250$ cal/g-atom
$\frac{1}{2}Cl_{2(\text{g})} = Cl_{(\text{g})},$	$\Delta H^0_{25} = +29{,}012$ cal/g-atom
$\frac{1}{2}Br_{2(\text{g})} = Br_{(\text{g})},$	$\Delta H^0_{25} = +23{,}040$ cal/g-atom
$\frac{1}{2}I_{2(\text{g})} = I_{(\text{g})},$	$\Delta H^0_{25} = +18{,}044$ cal/g-atom

Source: *Selected Values of Chemical Thermodynamic Properties*, as of July 1, 1953, edited by D. D. Wagman, National Bureau of Standards.

2–40
Enthalpy change of a chemical process at any temperature

Our discussions thus far have been limited to processes of chemical reaction, which began and ended at the same standard 25°C temperature. Obviously most chemical processes are not so restrained. A reaction may start at one temperature and products will be produced at quite another. We can solve the problem of determining the heat effect that accompanies such a general process by a computation involving both specific heats and the standard enthalpy of reaction.

Consider the process:

$$aA + bB \rightarrow cC + dD,$$

where A and B are initially at temperature T_1, which is different from 25°C, and C and D are at some final higher temperature T_2. The process occurs at constant pressure, and we wish to calculate the heat exchanged with the surroundings.

Figure 2–35 pictures the initial and final states of the system schematically, and on a temperature–enthalpy diagram.

The Conservation Law once again requires that:

$$Q_P = W + \Delta U = P\,\Delta V + \Delta U = \Delta H_{1-2} = H_{\text{final}} - H_{\text{initial}},$$

or the heat effect accompanying the process equals the change in enthalpy between the final and the initial states. Unhappily, we cannot determine ΔH_{1-2} simply by consulting the standard enthalpy tables, because reactants and products are not at the standard temperature at which these tables apply.

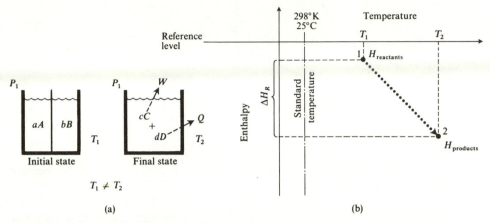

Fig. 2–35. The enthalpy change ($\Delta H_{\text{reaction}}$) that accompanies a chemical reaction process which takes place at temperatures other than 25°C.

We can find ΔH_{1-2} indirectly by devising a *calculation path* which uses the standard enthalpy of reaction (ΔH_R^0) as a *bridge* between products and reactants at 25°C. Such a path is indicated by points 1, e, f, and 2 in Fig. 2–36. It requires that the reactants be cooled to 25°C (298°K), converted to products at the same temperature, and that the products then be warmed from 25°C to T_2. Since enthalpy is a property, the difference in enthalpy between states (1) and (2) is independent of the path used to travel between them. We therefore find the desired enthalpy of reaction by summing the changes in enthalpy along the indicated path (subscripts refer to points in Fig. 2–36):

$$\Delta H_{1-2} = H_2 - H_1 = (H_2 - H_f) + (H_f - H_e) + (H_e - H_1).$$
$$(2\text{–}55)$$

The expression $(H_2 - H_f)$ is the change in enthalpy of the products when they are warmed from 298°K to T_2:

$$H_2 - H_f = c \int_{298°K}^{T_2} C_{P_C}\, dT + d \int_{298°K}^{T_2} C_{P_D}\, dT = \sum_{\text{products}} n_i \int_{298°K}^{T_2} C_{P_i}\, dT.$$

Similarly $(H_e - H_1)$ is the change in enthalpy of the reactants when they are cooled from T_1 to 298°K:

$$H_e - H_1 = a \int_{T_1}^{298°K} C_{P_A}\, dT + b \int_{T_1}^{298°K} C_{P_B}\, dT = \sum_{\text{reactants}} n_j \int_{T_1}^{298°K} C_{P_j}\, dT.$$

The remaining term $(H_b - H_a)$ in Eq. (2–55) is the standard enthalpy of reaction, ΔH_R^0.

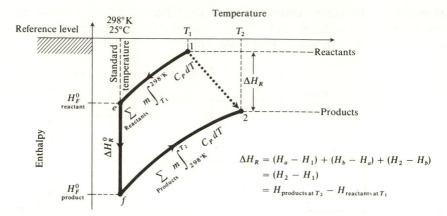

Fig. 2–36. Computation path $(1 \to e \to f \to 2)$ for computing $\Delta H_{\text{reaction}}$ for a chemical reaction process at nonstandard temperatures.

Thus the general expression for the change in enthalpy associated with any chemical reaction at temperatures other than 25°C is:

$$\Delta H_R = \Delta H_{1-2} = \Delta H_R^0 + \sum_{\text{reactants}} n_j \int_{T_1}^{298\,°K} C_{P_j}\, dT + \sum_{\text{products}} n_i \int_{298\,°K}^{T_2} C_{P_i}\, dT,$$

$$(2\text{–}56)$$

where n is the number of moles, and subscripts i and j refer to the products and reactants, respectively. ΔH_R is the enthalpy of reaction at *any* temperature, and must not be confused with ΔH_R^0, which is the *standard* enthalpy of reaction (reactants and products both at 25°C). The superscript zero designates standard conditions.

Note that the limits of integration indicate the direction of the process path.

Example 2–31. Determine the heat evolved when 1 g-mol of H_2 is burned completely at constant pressure with a stoichiometric* amount of O_2 supplied as dry air. The air and H_2 are fed to the combustion chamber at 100°C and the combustion products discharge at 500°C.

Answer: Using 1 g-mol of H_2, we have

$$H_{2\,(g)} + \tfrac{1}{2}O_{2\,(g)} \to H_2O_{(\text{vapor})}.$$

Note that the H_2O that is formed leaves the process as vapor. The O_2 required is

---

* A *stoichiometric* amount of a substance is that amount theoretically required for complete reaction.

0.50 g-mol. Air is composed of 21% O_2 and 79% N_2. Therefore the air required to furnish this much O_2 is

$$\frac{0.50}{0.21} = 2.38 \text{ g-mol.}$$

The amount of N_2 supplied $= 2.38 - 0.50 = 1.88$ g-mol.

Because we are using air, the equation describing the process is

$$H_{2\,(g)} + 2.38 \text{ air}_{(g)} \rightarrow H_2O_{(g)} + 1.88\,N_{2\,(g)}.$$

Now we consult Table 2–5 to find the standard enthalpy of formation of gaseous H_2O,

$$\Delta H_R^0 = -57,798\ \frac{\text{cal}}{\text{g-mol}}.$$

We assume C_P values to be constant at the values given in Table 2–3. More precise calculation requires the use of C_P data in the form of Eq. (2–20). Consequently the change in enthalpy of the reactants as they cool from 100° to 25°C is as follows.

Reactant	moles	C_P	ΔT_{1-e}	ΔH
$H_{2\,(g)}$	1.0	6.98	−75	−523
Air$_{(g)}$	2.38	7.15	−75	−1276

$$\sum_i n_i C_{P_i}\, \Delta T_{1-e} = -1799 \text{ cal}$$
$$\text{reactants}$$

The change in enthalpy of the products as they are warmed to 500°C is as follows.

Products	moles	C_P	ΔT_{f-2}	ΔH
$H_2O_{(g)}$	1.0	8.41	475	3991
$N_{2\,(g)}$	1.88	7.09	475	6330

$$\sum_j n_j C_{P_j}\, \Delta T_{f-2} = 10,321 \text{ cal}$$
$$\text{products}$$

We add all these sums to obtain

$$\Delta H_R = -57,798 - 1799 + 10,321 = -49,276 \text{ cal.}$$

Since the process is at constant pressure,

$$\Delta H_R = Q = -49,276\ \frac{\text{cal}}{\text{g-mol } H_2}.$$

2–41
Adiabatic chemical processes: flame temperatures

Adiabatic reaction processes proceed *without heat effects*, that is, $Q = 0$. Adiabatic conditions are achieved when a chemical reaction is carried out in a perfectly insulated container, or is carried out so rapidly that the process is over before any heat can leak into or out of the system. Flame reactions approach this latter condition.

For a constant-pressure *adiabatic* chemical-reaction process, the Conservation Law becomes:

$$Q = P \, \Delta V_R + \Delta U_R = \Delta H_R = 0, \qquad (2\text{–}57)$$

or *the final enthalpy of a system must equal the initial enthalpy of the system* ($\Delta H_R = 0$). This means that the energy released by the reactants is nearly all accumulated in the products of the reaction. As a result, if the reaction is normally exothermic, the products get very hot.

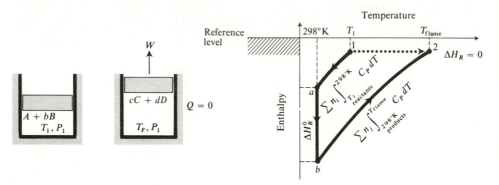

Fig. 2–37. Adiabatic chemical process: $aA + bB \to cC + dD$.

Figure 2–37 shows the process, and illustrates the calculation procedure. Again we employ Eqs. (2–55) and (2–56), but in the adiabatic process we use them to find the final temperature T_2, since the initial temperature T_1 is given, and $\Delta H_R = 0$. We call T_2 the ADIABATIC FLAME TEMPERATURE if the process examined is one of combustion. It represents the maximum temperature that can be achieved in a combustion process. Actual flame temperatures are lower than the adiabatic flame temperature because of heat losses, and also because of chemical equilibrium and dissociation effects, which absorb some of the energy released by combustion.

Problems

1. A body weighs 10 lb_f (as measured on an ordinary spring scale) on the moon, where the acceleration of gravity is 5.47 ft/sec².

a) What force is required to accelerate the body horizontally at a rate of 25 ft/sec² on the moon? on earth? Neglect friction.

b) What is the answer to (a) if the body weighs 10 pounds mass?

c) What is the answer to (a) if the body is accelerated vertically upward?

2. A 10-lb mass is weighed with a spring scale which has been calibrated for standard gravity. (A spring scale is a force meter rather than a mass meter.) What would you expect the reading to be on the moon, where the acceleration of gravity is 5.47 ft/sec^2? What would be the result if the weighing were performed on a double-pan balance (a mass meter) with weights calibrated on earth?

3. What absolute pressure in psia is equivalent to:

a) A Bourdon vacuum-gage reading of 21 in. Hg within a system, when the barometric pressure is 31.0 in. Hg?

b) A pressure-gage reading of 5 atm in a system? (1 atm $= 14.7$ lb$_f$/in^2 $= 29.9$ in. Hg.)

4. A gram of gas is allowed to expand slowly to atmospheric pressure inside a cylinder fitted with a frictionless piston 2 inches in diameter. The following data were obtained during the expansion process:

Distance of piston travel, mm	0	20	40	60	80
Absolute pressure within cylinder, mm Hg, 0°C	2600	1800	1200	900	760

a) What was the increase in the volume of the gas during this operation, in cubic feet? In liters? Plot $(V - V_0)$ versus P.

b) What was the work performed by the gas on the face of the piston in foot-pounds? Joules? Btu? Calories? Grams-rest mass?

c) What was the force exerted by the gas on the face of the piston?

d) What can you say about the temperature of the process? Is it isothermal?

5. A pot of water is boiling on an electric stove. Place boundaries around the pot so as to make it part of:

a) An open system.

b) A closed system.

c) An isolated system.

6. An ice cube is dropped into a heavily insulated jar of warm water and the jar is closed with a well-insulated cover. Draw boundaries within the closed jar which designate:

a) A system experiencing a positive heat effect.

b) A system experiencing a negative heat effect.

c) A system experiencing no heat effect.

7. A 20-ton freight car is rolling along a smooth horizontal track at a speed of 40 miles per hour.

a) How much more kinetic energy does the car have now than it had when it was standing still? Express your answer in kilowatt-hours.

b) If the track were to ascend a mountain, how high would the car climb if there were no friction? Express your answer in meters.

c) Repeat (b) for a car moving at 20 miles per hour.

8. A pure iron meteorite weighing 1 metric ton comes hurtling into the earth's atmosphere with an initial speed of 40,000 km per hour. Air friction reduces its speed to 2000 km per

hour, at which velocity it strikes the earth. The frictional heat is in part dissipated to the surrounding atmosphere and the remainder raises the temperature of the meteorite. What fraction of the frictional heat is dissipated, given that the temperature of the meteorite reaches 700°C (red hot) before impact? Assume that the meteorite warms up uniformly, i.e., without differences between internal and external temperatures. Also $C_{iron} = 0.12$ cal/gram·°C; the initial temperature of the meteorite is 50°K.

(Does this computation provide an insight into the magnitude of the problems of satellite re-entry?)

9. Count Rumford of Bavaria (Ben Thompson of Woburn, Mass.) found that the barrel of a rifle which fired a blank charge (powder but no bullet) was hotter immediately after firing than the barrel of a rifle that fired a bullet. Suggest an explanation for this curious phenomenon.

10. Captain Schultz, the star high diver of the circus, insists that his bath always be at exactly 98.6°F, and measures the temperature of the bath with a precision thermometer before relaxing in it at the end of his day's work. One terrible evening not so long ago the captain found his bath to be only 98.3°F. His noble countenance glowed with fury as he ordered his trembling valet to move his portable 20-gallon tub from his private dressing tent to the foot of the high-dive ladder. For a moment he paused, deep in thought, and then, ignoring the murmurs of the gathering crowd, he mounted the vertical ladder, counting the one-foot rungs as he climbed. He stopped at what was obviously a critical height. Disdainfully he let his dressing gown fall to the spellbound throng far below, and launched himself into the air, landing in his tub with such infinite precision that nary a drop of water was lost from the tiny splash. A smile lighted his features as he relaxed in his 98.6°F tub, while his valet quickly wheeled him back to the star's tent. Captain Schultz is 5 feet 10 inches tall, has blond hair, blue eyes, weighs 15 stone, and neglects air resistance. How high did he climb?

11. James Joule had difficulty convincing some of his scientific colleagues of the veracity of his results, because the effects he measured were very small. How much temperature rise would he measure in allowing a 20-lb mass dropping 30 ft to expend its energy in 5 lb_m of water? Discuss in reasonable detail some of the experimental problems he must have had to overcome and outline a probable experimental procedure. The specific heat of water was known to Joule. Should he have used the value for C_P or C_V?

12. A vertical cylinder is perfectly insulated, and fitted with an insulated piston which weighs 50 lb_m. The piston is 3 inches in diameter. No heat can enter or leave the system.

The gas in the cylinder, which may be considered to behave ideally, is initially at 200°F and 50 psia. The piston is held by a latch. The upper portion of the piston is exposed to the atmosphere. The latch is released and the piston moves rapidly upward 6 inches, after which it comes to rest, due to friction, with the gas in the cylinder at a pressure of 26 psia. The gas has a constant-volume molar specific heat of 5.0 Btu/lb mol·°F.

a) Discuss the work and heat effects in this process.
b) What is the final temperature of the gas?
c) Repeat (a) and (b) for the case in which the cylinder is turned upside down and the gas expands as before to 26 psia. (In this case the piston is locked in place after the pressure has dropped to 26 psia.)

13. The compressibility coefficient for liquid water at 20°C is approximately 1.1×10^{-6} in^2/lb_f. How much work is required to compress 10 lb of liquid water from 15 to 2000 psia at 20°C?

14. Compute the minimum tension–elongation work $(-\int \tau \, dL)$ needed to raise the tension in a mild steel rod, which has a cross section of 2 in^2, to 10 tons. Young's modulus [the relationship of stress (lb/in^2) to strain (inches of elongation/inch of original length)], at the temperature of the bar is 2.97×10^7 psia.

15. A cubic foot of liquid water at 1 atm pressure is heated from 0° to 50°C in an ideal piston-cylinder machine. The density of water at 0°C is 0.97 g/cm^3. The specific volume of liquid water depends on temperature (°C) as follows:

$$\bar{V}_t = \bar{V}_0 (1 - 0.0643 \times 10^{-3}t + 8.505 \times 10^{-6}t^2 - 6.79 \times 10^{-8}t^3), \text{ where } \bar{V}_0 =$$
specific volume at 0°C. How much work does the water do on its surroundings?

16. A certain gas undergoes an isobaric process such that $T_1 = 100°F$, $V_1 = 20$ ft^3 per pound, $P_1 = 1$ atm, and $T_2 = 300°F$, $P_2 = 25$ ft^3 per pound. The specific heat of this gas is temperature dependent, and is given by $C_P = (0.25 + 40/T°K)$ cal/gram°K. Calculate Q, W, ΔH, and ΔU for the process.

17. Carbon dioxide gas expands in a piston-cylinder machine. The machine is not ideal because of piston friction.

P	psia	50	40	30	20
V	ft^3/lb	2.0	2.4	3.0	4.3

a) Calculate the work done *by the gas* per pound of CO_2.
b) Suppose that the friction between the cylinder and the piston is equivalent to $0.2P$. How much of the work done by the gas can leave via the piston rod to do work outside the machine?

18. Assume that 7.0 ft^3 of a gas, which is not necessarily in an ideal state, is held in a piston-cylinder machine. Initially the gas is at 100 psia and 200°F. Then it expands until the pressure is 25 psia. The expansion is carried out so that the relationship between P and V is

$$PV^m = \text{constant}.$$

Calculate the work leaving the gas, given that (a) $m = 1.0$, (b) $m = 1.4$, and (c) $m = 1.6$. Plot the expansion curves for each value of m on a PV plane.

19. You are to design a motorless electric jack which will consist of a low-friction leak-proof piston-cylinder machine containing 1 ft^3 of nitrogen and an internal electric coil which can heat the gas from 70°F up to 1000°F. The jack is to be fitted with a piston 5 inches in diameter, and should be capable of raising 300 lb 1 ft.

a) Assuming that the jack is perfectly insulated, what is the least number of kilowatt-hours of electrical energy needed to meet the design load raising capacity? The internal pressure and temperature of the jack are initially 14.6 psia and 70°F.

b) Actually, the jack will lose heat energy at a rate

$$Q = 20(T_{\text{cylinder}} - 70°) \frac{\text{Btu}}{\text{hour}}.$$

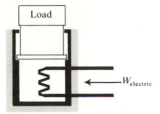

How much electric power is needed at design load and extension?

20. Prove that the work done by 1 g-mol or 1 lb-mol of an ideal gas with constant specific heats during a reversible adiabatic expansion is equal to:

a) $\overline{W} = C_V(T_1 - T_2)$

b) $\overline{W} = \dfrac{P_1\overline{V}_1 - P_2\overline{V}_2}{\gamma - 1}$

c) $\overline{W} = \dfrac{RT_1}{\gamma - 1}\left[1 - \left(\dfrac{P_2}{P_1}\right)^{(\gamma-1)/\gamma}\right]; \quad \gamma = C_P/C_V$

21. Helium is to be cooled by a series of adiabatic reversible expansions and isothermal compressions. The discharge pressure of the compressor is 100 psia. The high-pressure helium is expanded adiabatically to 50 psia in a special reversible turbine, and then returned to the compressor, to be recompressed isothermally to 100 psia. How many expansions are needed to cool the helium from 80°F to below −300°F? What are the heat and work effects in each step of the process? Assume ideal gas behavior. Initial inlet pressure of the compressor is 50 psia.

22. A horizontal cylinder, closed at both ends and insulated on all but the left end, contains a frictionless nonconducting piston. On each side of the piston is 1 ft³ of an inert monatomic ideal gas at 1 atm and 80°F. Heat is slowly supplied to the gas on the

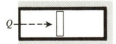

left side until the piston has compressed the gas on the right side to 1.6 atm.

a) How much work is done on the gas on the right side?
b) What is the final temperature of the gas on the right side?
c) What is the final temperature of the gas on the left side?
d) How much heat is added to the gas on the left side?

23. A frictionless piston which weighs 144 lb$_m$ and which has a face area of 1 ft² contains 2 ft³ of O_2 gas at 18 psia and 400°F. Initially the piston presses against the upper stop. As coolant circulates through the coil, the gas is cooled, and the piston descends and settles against the lower stop, having moved 0.1 ft. How much energy (Btu) must be removed to reduce the gas pressure to 14 psia? The pressure in the surrounding atmosphere is 14.7 psia. Draw the process on a P–V plane.

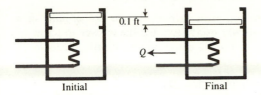

Initial   Final

24. An ideal monatomic gas (0.1 lb-mol) is housed in an ideal friction-free piston-cylinder machine. (The piston floats on gas.) A mass of 100 lb rides on the piston. The area of the piston is 10 in^2; atmospheric pressure is 15 lb/in^2. Suppose that heat is added to the system (the gas) so that its temperature increases from 70°F to 170°F. Compute Q, W, ΔH, ΔU, C_P, C_V, ΔV.

25. a) Repeat the above problem for a piston-cylinder machine which dissipates energy in friction. The frictional force that must be overcome is equivalent to $0.1P$.
b) How would the property values change if the piston were locked in its initial position but the ΔT for the process remained the same?

26. A diatomic ideal gas in a frictionless piston-cylinder machine at an initial pressure of 10 atm (absolute) and temperature of 300°C expands reversibly and adiabatically to a pressure of 5 atm (absolute). Calculate W, Q, $\Delta \bar{U}$, $\Delta \bar{H}$, ΔT, and $\Delta \bar{V}$ for the process, per g-mol of gas.

27. An ideal gas at an initial pressure of 10 atm (absolute) and temperature of 300°C goes through an irreversible adiabatic free expansion into a vacuum ($W = 0$). The pressure of the gas at the end of the process is 5 atm. Determine W, Q, $\Delta \bar{U}$, $\Delta \bar{H}$, ΔT, and $\Delta \bar{V}$ for the process, per g-mol of gas.

28. A solid rubber ball, 3 inches in diameter and weighing 415 grams, is rolling along a level floor at 60.0 ft/sec. The ball hits a small bump on the floor and flies into the air. At the maximum height of its arc, it is 27.8 ft above the floor and is traveling 16.7 ft/sec. There is no heat transfer between the ball and its surroundings, so that the temperature of the ball is raised 0.0215°C by internal friction after it hits the bump. What is the angular velocity of the ball after it leaves the floor? The heat capacity of the rubber is 0.271 Btu/lb·°F.

29. Suppose that 0.02 lb-mol of air at 120.0°F and 3 atm pressure is contained in a well-insulated vertical cylinder fitted with an insulated piston weighing 98.6 lb. The piston is unlatched, and rises quickly to a new position, about which it oscillates, and then stops. The final temperature of the air inside the cylinder is 105.7°F. The C_V of the air is 5.00 Btu/lb-mol·°F. What is the area of the piston? Be explicit about any assumptions you make in solving this problem.

30. A box of sand having a mass of 33.6 lb is standing on the surface of a frozen lake. A hunter happens along and shoots his high-powered rifle at the box. After the lead bullet has become embedded in the sand, the box has a velocity of 3.08 ft/sec and the sand has risen in temperature by 0.0109°F. What was the mass and velocity of the lead bullet shot into the sand? For lead, C_V = 0.0305 Btu/lb·°F; for sand, C_V = 0.194 Btu/lb·°F. Assume the surface of the frozen lake to be frictionless.

31. How much isothermal reversible work can be done by 10 lb of steam at 800°F expanding as a closed system from 1000 psia to atmospheric pressure?

32. What is the minimum work needed to compress carbon dioxide isothermally in a steady-flow system at 120°F from 80 psia to 800 psia? Express the answer in terms of foot-pounds per pound of fluid compressed. Assume elevation and kinetic effects to be negligible.

a) Assume that the carbon dioxide behaves like an ideal gas.
b) Assume that the pressure, volume, and temperature of CO_2 are related as follows:

$$\left(P + \frac{a}{\bar{V}^2}\right)(\bar{V} - b) = RT;$$

$$a = 2.70 \text{ atm} \frac{(\text{liters})^2}{\text{g-mol}}, \qquad b = 0.043 \frac{\text{liters}}{\text{g-mol}}.$$

(The equation in part (b) is called the *van der Waals equation*.)

33. Re-examine the operation of the electric jack in Problem 19, assuming that the working fluid of the jack is 0.1 lb of water initially at 14.6 psia and 70°F. Use data from the Steam Table.

Compare relative advantages and disadvantages of using water and N_2. How does the efficiency (that is, the ratio of work output to heat input) of the volatile liquid jack compare with the N_2 jack?

34. An ideal diatomic gas is housed in an insulated cylinder-and-piston apparatus. The internal volume of the cylinder is initially 2 ft³. A thin diaphragm is sealed to the inner walls of the cylinder parallel to the piston; the diaphragm is so located as to divide the volume of the cylinder in half. The part of the volume of the cylinder between the diaphragm and the piston is evacuated. The gas on the side of the diaphragm away from the piston is at a pressure of 10 atm, a temperature of 200°C, and a volume of 1 ft³.

Now suppose that the diaphragm ruptures suddenly and the gas rushes into the vacuum. It then continues to expand reversibly and adiabatically until its volume is 4 ft³. Calculate W, Q, ΔU, ΔH, and ΔT for each step of the process and for the total process.

35. Devise a reversible process or group of processes that will bring the gas in the above example back to its initial state (10 atm, 200°C, 1 ft³). Calculate W, Q, ΔU, ΔH, and ΔT for this return process.

36. Suppose that a stream of liquid water at 70°F is being pumped from an absolute pressure of 1 atm to an absolute pressure of 100 atm by a pump that operates reversibly and adiabatically. What is the work required, in Btu/lb$_m$? The specific volume of water may be assumed insensitive to pressure, in this pressure range.

37. Air (ideal gas) at 150 psia and 200°F is to be used to power a reversible turbine whose discharge pressure is 15 psia. Compare the work output of the turbine:

a) When it is operated adiabatically.
b) When it is operated isothermally.
c) Suggest a system design which uses the turbine for cooling the air in a room.

38. An insulated tank contains an ideal gas at room temperature (T_1) equal to 70°F, and at an absolute pressure (P_1) equal to 1.2 atm. The tank is fitted with an outlet valve and a pressure gage. The valve is opened and gas is allowed to escape into the atmosphere until the pressure (P_2) in the tank equals atmospheric pressure, at which point the outlet valve is closed quickly.

The insulation is then stripped off the tank, and the contents of the tank (whose temperature has dropped as a result of the initial process) are permitted to warm up to room temperature, at which point the pressure in the tank is found to have reached a value (P_3) of 1.057 atm (absolute). Prove that

$$\frac{P_1}{P_2} = \left(\frac{P_1}{P_3}\right)^{C_P/C_V}.$$

Can you say whether the gas is monatomic or diatomic? [*Hints:* Consider the system as being that amount of gas which remains inside the tank during the entire operation. Imagine that it (the gas system) is separated from the gas that escapes from the tank by a thin, frictionless, weightless piston.]

39. A Venturi meter in a horizontal water pipeline (water density: 62.4 lb_m per ft^3) has a throat area equal to two-thirds of the area of the pipeline. The pressure at the throat of the meter is 1.0 lb/in^2 less than the pressure in the pipeline. How fast is the water in the pipeline flowing? Give your answer in units of ft/sec.

40. A steam turbine is supplied with 1000 lb_m of steam per hour. The steam entering the turbine has an enthalpy of 1271.5 Btu/lb. When the steam leaves, its enthalpy is 1190 Btu/lb_m. The insulation on the turbine is faulty, so that heat losses to the atmosphere amount to 5000 Btu/hr. Assuming that changes in kinetic and potential energy are negligible, how much shaft work can the turbine deliver?

41. A pipe is carrying water through the system shown in the figure at the rate of 20 lb/min. Neglecting friction and other irreversible effects, what is the difference in pressure between points (1) and (2)? At the process conditions, $\bar{V}^{-1}$ of water is 62.4 lb_m/ft^3.

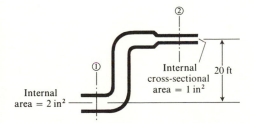

42. Air is heated by passing through a frictionless heat exchanger; it then flows through a reversible adiabatic turbine. The air enters the heat exchanger at 310 psia and 82°F. Only 0.238 Btu of shaft work is produced per Btu of heat added in the heat exchanger.

a) Specify the discharge state of the turbine.
b) What is the temperature of the inlet air to the turbine? Assume that air is an ideal gas, and that the specific heat ratio $\gamma = 1.4$.
c) Specify turbine operating conditions such that $W_s/Q_{exchanger} = 1:1$.

43. Saturated steam at 500 psia flows from a large main into a large evacuated vessel having a capacity of 81.9 ft^3. By the time the pressure in the tank reaches the pressure in the pipeline, 93.7 lb of steam have flowed into the tank. How much heat is lost in the process?

44. An insulated vertical cylinder is closed at both ends. The cross-sectional area of the cylinder is 0.785 ft^2 and it is 2.25 ft long (these are internal dimensions). The cylinder is fitted with an inlet tube in the top, and an insulated frictionless leakproof piston which is

3 in. long. The piston is initially at the top of the cylinder. The space below the cylinder is filled with saturated CO_2. Gas from a CO_2 supply line at 1000 psia and 120°F is allowed to flow into the space above the piston. When the piston reaches the midpoint of the cylinder, the flow is stopped. The temperature of the CO_2 below the piston is now 100°F. How much CO_2 has been passed into the upper chamber of the cylinder? (Thermodynamic properties of CO_2 may be found in chemical or refrigeration engineering handbooks.)

45. An insulated cylinder is closed at one end and is fitted with two insulated weightless pistons. The upper piston is initially latched, and the lower piston is free to move. Between the two pistons is 1.873 mol of air at 2 atm pressure and 205°F. Beneath the lower piston is 0.911 mol of methane at 119°F. The upper piston is now unlatched and is allowed to deliver work reversibly to its surroundings and to the atmosphere. When both pistons come to rest, $P_{air} = 1$ atm. What are the temperatures of the gases? Assume ideal gas behavior; $C_{P \text{ air}} = 7.0$, $C_{P \text{ methane}} = 11.0$.

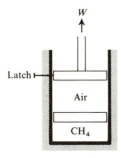

46. A flowing stream of gas at 300 psia and 525°F is to be used to run an adiabatic reversible turbine. The exit from the turbine is to be at -20°F. How much work is obtainable per mole of entering gas? The gas is an ideal gas, with $C_P = 4.9 + 0.0102T$, where T is in °R and C_P is in Btu/lb-mol·°R.

47. Superheated steam at 1400 psia is contained in an insulated cylinder. This steam is slowly released through a throttle valve to the atmosphere. When the pressure in the cylinder has reached atmospheric pressure, the steam in the cylinder has a quality of

88.5%. (*Quality* is the number of pounds of dry saturated steam per pound of steam containing saturated liquid mist.)

a) What is the specific enthalpy of 88.5% quality wet saturated steam?
b) At what cylinder pressure does liquid first form anywhere in the throttle valve?
c) How much heat, in Btu/lb of 1400-psia steam, would have to be added before throttling in order that the same process would give dry saturated steam as the final content of the cylinder?

   Assume that the cylinder contents are well mixed at all times.

48. An adiabatic, reversible turbine is being driven by two streams of compressed air. One stream is at 550°F and 20 atm and the other is at 1070°F and 11.3 atm. The flow rates are 5.00 and 15.00 mol per hour, respectively. The low-pressure stream of air is introduced at a stage in the turbine at which the pressure is 11.3 atm and mixes with the gas already in the turbine. The output pressure of the turbine is 15.0 psia.

   What is the power rating of the turbine? Assume that air is an ideal gas, with the specific heat ratio $\gamma = 1.40$. Compare the power output of this turbine with that of two turbines operating separately on each of the air-feed streams between the same pressure limits.

49. A well-insulated closed tank with a capacity of 7 ft³ contains saturated steam at 130 psia. To this tank is added steam from a source at 500 psia and 560°F. The final contents of the tank consist of saturated steam at 440 psia. What was the quality of the original contents of the tank?

50. Work is to be generated by operating an adiabatic, reversible turbine between two tanks. The first tank has a volume of 11.3 ft³ and contains air at 2000 psia and 300°F. The second tank contains air at 130 psia and 143°F, and has a volume of 140 ft³. Both tanks are well insulated. What is the maximum work the turbine can deliver? (Assume that air is an ideal gas with $\gamma = 1.40$.)

51. Dry saturated steam at 100 psia throttles into an insulated tank containing 500 pounds of saturated liquid water and an equal volume of dry saturated steam at 90 psia. How much steam has entered the tank by the time the tank pressure reaches the same pressure as the pipeline? Where might such a storage tank be used?

52. Can you justify the following statement?

   Every change in the extent of a system's U is indicative of a change in the system's thermodynamic state; however, not every change in thermodynamic state alters the system's U.

53. Compare the heat given off during the isothermal isobaric combustion of one pound of TNT [2,4,6 trinitrotoluene: $C_7H_5N_3O_6$ (solid), molecular weight 227] with the heat given off by combustion of one pound of graphitic carbon. Why is TNT a better explosive? Compare the volumes of products (products given by Table 2–6) per pound of TNT and per pound of carbon.

54. What percent of mass (rest mass) is lost from the system $C_{(graphite)} + O_2 \rightarrow CO_2$, as a result of the heat loss during a constant standard-pressure-and-temperature combustion process?

55. Compute the difference in the heat given off by processes (a) and (b):

a) CO (gas) + $\frac{1}{2}O_2$ (gas) at 1 atm and 25°C go through a process which yields CO_2 (gas) at 1 atm and 25°C.

b) CO (gas) + $\frac{1}{2}O_2$ (gas) at 1 atm and 25°C go through a process which yields CO_2 (gas) at 1 atm and 500°C.

56. Assume that in a lead storage cell the reaction is $Pb + PbO_2 + 2H_2SO_4 = 2PbSO_4 + 2H_2O$; $\Delta H_{18°C, 1 atm} = -121$ kcal/g-mol.

A series of 12 such cells is operating a 24-volt, $\frac{1}{2}$-hp motor having an efficiency of 90%. Estimate, per horsepower-hour of work performed by the motor, the heat evolved by the motor, assuming the temperature of the motor to be constant at 18°C.

57. A possible source of hydrogen for industrial use is the pyrolysis of methane to carbon and hydrogen.

a) Using data from Tables 2–5 and 2–6, estimate the enthalpy of cracking of methane:

$$CH_{4\ (g)} \rightarrow C_{(graphite)} + 2H_{2\ (g)}.$$

b) If all the cracking enthalpy is supplied by burning some of the methane feed, what is the maximum number of moles of hydrogen produced per mole of methane feed in a pyrolysis unit?

58. Calculate the standard enthalpy of formation of Fe_2O_3 from the following data, and compare with the value given in Table 2–5.

$$3C_{(s)} + 2Fe_2O_{3\ (s)} = Fe_{(s)} + 3CO_{2\ (g)}; \quad \Delta H_{25°C} = 113,600\ cal,$$
$$C_{(s)} + O_{2\ (g)} = CO_{2\ (g)}; \quad \Delta H_{25°C} = -94,500\ cal.$$

59. Calculate the standard enthalpy change for the reactions:

a) $Na_2CO_{3\ (s)} + 2HCl_{(g)} = 2NaCl_{(s)} + CO_{2\ (g)} + H_2O_{(l)}$
b) $Fe_2O_{3\ (s)} + CO_{(g)} = CO_{2\ (g)} + 2\ FeO_{(s)}$
c) $3C_2H_{2\ (g)} = C_6H_{6\ (l)}$
d) $HgO_{(s)} + 2HCl_{(g)} = H_2O_{(l)} + HgCl_{2\ (s)}$

60. Find the enthalpy of formation of ammonia at 1000°K from the following data, and from information given in Table 2–5 and in Table A, Appendix I:

$$NH_{3\ (g)}: C_P = 6.189 + 7.887 \times 10^{-3}T + 7.28 \times 10^{-7}T^2.$$

61. A water-cooled gasoline engine is consuming liquid fuel (octane, C_8H_{18}) at 70°F at a steady rate of 30 lb per hr. Assume that just enough air at 50°F and 1 atm enters to allow complete combustion of the fuel to carbon dioxide and water vapor. The exhaust gases leave at 500°F and 1 atm. The engine is producing work at a rate of 50 hp.

At what rate (in pounds per hour) must water circulate through the cooling jacket of the engine, given that the water enters at 60°F and leaves at 180°F? The data are:

$$C_P\ N_2 = 7.00\ Btu/lb\text{-}mol\cdot°F$$
$$C_P\ air = 7.00\ Btu/lb\text{-}mol\cdot°F$$
$$C_P\ CO_2 = 10.00\ Btu/lb\text{-}mol\cdot°F$$
$$C_P\ H_2O_{(g)} = 8.25\ Btu/lb\text{-}mol\cdot°F$$
$$C_P\ C_8H_{18\ (l)} = 81.92\ Btu/lb\text{-}mol\cdot°F$$

62. Hydrated lime, $Ca(OH)_2$, is being produced by reacting calcium oxide (CaO) with steam. The CaO enters the reactor at 68°F, while the steam enters at 25 psia superheated 20°F. The product, hydrated lime, after leaving the reactor, passes through a cooler in which it is cooled to 68°F. The whole process operates continuously and steadily with no excess steam. Per ton of hydrate produced, how much heat must be removed in the reactor and the product cooler?

63. Methane (CH_4) is burned at 1 atm pressure with pure oxygen according to the reaction:

$$CH_{4\,(g)} + 2O_{2\,(g)} = CO_{2\,(g)} + 2H_2O_{(g)}$$

after both reactants have been heated to 500°F.

a) How much heat (in Btu per pound-mole of methane) must be removed from the chamber, given that the products leave at 500°F?

b) What is the final temperature of the products, given that the process occurs in an insulated constant-pressure vessel?

64. A piston-cylinder machine containing an ideal-gas working fluid operates reversibly through the following series of processes which combine to form a cyclical process:

a) The gas expands isothermally from pressure P_1 to lower pressure P_2 at temperature T_1.

b) The gas then expands adiabatically from P_2 to P_3, which is less than P_2. During this adiabatic expansion the temperature drops to T_3.

c) The gas is then compressed isothermally from P_3 to P_4.

d) Finally the gas is subjected to an adiabatic compression, which raises the pressure to P_1 and the temperature to T_1.

Draw the cycle on a PV plane. Evaluate the heat and work effects for processes (a), (b), (c), and (d). Write an expression for the net amount of work performed during one complete cycle; for the ratio of net cycle work to heat input. Express the latter ratio in terms of temperature only.

Given that $T_1 = 75°F$ and $T_3 = 300°F$, find suitable values of P_1, P_2, and P_4 if $P_3 = 1$ atm.

What additional information do you need to compute W_{cycle}? Assume this information and compute W_{cycle}.

65. Calculate the adiabatic flame temperature for the combustion of CO with a stoichiometric amount of O_2 starting at 298°K. Neglect dissociation.

66. In the steel industry, carbon monoxide is used extensively as a fuel. It is almost always preheated before combustion in order to raise the temperature of the flame and furnace. Draw an enthalpy–temperature diagram similar to Fig. 2–37, comparing an adiabatic combustion process with and without preheating of the fuel. Compute the adiabatic flame temperature for CO combustion, given that the CO and $\frac{1}{2}O_2$ are preheated to 1000°K.

67. The Random House *Dictionary of the English Language* (1967) defines energy as "the property of a system which diminishes when the system does work on any other system, by an amount equal to the work so done."

In light of your reading, this definition should strike you as inadequate. Why?

RUDOLF CLAUSIUS, 1822–1878

Section 3
ENTROPY AND THE SECOND LAW
OF THERMODYNAMICS

3–1
Work from heat

Section 2 emphasized the equivalence of all forms of energy. We paid particular attention to the equivalence of work and heat, and to Joule's measurement of the amount of mechanical work energy equivalent to a unit of heat energy. We shall now examine the curious fact that, although heat and work are indeed equivalent energy forms, they are *not completely interconvertible*.

Work may be easily, completely, and continuously converted into heat by means of friction. However, the conversion of heat into work is much more difficult, requiring the use of a HEAT ENGINE, which is a machine specifically designed and constructed to effect this conversion. Furthermore, we shall find that even with the best of heat engines, *it is impossible to feed heat indefinitely into a heat engine and have it all emerge as work*. In other words, we shall find that it is impossible to build any continuously operating device which completely transforms heat into work. Part of the heat energy fed into any heat engine is always wasted. Surprisingly, this inevitable wastage of heat energy will be shown to be a necessary, deductive consequence of the premise stated in the introduction: that heat cannot flow spontaneously from a cold body to a hot body.

The wastefulness during the conversion of heat to work is also related to the fact that nature provides a singular direction to spontaneous changes. For example:

Apples, when they leave a tree branch, fall *down*, not up.

Hot coffee *cools* in a cup.

A raw egg dropped on a hard floor experiences a morphological change; the resulting pieces never recreate the original morphology. (Ref.: M. Goose, *Humpty-Dumpty*, Boston: Thomas Fleet, 1719.)

A zinc strip dissolves in hydrochloric acid and liberates hydrogen. The gas and solution of zinc chloride, even though kept in intimate contact, never spontaneously regain the original reactant forms.

3–2
The second law of thermodynamics

The principle underlying the directionality of spontaneous change and the wastefulness or inefficiency of a heat engine is the stuff of what is called the SECOND LAW OF THERMODYNAMICS. This law may be stated in various equivalent ways. Among the simpler statements are the following:

*Heat flows spontaneously from a hot object to a cooler object, and not in the reverse direction.**

All possible spontaneous changes always increase the disorder of the Universe.

---

* This corresponds to the Clausius statement of the Second Law.

114

In this section, we shall examine the consequences of the first statement. Discussion of the second statement will be reserved for the section on statistical thermodynamics (Section 5).

Although the word spontaneous is intuitively appreciated by all of us, we here state explicitly that:

A SPONTANEOUS flow or process is one that can occur in an isolated system, or between interacting systems at a finite (rather than infinitesimal) rate without effects on, or help from, or interaction with, any other part of the universe.

Furthermore—and this is the essential point—at the end of a spontaneous process the properties of the isolated or interacting systems have changed and cannot be restored to their initial condition by another spontaneous process.

3–3
Heat reservoirs

A HEAT RESERVOIR is any object or system which can serve as a heat source or sink for another system.

Heat reservoirs usually have accumulated energy capacities which are very, very large compared with the amounts of heat energy they exchange. They therefore may operate at constant temperature. Examples of large-capacity, constant-temperature heat reservoirs which make convenient heat sources and sinks are: a thermostated electric hot plate; a flame fed from a constant-pressure fuel line; a large quantity of condensing steam; a large block of melting ice; a river or large body of water; the atmosphere. We shall find it convenient to use heat reservoirs in describing the operation of heat engines.

3–4
Reversibility

We shall now enlarge on the concept of *reversibility* introduced in Section 2 in connection with work–energy interactions. Reversibility is a condition that is prerequisite to all ideally efficient processes of energy conversion. The concept owes its existence to Lazare Carnot and his son Sadi Carnot. Lazare introduced the idea of mechanical reversibility in 1803 (Fig. 2–2). Son Sadi laid the cornerstone of thermodynamics with a little book,* *Reflections on the Motive Power of Fire*, published in 1824, in which he extended the reversibility concept to the processes of energy conversion and heat transfer.

In his book Sadi Carnot reasoned that because matter changes its dimensions when it is alternately heated and cooled, any temperature difference can be used to produce work. (See Fig. 3–1.) For example, two heat reservoirs whose temperatures differ by a *finite* amount can be used to alternately expand and contract a

---

* Available in English translation from Dover Publications in New York (1960).

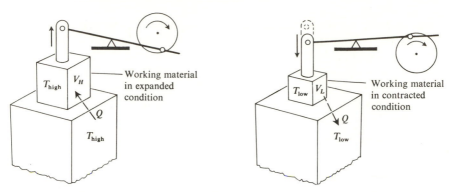

Fig. 3–1. "This . . . fact . . . appears evident as soon as we reflect on the *change of volume occasioned by heat*; wherever there exists a difference of temperature, motive power can be produced." (Sadi Carnot, "Reflections on the Motive Power of Fire.")

suitable working material. The working material will repeatedly absorb heat from the higher-temperature reservoir and reject heat to the lower-temperature reservoir. The resulting dimensional changes in the working material can be exploited by appropriate linkages, cranks, and couplings to raise weights or pump water, or in general, to do work on the surroundings.

Hence heat transferring from a high-temperature reservoir to one at a lower temperature can always produce work if the transfer is made by means of a device containing a working material and appropriately arranged couplings. Such a device is called a HEAT ENGINE. If the heat transfer occurs without the intercession of a heat engine, then the work that this engine could have performed is lost. Young Carnot (aged 25) therefore concluded that the maximum conversion of heat into work requires that all heat transfers (without the intercession of a heat engine) occur over infinitesimal temperature differences. Such a heat transfer is called *reversible* because an infinitesimal change in the temperature of either the heat receiver or the heat source will change the direction of heat flow.

3–5
Characteristics of reversible and irreversible processes

The distinguishing and most general characteristic of all reversible processes is: *The system and surroundings that have completed a reversible process can be restored to their initial states without any finite changes or effects in the rest of the universe.*

All processes that do not share this characteristic are irreversible. Thus *all spontaneous processes are irreversible.* The system and immediate surroundings that interact spontaneously can be restored to their initial state by absorbing work from the universe outside. (The work required for restoration may be large and highly specialized. For example, Humpty Dumpty's reconstruction has resisted the combined efforts of the government's military establishments. It can at best be accomplished by a hen fed with the remains of the original.)

3-6
Converting heat into work: noncyclical process

Small amounts of heat can be extracted from a heat reservoir and completely converted into work. However, the conversion of heat into work *cannot continue indefinitely* unless the engine operates in a *cyclical* process and unless some heat is *rejected* to a sink whose temperature is *lower* than the reservoir serving as a heat source. We shall demonstrate this unhappy fact using as our engine the thermodynamicist's *"gedanken apparat"* (imaginary, idealized apparatus): the ideal gas in a frictionless piston-cylinder machine (Fig. 3–2b). The conclusions we shall draw will then be shown to apply to all reversible engines operating between heat reservoirs having the same temperatures.

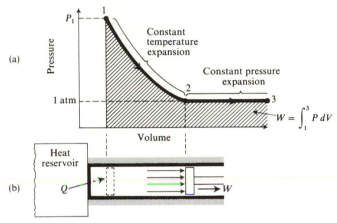

Fig. 3–2. (a) Pressure-volume diagram for (b). (b) Heat being converted into work by the expansion of a gas inside a frictionless piston-cylinder machine.

Let the gas in the cylinder be at an elevated pressure P_1 (Fig. 3–2a). It then expands isothermally to a lower pressure P_2, and does work on its surroundings. The gas temperature T_1 is kept constant during the expansion by adding heat to the cylinder from a heat reservoir whose temperature is differentially higher than T_1 (Fig. 3–2a, path 1–2). (To deliver the maximum amount of work to its surroundings, the piston must be opposed throughout the expansion by a force differentially smaller than the force exerted by the gas on the piston; in other words, the expansion process must occur reversibly.)

Isothermal expansion can continue so long as heat flows into the cylinder and the pressure within the cylinder is above atmospheric pressure (point 2 in Fig. 3–2a). Up to point 2, the work energy leaving the engine equals the heat energy entering it (see Exercise 3–1). Heat input may continue beyond point 2 and the gas will continue to expand, but expansion beyond 2 occurs at constant pressure as the gas temperature and volume increase (path 2–3). To maintain an indefinite heat input to the cylinder, the temperature of the heat source will have to rise indefinitely.

Also, to allow for indefinite expansion of the gas, the length of the cylinder must be unlimited. Obviously the heat input cannot continue indefinitely. The process must stop before the cylinder melts or the piston falls out of the cylinder.

Exercise 3–1. Prove that the work leaving the ideal-gas system in Fig. 3–2b during process 1–2 equals the heat energy supplied to the system.

Exercise 3–2

a) Having proved that $Q_T = W_T$ in Exercise 3–1, how do you reconcile it with the statement on the first page of this section that "it is impossible to feed heat indefinitely into a heat engine and have it all emerge as work"?

b) How does the (work to heat) conversion ratio during the process 2–3 compare with that during process 1–2?

3–7
Converting heat into work; Carnot cycle

The process shown in Fig. 3–2b cannot convert heat into work indefinitely. If the conversion of heat to work is to continue, and if the size of the engine as well as the temperature of the heat source is to be kept within reasonable bounds, some sort of cyclical operation must be used to bring the engine back to its initial condition.

As an example of cyclical operation, consider Fig. 3–3. Let the initial gas pressure be P_1. The gas expands *isothermally* along path 1–2 (Fig. 3–3a). Heat enters the system reversibly from a reservoir at a temperature which is infinitesimally higher than the temperature of the gas. At point 2, the flow of heat to the system is cut off by insulating the cylinder. The gas is allowed to continue its expansion *adiabatically* along path 2–3 (Fig. 3–3b). Pressure and temperature decrease and volume increases. During the isothermal process 1–2, the system has converted heat into work:

$$Q_{1-2} = W_{1-2}, \qquad \Delta U_{1-2} = 0.$$

More work is obtained when the gas expands adiabatically, along path 2–3, at the expense of the internal energy of the gas:

$$W_{2-3} = -\Delta U_{2-3}.$$

Therefore the temperature of the gas falls. At point 3, the expansion ceases, either because of the limitation of the size of the equipment or because the pressure of the system has reached the pressure of its surroundings. To convert more heat into work, we must restore the system to state 1, and repeat the expansion process. The gas must be compressed and its temperature raised. But how?

If we do this by simply reversing processes 1–2–3, the work and heat flows will be equal and opposite to those obtained during the expansion and the net effect

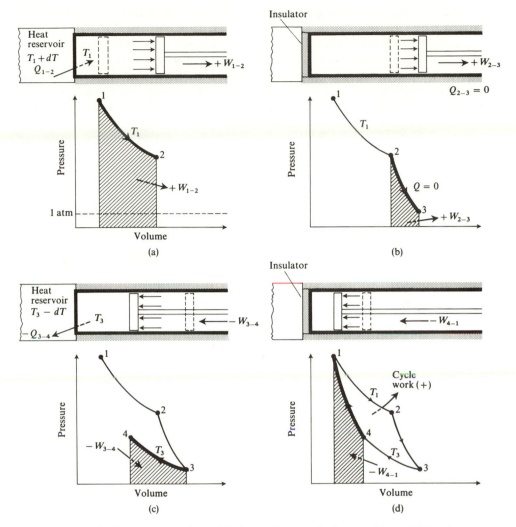

Fig. 3–3. Cyclical operation of an ideal-gas piston-cylinder machine. (a) Isothermal expansion. (b) Adiabatic expansion. (c) Isothermal compression. (d) Adiabatic compression.

will be zero. Nothing lost, nothing gained. No net conversion. Obviously we wish to return to state 1 (complete the cycle) in a less trivial fashion.

If the process is to deliver a net amount of work to its surroundings during a cycle, the work required to compress the gas must be less than the work delivered during the expansion. The cyclic process must have a net positive area when represented on a plot of pressure versus volume. We can accomplish this by compressing the gas at state 3 isothermally (rather than adiabatically) along path 3–4 in Fig. 3–3c. During the isothermal compression 3–4, W_{3-4} is negative, and so, therefore, is Q_{3-4}. By our sign convention, this means that *heat* is *rejected* by the system.

Now if the rejected heat Q_{3-4} could flow back into the supply reservoir which served process 1–2, there would be no waste or inefficiency. Unhappily, however, the supply reservoir is at temperature T_1 which is higher than T_3, and we have stated as a premise that heat flow occurs only down a temperature gradient (from hot to cold). Therefore the heat rejected must go into another heat reservoir or heat sink whose temperature is at least differentially lower than T_3, and this heat energy is of no avail to the system when it again needs heat at T_1 during a repetition of expansion process 1–2.

All cyclic thermal engines require at least two heat reservoirs, one at a high temperature to act as a *heat source* and one at a lower temperature to act as a *heat sink*.

The heat sink can be removed when the pressure reaches P_4. The cycle is completed by compressing *adiabatically* from P_4 to P_1 (Fig. 3–3d). The work done on the system, W_{4-1}, raises the internal energy and the temperature of the system to T_1. The cycle is completed.

The work energy delivered by the system in processes 1–2 and 2–3 is greater than the work energy which was supplied to the system in restoring it to its starting condition via processes 3–4 and 4–1. The net amount of work performed in one cycle, W_{cycle}, is the area bounded by processes 1–2–3–4 (Fig. 3–3d). The net heat is the difference between the heat absorbed isothermally at T_1 and rejected isothermally at T_3.

If we apply the law of conservation of energy to any system after the completion of one or more full cycles, we find that

$$Q_{\text{cycle}} = W_{\text{cycle}} + \Delta U.$$

But ΔU must be zero because the system has gone through a full cycle and is back at its initial state. (U depends on state and is independent of path.) Therefore

$$W_{\text{cycle}} = Q_{\text{cycle}}$$
$$= Q_{1-2} + Q_{3-4}. \tag{3-1}$$

Remember that Q_{1-2} is an *input* and its sign differs from that of Q_{3-4}, which is an *output*.

Exercise 3–3

i) Assume that the above engine operates on one mole of ideal gas and express Q_{1-2}, Q_{3-4}, W_{1-2}, W_{2-3}, W_{3-4}, W_{4-1}, and W_{cycle} in terms of P_1, P_2, P_3, P_4, T_1, T_3, and C_V.

ii) Using Eqs. (2–36) and (2–37), prove that:

$$\frac{W_{\text{cycle}}}{Q_{1-2}} = \frac{T_1 - T_3}{T_1}.$$

The reversible cycle just described, consisting of two adiabatic processes joined by two isothermal processes, is called a CARNOT CYCLE. An engine operating on this cycle is called a CARNOT ENGINE.

No practical engine operates on the reversible Carnot cycle. We shall give descriptions of practical engine cycles in Section 4.

3–8
Thermal efficiency of a heat engine

The *thermal efficiency* of a process is defined as the ratio of the net work energy produced to the heat energy fed into the system. Thus for the above cycle:

$$\text{Thermal efficiency} = \eta = \frac{W_{\text{cycle}}}{Q_{\text{in}}} = \frac{Q_{\text{in}} + Q_{\text{out}}}{Q_{\text{in}}}. \qquad (3\text{–}2)$$

We stated previously that a process delivers maximum work only when it occurs reversibly. Similarly, the minimum work required to cause a change in the state of a system is work supplied to the system reversibly. Hence the cyclic engine that delivers work via reversible processes and is restored to its initial state via other reversible processes will perform the maximum possible cycle work.

3–9
Refrigeration cycle

If the engine in Fig. 3–3 is operated in the reverse direction, that is, 1–4–3–2–1, the engine becomes a REFRIGERATION device. It rejects heat, during the isothermal

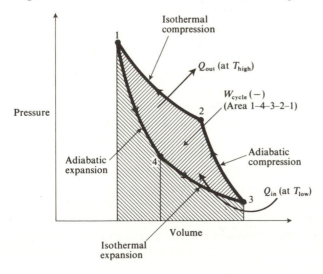

Fig. 3–4. Reversible refrigeration or heat-pump cycle.

compression process 2–1, to the high-temperature reservoir at T_1, and absorbs heat during the isothermal expansion process 4–3 from the low-temperature reservoir at T_3. The work done *on* the system during processes 3–2 and 2–1 exceeds the work done *by* the system during processes 1–4 and 4–3. Figure 3–4 shows this reversible refrigeration cycle.

The net effect of the cycle is the *absorption of work energy and the transfer of heat energy from the low-temperature* heat reservoir *to the high-temperature* heat reservoir. This does not violate the statement of the second law of thermodynamics because the movement of heat energy from a cold body to a hot one has not occurred spontaneously, but has been accomplished at the expense of work energy absorbed from the surroundings (Fig. 3–5).

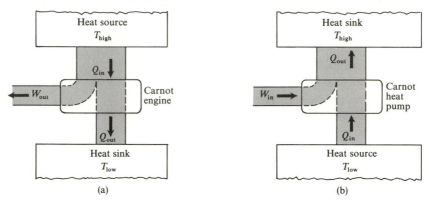

Fig. 3–5. (a) A Carnot engine. (b) A Carnot refrigerator or heat pump (engine operated in reverse.)

The *heat pump* and the *refrigerator* are important engineering devices whose operation depends on reversed cycles similar to that mentioned above. A refrigerator is a work-absorbing (i.e., motor-driven) device that compresses and expands a vapor in a cyclic process. The vapor, contained in a tubing loop which runs both inside and outside the refrigerator, absorbs heat from the refrigerator interior at the low refrigerator-interior temperature and rejects this heat to the high-temperature heat reservoir represented by the atmosphere surrounding the refrigerator.

A *heat pump* is any device that uses work energy to transfer heat energy from a cold reservoir to a hot reservoir. A refrigerator is one form of heat pump. Heat pumps may be designed to maintain an enclosed region at a temperature *above* rather than below surrounding temperatures. Thus heat pumps are employed to absorb heat energy from the cold atmosphere surrounding a dwelling and reject heat to the warmer interior of the dwelling. The heat energy rejected by the heat pumps to the interior of the dwelling is equivalent to the sum of the work and heat energy absorbed by the pump.

Figure 3–5 compares the energy flows in reversible refrigeration and engine cycles operating between reservoirs at the same temperature. The magnitudes of the energy streams (represented by the width of the bands entering and leaving the engines) are the same. Only their directions are reversed.

    We shall discuss practical refrigeration cycles briefly in Section 4.

Exercise 3–4

a) Show that the cycle work energy needed to run a reversible refrigeration cycle is related to the input and output of heat energy as follows:

$$W_{cycle} = Q_{in} + Q_{out}. \qquad (3\text{–}3)$$

b) How does Eq. (3–1) differ from Eq. (3–3)?

c) The effectiveness of a refrigerator is expressed by the ratio Q_{in}/W_{cycle}, called the COEFFICIENT OF PERFORMANCE, or COP. The COP measures the amount of heat removed from a refrigerated compartment per unit of work supplied to the refrigerator. Is the COP larger or smaller than unity? Is a large COP desirable or a small one?

d) Devise a performance index (ratio) for a heat pump used for heating purposes. Is your index larger or smaller than unity? Comment on its significance.

e) Compare the relative advantages and disadvantages of heating the interior of a house by using a 3000-watt electric resistance heater and a heat pump whose drive motor absorbs the same quantity of electrical energy.

3–10
Maximum thermal efficiency

One question that may occur to you is whether all reversible engines operating between two constant-temperature reservoirs are equally efficient.* *They are*—if heat does not flow from a cool body to a hot body without absorption of work from the surroundings; that is, if the second law of thermodynamics is true.

    This may be demonstrated as follows: Picture two reversible thermal engines E and R operating between a high and a low constant-temperature reservoir, as in Fig. 3–6a. The engines are represented schematically as rectangles with rounded corners. The streams of energy passing through the engines are shown as shaded bands whose width suggests the relative magnitudes of the energies in each case. Assume that engine E has a higher thermal efficiency than engine R. Since the engines operate cyclically, the total energy entering must equal the total energy leaving (no accumulation). Therefore the work energy delivered by each engine is

$$W_E = Q_{EH} + Q_{EL}$$

and

$$W_R = Q_{RH} + Q_{RL}.$$

---

* An engine operating between a high-temperature heat source and a low-temperature heat sink and exchanging heat with these reservoirs during isothermal processes at the reservoir temperatures is called a *2T engine*. This title distinguishes it from an engine which exchanges heat with reservoirs at more than two temperatures.

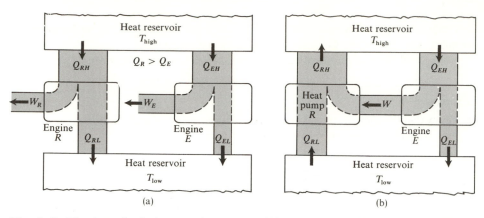

Fig. 3–6. The hypothetical case of two reversible engines which operate at different efficiencies.

(The subscripts are explained in Fig. 3–6.) The corresponding efficiencies are

$$\frac{W_E}{Q_{EH}} = \frac{Q_{EH} + Q_{EL}}{Q_{EH}} > \frac{W_R}{Q_{RH}} = \frac{Q_{RH} + Q_{RL}}{Q_{RH}}$$

and if $W_R = W_E$, then

$$Q_{RH} > Q_{EH}$$

and

$$Q_{RL} > Q_{EL}.$$

Now if the drive shaft on engine E is connected to engine R so as to drive the latter as a heat pump (Fig. 3–6b), then R will absorb Q_{RL} from the low-temperature reservoir and reject Q_{RH} to the high-temperature reservoir, while Q_{EH} leaves the high-temperature reservoir and Q_{EL} enters the low-temperature reservoir through the operation of engine E.

The two machines may now be considered a single system, the net effect of whose operation is the *unassisted flow of heat from the low-temperature to the high-temperature reservoir*. This defies the Second Law. Therefore we must conclude that all reversible $2T$ engines operating between the same two constant-temperature reservoirs have the same efficiency.

This conclusion must apply to all imaginable reversible engines, and is independent of the working material within the engine.

3–11
Temperature and the efficiency of a reversible engine

Having demonstrated that *all* reversible cyclical engines have the *same* efficiency, we shall now, in order to develop further information about the efficiency of reversible engines, examine a reversible engine using an ideal gas as its working material.

Referring again to Fig. 3–3, we sum up the work done by a mole of ideal gas in the reversible cycle shown as follows.

Figure 3–3a: Reversible isothermal expansion, path 1–2 (Eq. 2–30),

$$W_{1-2} = RT_1 \ln \frac{V_2}{V_1} = Q_{1-2}.$$

Figure 3–3b: Reversible adiabatic expansion, path 2–3,

$$W_{2-3} = -\Delta U_{2-3} = C_V(T_1 - T_3).$$

Figure 3–3c: Reversible isothermal compression, path 3–4,

$$W_{3-4} = RT_3 \ln \frac{V_4}{V_3} = Q_{3-4}.$$

Figure 3–3d: Reversible adiabatic compression, path 4–1,

$$W_{4-1} = -\Delta U_{4-1} = C_V(T_3 - T_1).$$

The cyclical work is the sum of the above terms, and the thermal efficiency is the ratio of the cyclical work to the heat input, Q_{1-2}. Therefore

$$\eta = \frac{W_{\text{cycle}}}{Q_{1-2}} = \frac{Q_{1-2} + Q_{3-4}}{Q_{1-2}} = \frac{T_1 \ln (V_2/V_1) + T_3 \ln (V_4/V_3)}{T_1 \ln (V_2/V_1)}. \tag{3-4}$$

Now during a reversible adiabatic process, volume is related to temperature by Eq. (2–37). Therefore

$$\frac{V_2}{V_3} = \left(\frac{T_1}{T_3}\right)^{C_V/R} = \frac{V_1}{V_4}, \tag{3-5}$$

which rearranges to

$$\frac{V_2}{V_1} = \frac{V_3}{V_4}. \tag{3-6}$$

Substituting Eq. (3–6) into Eq. (3–4), we obtain

$$\boxed{\eta = \frac{W_{\text{net}}}{Q_{1-2}} = \frac{T_1 - T_3}{T_1},} \tag{3-7}$$

which states that: *The efficiency of a reversible (2T) engine depends only on the temperatures of the reservoirs between which it operates.* Although Eq. (3–7) was derived for a reversible $2T$ engine which uses an ideal gas as a working fluid, it must apply to all $2T$ reversible engines. (Why?)

Equations (3–7) and (3–4) may be rearranged to yield

$$1 + \frac{Q_{3-4}}{Q_{1-2}} = 1 - \frac{T_3}{T_1}$$

or

$$\frac{Q_{3-4}}{T_3} + \frac{Q_{1-2}}{T_1} = 0. \tag{3–8}$$

Exercise 3–5. Show that the COP of a Carnot refrigerator is

$$\frac{Q_{\text{in}}}{W_{\text{cycle}}} = \frac{T_{\text{low}}}{T_{\text{high}} - T_{\text{low}}}.$$

[See Exercise 3–4, part (c).]

3–12
The thermodynamic temperature scale

Using the concept of the reversible engine, we can establish a THERMODYNAMIC TEMPERATURE SCALE, which is independent of the expansive properties or physical properties of any thermometric substance. Equal temperature intervals on the thermodynamic temperature scale are those intervals which produce equal amounts of work when reversible engines are operated between heat reservoirs at the temperatures of the interval terminals. The engines are arranged as in Fig. 3–7, so that all the heat leaving one engine flows into the engine in the next-lower temperature interval. The size of the temperature interval is quite arbitrary. For example, the interval between the triple point of water and absolute zero temperature might be divided into 273.16 intervals (as in the Kelvin absolute scale), or into 459.65 intervals (as in the Rankine absolute scale), or into another number of intervals.

The absolute temperatures (°Kelvin or °Rankine) used with the ideal-gas law are called *Thermodynamic Temperatures because equal temperature intervals on these scales produce equal amounts of work* when reversible engines are operated within the intervals, as in Fig. 3–7.

From Eq. (3–8), for any Carnot engine, we have

$$\frac{Q_{\text{in}}}{T_{\text{in}}} = -\frac{Q_{\text{out}}}{T_{\text{out}}} = \frac{|Q_{\text{out}}|}{T_{\text{out}}}.$$

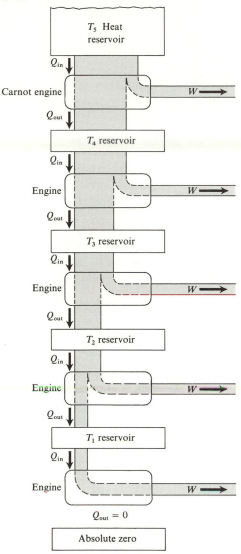

Fig. 3–7. The thermodynamic temperature scale. Equal temperature intervals on this scale are those which allow a reversible engine to produce equal amounts of work when operating between heat reservoirs at the temperatures of the interval terminals, given that the heat output of each engine is the total heat input to the engine in the next-lower interval. The minimum thermodynamic temperature—absolute zero—is the (unattainable) temperature of that low-temperature reservoir which enables a Carnot engine to operate at 100% efficiency.

($|Q|$ means *absolute value of Q*.) Therefore, for the engines in Fig. 3–7,

$$\frac{Q_5}{T_5} = \frac{Q_4}{T_4} = \frac{Q_3}{T_3} = \frac{Q_{in}}{T_{in}} = \text{constant.} \qquad (3\text{–}8a)$$

From Eq. (3–7), we have

$$T_5 - T_4 = W \times \frac{T_5}{Q_5},$$

$$T_4 - T_3 = W \times \frac{T_4}{Q_4}$$

$$T_n - T_{n-1} = W \times \frac{T_u}{Q_n}.$$

Now if W's are all equal, then by substituting Eq. (3–8a) in the above, we obtain

$$T_5 - T_4 = T_4 - T_3 = T_n - T_{n-1} = \frac{W}{Q_n/T_n},$$

and the temperature intervals are equal.

Equation (3–8a) indicates that the ratio of any two thermodynamic temperatures equals the ratio of the heat taken in to the heat rejected by a Carnot engine operating between reservoirs at these temperatures.

Exercise 3–6

a) Show that, for a Carnot engine,

$$\frac{W_{\text{cycle}}}{Q_{\text{out}}} = \frac{T_{\text{in}} - T_{\text{out}}}{T_{\text{out}}}.$$

b) Show that a Carnot engine which operates between 373°K and 273°K produces the same amount of work as one which operates between 492°R and 312°R and which is supplied with all the heat rejected by the first engine. Compare the efficiencies of the two engines under the two sets of conditions. How do these temperature intervals compare in absolute size? Are they equal Thermodynamic Temperature intervals? Why? What is the magnitude (in calories) of the work quantity per unit temperature interval on the thermodynamic Rankine scale, given that the Q leaving the 1°R scale is 1 calorie? How much larger is the quantity of work per °K on the same work basis?

3–13
Entropy

A *property of* a system is some calculable or measurable quantity of that system whose value depends only on the state of the system; i.e., it follows Eq. (2–39). If we reexamine the operation of a reversible engine, we shall find that the mathematical term dQ_{rev}/T behaves just this way:

$$\oint \frac{dQ_{\text{rev}}}{T} = 0.$$

This property was discovered in 1862 by Rudolf Clausius, who named it ENTROPY. It is the property that ties thermodynamics to the directionality of spontaneous change. A *spontaneous* change is always accompanied by an *increase* in the *total entropy* of the system and its surroundings. Phenomenologically, entropy is related to the molecular disorder of a system. We shall give this idea further attention in Section 5.

For the reversible heat engine cycle consisting of two isothermal processes connected by two adiabatic processes (recall the Carnot cycle diagrammed in Fig. 3–3),

$$\frac{Q_H}{T_H} + \frac{Q_L}{T_L} = 0. \tag{3-8}$$

(The subscripts H and L refer to the high- and low-temperature reservoirs previously referred to as 1 and 3.) We may also write Eq. (3–8) as

$$\sum_{\text{cycle}} \frac{Q_{\text{rev}}}{T} = 0. \tag{3-9}$$

The subscript "rev" emphasizes that the summation holds for reversible heat flow *only*.

We shall now show that the summation applies to *any* reversible cycle and not just to the Carnot cycle, and is therefore a property.

Consider the arbitrary cycle connecting states 1 and 2, shown on a P–V plane in Fig. 3–8. The cycle may be divided into differentially thin curved slices, each

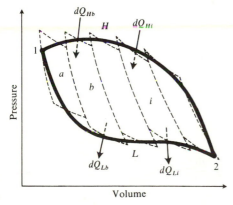

Fig. 3–8. A pressure-volume diagram of an arbitrary cycle passing through states 1 and 2.

slice being a Carnot cycle consisting of two tiny isothermal processes connected to two adiabatic processes.* A differential amount of heat, dQ_{H_i}, is absorbed by

---

* The differential cycles can be made to approach the overall arbitrary cycle as closely as desired by making the differential cycles sufficiently thin. It can also be shown that the heat transferred during any reversible process can always be duplicated by a suitable combination of reversible adiabatic and isothermal processes connecting the terminal states of the process. This is known as *Clausius' theorem.*

cycle i during the isothermal expansion at a temperature corresponding to the intersection of the cycle i with the upper part of the arbitrary cycle envelope. An amount of heat dQ_{L_i} is rejected at the low temperature corresponding to the state of the system at which the differential cycle i intersects the lower-cycle envelope.

For any one differential cycle i,

$$\frac{dQ_{H_i} + dQ_{L_i}}{dQ_{H_i}} = \frac{T_{H_i} - T_{L_i}}{T_{H_i}}$$

or

$$1 + \frac{dQ_{L_i}}{dQ_{H_i}} = 1 - \frac{T_{L_i}}{T_{H_i}}$$

or

$$\frac{dQ_{H_i}}{T_{H_i}} = -\frac{dQ_{L_i}}{T_{L_i}}. \tag{3–10}$$

Similarly, for cycles a, b, c, etc., we have

$$\frac{dQ_{H_a}}{T_{H_a}} = -\frac{dQ_{L_a}}{T_{L_a}}$$

$$\frac{dQ_{H_b}}{T_{H_b}} = -\frac{dQ_{L_b}}{T_{L_b}}$$

$$\vdots \qquad \qquad \vdots$$

$$\frac{dQ_{H_i}}{T_{H_i}} = -\frac{dQ_{L_i}}{T_{L_i}}$$

$$\vdots \qquad \qquad \vdots$$

$$\frac{dQ_{H_n}}{T_n} = -\frac{dQ_{L_n}}{T_n}$$

Summing the above, we obtain

$$\sum_{i=a}^{n} \frac{dQ_{H_i}}{T_{H_i}} = -\sum_{i=a}^{n} \frac{dQ_{L_i}}{T_{L_i}}.$$

If the slices are infinitesimals and the summation is carried to infinity, the $\sum$'s may be replaced with integral signs. Thus:

$$\int_1^2 \frac{dQ_H}{T_H} = -\int_2^1 \frac{dQ_L}{T_L} \tag{3–11}$$

The limits on the right-hand side of the equation are reversed because dQ_L is negative. The right-hand side of Eq. (3–11) sums dQ/T for the path 1–L–2, whereas the left-hand side sums dQ/T for the path 2–H–1. Both paths were arbitrarily chosen. Therefore $\int_1^2 dQ/T$ is independent of path and related to states 1 and 2 only. In other words, it is a property. We give the name *entropy* and the symbol S to this property, and restate Eq. (3–11) as a definition.

$$\oint dS \equiv \int_1^2 \frac{dQ}{T} + \int_2^1 \frac{dQ}{T} = \oint \frac{dQ_{\text{rev}}}{T} = 0, \qquad (3\text{–}12)$$

and

$$dS \equiv \frac{dQ_{\text{rev}}}{T}. \qquad (3\text{–}13)$$

Our earlier discussions emphasized that dQ_R is an inexact differential and that $\int_1^2 dQ_R$ is indeterminate unless the process paths linking states 1 and 2 are described. Even then the value of $\int dQ_R$ depends on the exact path traversed; $\int dQ_{\text{rev}}/T$, however, is *independent of path*. As a result, we can find the value of $\int_1^2 dQ_{\text{rev}}/T$ for a given change of state (from 1 to 2) by choosing *any reversible process* or processes which connect the terminal states, and along which Q_{rev} may be expressed as a function of temperature, or, when T is constant, in terms of other properties of the terminal state. We have used similar computational strategy in evaluating ΔH for any chemical reaction (recall Section 2–40).

3–14
Entropy and integrating factors

The statement that dQ is an inexact differential means that no mathematical function can be written whose derivative is equal to dQ. For example, the function

$$d\phi = y\,dx - x\,dy \qquad (A)$$

is an inexact differential because no $f(x, y)$ exists such that $d[f(x, y)]$ equals $y\,dx - x\,dy$. (Try finding one.) In order to evaluate the $\int_1^2 d\phi$, we need an additional equation (a functional relation) that expresses x in terms of y (or vice versa). This additional equation determines the path between 1 and 2.

A very interesting thing happens when we multiply equation (A) by $1/x^2$:

$$\int_1^2 \frac{d\phi}{x^2} = \int_1^2 \left(\frac{y\,dx}{x^2} - \frac{dy}{x}\right) = \int_1^2 d\left(-\frac{y}{x}\right) = \left(-\frac{y_2}{x_2}\right) - \left(-\frac{y_1}{x_1}\right). \qquad (B)$$

The $\int_1^2 d\phi/x^2$ depends only on the *points* y_1, x_1 and y_2, x_2, and is independent of the way that x changes with y. The term $1/x^2$ is called the *integrating factor* for $d\phi$. Integrating factors can always be found for an inexact differential in two variables. The integrating factor for dQ_R is $1/T$, and $dS \equiv dQ_R/T$ is therefore a point function or property.*

The mathematician Carathéodory† derived the entropy function without

---

* See Exercises 2–22 and 2–23 and the accompanying discussion.
† Reference: C. Carathéodory, "Grundlagen der Thermodynamik," *Math. Ann.* **67**, 355, 1909.

recourse to Carnot cycles, by proving that an integrating factor can always be found for an inexact differential of the form $dQ = C(T, V)\, dT + P(T, V)\, dV$.

3–15
Computing changes in entropy

a) For processes that are carried out at constant pressure and constant volume (per mole of material, having an average or constant C_P and C_V):

$$d\bar{S}_P = \left(\frac{dQ_{rev}}{T}\right)_P = \frac{dH_P}{T} = \frac{C_P\, dT}{T} = C_P d \ln T, \tag{3–14}$$

$$d\bar{S}_V = \left(\frac{dQ_{rev}}{T}\right)_V = \frac{dU_V}{T} = \frac{C_V\, dT}{T} = C_V d \ln T. \tag{3–15}$$

b) For processes that are carried out under isothermal conditions:

$$(dQ_{rev})_T = dW_T = P\, dV = -RTd \ln P. \qquad \text{(For ideal gases only)} \qquad \text{(3–15a)}$$

Therefore

$$d\bar{S}_T = \frac{1}{T}(dQ_{rev}) = -Rd \ln P. \qquad \text{(For ideal gases only)} \qquad \text{(3–16)}$$

c) For processes that are both adiabatic and reversible:

$$dQ_{rev} = 0.$$

Therefore

$$dS_{adiabatic-reversible} = \frac{dQ_{rev}}{T} = 0. \tag{3–17}$$

An adiabatic reversible process is therefore a *constant-entropy* or an ISENTROPIC process.

Example 3–1. Determine the change in entropy which takes place when 1 pound of N_2 (ideal gas) is:

a) compressed reversibly and isothermally from 150 psia to 200 psia at a temperature of 600°F (path 1–2 in Fig. 3–9),

b) heated reversibly from 600°F to 1200°F at a constant pressure of 200 psia (path 2–3 in Fig. 3–9),

c) cooled reversibly at constant volume until the pressure drops to 96 psia (path 3–4 in Fig. 3–9),

d) compressed adiabatically to a pressure of 150 psia (path 4–5), and

e) heated reversibly at 150 psia until the volume returns to V_1 (path 5–1).

Reversible heating or cooling at constant pressure or volume is more easily imagined than accomplished. It requires a *series* of heat reservoirs, each at a

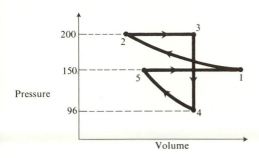

Figure 3–9

temperature differentially higher (or lower) than the preceding reservoir, and there must be a sufficient number of these reservoirs to cover the desired temperature interval. To heat a system reversibly from T_2 to T_3, one must bring the system into sequential contact with each of the heat reservoirs in the series. At any given instant the system receives heat from a reservoir only differentially warmer than itself, so that the heat transfer is always reversible. (An alternative scheme is to use a differential Carnot engine or heat pump. See Section 3–32.)

a) Process 1–2 is a reversible isothermal compression. We know that $Q = W + \Delta U$, and if we assume ideal-gas behavior, then $\Delta U = C_V \Delta T = 0$. Therefore

$$Q_{rev} = (W)_T = nRT \ln \frac{P_1}{P_2}, \qquad \text{(from Eq. 3–15a)}$$

and from Eq. (3–16),

$$\Delta S_{1-2} = \frac{Q_{rev}}{T} = nR \ln \frac{P_1}{P_2}$$

or

$$\Delta S_{1-2} = \frac{Q_{rev}}{T} = nR \ln \frac{P_1}{P_2} = \left(\frac{1}{28.02}\right)(1.98) \ln \frac{150}{200} = -0.0211 \text{ Btu/°R.}$$

b) Process 2–3 is a reversible isobaric expansion. From Eq. (3–14),

$$\Delta S_{2-3} = nC_P \ln \frac{T_3}{T_2}$$

$$= \left(\frac{1}{28.02}\right)(7.09) \ln \frac{1660}{1060} = 0.1134 \frac{\text{Btu}}{°R}.$$

c) Process 3–4 is a reversible, constant-volume cooling process. From Eq. (3–15),

$$\Delta S_{3-4} = \int_{T_3}^{T_4} nC_V d \ln T = nC_V \ln \frac{T_4}{T_3}.$$

Since the volume is constant,

$$\frac{T_4}{T_3} = \frac{P_4}{P_3} = \frac{96}{200};$$

also

$$C_{\bar{V}} = \frac{C_P}{1.4}. \tag{2-23}$$

Therefore

$$\Delta S_{3-4} = \left(\frac{1}{28.02}\right)\left(\frac{7.09}{1.4}\right) \ln \frac{96}{200} = -0.1327 \frac{\text{Btu}}{°\text{R}}.$$

d) Process 4–5 is a reversible adiabatic compression. Therefore, from Eq. (3–17),

$$\Delta S_{4-5} = 0.$$

e) Process 5–1 is a reversible isobaric expansion. We again apply Eq. (3–14):

$$\Delta S_{5-1} = \int_{T_5}^{T_1} nC_p d \ln T.$$

To find T_5, we must find T_4. From (c) above, we obtain

$$T_4 = T_3 \left(\frac{96}{200}\right) = 1660 \left(\frac{96}{2000}\right) = 796°\text{R}.$$

For an adiabatic expansion, Eq. (2–36) applies:

$$\frac{T_5}{T_4} = \left(\frac{P_5}{P_4}\right)^{R/C_P}, \qquad T_5 = 796 \left(\frac{150}{96}\right)^{0.286} = 904°\text{R}.$$

Therefore

$$\Delta S_{5-1} = \left(\frac{1}{28.02}\right)(7.09) \ln \frac{1060}{904} = 0.0404 \frac{\text{Btu}}{°\text{R}}.$$

This means that

$$\Delta S_{\text{total}} = \Delta S_{1-2} + \Delta S_{2-3} + \Delta S_{3-4} + \Delta S_{4-5} + \Delta S_{5-1}$$

$$= \oint dS$$

$$= -0.0211 + 0.1134 - 0.1327 + 0.0404 = 0.000.$$

Example 3–2. Using data from the steam table, repeat Example 3–1 for 1 pound steam. (See Fig. 3–10.)

a) Process 1–2 is a reversible isothermal compression of steam. The values of the

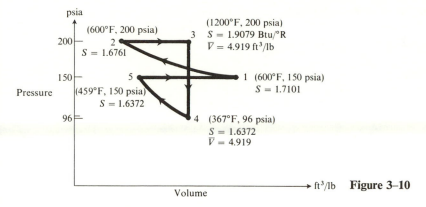

Figure 3–10

specific entropy are read from Steam Table 3 (Appendix II) for the given temperatures and pressures:

State 1: $T = 600°F$, $P = 150$ psia, $S_1 = 1.7101$ Btu/lb · °R,

State 2: $T = 600°F$, $P = 200$ psia, $S_2 = 1.6761$ Btu/lb · °R,

Therefore

$$\Delta S_{1-2} = S_2 - S_1 = 1.6761 - 1.7101 = -0.0340 \text{ Btu/lb · °R.}$$

b) Process 2–3 is a reversible isobaric expansion. We find from the steam table:

State 3: $T = 1200°F$, $P = 200$ psia; $S_3 = 1.9079$ Btu/lb · °R,
$\overline{V} = 4.919$ ft³/lb.

Therefore

$$\Delta S_{2-3} = S_3 - S_2 = 1.9079 - 1.6761 = +0.2318 \text{ Btu/lb · °R.}$$

c) Process 3–4 is a reversible, constant-volume expansion.

For state 4, $P = 96$ psia, $\overline{V} = 4.919$ ft³/lb. We can find the values of the temperature and entropy from the steam table by interpolating between 360° and 380°F and 95 and 100 psia in Steam Table 3:

$$T_4 = 367°F, \qquad S_4 = 1.6372 \text{ Btu/lb · °R.}$$

Therefore

$$\Delta S_{3-4} = S_4 - S_3 = 1.6372 - 1.9079 = -0.2707 \text{ Btu/lb · °R.}$$

d) Process 4–5 is a reversible adiabatic compression. Therefore

$$\Delta S_{4-5} = 0. \qquad\qquad (3\text{–}17)$$

Or:

$$S_4 = S_5 = 1.6372 \text{ Btu/lb · °R.}$$

For state 5, $P = 150$ psia, $S_5 = 1.6372$ Btu/lb · °R, we can find the temperature value from the steam table by interpolation: $T_5 = 459°F$.

e) Process 5–1 is a reversible isobaric expansion. Thus

$$\Delta S_{5-1} = S_1 - S_5 = 1.7101 - 1.6372 = +0.0729 \text{ Btu/lb} \cdot {}^{\circ}\text{R}.$$

Therefore

$$\Delta S_{\text{total}} = \Delta S_{1-2} + \Delta S_{2-3} + \Delta S_{3-4} + \Delta S_{4-5} + \Delta S_{5-1}$$

$$= \oint dS$$

$$= -0.0334 + 0.2318 - 0.2707 + 0.0729$$

$$= 0.3047 - 0.3047 = 0.000.$$

3–16
Temperature-entropy diagram: Carnot cycle

The Carnot cycle shown in Fig. 3–3 consists of two isothermal and two isentropic processes. When these are represented on a T–S (temperature-entropy) diagram, the cycle appears as a rectangle (Fig. 3–11) whose area ($T_H \, \Delta S - T_L \, \Delta S$) equals

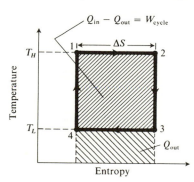

Fig. 3–11. T–S diagram of a Carnot cycle.

the net heat absorbed (or work done) by the cycle. The numbered states on the diagram correspond to the states in Fig. 3–3.

From Eq. (3–13) it is apparent that the area under the curve of a reversible process, connecting two states on a T–S plane, is equal to

$$\int_1^2 T \, dS = Q_{\text{rev}}, \tag{3–17a}$$

which is the heat exchanged during the reversible process.

3–17
Free expansion: entropy increase for an irreversible process

When a gas expands adiabatically into a vacuum (Fig. 3–12), both Q and W are zero because there is neither heat nor work exchanged with the surroundings. Therefore

$$Q = W = \Delta U = 0.$$

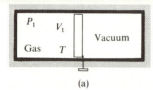 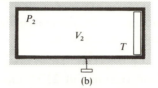

(a)   (b)

Fig. 3–12. Free expansion.

For an ideal gas:

$$\Delta U = C_V \, \Delta T = 0,$$

which means that the temperature of an ideal gas remains constant during a free expansion.

The process is completely irreversible, and we shall find that

$$\Delta S_{\text{universe}} > 0.$$

To calculate ΔS for this process, we must choose reversible processes or calculable paths connecting the terminal equilibrium states of the process.

The terminal states of an ideal-gas free expansion can be joined by a reversible isothermal expansion from the initial pressure to the final pressure. Therefore the change in entropy is the same for the irreversible free expansion as for the reversible isothermal expansion. Using Eq. (3–16) and realizing that $P_1 > P_2$, we obtain:

$$\Delta S_{\text{system}} = \frac{1}{T} \int_1^2 (dQ_{\text{rev}})_T = +R \ln \frac{P_1}{P_2} > 0.$$

Because the system is isolated, the process does not change the entropy of the surroundings. Therefore

$$\Delta S_{\text{universe}} = \Delta S_{\text{system}} + \Delta S_{\text{surroundings}}^{\ \ 0} = \Delta S_{\text{system}} > 0.$$

Exercise 3–7. A cubic foot of dry steam at 400°F and 247.3 psia doubles its volume in a free expansion. Determine the entropy change of the steam and of the universe, using data from the steam table and the fact that $\Delta U = 0$, during a free expansion.

3–18
Other irreversible processes

An irreversible process always increases the entropy of the universe. The entropy of a system may decrease during an irreversible process, but it will then be found that the surroundings have experienced a larger increase in entropy, so that the net change of the system plus surroundings (universe) is positive.

Example 3–3. Suppose that 200 g of water at 95°C cools to room temperature, 23°C. What is the change in entropy of the system (the water), the surroundings, and the universe, as a result of the water's cooling?

We choose the water as our system. The room and the atmosphere surrounding the water constitute a heat reservoir at 23°C, into which heat flows irreversibly from the water.

The change in entropy of the system (water) is that change that would occur if the water were cooled reversibly at constant pressure from 95°C to 23°C:

$$\Delta S_{\text{system}} = \int \frac{dQ_R}{T} = mC_P \int_{368°}^{296°} d \ln T$$

$$= 200 \text{ g} \times \frac{1 \text{ cal}}{\text{g} \text{ °K}} \times \ln \frac{296}{368}.$$

Therefore

$$\Delta S_{\text{system}} = -43.6 \frac{\text{cal}}{\text{°K}}.$$

Note that the entropy of the system decreases.

The temperature of the surroundings does not change. Therefore the change in entropy of the surroundings is the heat received from the water (positive) divided by the absolute temperature of the surroundings, 296°K. A constant-temperature reservoir always responds reversibly to a heat input or output:

$$\Delta S_{\text{surroundings}} = \frac{1}{296°} \int_{296°}^{368°} mC_P \, dT = +48.7 \frac{\text{cal}}{\text{°K}}.$$

Therefore

$$\Delta S_{\text{universe}} = \Delta S_{\text{system}} + \Delta S_{\text{surroundings}} = -43.6 + 48.7 = +5.1 \text{ cal/°K.}$$

The entropy of the universe has increased.

Example 3–4. A 10-kg mass falls from the top of the Tower of Pisa. What is the change in entropy of the system (the mass), the surroundings, and the universe? Temperatures remain constant at 20°C.

To find the change in entropy of the *system* (the mass), we must replace the

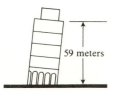

59 meters

irreversible fall with an arbitrary reversible process and find the change in entropy accompanying that reversible process. The mass can be brought to the foot of the tower reversibly by a pulley and string system (as in Fig. 2–2a). For the reversible process, the Energy Conservation Law states

$$Q_{rev} = W_{rev} + \Delta E.$$

Because no heat effects accompany the reversible process,

$$W_{rev} = -\Delta E$$

and

$$Q_{rev} = 0.$$

Therefore

$$\Delta S_{system} = 0.$$

But, if we apply the Energy Conservation Law to the actual irreversible process:

$$Q = W + \Delta E.$$

No work is done by the system in the actual process; therefore W equals zero. Also the accumulated gravitational potential energy decreases. Therefore

$$Q = \Delta E = -5790 \text{ joules} = -1385 \text{ cal} \qquad \text{(from Example 2–2).}$$

This amount of heat energy was rejected (negative sign) by the system, and must have flowed into the surroundings (positive heat input to the surroundings). Therefore

$$\Delta S_{surroundings} = +\frac{1385 \text{ cal}}{293°K} = +4.73 \frac{cal}{°K}$$

and

$$\Delta S_{universe} = \Delta S_{system} + \Delta S_{surroundings} = 0 + 4.73 = +4.73.$$

Therefore $S_{universe}$ has increased by 4.73 calories.

| Heat reservoir T_H | $\xrightarrow{Q}$ | Heat reservoir T_L |

Example 3–5. Some heat, Q, is transferred from a high-temperature heat reservoir at T_H to a low-temperature heat reservoir at T_L. Reservoir temperatures remain constant. Determine $\Delta S_{universe}$.

We approach the problem as follows:

$$\Delta S_{\text{universe}} = \Delta S_{\text{reservoir } H} + \Delta S_{\text{reservoir } L}.$$

Then, as a quantity of heat, Q, leaves the T_H reservoir,[*]

$$\Delta S_H = -\frac{Q}{T_H}.$$

As the same quantity of heat enters the T_L reservoir,

$$\Delta S_L = +\frac{Q}{T_L}.$$

Therefore we may say that

$$\Delta S_{\text{universe}} = Q\left(\frac{1}{T_L} - \frac{1}{T_H}\right) \gtreqless 0. \tag{3–18}$$

If the heat flows in the direction shown, then $T_H > T_L$ and $1/T_H < 1/T_L$. Hence $\Delta S_{\text{universe}}$ is positive.

Equation (3–18) reveals that $\Delta S_{\text{universe}}$ approaches zero as T_H approaches T_L: that is, *as the heat transfer becomes reversible.* Thus a condition for reversibility is that

$$\Delta S_{\text{universe}} = 0. \tag{3–19}$$

3–19
Entropy change during chemical reactions

A system that goes through a chemical-reaction process usually experiences a change in entropy. We can calculate the change in entropy that occurs when pure reactants are converted to pure products, each at a standard temperature and pressure, by subtracting the standard-state entropies of the reactants from those of the products, using values of Standard Absolute Entropies such as those found in Table 10–1. The calculation is similar to that used in finding ΔH_R^0 (Eq. 2–54c), except that entropy values are absolute rather than relative and entropies of elements must be included. We shall say more about this matter in the chapter on chemical equilibrium (Section 10).

---

[*] Although this process is irreversible, the irreversibility occurs between the reservoirs, and is external to the reservoirs themselves. The Q entering or leaving each is indistinguishable from the Q_{rev} that each reservoir might exchange with a system either infinitesimally warmer than or cooler than itself. Therefore

$$\Delta S_H = -\frac{Q}{T_H}; \qquad \Delta S_L = \frac{Q}{T_L}.$$

3–20
Entropy of phase change

Pure materials generally undergo changes of phase at constant temperatures and pressures, and this is usually accompanied by a sizable exchange of heat with their surroundings. You are already familiar with the enthalpy of fusion and vaporization. The change in entropy associated with an equilibrium change in phase is simply

$$\Delta S_{\text{phase change}} = \frac{\Delta H_{\text{phase change}}}{T}. \tag{3–20}$$

For example: For pure water at 1 atmosphere,

$$\Delta H_{\text{fusion}} = +1436 \frac{\text{cal}}{\text{mol}} \text{ at } 273°\text{K},$$

and

$$\Delta H_{\text{vaporization}} = +9729 \frac{\text{cal}}{\text{mol}} \text{ at } 373°\text{K}.$$

Therefore

$$\Delta S_{\text{fusion}} = +\frac{1436}{273} = +5.28 \frac{\text{cal}}{\text{mol} \cdot °\text{K}}$$

and

$$\Delta S_{\text{vaporization}} = +\frac{9729}{373} = +26.1 \frac{\text{cal}}{\text{mol} \cdot °\text{K}}.$$

The processes of phase change are reversible when carried out at equilibrium melting and boiling points.

3–21
Trouton's law

The entropy of vaporization of many organic liquids is approximately 21 entropy units (cal/g-mol · °K). This is the basis of an empirical rule called *Trouton's Law* (Eq. 3–20a), which may be used to estimate the enthalpy of vaporization of a *nonpolar* liquid on the basis of its boiling point when other data are lacking:

$$\frac{\Delta H_{\text{vaporization}}}{T_{\text{boiling point}}} \doteq 21 \text{ cal/g-mol} \cdot °\text{K}. \tag{3–20a}$$

3–22
Third law of thermodynamics: absolute entropy

In 1906 Walther Nernst proposed that: *The entropy of a pure perfect crystal be taken as zero at absolute zero temperature.* This means that all molecular random motions disappear at absolute zero. Both experimental and theoretical findings

have supported this proposal, which is now known as the THIRD LAW OF THERMO-
DYNAMICS. It is therefore possible to assign a Standard Absolute Entropy to a pure
substance at 298°K by computing the entropy gained when the substance is
warmed from 0°K to 298°K at 1 atm. If the substance experiences no phase changes,
then

$$\Delta S^{\circ}_{\text{absolute}} = \int_{0°K}^{298°K} C_P \frac{dT}{T}.$$

If there are phase changes, then the computation equation takes the form

$$\Delta S^{\circ}_{\text{absolute}} = \sum \int C_P \frac{dT}{T} + \sum \frac{\Delta H_{\text{phase change}}}{T_{\text{phase change}}}, \tag{3-21}$$

where the first term is the sum of the entropies gained by heating between phase
changes and the second term sums the entropies associated with each phase change.
(Absolute entropies may also be calculated from spectroscopic data, using the
methods of statistical thermodynamics.)

3–23
Absolute entropy and molecular structure

The absolute entropy of a substance is related to its molecular structure. Solids
have lower entropies than liquids, and liquids have lower entropies than gases,
reflecting the relative disorder of molecular motion in each of these forms of matter.
Soft solids, whose atoms are held loosely in the crystal lattice, generally have
higher entropies than hard solids, whose hardness reflects a more rigid crystal
lattice. For example, graphite has a higher absolute entropy than diamond. (See
Table 10–1.)

3–24
Work functions

The enthalpy function ($H \equiv U + PV$) was devised as a convenient way to combine thermodynamic properties that occur together in equations describing processes of flow and constant pressure. Other important functions have been in-
vented or defined because, like H, they are convenient combinations of already
familiar fundamental properties. These new defined functions depend on the *state*
of the system (P, T, and composition if the only work that is possible is expansion
work), and are independent of path or processes by which the system moves from
one state to another. Hence they are all properties. As a group they are called
WORK FUNCTIONS because they represent the *maximum work* a system can exchange
with its surroundings under various restraints.

3–25
Gibbs free energy

$$G \equiv U + PV - TS.$$ (3–22)*

The function defined above has various names: Gibbs Function, Gibbs Free Energy, or simply FREE ENERGY, which is the term we shall use. Its inventor was Josiah Willard Gibbs, one of the greatest intellects America ever produced. He developed vector analysis, statistical mechanics, and the phase rule. From 1871 to 1903, he was professor of mathematics at Yale University.

The free energy G is a property defined in terms of other properties. If we substitute Eq. (2–16) in Eq. (3–22), the definition reduces to

$$G = H - TS.$$ (3–22c)†

The importance and utility of the free-energy function lie in the fact that it can furnish a general criterion for equilibrium and information on equilibrium conditions in systems whose composition can change (i.e., chemical-reaction systems).

The function also provides a convenient way of separating out or marking those reversible work effects which are *not* expansion, $P\,dV$, work (or acceleration or elevation work). For example, if a system engages in many kinds of reversible work,

$$dU = dQ_{\text{rev}} - \sum dW_{\text{rev}},$$ (3–23)

where $\sum dW_R$ represents $P\,dV$ and all other independent reversible kinds of work (surface, $-\Gamma\,d\mathscr{A}$; electrical, $-\mathbf{V}\,dq$; elongation, $-\tau\,dL$; $F_i\,dX_i$,‡ etc.) that a given system may engage in. Letting dW_{other} represent non-$P\,dV$ works, and using Eq. (3–13) in (3–23), we obtain

$$dU = T\,dS - P\,dV - dW_{\text{other}}.$$ (3–23a)

---

* The definition can also be given as

$$G \equiv E + PV - mgx - \frac{mu^2}{2} - TS.$$ (3–22a)

Because $E = U + \sum E_{\text{external}}$, Eq. (3–22a) is the same as (3–22) when there are no relativistic effects.

Some references define the Gibbs free energy as

$$G' \equiv E + PV - TS.$$ (3–22b)

When gravitational and kinetic-energy effects are important, G' differs from G.

† In this text, H is always equal to $U + PV$. In circumstances in which a system does no $P\,dV$ work but does other kinds of reversible work, $F_i\,dX_i$, there may be a need to define a special enthalpy function, $H' \equiv U + F_iX_i$. Equation (3–22c) does *not* include such special enthalpies.

‡ We exclude $mu\,du$ and $mg\,dx$; F_i and X_i are properties of the system, and not of the surroundings. We shall discuss Eqs. (3–23) and (3–23a) further at the beginning of Section 6.

Rearranging, we have

$$-dW_{\text{other}} = dU + P\,dV - T\,dS. \tag{3–24}$$

We can obtain the same result by differentiating Eq. (3–22) at *constant T and P*:

$$dG_{T,P} = dU + P\,dV - T\,dS.$$

Hence

$$\boxed{dG_{T,P} = -dW_{\text{other}},} \tag{3–25}$$

when the system can do work other than $P\,dV$ work. Equation (3–25) states that for a constant-temperature-and-pressure process, the change in the system's free energy (a property) is equal to (minus) the reversible work, other than $P\,dV$ work, that can be done by the system. dW_{other} is sometimes called *useful work* because it represents work delivered by a system in excess of that expended in displacing the atmosphere. For systems that do $P\,dV$ work *only*,

$$dG_{T,P} = 0. \tag{3–25a}$$

3–26
dG for any system

The total differential of G (Eq. 3–22) is

$$dG = dU + P\,dV + V\,dP - T\,dS - S\,dT. \tag{3–26}$$

On substituting the general expression for dU (Eq. 3–23a) in Eq. (3–26), we obtain

$$\boxed{dG = V\,dP - S\,dT - dW_{\text{other}}.} \tag{3–27}$$

Clearly if P and T are constant, Eq. (3–27) reduces to Eq. (3–25). If in Eq. (3–27) we write

$$dW_{\text{other}} = F\,dX,$$

where F and X are the generalized force and displacement (other than P and V) yielding dW_{other}, then we have

$$dG = V\,dP - S\,dT - F\,dX, \tag{3–28}$$

which is the general expression for the change in free energy in any process or system.

If T and X are constant during a process,

$$dG_{T,X} = V\,dP, \qquad \text{(isothermal process)} \tag{3–29}$$

and if T, P and X are constant, then

$$dG_{T,P,X} = 0. \tag{3–30}$$

3–27
Applications of free-energy equations

A. Isothermal Process: Ideal Gas

A pure gas generally can do only expansion work. For an isothermal process in which dW_{other} is zero, Eq. (3–27) becomes

$$\boxed{dG_T = V\,dP. \qquad (P\,dV \text{ work only})} \qquad (3\text{–}29a)$$

If the gas is ideal, $PV = nRT = \text{constant}$, and therefore

$$dG_T = nRT\,\frac{dP}{P},$$

or, for 1 mol of gas:

$$\boxed{d\bar{G}_T = RT\,\frac{dP}{P} = RT\,d\ln P. \qquad (\text{Ideal gas only})} \qquad (3\text{–}29b)$$

Exercise 3–8
a) Write the general expression for dG_T. How does it differ from Eq. (3–29a)?
b) Prove that for an ideal gas (doing $P\,dV$ work only), $dW_{reversible} = -dG_T$.
c) Is the above equation the same as Eq. (3–25)?

B. Isothermal Reversible Steady-Flow Process Involving Shaft Work

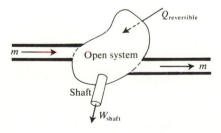

Figure 3–13

The law of conservation of energy for a steady-flow process (Fig. 3–13) is

$$\bar{Q} - \bar{W}_S = \Delta\bar{U} + \Delta(P\bar{V}) + \frac{\Delta u^2}{2} + g\,\Delta x. \qquad (2\text{–}45)$$

Also at constant temperature we may write Eq. (3–22) as

$$\Delta\bar{G}_T = \Delta\bar{U} + \Delta(P\bar{V}) - T\,\Delta\bar{S}. \qquad (3\text{–}31)$$

If the process of isothermal flow is reversible, then

$$\bar{Q}_{\text{reversible}} = T \, \Delta \bar{S} \tag{3-13}$$

and on substituting Eqs. (3–31) and (3–13) in (2–45), we obtain

$$\Delta \bar{G}_T = -\bar{W}_S - \frac{\Delta u^2}{2} - g \, \Delta x. \tag{3-31a}$$

When kinetic and potential energy changes are negligible, Eq. (3–23b) becomes

$$\boxed{\Delta \bar{G}_T = -\bar{W}_S.} \tag{3-31b}$$

The change in free energy per unit mass of fluid passing through an isothermal steady-flow process equals (minus) the maximum (reversible) shaft work obtainable per unit mass of fluid (when changes in elevation and velocity are negligible).

Exercise 3–9. Equations (3–29a) and (3–31a) combine to yield an equation used in Section 2. Find the equation and equation number.

Example 3–6. Consider again the reversible isothermal compression process (1–2) described in part (a) of Example 3–1, page 132.

a) Does the energy accumulated in the system (1 lb of N_2 in an ideal-gas state) change as a result of the compression?
b) Does the free energy of the system change?
c) What is the minimum work required to compress the gas from 1–2?

Answers

a) No; The process is isothermal, and therefore the energy accumulation (ΔU of an ideal gas) must be zero.
b) Yes:

$$\Delta G_T = \int_1^2 dG_T = \int_1^2 dH - \int_1^2 T \, dS.$$

But for an isothermal ideal-gas process $dH = 0$ and $dQ = dW$. Therefore

$$\Delta G_T = -T \int_1^2 dS = -T \int_1^2 \frac{dQ_{\text{rev}}}{T} = -\int_1^2 dQ_{\text{rev}} = -\int_1^2 dW_T$$

or

$$\Delta G_T = +nRT \int_1^2 d \ln P = nRT \ln \frac{P_2}{P_1}$$

$$= \frac{1.987}{28.02} \times 1060 \ln \frac{200}{150} = 21.6 \text{ Btu .}$$

c) The minimum work required equals $(-) \Delta G_T$, or $W_{\text{rev}} = -21.6$ Btu.

Example 3–7. *Equilibrium Vaporization*

a) What is the change in free energy associated with the vaporization of 1 lb-mol of water at 212°F and 1 atm pressure? How much "other work" is performed by the pound-mole of vapor?

   To answer the question, let us picture the vaporization process as occurring in the ideal piston-cylinder machine shown in Fig. 3–14. Not only is the piston frictionless and leakproof, but it is also massless. We might think of the piston as the boundary between the expanding water vapor and the surrounding atmosphere pushed back by this vapor. Vapor and liquid are in equilibrium.

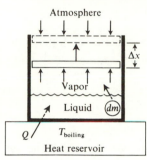

Fig. 3–14. Equilibrium phase change at 1 atm.

   To compute ΔG, we assume that the process occurs reversibly at constant temperature and pressure. For the vaporization of dm pound-moles of liquid to vapor,

$$dG_{T,P} = dH - T\,dS,$$

but

$$dH = \Delta\bar{H}_{\text{vaporization}} \cdot dm$$

and

$$dS = \frac{\Delta\bar{H}_{\text{vaporization}} \cdot dm}{T}. \tag{3–20}$$

Therefore, for an *equilibrium phase change*,

$$dG_{T,P} = 0. \tag{3–25a}$$

   Since the only work the system (liquid plus vapor) performs is $P\,dV$ (expansion) work, the "other work" performed (Eq. 3–25) is zero.

b) What is ΔG for the vaporization of one pound of water for the system (water + water vapor) shown in Fig. 3–15?

   The system is now at a higher pressure than in (a):

$$P_{\text{H}_2\text{O}} = P_{\text{atmosphere}} + \frac{\text{weight of mass}}{\text{area of piston}}.$$

Therefore the reservoir supplying the heat energy for vaporization must be at a higher temperature than in (a). The vaporization process at this higher

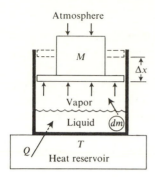

Fig. 3–15. Equilibrium phase change at an elevated pressure and temperature.

pressure and temperature is reversible. The system consists of liquid and vapor in equilibrium inside the cylinder. Although $\Delta H_{\text{vaporization}}$ and T are different, the free-energy effect is the same as in (a):

$$dG_{T,P} = dH - \frac{T\, dH}{T} = 0.$$

The examples demonstrate that *free energy does not change during an equilibrium phase change*. Thus: *The free energy of a vapor is always the same as that of the equilibrium liquid in contact with the vapor*. All equilibrium chemical or phase changes involve *no* change in free energy.

Example 3–8. *Free-Energy Change Accompanying the Isothermal Compression of a Liquid or a Gas*

a) A pound of liquid water at 212°C and 1 atm is compressed isothermally to 50 psia. Calculate $\Delta \bar{G}$.
b) Calculate $\Delta \bar{G}$ for the case in which the liquid is compressed to 1000 psia.
c) How large an isothermal change in pressure would create the same ΔG as in (b) in a pound of water vapor, initially at 14.7 psia and 220°F, assuming that the water vapor behaves like an ideal gas?

Answer
$$\Delta \bar{G}_T = \int_{14.7}^{50} \bar{V}\, dP. \qquad \text{(from 3–29a)}$$

a) Since the volume of the liquid water does not change appreciably during the compression, we may consider $\bar{V}$ as constant and move it outside the integral sign. Therefore

$$\Delta \bar{G}_T = \bar{V} \int_{14.7}^{50} dP = \bar{V}(\Delta P).$$

We find, from Appendix II, that $\bar{V} = 0.0167$ ft³/lb at 212°F and 1 atm.

And thus we obtain

$$\Delta \bar{G}_T = 0.0167 \frac{\text{ft}^3}{\text{lb}_m} (50 - 14.7) \frac{\text{lb}_f}{\text{in}^2} \times \frac{144 \text{ in}^2}{\text{ft}^2}$$

or

$$\Delta \bar{G}_T = 84.8 \text{ ft-lb}_f/\text{lb}_m.$$

b) Again neglecting the change in volume of the liquid water, we may write

$$\Delta \bar{G}_T = 0.0167(1000 - 14.7)144 \text{ ft-lb}_f/\text{lb}_m = 2370 \text{ ft-lb}_f/\text{lb}_m.$$

c) For the isothermal compression of an ideal gas, we obtain

$$\Delta \bar{G}_T = \int_{14.7}^{P_{\text{final}}} V \, dP = nRT \ln \frac{P_{\text{final}}}{14.7} = 2370 \text{ ft-lb}_f/\text{lb}_m.$$

Therefore the gas pressure needed to produce a ΔG of 2370 ft-lb$_f$ is

$$\ln \frac{P_{\text{final}}}{14.7} = \frac{2370 \text{ ft-lb}_f}{\frac{\text{lb-mol}}{18} \times \frac{1545 \text{ ft-lb}_f}{\text{lb-mol °R}} \times 680°\text{R}} = 0.041.$$

And thus we have

$$\frac{P_{\text{final}}}{14.7} = 1.042, \quad \text{or} \quad P_{\text{final}} = 15.3 \frac{\text{lb}}{\text{in}^2}.$$

This example illustrates that the free energy of a gas is vastly more sensitive to a change in pressure than is the free energy of an incompressible material.

3–28
Effect of total pressure on vapor pressure: the Poynting effect

Let the vapor space below the piston in Fig. 3–14 contain helium or some other noncondensing insoluble gas, in addition to the vapors of the liquid, all at a pressure slightly above the normal vapor pressure of the liquid. Now let the piston be loaded so as to isothermally compress the contents of the cylinder to a high pressure. In the compressed state, the gas above the liquid contains liquid vapors to an extent determined by the vapor pressure of the liquid. (In the absence of the noncondensable gas, isothermal compression would have caused the vapor space to disappear, because the vapor would condense into the liquid phase.)

The free energy of the liquid increases because of the increase in total pressure (Example 3–8) to an extent:

$$\Delta G_T = \int V \, dP = V \, \Delta P = V(\mathbb{P} - P_L),$$

where $\mathbb{P}$ = final total pressure in the cylinder, and P_L = initial pressure in the cylinder, which has the same magnitude as the equilibrium vapor pressure of the pure liquid at the cylinder temperature.

Because the vapor in the gas space is in equilibrium with the liquid, it must have the same free energy as the liquid (Example 3–7). Therefore it must have experienced the same ΔG, which means that the *vapor pressure must have increased* slightly above the pressure that would be normal at the system's temperature. We can find the extent of the increase by using the fact that

$$\Delta G_{\text{vapor}} = \Delta G_{\text{liquid}}, \qquad (3\text{--}32)$$

and assuming that the vapor is an ideal gas (Example 3–8c):

$$RT \ln \frac{P'_L}{P_L} = \overline{V}(\mathbb{P} - P_L),$$

or

$$\ln \frac{P'_L}{P_L} = \frac{\overline{V}}{RT}(\mathbb{P} - P_L), \qquad (3\text{--}33)$$

where P'_L = the new augmented vapor pressure at $\mathbb{P}$ and T. The effect of total pressure on vapor pressure is called the *Poynting effect*. It is appreciable only when $\mathbb{P}$ is very high.

3–29
The Helmholtz function

$$\boxed{A \equiv U - TS.} \qquad (3\text{--}34)^*$$

THE HELMHOLTZ FUNCTION, A, plays a role in closed systems and constant-volume processes similar to the role that the Gibbs free energy plays in open systems and constant-pressure processes. (The function is also called the Arbeit function, the work function, and the Helmholtz free energy.)

The total work (other than acceleration and elevation work) done by an isothermal reversible process in a closed system is

$$dW_{\text{rev}} = dQ_{\text{rev}} - dU,$$

$$dW_{\text{rev}} = T\,dS - dU = (-)\,(dA)_T. \qquad (3\text{--}35)$$

Compare Eq. (3–35) with Eq. (3–25).

The change in Helmholtz function is $(-)$ the maximum total work ($P\,dV$ plus all other dW_{rev}) obtainable from an *isothermal process*.

---

* Note that the Helmholtz function (Eq. 3–34) is here defined in terms of accumulated internal energy U. Some authors define this property in terms of E, or $U + E_{\text{external}}$.

If the system can do expansion work only,

$$dA_T = -dW_T = -P\,dV.$$ (3-36)

Exercise 3-10. Write an equation: for G in terms of A; for $dA_{T,V}$.

3-30
Equilibrium and spontaneous change

We have previously demonstrated that spontaneous processes, such as free expansions, falling of masses, and the cooling of liquids (Examples 3-3, 3-4, and 3-5) all result in an increase in the $\Delta S_{\text{universe}}$. All spontaneous changes have similar entropy effects, and we now state this as an equivalent form of the Second Law of Thermodynamics. Thus:

> *A spontaneous process is always similar in entropy effect to the flow of heat from a warm object to a cooler object; that is, it produces an increase in the entropy of the universe.*

An isolated system in effect comprises its own universe, and the only process that can occur in an isolated system (other than equilibrium changes, which cause no net change in properties) is a spontaneous process. It therefore follows that:

> *The entropy of an isolated system can only increase.*

This last statement has disturbing philosophical and material implications, because our universe is by definition an isolated system, and we shall see in a later section that entropy is a measure of disorder.

Example 3-9. *Spontaneous Compression*

The isolated system (ideal gas) in Fig. 3-12 is initially in the state shown in part (b) (P_2, V_2) and goes through a process which brings it to the state shown in part (a) (P_1, V_1).

a) What is $\Delta S_{\text{universe}}$?
b) Does the process violate the First Law of Thermodynamics?
c) Does the process violate the Second Law of Thermodynamics?
d) Is the process possible?

Answer

a) $\Delta S_{\text{universe}} = \Delta S_{\text{isolated system}}$. Since the initial and final temperatures are the same and $P_1 > P_2$:

$$\Delta S = R \ln \frac{P_2}{P_1} < 0; \qquad \Delta S_{\text{universe}} \text{ decreases.}$$

b) $$Q = W + \Delta U = 0; \qquad W = 0; \qquad \Delta U = 0.$$

Therefore the First Law is not violated. Energy is conserved.

c) A superficial examination of the process might lead one to conclude that there is also no violation of the Second Law, since the process is isothermal and involves no heat flow. However, if the process is possible, then the gas in Fig. 3–12a can be placed in contact with a heat reservoir at T, and, by suitable couplings, made to expand isothermally to its initial state (Fig. 3–12b). In so doing, the system will absorb heat, Q_A, from the reservoir and convert it into an equivalent amount of work, W_A, in a cyclic process. The spontaneous compression process from Fig. 3–12b to Fig. 3–12a could then be repeated, and the system would again be able to convert more heat completely into work.

   This looks very suspect. To show that the process does in fact violate the Second-Law statement about the direction of spontaneous heat flow, we need only use the work just obtained to operate a reversible heat pump (see figure). This pump will absorb heat Q_B from a low-temperature reservoir and discharge heat $Q_B + W_A = Q_B + Q_A$ to the reservoir supplying the engine. If the engine and heat pump are considered as a single system, then this combined system does nothing but transfer Q_B from the low-temperature reservoir to the high-temperature one. Hence it violates the Second Law.

d) We conclude that, if the Second Law holds, the process from (b) to (a) of Fig. 3–12 is impossible, and $\Delta S_{\text{isolated system}}$ does not decrease.

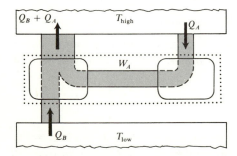

Exercise 3–11. Consider an isolated system consisting of 120 grams of pure liquid water at 1 atm and 0°C. Imagine that a process occurs within the isolating boundaries of the system that results in the formation of 20 grams of ice at 0°C, and 100 grams of liquid water at 15.96°C. ($\Delta U_{\text{fusion}} = 79.8$ cal/g)

a) Compute $\Delta S_{\text{universe}}$.
b) Does the process violate the law of conservation of energy?
c) Does the process violate the Second Law of Thermodynamics? Explain.
d) Is the process possible?

   A system is in an EQUILIBRIUM STATE at the end of a spontaneous process when its properties become time invariant. Such states are characterized by the absence of any unequal or unbalanced intensity factors (temperature, force, pressure, concentration, voltage, etc.) both within the system and between the system and its surroundings, if the system can interact with its surroundings. An equilibrium state may be considered the *terminal state of a spontaneous change*. When a system is given a small displacement from an equilibrium state, it tends to return spontaneously to that equilibrium state.

Since movements *toward* equilibrium states are spontaneous ($\Delta S_{universe} > 0$) and movements away are not, we may state a *criterion for equilibrium* in terms of entropy as follows:

A system is in equilibrium with its immediate surroundings if, for all conceivable interactions* between the system and its immediate fixed surroundings,

$$\Delta S_{universe} \leqslant 0. \qquad \text{(At equilibrium)} \qquad\qquad (3\text{--}37)$$

Equation (3–37) implies that the entropy of the universe *is at a maximum at equilibrium*.

At this point we need to realize that *equilibrium states exist relative to constraints* inherent in the system and in the surroundings. For example, a bounded system can be in equilibrium only with that *limited* part of the universe (the immediate surroundings) with which the system can effectively interact. We implicitly construct an isolating boundary around both system and interacting surroundings whenever we state that a system is in equilibrium with its surroundings, and make of the interacting entities a *composite isolated system*. The properties of the equilibrium state depend on the constraints imposed by the isolating boundaries, as well as those imposed by any internal constraints, such as container walls. For example, a cylinder of high-pressure gas can be in equilibrium with the surrounding atmosphere, given the constraint imposed by the steel cylinder walls. If a system consisting of a block of ice interacts with surroundings consisting of a body of warm water, the final equilibrium state depends very much on the volume of the surroundings. In other words, it depends on the constraints (volume) of the isolating

---

* The interactions must be restrained by the First Law. For example, a situation wherein a perfectly *isolated* jug of water at a uniform, time-invariant temperature and pressure suddenly changes into a jug of steam is conceivable, and increases the universe's entropy. However, it also violates the First Law, and thus is not fair game. In addition, we *cannot alter* the internal or external *constraints* on the system. For example, if the system is a high-pressure gas inside a vertical cylinder sealed with a heavy frictionless piston, all in equilibrium with the immediate external atmosphere, Eq. (3–37) does not apply to processes in which holes are drilled through the cylinder wall or in which the mass of the piston is reduced, or in which the atmosphere changes temperature. Such processes allow the entropy of the universe to increase, but they are outside the purview of Eq. (3–37) because they involve changes in the constraints on the system, or changes in the state of the surroundings.

boundary for the composite of system (ice) and interacting surroundings (warm water).

We can therefore also write Eq. (3–37) as

$$\Delta S_{\text{isolated equilibrium system}} \leqslant 0, \tag{3–37a}$$

by which we mean that the entropy of an isolated system that is at equilibrium is a maximum. The magnitude of the entropy maximum depends on the constraints imposed by the isolating boundaries. If the constraints change, the system may seek a new equilibrium state and a new maximum entropy consistent with the new constraints.

The equals portion of the $\leqslant$ sign in Eqs. (3–37) and (3–37a) takes equilibrium processes into account. For example, if, in an isolated system consisting of 10 grams of ice in equilibrium with water at 0°C, 1 gram of the ice melts to form water and 1 gram of water freezes to form ice, then $\Delta S_{\text{isolated}} = 0$, and the system has not moved from its entropy maximum.

3–31
Work functions and equilibrium

The work functions G and A (Eqs. 3–22 and 3–34) also provide criteria for equilibrium. These criteria are equivalent to the entropy criterion (Eq. 3–37), but are more convenient in that they do not require us to investigate separately the entropy effects in the system and in its interacting surroundings.

Let us first consider the Helmholtz function, A. From Eq. (3–34), we know that, for a differential process at constant temperature and volume,

$$dA_{T,V} = dU - T\,dS = T\left(\frac{dU}{T} - dS\right), \tag{3–38}$$

where all the properties refer to the *system*.

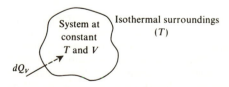

The system (which is closed) can exchange heat with its surroundings, and is at the temperature of its surroundings. (The surroundings act as a thermostat.) From the first law

$$dU = dQ - dW.$$

If $P\,dV$ work is the only kind of work that can be exchanged, then, at constant V, $dW = 0$, and therefore $dU_V = dQ$. Substituting this into Eq. (3–38), we obtain

$$dA_{T,V} = T\left(\frac{dQ}{T} - dS\right). \tag{3–39}$$

But the dQ *entering* the system equals the $(-)\,dQ_{\text{rev}}$ *leaving* the surroundings. Consequently,

$$dS_{\text{surroundings}} = -\frac{dQ_{\text{rev}}}{T}.$$

Equation (3–39) becomes

$$dA_{T,V} = T(-dS_{\text{surroundings}} - dS_{\text{system}}) = -T\,dS_{\text{universe}}.$$

Therefore, from Eq. (3–37), it follows that, in a $P\,dV$ work system,

$$\boxed{dA_{T,V} \geqslant 0. \qquad \text{(At equilibrium)}} \tag{3–40}$$

Next let us consider the Gibbs free energy, G. We know from Eq. (3–22) that, for a process that takes place under conditions of constant pressure and temperature,

$$dG_{T,P} = dU + P\,dV - T\,dS = T\left(\frac{dU + P\,dV}{T} - dS\right). \tag{3–41}$$

But for a process that does $P\,dV$ work *only*,

$$dQ = dU + P\,dV.$$

Therefore Eq. (3–41) becomes

$$dG_{T,P} = T\left(\frac{dQ}{T} - dS\right). \tag{3–41a}$$

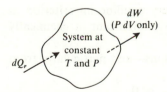

But dQ, the heat entering the system at T, must equal $(-dQ_{\text{rev, surroundings}})$, which is the heat leaving the surroundings (also at T). Therefore

$$dG_{T,P} = T\left(-\frac{dQ_{\text{rev}}}{T_{\text{surroundings}}} - dS_{\text{system}}\right)$$

$$= T(-dS_{\text{surroundings}} - dS_{\text{system}})$$

$$= T(-dS_{\text{universe}}).$$

And a consequence of Eq. (3–37) is

$$dG_{T,P} \geqslant 0. \qquad \text{(At equilibrium)} \tag{3-42}$$

Equations (3–40) and (3–42) show that equilibrium corresponds to *minimum* values of the Gibbs free energy and the Helmholtz functions.

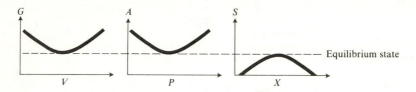

Equilibrium state

3-32
Equilibrium criteria for systems that do work other than expansion work

For systems which do work, $\sum dW_{\text{other}}$, which is other than $P\,dV$ work (let us, for example, assume that they can engage in $\sum_i F_i\,dX_i$ kinds of work), both Eqs. (3–39) and (3–41a), when combined with Eq. (3–23a), become:

$$dG_{T,P} = dA_{T,V} = T\left(-\frac{dQ_{\text{surroundings}}}{T} - \frac{\sum_i F_i\,dX_{i,\,\text{system}}}{T} - dS_{\text{system}}\right)$$

$$= T\left(-\,dS_{\text{universe}}\right) - \sum_i F_i\,dX_{i,\,\text{system}}\,. \tag{3-42a}$$

Therefore, in such other-work systems, Eq. (3–42a) is the equivalent of the basic entropy equilibrium criterion, Eq. (3–37), only if

$$\sum_i F_i\,dX_i = \sum dW_{\text{other}} = 0. \tag{3-43}$$

Equation (3–43) is therefore a necessary supplementary equilibrium criterion in $\sum dW_{\text{other}}$ systems. We shall see that it is particularly important in chemically reactive systems (Sections 9 and 10).

It should be apparent from the foregoing discussions—and particularly from Eq. (3–42a)—that

$$dA_{T,V,X_i} \geqslant 0 \qquad \text{and} \qquad dG_{T,P,X_i} \geqslant 0$$

are equilibrium criteria equivalent to Eq. (3–37).

Exercise 3–12

a) Calculate $\Delta S_{\text{universe}}$ for the reversible vaporization of 1 pound of toluene (which is the system) at its normal boiling point in contact with a suitable heat reservoir. What properties of the system remain constant during the process?
b) Calculate ΔG for the process in (a).

3–33
Carnot engine and the measurement of available work energy

One of the basic ideas conveyed by this section is that heat energy is not completely available for the performance of work. The Carnot engine may be used to gauge the maximum amount of work energy that can be extracted from any heat source. The magnitude of this available work energy depends on the temperature of the source and the temperature of the heat sink to which the engine rejects heat.

Example 3–10. What is the maximum work that can be obtained from heat energy transferred from a 100-pound mass of water at 1 atm and 210°F if the only heat sink available is the atmosphere at 70°F?

Answer: The maximum work obtainable from any heat source is the work obtained by feeding this heat to a Carnot engine. In this example, the heat source is *finite* and will cool down as it sends heat into the engine. We therefore employ a tiny Carnot engine disposed so as to absorb heat from the water mass and reject it to the atmosphere. This engine differs from the engines we have used previously, in that it performs only a *differential* amount of work, dW, in each cycle (see Fig. 3–16).

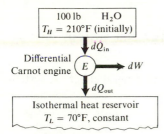

Figure 3–16

During a single cycle: The amount of heat absorbed by the engine (which is the same as the heat lost by the hot water) is

$$dQ_{in} = -mC_P \, dT_{H_2O}.$$

The efficiency of the engine, by Eq. (3–7), is

$$\frac{dW}{dQ_{in}} = \frac{T_{H_2O} - T_L}{T_{H_2O}}.$$

Therefore the work is

$$dW = \left(1 - \frac{T_L}{T_{H_2O}}\right) dQ_{in} = mC_P \left(T_L \frac{dT_{H_2O}}{T_{H_2O}} - dT_{H_2O}\right).$$

The maximum work available from the heat transferred during the cooling of the water from 210°F to 70°F (the temperature of the atmosphere) is

$$W = mC_P \int_{T_{H2O}}^{T_L} (T_L \, d \ln T - dT)$$

$$= 100 \, lb_m \times \frac{1 \, Btu}{lb_m \cdot °R} \left(530°R \ln \frac{530}{670} + 140°R \right) = 1580 \, Btu.$$

Exercise 3-13. How does the available work energy in the above example change if heat is rejected to a stream of cool water at 40°F?

Note that the differential Carnot engine of Example 3-10 provides us with means for cooling and/or heating a system reversibly while the temperature of the system is changing.

Exercise 3-14. What is the least amount of work energy needed to heat a pound of water at constant pressure from 100°F to 200°F, when the heat reservoir available in the surroundings is at 70°F? How high would this amount of work energy heat the water if it were expended as in the Joule experiments?

3-34
Supplementary remarks on reversible work, energy and entropy change, lost work, and the Clausius inequality

We conclude this section by considering a very simple and general effect. A system is in an equilibrium state and the constraints on the system are changed so that the system moves to another nearby equilibrium state, with an accompanying change in its total accumulated energy of dE. The magnitude of dE is, of course, always given by

$$dE = dQ - dW. \tag{2-13}$$

Equation (2-13) applies to any and all processes that can move the system between its initial and final states, irrespective of whether they are reversible or irreversible. The size of the heat and work effects are different in reversible and irreversible processes. However, the difference between the heat and work effects in each kind of process is always the same.

An important idea that you should grasp is that the *heat and work effects* during a *reversible* process *are related to the properties of the system* and may be evaluated in terms of these properties (for example, $dQ_{rev} = T \, dS$ and $dW_{rev} = P \, dV$, or more general $F_i \, dX_i$). By contrast, dQ and dW for irreversible processes bear no special relation to the system's properties and may be evaluated only from changes that have occurred in the surroundings.

Because E is a state property,

$$dE = dQ_{actual} - dW_{actual} = dQ_{rev} - dW_{rev}. \tag{3-44}$$

If we substitute $T\,dS$ for dQ_{rev} in the above, then:

$$dQ_{actual} - dW_{actual} = T\,dS - dW_{rev},$$

which rearranges to

$$dS = \frac{dQ_{actual}}{T} + \frac{dW_{rev} - dW_{actual}}{T}. \qquad (3\text{--}45)$$

The term $(dW_{rev} - dW_{actual})$ is always positive ($\geqslant 0$) and is called the LOST WORK. It is the work dissipated because of irreversibilities during an actual process. As a consequence of the sign of the lost-work term, Eq. (3–45) may be restated as

$$dS \geqslant \frac{dQ}{T}. \qquad (3\text{--}46)$$

Equation (3–46) is called the CLAUSIUS INEQUALITY equation. It says that the change in entropy is always greater than dQ/T, and, in the limit of reversible processes, equals dQ_{rev}/T.

Exercise 3–15. A real fluid dissipates or loses work by fluid friction whenever it flows. Let this lost work in a fluid-flow system be represented by dF, and combine Eq. (3–45) with Eq. (2–45a) so as to obtain Eq. (2–50).

Exercise 3–16. Show that for *any* adiabatic process

$$dS_{adiabatic\ process} \geqslant 0. \qquad (3\text{--}47)$$

Summary of Concepts

A reversible process is a process that can be completely undone without causing changes in the universe outside the system and its interacting surroundings.

Maximum work is obtained from systems when they follow reversible processes.

Heat engines

Work of heat engine cycle = Heat absorbed − Heat rejected.

$$\text{Efficiency of any heat engine} = \eta = \frac{W_{cycle}}{Q_{in}} = \frac{\text{Work output}}{\text{Heat input}}$$

$$= \frac{Q_{in} - |Q_{out}|}{Q_{in}}.$$

$$\text{Efficiency of a } 2T \text{ reversible engine} = \eta = \frac{T_H - T_L}{T_H}.$$

All reversible $2T$ engines have the same efficiency when they are operating between the same two constant-temperature heat reservoirs.

$$Entropy \equiv \int_1^2 \frac{dQ_{rev}}{T} \equiv \Delta S_{1-2} \,.$$

A property of the state of a system, ΔS_{1-2}, depends only on states 1 and 2, and therefore may be evaluated over any reversible process or set of processes connecting states 1 and 2.

Refrigeration cycle or heat pump

A heat engine run in reverse becomes a refrigerator or heat pump. These are devices which absorb *work* from their surroundings and use it to move *heat* from a low-temperature body to a high-temperature body.

The quantity of heat delivered to the high-temperature reservoir equals the sum of the heat energy absorbed at the low temperature plus the work energy absorbed.

$$\text{COP} = \text{coefficient of performance} = \frac{Q_{in}}{W} \,.$$

Gibbs free energy

$$G \equiv U + PV - TS,$$
$$dG_{T,P} = -dW_{other},$$
$$dG_T = -dW_{shaft} - g\,dx - u\,du, \qquad \text{isothermal reversible steady-flow process.}$$
$$dG_T = V\,dP, \qquad \text{any isothermal process, expansion work only.}$$

Helmholtz function

$$A \equiv U - TS,$$
$$dA_T = -\sum dW_{reversible}; \qquad \text{isothermal process.}$$

Second Law of Thermodynamics

Heat flows spontaneously from a hot object to a cooler object.

Alternate Statements or Consequences

Heat energy is not wholly convertible to work in a continuous process.

No engine can be built which operates in a cycle and produces work from heat absorbed from and rejected to a single temperature heat reservoir.

The entropy of an isolated system (or the universe) can never decrease.

Third Law of Thermodynamics: Absolute Entropy

The entropy of a pure perfect crystalline material approaches zero as the temperature approaches absolute zero.

Spontaneous processes

- occur in finite time in an isolated system or between a system and its immediate surroundings, with no effects or assistance from the surroundings, and
- *cannot be undone by other spontaneous processes.*

- always have the same effect on the entropy of the universe as the spontaneous flow of heat; that is,
- always increase the entropy of the universe.
- decrease G at constant T and P.
- decrease A at constant T and V.
- move a system *toward an equilibrium state*.

Equilibrium states involve systems which can interact only with their immediate surroundings (composite isolated systems) or simple isolated systems. If a composite or simple isolated system is in an equilibrium state, any process occurring within the system will make

$$\Delta S_{\text{universe}} \leqslant 0 \qquad \text{(All systems)}$$

$$\left. \begin{array}{l} \Delta G_{T,P} \geqslant 0 \\ \Delta A_{T,V} \geqslant 0 \end{array} \right\} \begin{array}{l} \text{Systems doing} \\ \text{expansion work only} \end{array}$$

Clausius inequality

$$dS \geqslant \frac{dQ}{T} .$$

Problems

1. Dante observed that the temperatures of the Inferno (hell) become increasingly intense as one descends into its depths. If this design is intentional—and it cannot be otherwise—then it is evidence of considerable engineering sophistication on the part of the local authorities. Suggest why this is so, and how the condition observed by Dante is probably exploited by the indigenous authority.

2. The patent office received an application for a patent on an "Electro-Optical Micro-Torch." The maker claims that this torch is capable of producing temperatures above the melting point of tungsten. The device consists of a 1000-watt incandescent projection bulb which has a tungsten element spread over an area of 1.0 in², plus an elaborate lens system which projects a concentrated, tiny (0.005 in²), hot image of the bulb filament on the spot to be heated. Is the maker's operating claim valid?

3. The ocean off Dakar, in West Africa, has a surface temperature of 85°F. The temperature 200 feet below the surface is 40°F. Some attempts have been made to exploit this temperature difference to generate power. What is the maximum thermal efficiency that can be obtained from the operation? What is the minimum rate of heat input required to generate 1 megawatt of power?

4. Can the proof of Exercise 3–1 be performed if the cylinder contains saturated (dry) steam, and P_1 is 200 psia? Does the steam do more or less work per Btu of heat supplied during isothermal expansion?

5. Reversible engines all have the same efficiency and therefore cannot operate as shown in Fig. 3–6b. Suppose that engines E and R are now irreversible (have less than the maximum thermal efficiency) and that E is more efficient that R. What is the net effect of operating them as in Fig. 3–6b?

6. A company that makes automobile engines claims that it is developing a new internal combustion engine which will convert 90% of its fuel energy into useful power. How do you view this claim?

7. Show that the efficiency of the nth engine in Fig. 3–7 is a function of T_R and W only, where T_R is the temperature of the reservoir supplying heat to the nth engine and W is the work produced by the engines in each temperature interval.

8. Carnot described a reversible engine cycle which used a kilogram of water as its working fluid. The fluid at an elevated pressure was converted into steam and the steam expanded adiabatically. The steam was then condensed and the cycle repeated. Draw a $P–V$ diagram for such a cycle operating between 1 atm and 30 psia. Calculate the work of this cycle and compare it with an engine using 1 kg of air as a working fluid. You can find the volume of steam and liquid water at saturation conditions (boiling point) in Appendix II. (Assume that air behaves as an ideal gas.)

9. Water flows over the American falls at Niagara at the rate of 4 million gal/hr and drops 197 ft. Is the rate of increase of the entropy higher in the summer or in the winter?

10. If the human body operated as a heat engine, what would be the minimum number of calories a 160-pound man would have to ingest in order to climb a 100-foot tower on a 25°C spring day? Normal body temperature is 98.6°F.

11. Evaluate and tabulate Q, $\Delta \bar{S}$, $\Delta \bar{H}$, $\Delta \bar{U}$, $\Delta \bar{G}$, and W for each of the paths connecting states 1 and 2 in the diagram. Assume that the working fluid is an ideal gas.

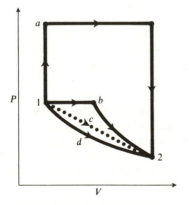

1–a–2 Reversible constant-volume compression, followed by isobaric expansion, followed by constant-volume pressure drop.

1–b–2 Reversible isobaric expansion followed by isentropic expansion

1–c–2 Irreversible, adiabatic free expansion.

1–d–2 Reversible isothermal expansion.

$P_1 = 2\,(P_2) = 10$ atm; $P_a = 20$ atm; $T_1 = 300°$K

12. Plot $\Delta S_{\text{universe}}$ versus mass of B for the machine in Fig. 2–2b as the mass of B varies from 0 to A in steps of $0.1A$. Also plot entropy versus work done and entropy versus power (work/working time).

13. What is the change in entropy resulting from Captain Schultz's dive (Section 2, Problem 10)?

14. A 10-g ice cube is placed in 100 cm³ of water, initially at room temperature (25°C), in an insulated container. What is the final temperature and $\Delta S_{\text{universe}}$? Neglect the effects of heat on the container.

15. Repeat Exercise 2–25, using entropy values taken from the steam table instead of graphical integration to compute Q and W.

16. Steam at 160 psia and 400°F flows into a reversible engine, where it expands adiabatically to 30 psia. What is the maximum work this engine can produce? What are the final temperature and the final quality?

17. In a Carnot cycle operating on nitrogen, the heat supplied is 40 Btu, and the adiabatic expansion ratio is 15.6. The sink temperature is 60°F. Determine (a) the efficiency, (b) the work output, and (c) the heat rejected.

18. One lb-mol of gaseous nitrogen actuates a Carnot cycle in which the respective volumes at the beginning of the isothermal expansion are $V_1 = 0.3565$ ft^3, $V_2 = 0.5130$ ft^3, $V_3 = 8.0$ ft^3, and $V_4 = 5.57$ ft^3. What is the efficiency? Given that the heat input is 4 Btu, calculate the work output and determine the pressure and temperature for each state.

19. A Carnot engine operating between 900°F and 90°F produces 80,000 ft-lb of work. Determine (a) the heat supplied, (b) the change in entropy during heat rejection, and (c) the efficiency of the engine.

20. A reversible engine operating on the Carnot cycle receives heat from a large reservoir containing a mixture of liquid water and water vapor in equilibrium at 1 atm, and discharges 5000 Btu of heat per hour to another large reservoir of water. The power developed is 536 watts. Determine the number of Kelvin degrees separating the high-temperature and low-temperature reservoirs.

21. The efficiency of a particular engine operating on an ideal cycle is 35%. Determine (a) the heat supplied per 1200 watt-hours of work developed, (b) the ratio of heat supplied to heat rejected, and (c) the ratio of work developed to heat rejected.

22. A Carnot engine operating between 800°F and 100°F develops 5 hp. Determine (a) the efficiency, (b) the heat supplied and the heat rejected per second, and (c) the change in entropy each second during heat intake and heat rejection.

23. What is the thermal efficiency of the reversible engine (whose working fluid is air) that operates on the cycle shown in the diagram? How does its efficiency compare to that of the cycle in Fig. 3–3 operating between the same temperature *limits*? What are the temperature limits?

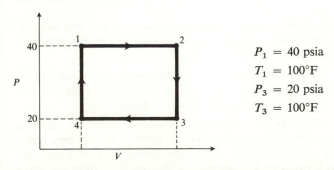

$P_1 = 40$ psia
$T_1 = 100°F$
$P_3 = 20$ psia
$T_3 = 100°F$

Draw the cycle on a chart having temperature as ordinate and entropy as abscissa. Is this a $2T$ engine?

24. The entropy of an equilibrium phase change is given by Eq. (3–20). Does it therefore follow that the entropy change of the chemical reaction process wherein 1 mol of A at

STP is converted to 1 mol of B at STP, A → B, is

$$\Delta S^\circ_{\text{reaction}} = \frac{\Delta H^\circ_{\text{reaction}}}{T}?$$

Explain, keeping in mind that the ΔS between any two equilibrium states can be computed via any reversible path connecting the states.

25. How much work can be performed by a Carnot engine operating between a mass M_1 initially at T_H and another mass M_2 initially at T_L, given that the specific heats of each mass are C_{P1} and C_{P2}, respectively? [*Hint:* Use a differential Carnot engine.]

What is overall thermal efficiency of this engine? How does it compare with the η of an engine operating between constant-temperature heat reservoirs? What is the final temperature of mass M_1 and M_2? How are the answers to the foregoing questions changed if M_1 equals M_2?

26. How high an external pressure must be imposed on liquid benzene at 80.1°C to raise its vapor pressure 5%? ($P_{\text{benzene}} = 760$ mm at 80.1°C.)

27. What are the changes in entropy of M_1 and M_2 and $\Delta S_{\text{universe}}$ in Problem 25?

28. We have all heard of the little boy who decided to cool the family kitchen by keeping the refrigerator door open. How effectively would a conventional 500-watt refrigerator with its door open cool a 900-ft³ room when the outside temperature was 90°F? What change in temperature would take place in 5 hours, assuming that the room was perfectly insulated, and neglecting the specific heat of the walls?

29. Many schemes have been proposed for using the accumulated internal energy of the seas to produce work. One such scheme involves the construction of a large ship whose engines would be driven by heat absorbed from the sea water through which it passed. The ship would leave behind it a trail of cold water, or even ice (depending on speed and operating conditions). The ship's engines (turbines) would be designed to use Freon 12 as a working fluid. The Freon 12, which boils at −21°F, would be converted into high-pressure vapor simply by passing it through heat exchangers "warmed" by sea water. The resulting vapor would then drive the ship's turbines very much as conventional steam does. Because of the low boiling point, the ship could operate in all latitudes, including the extreme polar seas.

This scheme, though worked out in great detail, has never been put into practice. Some claim that it is being suppressed by conventional shipbuilders in collaboration with the major fuel-distribution companies. Might there be other reasons?

30. Which produces the greater increase in efficiency in a heat engine, an increase in the hot-reservoir temperature or an equal-magnitude decrease in the cold-reservoir temperature? What are the corresponding effects on a Carnot refrigerator?

31. Prove that two curves—depicting two different processes of reversible adiabatic expansion—drawn on a P–V plane can never intersect, by showing that an intersection would make possible cyclical processes which violate the Second Law of Thermodynamics. Your proof is also a proof of Carathéodory's theorem that there are states adjacent to every equilibrium state which cannot be reached by adiabatic processes. Do you agree?

32. How would you modify a household refrigerator so as to be able to use it as a room air-conditioner?

33. How much did James Joule, in his famous paddle-wheel experiment, increase the entropy of the universe when he expended 778 ft-lb of work energy in 5 lb of 21°C water? What would the increase in entropy of the universe have been had he expended 778 ft-lb of heat in the 5 lb of water?

34. Which contains more usable *work* energy: a ton of copper at 700°F or 500 lb of liquid water at 500°F? A heat sink is available in a nearby river at 50°F. The specific heat of copper is 0.102 Btu/lb·°F.

35. Show, by the criterion given in Section 3–5, that transfer of heat from a high-temperature system to a finitely lower-temperature system is irreversible; that is, that restoration of the systems to their initial states requires absorption of work from outside the two systems.

36. An adiabatic reversible turbine is supplied with steam at 200 psia and 700°F. (a) Part of this steam is bled from the turbine when its pressure has dropped to 40 psia. How much work is obtained from each pound of this steam? What is the steam temperature? See Example 2–19. (b) The remainder of the steam expands down to 10 psia. Calculate the final temperature; the final quality; the work/pound of steam.

37. Determine ΔS, ΔG, and ΔA for the process described in Problems 2–26 and 2–27.

38. Compute ΔG and ΔA for a process that takes a pound of steam from its dry saturated state at 700°F to 500°F and 500 psia. Could this process occur in an isolated system?

39. Determine ΔG and ΔA for 1 lb of steam for the free-expansion process described in Exercise 3–7.

40. Calculate ΔS_{gas} and $\Delta S_{universe}$ in Problems 2–34 and 2–35.

41. In an ideal steam-power plant, water passes continuously through the following sequence of states:

1) Liquid water, 100°F saturated
2) Liquid water, 100°F and 100 psia
3) Liquid water, saturation temperature at 100 psia
4) Dry saturated steam at 100 psia
5) Steam at 100 psia and 500°F
6) Steam at the same entropy as (5) but at 100°F and saturation pressure

    a) Calculate ΔG for each state change.
    b) Compute the minimum work done on the water in going from 1 to 2.
    c) Which of the above processes could occur spontaneously and which could not? What is the sign of ΔG for these spontaneous processes?

42. A closed cylinder is insulated on all sides but one, as shown in the diagram, and contains a frictionless insulated piston which divides the contents (an ideal gas) into a volume ratio of 1:4. Initially temperature and pressure are uniform. Heat is added to the lower, larger section of the cylinder, causing the piston to rise and compress the gas in the upper section. What is the relationship between temperature in the upper and lower sections? Can $T_{upper} > T_{lower}$, and if so, is this a violation of the Second Law of Thermodynamics?

43. Two containers of equal volume—one containing 1 mol of hydrogen and the other 1 mol of helium—are brought into *thermal* contact, and exchange heat only with each other. Initially, one of the gases is at 1000°R and the other at 600°R. After a period of time, the temperatures in each vessel are equal, as are the pressures (1 atm). The gases are both ideal. What are the maximum and minimum changes in entropy of the universe that accompany the above process? Neglect the effects of changes in the temperature of the container walls.

44. For the above system, compute the maximum work that can be obtained using the gases as heat reservoirs. Which should be the low-temperature reservoir, the helium or the hydrogen?

45. Compute $\Delta S_{universe}$ for the steam-storage process described in Problem 2–51.

46. Insulated tank (1) at P_1 and T_1 is connected through a small valve to an insulated tank (2) at P_2 and T_2. The valve is opened momentarily and M pounds of gas flow from (1) to (2). The value M is small compared to the tank capacities, and there is no change in T or P as a result of the flow. Compute $\Delta S_{universe}$.

47. A walrus drops a tear into the ocean. Prove that the entropy of the universe has increased by an amount

$$\Delta S_{universe} = mC_{P,\,tear}\left(\ln\frac{T_o}{T_w} - \frac{T_o - T_w}{T_o}\right) + \frac{mgh}{T_o},$$

which is always positive, irrespective of whether the sea is boiling hot or freezing cold. Assume that the tear and the sea water have the same composition, and that T_o = ocean temperature, T_w = walrus tear temperature, m = mass of tear, and h = height of walrus' eye.

Original illustration by John Tenniel from Lewis Carroll's *Through the Looking Glass*, Macmillan Publishing Co., London.

RUDOLF DIESEL, 1858–1913

Section 4
POWER AND REFRIGERATION CYCLES

It is usually neither possible nor practical to build real heat engines in such a way that they operate on the Carnot cycle. Practical heat engines operate on cycles which are a compromise between the Carnot cycle and the limitations imposed by design considerations.

Real engine cycles are either *open* or *closed*. An open-cycle engine discards its working fluid after each cycle and replenishes it with fresh fluid. A closed-cycle engine recirculates and reuses its working fluid. Air-breathing engines—such as gasoline and diesel engines, gas turbines, and jet engines—are open-cycle engines. Condensing steam engines and turbines continuously reuse the same water as working fluid, and hence are closed-cycle engines.

The terms *internal* and *external combustion* are applied to heat engines powered by chemical fuels, depending on whether heat is obtained from combustion carried out *within* the engine—that is, within and in direct contact with the working fluid— or *outside* the engine, with the heat from the combustion being transferred to the working fluid through a surface such as the metal walls of a boiler or a similar heating device. Thus gasoline and diesel engines that burn fuel inside the engine cylinder are internal-combustion engines, whereas steam engines with oil- or coal-fired boilers are external-combustion engines.

The following paragraphs describe in idealized fashion the cycles in which common types of real engines operate. Friction, acceleration, and heat losses present in all real engines are neglected in the descriptions.

4–1
Gasoline engines: the Otto cycle

The ordinary automobile gasoline engine operates on the cycle shown in Figs. 4–1 and 4–2 in an idealized version known as the AIR-STANDARD OTTO CYCLE.*

Process (1–2) in Fig. 4–2 represents the *intake stroke*, during which fresh air and gasoline vapor are drawn into the engine cylinder. At (2), the intake valve closes

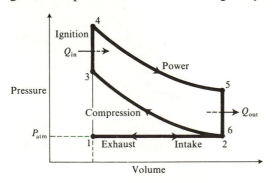

Fig. 4–1. Pressure–volume diagram of the air-standard Otto cycle.

---

* The idealized cycles of engines using pure air as a working fluid are called *air-standard cycles*. The Otto cycle takes its name from N. A. Otto, a German engineer who, in 1866, built the first successful 4-cycle internal combustion engine.

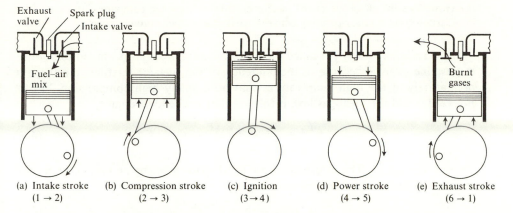

(a) Intake stroke (b) Compression stroke (c) Ignition (d) Power stroke (e) Exhaust stroke
   $(1 \rightarrow 2)$ $(2 \rightarrow 3)$ $(3 \rightarrow 4)$ $(4 \rightarrow 5)$ $(6 \rightarrow 1)$

Fig. 4–2. The 4-stroke cycle of a gasoline engine.

and the compression stroke begins. The fuel–air mixture is compressed adiabatically (2–3).

At (3) the compressed fuel–air mixture is ignited by a spark and the heat from the resulting combustion causes a rapid increase in pressure, ideally at constant volume (3–4). Process (4–5) is the *power stroke*, during which the hot gases expand adiabatically, driving the engine crankshaft. At (5), the exhaust valve opens and the pressure drops to atmospheric (6), while the volume of the cylinder remains constant. Process (6–1) is the exhaust stroke, during which the burned gases are expelled from the cylinder to prepare it for a fresh charge of working fluid and fuel, and the start of a new cycle. The cycle is completed in four passes or strokes of the piston, and hence is called a *four-stroke cycle*.

We may compute the efficiency of an idealized Otto-cycle engine by assuming that the engine contains a fixed mass (1 mol) of pure air in an ideal-gas state, and that heat is added at constant volume during (3–4) and rejected at constant volume during (5–6). In this analysis we neglect the exhaust and intake strokes and all irreversible effects:

$$Q_{in} = Q_{3-4} = C_V(T_4 - T_3), \qquad Q_{out} = Q_{5-6} = C_V(T_6 - T_5).$$

Therefore the efficiency η (eta) of the cycle is

$$\eta = \frac{W_{cycle}}{Q_{in}} = \frac{Q_{3-4} + Q_{5-6}}{Q_{3-4}}$$

or

$$\eta_{\text{ideal Otto}} = 1 - \frac{T_5 - T_6}{T_4 - T_3}. \tag{4–1}$$

Exercise 4–1

  i) Is the idealized air-standard Otto-cycle engine reversible?
 ii) Is it a $2T$ engine?

iii) How does the efficiency of the idealized Otto-cycle engine compare with that of the Carnot-cycle engine operating between the same temperature extremes?

The ratio of V_2 to V_3 is called the *compression ratio, r*. It is limited to about 9 in gasoline engines because of the tendency of the fuel–air mixture to self-ignite prematurely at the high temperatures that accompany high compression. From Eq. (2–35) and the ideal-gas law, it follows that in the ideal engine,

$$\frac{P_4}{P_5} = \frac{P_3}{P_2} = \left(\frac{V_2}{V_3}\right)^{\gamma} = r^{\gamma}, \tag{4–2}$$

where $\gamma = C_P/C_V$. The efficiency of the ideal Otto-cycle engine may be expressed in terms of the compression ratio as

$$\eta_{\text{Otto}} = 1 - r^{(1-\gamma)}. \tag{4–3}$$

Exercise 4–2

a) How does an increase in compression ratio affect the efficiency of an Otto-cycle engine?

b) Derive Eqs. (4–1), (4–2), and (4–3).

Figure 4–3 shows the pressure-volume diagram (called an *indicator diagram*) for an actual gasoline-engine cylinder. This diagram, which emphasizes the idealizations made in Fig. 4–1, has numbers which correspond to the events in Fig. 4–2.

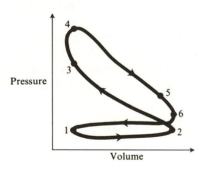

Fig. 4–3. Indicator diagram for actual gasoline-engine (spark-ignition) cycle.

4–2
The diesel engine cycle

The *diesel engine** differs from the gasoline engine in that it has no spark plug, and fuel is injected into the cylinder at the end of the compression stroke. Its compression ratios can be higher (~ 15) than those of an Otto-cycle engine because the

---

* Named after its inventor, Rudolf Diesel, a German engineer who patented the engine design in 1892.

cylinder contains only air during the compression stroke [process (2–3) in Fig. 4–4]. Fuel is injected into the hot compressed air during process (3–4). Because the compression stroke is adiabatic, or nearly so, air temperatures are high enough to ignite the fuel as it is sprayed into the cylinder. The combustion is idealized as a constant-pressure process (3–4). Process (4–5) is the adiabatic work-producing expansion. Events (5), (6), and (7) are the opening of the exhaust valve and the expulsion of exhaust gases, and are the same as in the Otto cycle.

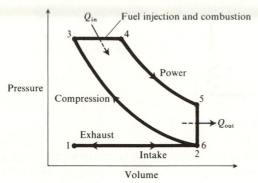

Fig. 4–4. Air-standard diesel cycle (4-stroke).

We may determine the efficiency of the idealized air-standard engine by assuming:

A fixed mass (1 mol) of ideal gas in the cylinder.

Process 2–3, adiabatic reversible compression.

Process 3–4, isobaric reversible heat addition.

Process 4–5, adiabatic reversible expansion.

Process 5–6, isovolumetric heat rejection.

We neglect the work required for the exhaust and intake strokes. The result is that:

$$\eta_{\text{ideal diesel}} = 1 - \frac{V_5}{\gamma} \frac{\left(\dfrac{V_4}{V_5}\right)^\gamma - \left(\dfrac{V_3}{V_2}\right)^\gamma}{V_4 - V_3}. \tag{4–4}$$

Exercise 4–3. Derive Eq. (4–4).

4–3
Steam engines: the Rankine cycle

Many heat engines have been devised which use a condensable vapor (usually steam) as their working fluid. Such engines may operate in a closed cycle, as shown schematically in Fig. 4–5. Heat from the combustion of chemical or nuclear fuel

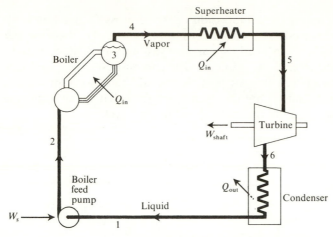

Fig. 4–5. Steam power cycle.

enters the cycle through the metal walls of the boiler, which contains liquid (water) at a high pressure. The liquid is converted into high-pressure vapor (steam), which then flows through pipes called *superheaters*, which are further heated by hot fuel gases. The superheated high-pressure vapor then passes to a cylinder or a turbine, where part of its accumulated energy is extracted as work during an adiabatic expansion. The expanded low-pressure vapor then exhausts from the turbine or cylinder to a condenser, which converts it back into liquid. The pressure in the condenser may be considerably below atmospheric pressure. A boiler-feed pump then returns the low-pressure condensed vapor back to the high-pressure boiler for another cycle.

The *Rankine cycle* (named for William J. M. Rankine, 1820–1872, Scottish engineer and author of many early works on thermodynamics, whose name is also associated with the absolute Fahrenheit temperature scale) serves as the ideal standard cycle for heat engines that operate with condensable-vapor working

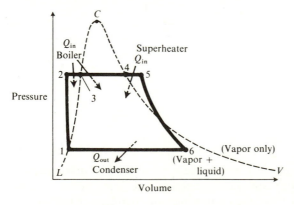

Fig. 4–6. Pressure–volume diagram of the Rankine cycle.

fluids. The cycle for a unit mass of steam is shown on a *PV* diagram in Fig. 4–6 by the closed path 1–2–3–4–5–6–1.

The dashed line *CV* is the saturation curve encompassing the region of values of pressure and specific volume within which liquid water is in equilibrium with water vapor or steam. Water exists as a pure liquid to the left of *LC* and as a dry vapor to the right of *CV*. The region under *LCV* represents mixtures of liquid and vapor. The numbers in Fig. 4–6 correspond to those in Fig. 4–5. Liquid water at condenser temperature and pressure (1) is pumped into the high-pressure boiler with an almost-negligible change in specific volume (process 1–2). The water is heated to its boiling point at boiler pressure (2–3). Addition of heat to the boiler continues at constant pressure and temperature until vaporization is complete (3–4). The vapor is then superheated (4–5), and expanded adiabatically in a suitable cylinder or turbine machine. At (6), the low-pressure exhaust steam, which may now contain suspended droplets of water, enters the condenser, where it gives off its enthalpy of vaporization at a constant temperature and pressure (6–1) and returns to the liquid phase. The condenser, which is the heat sink for the cycle, is at a much lower temperature than the boiler (the heat source).

The ideal Rankine cycle resembles the Carnot cycle in that—except for the processes of feed-water preheating (2–3) and superheating (4–5)—it consists of two isotherms and two adiabatic processes. Figure 4–7 shows the cycle plotted on a temperature–entropy (*TS*) diagram. Process (5–6) on the diagram represents the isentropic expansion of the steam. If the expansion is allowed to proceed to low temperatures (or pressures), state (6) lies below the saturation envelope (line *LCV*) and represents a state in which exhaust steam contains suspended droplets of saturated liquid water. Such steam is called *wet steam*. The term *steam quality* indicates the amount of suspended liquid water in steam; it is equal to the pounds of dry (liquid-free) saturated vapor present in a pound of wet saturated steam, expressed as a percentage. Thus 98% quality steam is steam containing 0.98 pound of dry saturated steam per pound of wet steam.

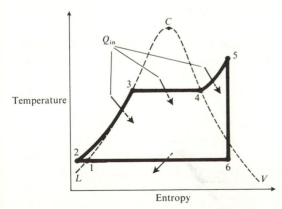

Fig. 4–7. Temperature–entropy diagram of the Rankine cycle.

Exercise 4–4. In the ideal Rankine cycle, all heat is added to and removed from the working fluid at constant pressure, and therefore the heat effects for each step of the cycle equal the changes in enthalpy, H. Assume that $(H_3 - H_2)$ (Figs. 4–6 and 4–7) is essentially identical to $(H_3 - H_1)$, and use Steam Table data to compute the thermal efficiency of a Rankine-cycle engine which isentropically expands steam at 300 psia from (a) dry saturated conditions to 10 psia; (b) 500°F superheated conditions to 10 psia.

Draw cycles (a) and (b) on the same PV plane and estimate the quality of the exhaust steam.

[*Note:* Problem 12 in Section 6 justifies the assumption regarding H_1 and H_2.]

Why is superheated steam used in most steam power cycles?

4–4
Gas turbines: the Joule cycle

Jet-aircraft engines are open-cycle gas turbines. The cycle in which gas turbines operate is called the *Joule* or *Brayton cycle* (shown in idealized air-standard form in Fig. 4–8), and consists of a pair of isentropic and isobaric processes.

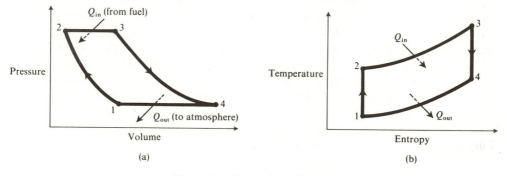

Fig. 4–8. Gas turbine (Joule) cycle.

Air is compressed isentropically (1–2) and heated isobarically (2–3). It then passes to a turbine where it expands isentropically (3–4). The low-pressure turbine exhaust is cooled along (4–1). At (1), the cooled air enters a compressor and is compressed adiabatically (1–2) until it attains the pressure of the heater inlet (P_2) and the cycle repeats.

Most gas turbines are open-cycle, internal-combustion devices. This means that the constant-pressure heating process (2–3) occurs in a combustion chamber in which the air comes in direct contact with burning fuel. In addition, the cooling process represented by (4–1) effectively occurs in the atmosphere outside the engine. The engine continuously draws in fresh air, compresses it, and heats it by direct injection of burning fuel. The hot high-pressure air and the burnt fuel gases expand through a turbine and exhaust to the atmosphere (Fig. 4–9).

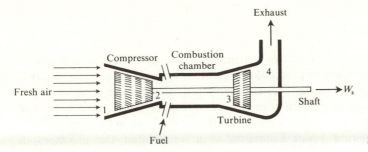

Fig. 4–9. Open-cycle gas turbine.

A gas turbine used for conventional power purposes has a shaft extension by means of which the work energy expended by the hot high-pressure gas in the power turbine is brought outside the turbine casing. Jet engines, on the other hand, have no external shaft, although their internal components have forms and functions similar to those of a power-generating gas turbine. The turbine in a jet engine extracts from the combustion gases only enough energy to drive the compressor feeding high-pressure air to the combustion chamber. The free energy remaining in the hot gases as they leave the turbine is converted into kinetic energy by passing the gases through a nozzle that accelerates them to supersonic velocities. It is this high-velocity gas jet, directed out the rear of the engine, which drives the aircraft forward. Ideally the process of expansion in the nozzle is isentropic, and the cycle appears as in Fig. 4–4.

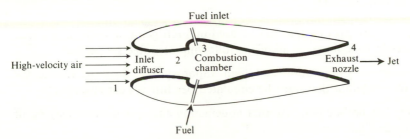

Fig. 4–10. A ramjet engine.

The *ramjet* is an open-cycle jet engine completely devoid of moving parts. It is, in essence, a hollow pipe whose inside configuration has the shape of a compound nozzle (Fig. 4–10). It operates on the Joule cycle, and achieves compression (1–2) by ramming high-velocity air into its combustion chamber. The heated air expands and accelerates through an exhaust nozzle, and is discharged at a velocity greater than the velocity of the vehicle, thus driving the vehicle forward. Ramjet operation, depending as it does on ram compression of high-velocity air, is effective only after a vehicle has attained a high (near-sonic) speed. In spite of its great mechanical simplicity, the ramjet has found few applications.

Many variations of steam- and gas-turbine cycles are used. You will find more complete description of them in standard texts on turbine design.

4–5
Refrigeration cycles

As indicated in Section 3, we obtain refrigeration by operating a heat-engine cycle backward. The variety of practical refrigeration cycles is perhaps not as great as the variety of practical power cycles. Refrigeration cycles use a gas or vapor working fluid in a sequential chain of compression, heat rejection, expansion, and heat absorption processes. The working fluid is called the *refrigerant*. Most frequently the refrigerant vapor condenses to a liquid after the processes of compression and heat rejection, and re-evaporates to the vapor phase during the heat-absorption process. The large enthalpy demand during the process of refrigerant vaporization makes it possible for a large amount of heat to be absorbed in the refrigerator cold box using a minimum amount of refrigerant (working fluid). Because of the cost and special properties of most refrigerants, refrigerators usually operate on a closed cycle.

The cooling capacity of refrigerators is frequently expressed in "tons," which is approximately equal to the tons of ice the machine can produce per day from 32°F water under certain standardized conditions. A machine with a capacity of one "ton" can absorb 200 Btu/min or 288,000 Btu/24 hr.

4–6
Properties of refrigerants

Although almost any gas or vapor can be used in a refrigeration cycle, a practical refrigerant should have many or all of the following desirable properties:

1. High enthalpy of evaporation and high vapor specific heat (to reduce the quantity of refrigerant that must be circulated per Btu of heat absorbed).

2. Reasonably low boiling point (so that liquefaction and vaporization may occur at low pressures and temperatures).

3. High critical temperature (so that vapor may always be liquefied in the temperature range of the refrigeration cycle) and low freezing point (so that refrigerant will not solidify and plug the refrigerant conduits in the cold end of the cycle).

4. Low cost.

5. They must, in addition, be chemically inert, nonflammable, nonexplosive, noncorrosive, nontoxic, and nonirritating.

The last group of properties are important for the sake of safety in the event of a leak in the refrigerant conduits, particularly in light of the fact that refrigeration is most commonly used to cool foodstuffs, and in completely enclosed spaces.

The refrigerants most commonly used today are the fluorochloro compounds of methane and ethane,* and ammonia (NH_3). The latter is used only in large commercial installations. Before the advent of the Freons, sulfur dioxide (SO_2) was used in household refrigerators.

4–7
Vapor-compression cycle

Almost all commercial refrigeration devices employ a condensable refrigerant in a vapor-compression cycle. Figure 4–11 shows the mechanical components of an idealized cycle schematically and Fig. 4–12 shows the same cycle in *PV* and *TS* diagrams. An interesting feature of this cycle is that it employs a compressor to feed work into the cycle but *no* device to produce expansion work. The work to be

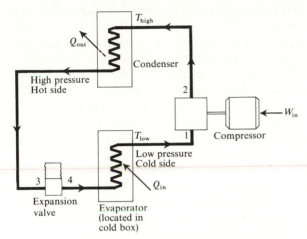

Fig. 4–11. An idealized vapor-compression refrigeration cycle.

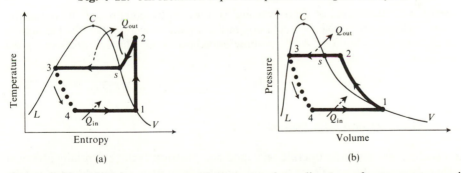

Fig. 4–12. Temperature–entropy and pressure–volume diagrams of a vapor-compression refrigeration cycle.

* Sold under the trade name Freon. For example, Freon-12, dichlorodifluoro methane, is nonflammable, nontoxic, and has a boiling point of 28°F.

obtained by the expansion of the liquefied refrigerant from the high (condenser) pressure to the low (evaporator) pressure is too little to justify the expense of using an expansion turbine or cylinder, and this potential expansion work is simply wasted by throttling the liquid irreversibly through an expansion valve (process 3–4).

The refrigerant enters the compressor as a low-pressure cold saturated vapor (state 1) and is raised isentropically to state (2). Now, as a high-pressure super-heated vapor, it flows into the condenser, where it is cooled at constant pressure and gives up its enthalpy of vaporization to a suitable external coolant (2–3). (The condenser in a home refrigerator is a long tubing coil, located outside and behind or beneath the cold box, cooled by the room air outside the refrigerator. In practice the operating temperature of the condenser is higher than room temperature.)

The condensed refrigerant, now a saturated liquid at high pressure, is throttled through the expansion valve (3–4) at the inlet to the low-pressure evaporator coils. Depending on the design of the system, all or part of the liquid, on exposure to the low pressure in the evaporator coil, flashes—or evaporates explosively—into cold vapor and flows into the coils of the evaporator, which is the heat-absorbing (cooling) device, located inside the refrigerator. In Fig. 4–12, the throttling process (3–4) is shown as a broken line to indicate that it is irreversible. State (4) is shown beneath the saturation curve (*LCV*) in the liquid–vapor equilibrium region, which means that the liquid refrigerant in this particular cycle is only partially converted into vapor by the throttling process. Since throttling occurs at constant enthalpy (Eq. 2–51), the gain in enthalpy due to vaporization must be offset by a large drop in temperature ("sensible" heat) of the refrigerant. Therefore at state (4) the refrigerant is very much colder than it is at state (3). The remainder of the liquid in the coils of the evaporator is vaporized at constant pressure and tem-perature by heat absorbed from the refrigerator cold box (process 4–1).

Exercise 4–5. What kind of controls would you use on the compressor and expansion valve to maintain the refrigerator cold box at a constant temperature? What is point *s* in Fig. 4–12? Modify Fig. 4–12 to show superheating at constant pressure in the evaporator.

4–8
Air conditioners and heat pumps

Air conditioners generally operate on vapor-compression cycles essentially identical to that shown in Fig. 4–11. The small window-mounted air conditioner has its evaporator or refrigeration coil on the room side of the window and has a blower to direct room air over the cold coil. An insulation panel separates these room-side components from the compressor and condenser coil, located outside the window. Heat picked up inside the room is rejected to the air outside. Figure 4–13 shows a circuit for refrigeration fluid which makes it possible to cool or heat a room by positioning a rotary valve in the refrigerant circuit. The valve interchanges the evaporator and condenser coils so that the device can operate either as a heat pump in cool weather or as an air conditioner in warm weather.

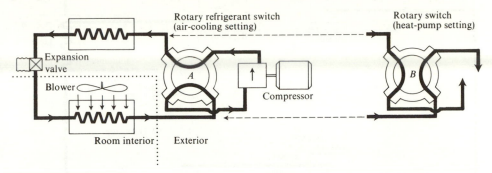

Fig. 4–13. Fluid circuit for combined operation of an air conditioner and heat pump. Rotating the refrigerant valve 90° interchanges the evaporator and condenser coils and makes possible either cooling (A position) or heating (B position) of room air.

4–9
Absorption refrigeration

In vapor-compression refrigeration, a volatile saturated liquid passes from a high- to a low-pressure region, so that a cooling effect is produced by rapid (flash) evaporation of the liquid. The cycle is completed by compressing and cooling the vapor so as to reconvert it into a high-pressure liquid. The mechanical work required to maintain a low pressure in the evaporator and to transfer the refrigerant from the low- to the high-pressure side can be reduced to a very small quantity by taking advantage of the ability of certain liquids to absorb large quantities of refrigerant vapor.

The materials most frequently used in an absorption refrigeration cycle are ammonia and water. Figure 4–14 shows the essential features of the absorption refrigeration cycle. The part of the diagram to the left of the vertical dashed line is identical to the vapor-compression cycle in Fig. 4–11. In effect, the low-pressure ammonia vapor from the evaporator (refrigerator) coils is transferred to the high-pressure side of the cycle by dissolving it in water and pumping the water into the high-pressure regenerator, where the ammonia vapor is driven out of solution by heating.

The work required to raise the pressure of the water solution is very small (why?), and may be negligible compared with Q_L and Q_R (symbols refer to Fig. 4–14). In the domestic Servel Electrolux refrigerator, which is an absorption-cycle machine, the mechanical pump is completely eliminated. The concentrated ammonia solution is transferred from the absorber to the regenerator by a thermal

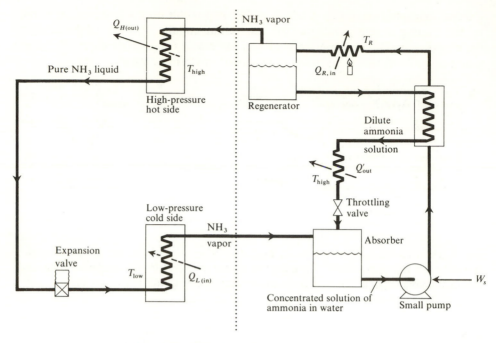

Fig. 4–14. The absorption refrigeration cycle.

siphon, similar to the way water is pumped in a coffee percolator. The refrigerator is powered entirely by a small gas or kerosene flame; $W_s = 0$.

The heat supplied to the regenerator, Q_R, may be looked on as being equivalent to the heat that would have been required to operate a heat engine supplying work to a compressor in a conventional vapor-compression refrigeration cycle.

Exercise 4–6

i) Show that the minimum work needed to operate a refrigerator which absorbs Q_L calories from a cold box at T_L, and with surroundings at T_H, is

$$W = \frac{T_H - T_L}{T_L} Q_L, \tag{a}$$

and that the minimum heat required to produce this quantity of work (with a Carnot engine) is

$$Q_R = W \frac{T_R}{T_R - T_H}, \tag{b}$$

where T_R is the temperature of the heat source (that is, the temperature of the regenerator) and T_H is, as before, the temperature of the surroundings (i.e., the high-temperature sink to which heat is rejected).

Then show that, as a consequence of (a) and (b), the maximum amount of refrigeration obtainable per calorie of heat available at T_R must be

$$\frac{Q_L}{Q_R} = \frac{T_L}{T_H - T_L} \cdot \frac{T_R - T_H}{T_R}.$$ (4-5)

ii) Show that the heat absorbed by an ideal absorption refrigeration cycle is

$$|Q_L| = |Q_H| + |Q'_{out}| - |Q_R| - |W_S|.$$

The symbols refer to Fig. 4–14.

iii) Does an absorption refrigeration cycle in which W_s is reduced to zero (as in the Servel refrigerator) violate the Second-Law statement about heat flowing from a cold to a hot reservoir?

4–10
Steam jet refrigeration

Industrial requirements for large quantities of chilled water are frequently met by using steam ejectors (steam-actuated aspirating pumps) to maintain a low pressure in large water tanks. The resulting rapid evaporation chills the remaining water, which is then circulated for purposes of air cooling and similar applications. The process, when justified by large-scale operations and the availability of steam, is both simple and economical.

Exercise 4–7. Water at 30 psia and 80°F flows through a throttle valve into a large insulated tank whose pressure is maintained at 0.1 psia. What is the lowest temperature to which the water will cool? What fraction of the liquid water vaporizes if this temperature is reached?

4–11
Air liquefaction

Liquefaction of air and similar noncondensable gases is accomplished by cooling gases to temperatures below their boiling points, using the general principles of mechanical refrigeration. Heat exchange, throttling, and adiabatic expansion engines are used in various combinations to achieve the extreme cold required. (The boiling point of $N_2 = -196°C$, of $O_2 = -183°C$, of $H_2 = -253°C$, and of $He = -269°C$.)

Figure 4–15 is a schematic diagram of a Joule-Thomson or Linde-type air-liquefaction process. Dry, CO_2-free air is compressed (1–2), cooled by passage through a heat exchanger (2–3), and further cooled by throttling through an expansion valve (3–4). Throttling cooling occurs not as a result of flash evaporation (as in the vapor-compression cycle described in Section 4–7), but because the air at high pressure behaves as a non-ideal gas whose temperature depends on its pressure,

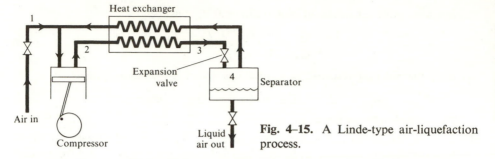

Fig. 4–15. A Linde-type air-liquefaction process.

even though the enthalpy does not change. The rate at which temperature changes with pressure during a throttling (constant-enthalpy) process, $(\partial T/\partial P)_H$, is called the *Joule-Thomson coefficient*.

Because the Joule-Thomson coefficient is positive for air at ordinary temperatures, the temperature of the air drops when it is throttled through the expansion valve into the low-pressure separator chamber. (Compressor discharge pressures may range from 2 to 100 atm, depending on the process used. Recent advances have tended to reduce pressure requirements.) The cooled, low-pressure air then passes back through the heat exchanger countercurrent to the high-pressure gas flowing to the expansion valve, and is recompressed and recycled. As the recycling continues, the temperature of the air leaving the throttling valve drops to below its boiling point. Liquid air then begins to condense in the separator, and additional air is admitted to the compressor to make up for the fraction liquefied. The process then operates at a steady state, discharging liquid air at the same rate that fresh air is fed to the compressor.

Some gases (H_2, He) have negative Joule-Thomson coefficients at ordinary temperatures, and therefore cannot be cooled by throttling. (The temperature at which the Joule-Thomson coefficient changes sign is called the *J-T inversion temperature*. For H_2 it is 100°K and for He it is 20°K.) These gases are best cooled by adiabatic expansion, in processes similar to that shown in Fig. 4–15, but with an expansion engine replacing the throttling valve (process 3–4).

Exercise 4–8. What is the loss in enthalpy of a flowing stream of gas expanding in a reversible adiabatic turbine located between (3) and (4) in Fig. 4–15, given that the turbine delivers W_s Btu of work? Would the process of Fig. 4–15 work with an ideal gas?

Liquefaction processes which employ an expansion engine (called Claude-type processes) generally operate at lower pressures than comparable throttling-type processes. It is frequent practice to combine both expansion-engine and Joule-Thomson cooling in tandem. The expansion engine takes the gas down to a temperature and pressure such that liquefaction will occur in the subsequent throttling process (i.e., below the J-T inversion temperature). The advantage of this arrangement is that it avoids problems in the operation of an expansion engine that result from liquid condensing in the engine.

Problems

1. Draw a cycle on a PV plane for an Otto-cycle engine which follows the sequence of operations shown in Fig. 4–2, except that the exhaust stroke occurs at a constant pressure slightly above atmospheric and the intake stroke occurs at a constant pressure slightly below atmospheric. Discuss why such intake and exhaust conditions might or might not occur in a real engine. What area on the diagram represents the net work output of the engine? What effect would an obstruction in the exhaust line have on the cycle and on the operation of the engine? What effects would supercharging have on the cycle? (*Supercharging* is the providing of moderately compressed air to the cylinder during the intake stroke.) Use a PV diagram to illustrate the effects of supercharging.

2. The *desert cooler* is an inexpensive device for cooling dry warm air. It consists, in essence, of a fan which blows warm air over a moist, porous pad. Water evaporating from the pad cools the air, and a slow trickle of water onto the pad replenishes the moisture that is evaporated. Is this device a heat pump? Can you describe a working cycle for the coolant? Is this an open- or a closed-cycle device?

3. An ordinary household refrigerator operates at about 40°F, whereas home freezer units may operate at temperatures of 0°F.

a) Compare the minimum amounts of work required to cool a given mass of food material from room temperature (70°F) to the operating temperatures of the refrigerator and of the freezer. (*Hint*: See Section 3–33.)

b) Refrigerators and freezers operate at a steady state for long periods of time, the refrigeration unit in each machine serving to remove from the cold chamber only that amount of heat which leaks into the cold chamber through the chamber's imperfectly insulated walls. Assuming that heat leaks into the freezer 10% faster than into the refrigerator, compare the least amount of power needed to operate each at the previously specified temperatures.

4. Draw a TS and PV diagram for the following refrigeration cycles:

a) A vapor-compression cycle in which the liquid passing through the expansion valve is all converted to dry saturated vapor.

b) A compression cycle in which the refrigerant does not condense to a liquid, and which uses an expansion valve to admit the high-pressure refrigerant gas into the coils of the cold box.

c) Same as (b), except that a reversible adiabatic expansion engine replaces the expansion (throttling) valve.

5. The equilibrium water-vapor pressure of certain aqueous solutions of salts is considerably lower than that of pure water, and is strongly temperature dependent. This phenomenon is exploited in the lithium bromide absorption refrigeration cycle which is used in air-conditioning applications.

   Use the following data to sketch an absorption refrigeration cycle and estimate its coefficient of performance (COP). What is the refrigerant in this cycle?

LiBr solution			Temp. at which water has same vapor pressure
Temp.	Wt. % LiBr	Vapor pressure	
100°F	60%	0.23 in. Hg	38°F
226°F	65%	3.50 in. Hg	120°F

6. A universal engine is being designed with a compression ratio of 10. What are the relative advantages of running it on the Otto or the diesel cycles?

7. Aircraft cabins are frequently cooled by open-cycle refrigeration which uses air bled from the jet engine turbocompressor as a refrigerant. The compressed air is passed through a heat exchanger, where it gives off heat directly to the outside atmosphere. It then expands as it passes through a miniature adiabatic turbine which exhausts directly into the cabin interior. Sketch the cycle on a PV and a TS plane and write an expression for a suitable COP. Draw a sketch showing how an open refrigeration cycle might be adapted for domestic air-conditioning. What advantages and disadvantages would such a system have compared with that of a closed-cycle air-conditioner?

8. Draw ideal air-standard diesel and Otto cycles on TS planes and indicate which portions of the diagrams represent heat flows into the cycle and which represent heat flows out of the cycle.

9. The percent clearance volume in an internal combustion engine is the ratio V_1/V_2 in Fig. 4–4 or 4–1. Express the efficiency of the ideal air-standard diesel cycle in terms of the percent clearance.

10. From Eq. (4–4), determine how the efficiency of an ideal (air-standard) diesel cycle is related to the temperatures of the cycle. How much less efficient is this engine than a $2T$ engine operating over the same T limits?

11. An air-standard *dual* cycle is a cycle which receives heat both at constant volume and at constant pressure (i.e., it combines the air-standard diesel cycle and the Otto cycle). Compare the efficiency of such a cycle with those of air-standard diesel and Otto cycles operating at the same compression ratio, given that the isobaric expansion occupies 20% of the length of the expansion or power stroke.

12. An idealized Otto-cycle engine contains $\frac{1}{2}$ lb-mol of air. At the start of the compression stroke, the pressure is 12 psia and the temperature is 100°F. The compression ratio is 9:1. When the fuel ignites in the cylinder, this is equivalent to adding 400 Btu of heat. Assuming adiabatic expansion and compression, compute:

a) the cylinder pressure at the end of the power stroke,
b) the heat discharged to the atmosphere,
c) the work done during each cycle,
d) the thermal efficiency of the engine.

13. Consider an idealized steam power plant, operating on a cycle similar to that shown in Fig. 4–5. The working substance is water, whose states at the numbered points in the figure are specified below:

1) Liquid water, 100°F and saturated
2) Liquid water, 100 psia, 100°F
3) Liquid water at 100 psia and saturation temperature
4) Dry saturated steam, 100 psia
5) Steam, 100 psia and 500°F
6) Steam at same entropy as state (5) and at condenser saturation pressure corresponding to 100°F.

Neglecting all friction and any changes in kinetic or potential energy, determine the following:

a) The heat Q per pound of water circulated, added in the boiler and superheater, and removed in the condenser.
b) The quality of the steam entering the condenser.
c) The work W of the turbine and pump (both turbine and pump are adiabatic).
d) The efficiency η of the energy conversion of this plant:

$$\eta = \frac{W_{net}}{Q_{in}}$$

14. In the preceding problem, what effect would raising the temperature of the super-heater by 200°F have on the efficiency?

15. In the idealized power plant in Problem 13, what effect would the following modifications of the cycle have on the efficiency of the energy conversion and on the cycle conditions?

a) Removal of the superheater.
b) Removal of the condenser and allowing the turbine to exhaust to the atmosphere (open-cycle operation).

Show each of the above cycles on PV and TS diagrams.

16. In a vapor-compression refrigeration cycle, the temperature of the refrigerant upstream of the expansion valve is 80°F and in the evaporator is 40°F. The refrigerant leaves the condenser as saturated liquid. Compute the minimum rate of circulation of the refrigerant for 20 tons of cooling capacity for the following refrigerants: (a) NH_3, (b) Freon-12, (c) SO_2, (d) H_2O. Consult the "ASHRAE Guide" for thermodynamic data not found in Section 7 or Appendix II.

17. Solid CO_2 is made in an apparatus similar to that in Fig. 4–15 when the discharge temperature at the expansion valve is below the CO_2 triple point. (The valve discharges a mixture of solid CO_2 snow and gaseous CO_2.) In steady-state operation, CO_2 enters the compressor at 1 atm and 70°F and leaves it at 800 psia and 90°F (water coolers are built into the compressor assembly). Gaseous CO_2 enters the heat exchanger at 90°F, and is then throttled to 1 atm and partially solidified. Vapor returns to the compressor via the heat exchanger, where it is warmed to 70°F. Compute:

a) The minimum work W per pound of CO_2 made.
b) The number of pounds of solid CO_2 per pound of CO_2 leaving the compressor.
c) The temperature upstream of the throttle valve.

18. A steam-jet refrigeration system operates at 0.362 in. Hg (absolute) pressure in its water evaporator. The chilled-water refrigerant returns to the evaporator at 70°F after doing 50 tons of cooling. Makeup water enters the system at 80°F. What weight and volume of water vapor must be removed from the evaporator per minute to maintain the refrigeration rate?

19. Examine the feasibility of operating a heat-pump cycle that uses rubber bands as a refrigerant, taking advantage of the well-known facts that rubber bands cool when they are allowed to contract adiabatically and that they have negative temperature coefficients of expansion (they shrink when warmed).

The cycle might operate as shown in the figure. Process (1–2) is an adiabatic reversible contraction, (2–3) is an isothermal contraction, (3–4) is an adiabatic stretching, and (4–1) is an isothermal stretching.

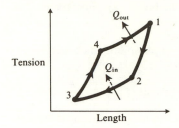

Determine the COP of this cycle in terms of T_{in} and T_{out}. Estimate the rubber circulation rate per ton of refrigeration between 0° and 50°C.

The specific length and tension are related by

$$\tau = K(L - L_0)T,$$

where τ = tension, g/cm^2 of rubber cross section; L = specific length, cm/g; L_0 = specific length at 0 tension at 300°K = 5 cm/g; T = temperature, °K; C_L = constant-length specific heat = 0.07 cal/g·°K; and K = constant = 0.12 g^2/cm^3·°K, for the band whose L_0 is given above. Also assume that $U = U(T$ only).

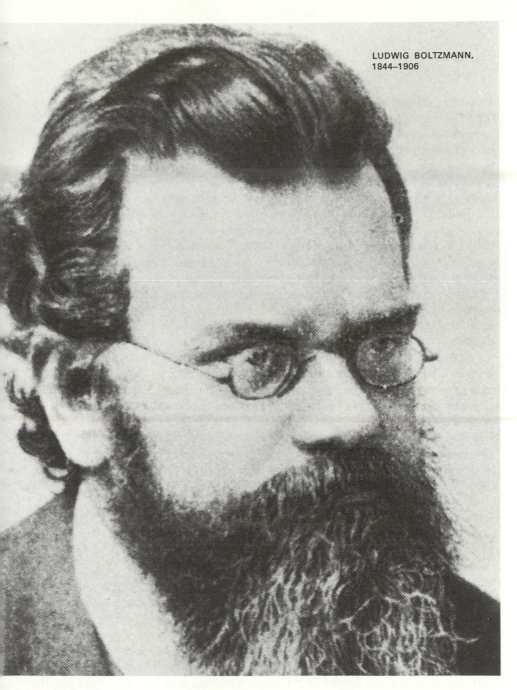
LUDWIG BOLTZMANN,
1844–1906

Section 5
STATISTICAL THERMODYNAMICS

5–1
Connecting the microscopic and macroscopic

The thermodynamics that we have been describing so far has dealt with energy and other properties of pieces of the universe (that is, systems), while ignoring the internal structure of systems, particularly the fact—or the universally accepted conjecture—that matter consists of atoms, molecules, etc. Our point of view has been *macro*scopic. We have dealt with systems that are large in comparison with molecules and have treated these as entities whose intensive properties are uniformly and continuously distributed. By watching the boundaries of a system, and the traffic of energy across these boundaries, we have been able to draw useful conclusions about the system, its surroundings, and the interactions between the two. We have not worried about the so-called "fine structure" of the systems under scrutiny. We have developed a useful science of the energetic interactions of matter without constructing theories about the nature of matter and energy.

However, theories on the nature of matter and energy—on the fine structure of matter and energy—do exist, and in fact state that matter and energy are discontinuous, and that they consist of atoms and molecules and discrete energy levels.

Statistical thermodynamics developed out of the need to reconcile the conjectures about and knowledge of the fine structure of matter and energy with the measurable thermodynamic properties of large chunks of matter. In a sense, statistical thermodynamics is the bridge between the *micro*scopic theory and the *macro*scopic observables of matter and energy. Out of the union of the microscopic and macroscopic points of view comes a more widely applicable thermodynamics, which gives us a physical understanding of the nature of entropy, which enables us to determine thermodynamic properties from spectroscopic data, to derive equations of state, and to do other useful things.

5–2
Allowed energies

Speculations about the microscopic nature of matter have led to the assertion that matter is ultimately discontinuous and that it consists of discrete entities called molecules, atoms, electrons, etc. More recent speculations, via the quantum theory, have led to the conceptually more difficult assertion that the energy stored by matter (i.e., the accumulated internal energy) is also ultimately discontinuous. The quantum theory says that the tiny packets of energy carried by the ultimate particles of matter are limited to certain discrete magnitudes. For example, the exchangeable accumulated internal energy of a pure, ideal, monatomic gas at ordinary temperatures, in the absence of potential fields (the effect of gravity on gases can usually be neglected), is the sum of the translational kinetic energy,

$mu^2/2$, of its constituent atoms. Quantum theory* allows x, y, and z components of this kinetic energy of individual atoms only those magnitudes given by:

$$\epsilon_i = \frac{mu_i^2}{2} = \frac{h^2}{8mV^{2/3}}\,(i)^2 \qquad\qquad (5\text{–}1)$$

where

h = Planck's constant = 6.6256×10^{-27} g $\cdot$ cm^2/sec,

ϵ_i = component (x, y, or z) energy of a particle (atom),

m = mass of the atom,

u_i = the velocity corresponding to quantum number i,

V = volume of container in which the atom is located,

i = quantum number, an integer which may have value 0, 1, 2, 3, ... etc.

For a given volume and kind of gas, all the terms on the right-hand side of Eq. (5–1) are fixed, with the exception of i. The term i is called a QUANTUM NUMBER. It is a dimensionless integer that may have any value between zero and infinity. Each value of i determines the magnitude of a component of the ALLOWED ENERGY of a gas atom. This energy is translational kinetic energy, $mu_i^2/2$, and we may therefore use the middle term in Eq. (5–1) to calculate corresponding allowed component velocities. Figure 5–1 is a plot of allowed energy levels versus quantum numbers,

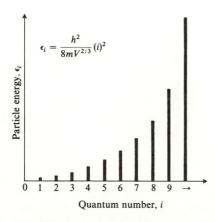

Fig. 5–1. Translational kinetic energy levels of an ideal-gas particle moving parallel to the x-axis.

---

* Equations such as (5–1), (5–2), and (5–3) are obtained by solving a fundamental equation of quantum mechanics called the *Schrödinger equation*. For our purposes we need only realize that this equation makes possible, in principle if not always in practice, the exhaustive enumeration of all possible microscopic states of a system.

as determined by Eq. (5–1). The gas particle represented by this plot moves parallel to a coordinate axis.

The total kinetic energy of a gas particle is the sum of its component energies:

$$\epsilon_{ijk} = \epsilon_i + \epsilon_j + \epsilon_k = \frac{h^2}{8mV^{2/3}} (i^2 + j^2 + k^2), \qquad (5\text{–}1a)$$

where i, j, and k are the quantum numbers in the x, y, and z coordinate directions. A set of i, j, k quantum numbers specifies the QUANTUM STATE of an ideal-gas atom.

Gases which are molecular rather than atomic can store internal energy in rotational and vibrational forms *in addition to* translational forms (Fig. 5–2). These energy forms are also limited to discrete values which differ only by a quantum number.* Thus, in a *perfect diatomic gas*, allowed *vibrational* energies satisfy the limitation:

$$\epsilon_l = hv(l + \tfrac{1}{2}), \qquad (5\text{–}2)$$

where v is the vibration frequency and l is a *vibrational quantum number*, equal to $0, 1, 2, 3, \ldots$, etc.

Rotational energies are quantized according to

$$\epsilon_J = \frac{h^2}{8\pi^2 I} (J)(J+1), \qquad (5\text{–}3)$$

(a)

Fig. 5–2. A monatomic gas particle (a) stores translational kinetic energy. A diatomic gas particle stores kinetic energy in translational (b), rotational (c), and vibrational (d) forms.

(b)

(c)

(d)

---

* There are also electronic, nuclear, intermolecular, and other microscopic modes of storing energy. At moderate temperatures, electronic and nuclear modes do not participate in energy interactions. Intermolecular modes may contribute in real materials, but not in ideal gases. The point to be emphasized is that all microscopic energy modes are quantized into discrete levels identified by one or more quantum numbers. As the complexity of a molecule grows or as the temperature increases, so do the number of modes of energy storage and the number of quantum numbers needed to specify its quantum state. In principle, all allowed states can be specified, although the complexity and size of this specification may be astronomical.

where I is the molecular moment of inertia and J is the *rotational quantum number*, which can be 0, 1, 2, 3, . . .

Consequently we specify the quantum state of a diatomic ideal-gas molecule by assigning values to each of its 5 quantum numbers (i, j, k, l, J). (A diatomic molecule is therefore said to have 5 *degrees of freedom*, in contrast to an atomic particle, which, at ordinary temperatures, has only 3 degrees of freedom.)

Exercise 5–1. Plot the x-component of the allowed atomic velocities versus the quantum number i, for i between 0 and 20, for a He atom confined to a 1-cm^3 volume which is surrounded by a 300°K heat reservoir, given that

$$m = 7 \times 10^{-24} \text{ g}, \qquad h = 6.6256 \times 10^{-27} \text{ g-cm}^2/\text{sec}.$$

Do the magnitudes of the allowed velocities and energies depend on temperature?

5–3
Multiparticle quantum states

Now the energy of a macroscopic system must equal the sum of the energies of all of its microscopic particles, that is the total energy of all of its atoms, or its molecules. Because a macroscopic system has a large number of microscopic particles and an even larger number of particle quantum states there is an immense number of differing arrangements of particles in quantum states which have the same total energy. We shall refer to an arrangement of particles in quantum states as a *multiparticle quantum state*. Therefore the next-to-last sentence may be restated as: An immense number of multiparticle quantum states is completely consistent with the energy of the macroscopic system.

Consider, for example, the exceedingly simple system shown in Fig. 5–3, consisting of 4 identical particles, each of which can be in any of 4 single-particle quantum states $(i = 0$ or 1 or 2 or 3), having energies $\epsilon_0 = 0$, $\epsilon_1 = 1$, $\epsilon_2 = 2$, $\epsilon_3 = 3$ energy units.*

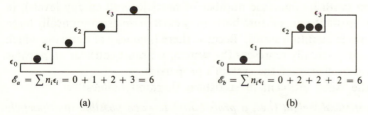

$$\mathcal{E}_a = \sum n_i \epsilon_i = 0 + 1 + 2 + 3 = 6 \qquad\qquad \mathcal{E}_b = \sum n_i \epsilon_i = 0 + 2 + 2 + 2 = 6$$

(a)   (b)

Fig. 5–3. Two possible quantum states of a 4-particle system which has a total energy of 6 units and in which the individual particles may be in any of four different single-particle states.

---

* These particles have one degree of freedom because the quantum state of a particle is specified by *one* quantum number.

The 4-particle system, as illustrated, has an energy of 6 units. In Fig. 5–3a, the system is in the multiparticle quantum state given by the *set* of its particle quantum numbers, that is, 0, 1, 2, 3. In Fig. 5–3b, it is in the multiparticle quantum state 0, 2, 2, 2. Other 4-particle quantum states which have an energy of 6 units, and which are therefore consistent with the system's energy, are as follows.

4-particle quantum states	$\mathscr{E}_6$
2, 2, 1, 1	$2 + 2 + 1 + 1 = 6$
3, 1, 1, 1	$3 + 1 + 1 + 1 = 6$
3, 3, 0, 0	$3 + 3 + 0 + 0 = 6$

If the system's total energy can *fluctuate*,* so that instead of having a precise value of 6 units, it has an *average* value of 6 units, then 30 additional 4-particle quantum states become consistent with the system's (*average*) energy. The system may conceivably spend time in any of these states. The available information (the system's energy, number of particles, and allowed particle energies) does not enable us to uniquely specify its quantum state.

Exercise 5–2. Tabulate all the 35 4-particle quantum states consistent with an *average* system energy of 6 units, assuming that the system of Fig. 5–3 fluctuates between all its allowed total energy states (0 to 12 energy units).

Note that there are 13 different energies among the quantum states considered in Exercise 5–2. (See Table 5–1.) The system is therefore said to have 13 ENERGY STATES. The number of quantum states in an energy state is called the DEGENERACY of that energy state. Thus the 6-unit energy state has a degeneracy of 5, and the 10-unit energy state has a degeneracy of 2. It should now be apparent that a macroscopic system (with its immense number of particles and energy levels), in equilibrium with a constant-temperature bath, may exist in an astronomically large number of multiparticle quantum states. Because there is *no* way of knowing which quantum state or states actually represent the system, we must consider *all possible states* when we wish to determine the system's properties.

This defines the basic problem in statistical thermodynamics:

To assign a statistical weight (i.e., a probability) to each possible multiparticle quantum state which reflects that quantum state's contribution to the properties of the macroscopic system.

---

* A completely *isolated* system must have a precisely definite energy because of the First Law. However, a system in a thermostated, constant-temperature bath has an energy that fluctuates about its average value (or *expectation value*).

Table 5–1. Quantum and Energy States of the Fig. 5–3 System of 4 Identical Particles Having 4 Single-Particle Quantum States

4-particle quantum states, l	Energy states, $\mathscr{E}_l$	Degeneracy of energy states
(Specified by a combination or set of 4 quantum numbers)	(Specified by total energy of 4 particles)	(Number of multiparticle quantum states in an energy state)
0,0,0,0	$\mathscr{E}_0 = 0$ energy units	$w_0 = 1$
0,0,0,1	$\mathscr{E}_1 = 1$	$w_1 = 1$
0,0,1,1 0,0,0,2	$\mathscr{E}_2 = 2$	$w_2 = 2$
0,1,1,1 0,0,1,2 0,0,0,3	$\mathscr{E}_3 = 3$	$w_3 = 3$
1,1,1,1 0,1,1,2 0,0,1,3 0,0,2,2	$\mathscr{E}_4 = 4$	$w_4 = 4$
1,1,1,2 0,1,1,3 0,1,2,2 0,0,2,3	$\mathscr{E}_5 = 5$	$w_5 = 4$
1,1,2,2 1,1,1,3 0,2,2,2 0,1,2,3 0,0,3,3	$\mathscr{E}_6 = 6$	$w_6 = 5$
1,1,2,3 0,2,2,3 0,1,3,3 1,2,2,2	$\mathscr{E}_7 = 7$	$w_7 = 4$
1,1,3,3 1,2,2,3 0,2,3,3 2,2,2,2	$\mathscr{E}_8 = 8$	$w_8 = 4$
2,2,2,3 1,2,3,3 0,3,3,3	$\mathscr{E}_9 = 9$	$w_9 = 3$
2,2,3,3 1,3,3,3	$\mathscr{E}_{10} = 10$	$w_{10} = 2$
2,3,3,3	$\mathscr{E}_{11} = 11$	$w_{11} = 1$
3,3,3,3	$\mathscr{E}_{12} = 12$	$w_{12} = 1$

Thus statistical thermodynamics seeks to assign a probability $\mathcal{P}_l$ to quantum state l such that $\mathcal{P}_l \, \mathcal{E}_l$ is the contribution of quantum state l to the energy of the macroscopic system. State l for a *multi*particle system is a possible combination of the single-particle quantum numbers of all particles in the system. $\mathcal{E}_l$ is the energy of the quantum state having the l set of quantum numbers.

The average or *expectation value** of the system's energy is the sum of the contributions from all quantum states, or

$$\langle \mathcal{E} \rangle = \sum_l \mathcal{P}_l \mathcal{E}_l \equiv U , \tag{5–4}$$

where

$\mathcal{P}_l =$ the probability of the existence of a multiparticle quantum state l,

$\mathcal{E}_l =$ the energy of state l, and

$\langle \mathcal{E} \rangle =$ the expected energy of the multiparticle system, which can be in any of the l states. $\langle \mathcal{E} \rangle \equiv U$, the system's accumulated internal energy.

The problem of assigning $\mathcal{P}_l$'s to all the possible quantum states is a problem in skillful *guessing*, or in making decisions without complete information. It is an act of desperation prompted by the fact that nothing else can be done,† and is made acceptable by using the laws of logic and statistical inference.

It is logically necessary that the assignment of probabilities to allowed energy states of a system neither neglects existing information nor implies the presence of nonexisting information. For example, we cannot assign certainty ($\mathcal{P}_l = 1$) or impossibility ($\mathcal{P}_l = 0$) to any allowed state, because these are unequivocal assignments and imply the existence of supporting information which does not exist. In addition, one state can be assigned a probability which is more or less than that of another only if information exists which supports unequal weighting. An assignment of probabilities made as above is said to be *free of bias* because it distributes probability values to states without prejudging and on the basis of available information only.

---

* The term "expectation value" is preferable to "average" to describe a summation of properties weighted by their corresponding probabilities [for example, Eq. (5–4)]. An average is obtained by operating on a set of measurements representing complete information about events in the *past*, whereas an "expectation value" represents a guess as to the outcome of many *future* events or measurements. Thus "expectation" is a guess based on incomplete information, whereas "average" is a way of summarizing actual experiences.

† There just is no way to determine the exact microscopic state of a macrosystem, first because of the immense number of simultaneous measurements needed to know the positions and velocities of all particles in a macrosystem ($6 \times 6.02 \times 10^{23}$ simultaneous measurements in a mole of noble gas), and second because the measuring process changes the energy of the microscopic particle being measured. The limit on the accuracy of measurement is given by the *Heisenberg uncertainty principle*, which states that the product of the errors in measuring the momentum and position of a particle cannot be less than $h/4\pi$.

5–4
Additional hypothesis

We need an additional hypothesis in order to select a bias-free distribution* of probabilities. We here state the hypothesis† in the form of a criterion for selection: *In a distribution of probabilities that is maximally free of bias, the property*

$$S \equiv -k \sum_l \mathscr{P}_l \ln \mathscr{P}_l \qquad (5\text{–}5)$$

is a maximum, subject to all the constraints on the system.

This criterion for choosing a distribution function is referred to as the Information Theory approach because Eq. (5–5) is identical to the property called *information* or EXPECTED UNCERTAINTY in the Mathematical Theory of Information.‡ (We shall see later that the *maximum S* is the same as the macroscopic property *entropy*.)

A qualitative justification for the use of the criterion is given at the end of this section. For the moment, suffice it to say that $-\sum_l \mathscr{P}_l \ln \mathscr{P}_l$ is the expectation value of $-\ln \mathscr{P}_l$, and maximizing S tends to *smooth out* (remove excessively high peaks or deep valleys) *an ordered set of $\mathscr{P}_l$'s, making them as uniformly small as the data allow*. The expectation value of any function $f(x)$ is defined as

$$\langle f(x) \rangle \equiv \sum_i \mathscr{P}_i f(x_i). \qquad (5\text{–}6)$$

5–5
Maximizing uncertainty

We shall now apply the maximum uncertainty criterion to a system about which we know only the following:

a) The system has N identical particles, and very many single-particle quantum states are accessible to these particles.

b) The system is immersed in a thermostated heat bath and therefore has an energy of U; (the energy fluctuates about $\langle \mathscr{E} \rangle$).

c) The system has a constant volume and temperature.

---

* A function which determines the value of the probability assigned to each state is called a *distribution function*. The set of probabilities corresponding to a set of states is called the *distribution*.

† Other hypotheses, some of which are described and compared at the end of this section, may also be used.

‡ See C. Shannon, *The Mathematical Theory of Information*, Urbana, Ill.: University of Illinois Press, 1949. See also M. Tribus, *Thermostatics and Thermodynamics*, Princeton, N.J.: Van Nostrand, 1960.

d) The N-particles and their accessible single-particle quantum states in combination form an immense number of N-particle quantum states.

Since it is possible in principle to describe *all possible* N-particle quantum states,* that is, since we know all possible states of the N-particle system, it necessarily follows that the system is *certain* to be in one of these states. This seemingly simple idea can be expressed as

$$\mathscr{P}_0 + \mathscr{P}_1 + \mathscr{P}_2 + \mathscr{P}_3 + \cdots + \mathscr{P}_l + \cdots = 1$$

or

$$\sum_l \mathscr{P}_l = 1, \tag{5-7}\dagger$$

where $\mathscr{P}_l$ is the probability of the system being in the l state. It also follows from item (b) above that

$$\sum_l \mathscr{P}_l \mathscr{E}_l = \langle \mathscr{E} \rangle = U, \tag{5-8}$$

where $\mathscr{E}_l$ is the energy of state l and $\langle \mathscr{E} \rangle$ is the expected value of the system's energy, which is the same as U, the accumulated internal energy.

Equations (5-7) and (5-8) summarize the available information on the system (it must be in some quantum state, and it has a certain expected energy), and are the constraints under which Eq. (5-5) is to be maximized.

We can perform the maximization, subject to constraints, by using a technique known as *Lagrange's Method of Indeterminate Multipliers*:

a) We multiply each equation of constraint (Eqs. 5-7 and 5-8 in this instance) by an arbitrary constant, called a *Lagrangian Multiplier*.

b) We set all the derivatives of Eq. (5-5) and the constant-augmented constraint equations equal to zero. The maximum condition will then be expressible in terms of the Lagrangian multipliers.

Therefore, from Eq. (5-5), by taking the derivative of each term in the summation, we obtain

$$dS = 0 = -k \sum_l \left(\ln \mathscr{P}_l \, d\mathscr{P}_l + \mathscr{P}_l \frac{d\mathscr{P}_l}{\mathscr{P}_l} \right)$$

or

$$0 = \sum_l (\ln \mathscr{P}_l + 1) \, d\mathscr{P}_l. \tag{5-9}$$

---

* We can do this by enumerating all the single-particle quantum-number combinations for N particles. This is possible in principle, but certainly not in practice, if we are dealing with 10^{23} particles and an even greater number of allowed single-particle states.

† Equation (5-7) can apply to any exhaustively enumerated set of states. For example, if applied to the throwing of a single die, it means that the die face will show either a 1, or a 2, or a 3, or a 4, or a 5, or a 6.

From Eq. (5-7), we have

$$0 = \sum_l d\mathcal{P}_l,$$

and multiplying by an arbitrary constant $(\psi - 1)$,

$$0 = \sum_l (\psi - 1)\, d\mathcal{P}_l. \tag{5-10}$$

From Eq. (5-8),*

$$0 = \sum_l \mathcal{E}_l\, d\mathcal{P}_l,$$

and multiplying by another arbitrary constant β, we obtain

$$0 = \sum_l \beta \mathcal{E}_l\, d\mathcal{P}_l. \tag{5-11}$$

Adding Eqs. (5-9), (5-10), and (5-11) yields

$$0 = \sum_l (\ln \mathcal{P}_l + \psi + \beta \mathcal{E}_l)\, d\mathcal{P}_l. \tag{5-12}$$

If the summation is always zero, the term in parentheses must be zero. Therefore:

$$\ln \mathcal{P}_l = -\psi - \beta \mathcal{E}_l \tag{5-13}$$

or

$$\boxed{\mathcal{P}_l = e^{-\psi - \beta \mathcal{E}_l},} \tag{5-14}$$

which says that the probability of the occurrence of the N-particle quantum state l is a function only of the energy of that state, and of two parameters of the system, the Lagrangian multipliers ψ and β.

Equation (5-14) is the answer we seek, the minimally biased distribution of probabilities consistent with our knowledge of the system. Figure 5-4 shows the equation plotted for two values of β.

Exercise 5-3. Show that the maximum uncertainty criterion applied to a system about which the only available information is

$$\sum_{l=1}^{w} \mathcal{P}_l = 1$$

---

* In a constant-volume system, the energy $\mathcal{E}_l$ of *quantum* state l is constant. (See Eq. (5-1).)

(where w is the total number of quantum states) leads to the conclusion that

$$\mathscr{P}_l = \text{constant} = \frac{1}{w}.$$

(Note that when we are applying to this system the procedure outlined by Eqs. (5–9) through (5–14), the information giving rise to Eq. (5–11) is not available.)

Exercise 5–4. Show that a result similar to that of Exercise 5–3 is obtained for an isolated system whose energy *cannot* fluctuate, but is limited to a precise value, U. ($U = \mathscr{E}_l = $ constant). Given that the system can be in any of w different quantum states, what is the probability of its being in each state, and the degeneracy of the system's *energy states*? Show that

$$\boxed{S = k \ln w.}$$ (5–5a)

Exercise 5–5. Imagine a system so arranged that both its energy and its particle content may *fluctuate* about certain expected values. An example of such a system might be a gram of liquid water in a large closed container filled with water vapor at the equilibrium vapor pressure of the water, the container being immersed in a thermostated bath. The number of particles in the system (the liquid) fluctuates because water molecules are continuously passing in both directions through the liquid–vapor interface. A quantum state l for this system has both an energy, $\mathscr{E}_l$, and a particle content, n_l, associated with it. Using uncertainty maximization, show that this kind of *open* system, which is constrained by

$$\sum_l \mathscr{P}_l = 1,$$

and by the system's expected energy,

$$\sum_l \mathscr{P}_l \mathscr{E}_l = \langle \mathscr{E} \rangle;   \text{(or } U)$$

and by the system's expected particle content

$$\sum_l \mathscr{P}_l n_l = \langle N \rangle$$

has a distribution function

$$\mathscr{P}_l = e^{-\psi' - \beta' \mathscr{E}_l - \alpha n_l},$$

where ψ', β', and α are Lagrangian multipliers.

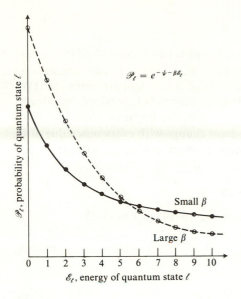

$$\mathscr{P}_\ell = e^{-\psi - \beta \mathscr{E}_\ell}$$

Small β

Large β

Fig. 5–4. Probability versus energy of a multiparticle quantum state; minimally biased distribution.

5–6
Partition functions

When we substitute Eq. (5–14) into the equation of constraint (5–7), we get

$$\sum_l \mathscr{P}_l = \sum_l e^{-\psi - \beta \mathscr{E}_l} = 1. \tag{5–15}$$

Here $e^{-\psi}$ is a constant and can be moved out of the summation sign:

$$e^{-\psi} \sum_l e^{-\beta \mathscr{E}_l} = 1$$

or

$$\boxed{e^{+\psi} = \sum_l e^{-\beta \mathscr{E}_l} \equiv Z.} \tag{5–16}$$

Equation (5–16) is called the PARTITION FUNCTION or the *sum-over-states function.* The abbreviation for it is the symbol Z.

On substituting Eq. (5–16) back into Eq. (5–14), we obtain

$$\mathscr{P}_l = \frac{e^{-\beta \mathscr{E}_l}}{\sum_l e^{-\beta \mathscr{E}_l}} = \frac{e^{-\beta \mathscr{E}_l}}{Z}. \tag{5–17}$$

On taking the logarithm of Eq. (5–16), we see that

$$\boxed{\psi = \ln \sum_l e^{-\beta \mathscr{E}_l} = \ln Z.} \tag{5–18}$$

5-7
Changing β and ψ

If the heat bath surrounding the system is changed, the energy U of the system changes. The individual quantum-state energies $\mathscr{E}_l$, however, remain constant so long as the volume or other external parameters of the system are constant. (We can see this in part by reexamining Eq. (5–1) and Exercise 5–1, in which the allowed energies are independent of temperature.) It is therefore apparent from Eqs. (5–4) and (5–14) that $\mathscr{P}_l$, and therefore ψ and β, must change with every new value of U.

From Eq. (5–18),* we can obtain the rate at which ψ changes with β:

$$\left(\frac{\partial \psi}{\partial \beta}\right)_V = \frac{\partial \ln Z}{\partial \beta} = \frac{\partial Z}{\partial \beta} \cdot \frac{1}{Z} = \frac{\partial(\sum_l e^{-\beta \mathscr{E}_l})}{\partial \beta \cdot Z} = \frac{\sum_l e^{-\beta \mathscr{E}_l}(-\mathscr{E}_l)}{Z}. \qquad (5\text{–}19)$$

Now, substituting Eq. (5–16) for Z and using Eq. (5–14) in Eq. (5–19),

$$\left(\frac{\partial \psi}{\partial \beta}\right)_V = \frac{\sum_l e^{-\beta \mathscr{E}_l}(-\mathscr{E}_l)}{e^{\psi}} = \sum_l e^{-\psi - \beta \mathscr{E}_l}(-\mathscr{E}_l) \qquad (5\text{–}19a)$$

or

$$\left(\frac{\partial \psi}{\partial \beta}\right)_V = -\sum_l \mathscr{P}_l \mathscr{E}_l. \qquad (5\text{–}20)$$

Then, recalling Eq. (5–4), we see that

$$\boxed{\left(\frac{\partial \psi}{\partial \beta}\right)_V = -U,} \qquad (5\text{–}21)$$

or the rate at which the ln partition function varies with β is equal to minus the system's energy!

All thermodynamic functions can be expressed in terms of the ln partition function. Hence the importance of the partition function.

5-8
The nature of β

For the two separate multiparticle systems A and B in Fig. 5–5a, whose energies are U_A and U_B, respectively, we can calculate the probability of allowed quantum states in each system by the technique just described. The expected uncertainty of system A,

$$S_A = -k \sum_l \mathscr{P}_{Al} \ln \mathscr{P}_{Al}, \qquad (5\text{–}22)$$

is maximized subject to the restraints:

$$\sum_l \mathscr{P}_{Al} = 1 \qquad (5\text{–}23)$$

---

* The differentiation is not as formidable as it may first appear. Remember that

$$\frac{\partial \ln y}{\partial x} = \frac{\partial y}{\partial x} \cdot \frac{1}{y}, \qquad \text{and that} \qquad \frac{d(e^{-ay})}{dy} = e^{-ay}(-a).$$

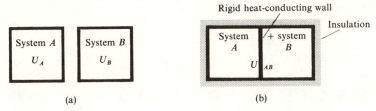

Fig. 5–5. (a) Two separate multiparticle systems. (b) The same two systems in thermal contact.

and

$$\sum_l \mathscr{P}_{A_l} \mathscr{E}_{A_l} = U_A. \tag{5–24}$$

Similar equations apply to system B:

$$S_B = -k \sum_l \mathscr{P}_{B_l} \ln \mathscr{P}_{B_l},$$

$$\sum_l \mathscr{P}_{B_l} = 1, \tag{5–25}$$

$$\sum_l \mathscr{P}_{B_l} \mathscr{E}_{B_l} = U_B. \tag{5–26}$$

The calculated distribution functions have identical forms, but different ψ and β parameters:

$$\mathscr{P}_{A_l} = e^{-\psi_A - \beta_A \mathscr{E}_{A_l}}, \qquad \mathscr{P}_{B_l} = e^{-\psi_B - \beta_B \mathscr{E}_{B_l}}.$$

If the two systems are now brought into thermal contact, as in Fig. 5–5b, so that they exchange energy without exchanging particles (contact is through a rigid heat-conducting wall), they each eventually achieve a new, stable set of quantum states consistent with their shared energies. We can find new distribution functions by maximizing the uncertainty of the *combined* system:

$$S_{AB} = -k \sum_l \mathscr{P}_{A_l} \ln \mathscr{P}_{A_l} - k \sum_l \mathscr{P}_{B_l} \ln \mathscr{P}_{B_l} \tag{5–27}*$$

---

* Implicit in Eq. (5–27) is the idea that S is an extensive function; that is,

$$S_{AB} = S_A + S_B.$$

From the definition, Eq. (5–5),

$$S_{AB} = -k \sum \mathscr{P}_{A_l B_l} \ln \mathscr{P}_{A_l B_l};$$

where $\mathscr{P}_{A_l B_l}$ is the probability of system A being in state A_l when B is in state B_l. The summation is over all *combinations* of A_l and B_l. Now

$$\mathscr{P}_{A_l B_l} = (\mathscr{P}_{A_l})(\mathscr{P}_{B_l}), \text{ if states of } A \text{ and } B \text{ are independent.}$$

Therefore

$$S_{AB} = -k \sum_{A_l} \sum_{B_l} \mathscr{P}_{A_l} \mathscr{P}_{B_l} \ln \mathscr{P}_{A_l} \mathscr{P}_{B_l}$$

$$= -k \sum_{A_l} \sum_{B_l} \mathscr{P}_{A_l} \mathscr{P}_{B_l} \ln \mathscr{P}_{A_l} - k \sum_{A_l} \sum_{B_l} \mathscr{P}_{A_l} \mathscr{P}_{B_l} \ln \mathscr{P}_{B_l}$$

$$= -k \sum_{A_l} \mathscr{P}_{A_l} \ln \mathscr{P}_{A_l} \left(\sum_{B_l} \mathscr{P}_{B_l} \right) - k \sum_{B_l} \mathscr{P}_{B_l} \ln \mathscr{P}_{B_l} \left(\sum_{A_l} \mathscr{P}_{A_l} \right)$$

or

$$S_{AB} = S_A + S_B. \qquad \text{Q.E.D.}$$

subject to the constraints:

$$\sum_l \mathscr{P}_{A_l} = 1, \tag{5-28}$$

$$\sum_l \mathscr{P}_{B_l} = 1, \tag{5-29}$$

$$\sum_l \mathscr{P}_{A_l}\mathscr{E}_{A_l} + \sum_l \mathscr{P}_{B_l}\mathscr{E}_{B_l} = U_A + U_B = U_{AB}. \tag{5-30}$$

Equation (5–30) takes cognizance of the fact that we no longer know the individual internal energies of systems A and B, only the combined energy of A and B together. The equation states that energy is conserved during the contacting process.

The initial condition (Fig. 5–5a) was described by four equations of constraint: Eqs. (5–23), (5–24), (5–25), and (5–26), whereas only three equations—Eqs. (5–28), (5–29), and (5–30)—are available for the final condition (Fig. 5–5b). Since each constraint gives rise to one Lagrangian multiplier in the maximization procedure, the final distribution functions have only *three* multipliers:

ψ_A, from Eq. (5–28), ψ_B, from Eq. (5–29), β, from Eq. (5–30).

The distribution functions for maximum S are found [by the procedures outlined in Eqs. (5–9) through (5–14)] to be

$$\mathscr{P}_{A_l,\,\text{final}} = e^{-\psi_A - \beta\mathscr{E}_{A_l}}, \tag{5-31}$$

$$\mathscr{P}_{B_l,\,\text{final}} = e^{-\psi_B - \beta\mathscr{E}_{B_l}}. \tag{5-32}$$

The parameter β has the *same value* in both systems. *β is therefore a distribution-function parameter common to systems in thermal contact which have each achieved a stable distribution of allowed states*, i.e., they are in thermal equilibrium.

We can obtain further insights into the nature of β by using Eq. (5–14) to examine the relative probabilities of two quantum states i and j of the same system:

$$\frac{\mathscr{P}_j}{\mathscr{P}_i} = \frac{e^{-\psi - \beta\mathscr{E}_j}}{e^{-\psi - \beta\mathscr{E}_i}} \tag{5-33}$$

or

$$\boxed{\frac{\mathscr{P}_j}{\mathscr{P}_i} = e^{-\beta(\mathscr{E}_j - \mathscr{E}_i)}.} \tag{5-34}$$

When $\mathscr{E}_j > \mathscr{E}_i$ (and β is positive), Eq. (5–34) shows that

$$\frac{\mathscr{P}_j}{\mathscr{P}_i} < 1,$$

or the *high-energy state is less probable than the low-energy state*. Equation (5–34) is known as the *Boltzmann factor*.

Again referring to Eq. (5–34), as β approaches infinity,

$$\frac{\mathscr{P}_j}{\mathscr{P}_i} \to 0, \tag{5–35}$$

which is to say that only the *lowest energy state is possible*. With only *one* energy state possible, the uncertainty (entropy) must go to zero (or, if the lowest energy level is degenerate, to a constant).

As β approaches zero, Eq. (5–34) approaches 1 and all states become equally likely. In general, the higher allowed energy states become more probable (or accessible) as β decreases. The two curves in Fig. (5–4) should now be reexamined in light of this discussion.

A relationship between β and *pressure* (Eq. 5–36) can be derived by considering the momentum transferred to the walls of a container by 1 mol of ideal-gas atoms. (See Problem 2 at the end of this section.)

$$P = \frac{\bar{N}}{\beta \bar{V}}, \tag{5–36}$$

where $\bar{N}$ is Avogadro's number: 6.02×10^{23} atoms per g-mol. When we compare Eq. (5–36) with the ideal-gas law, $P = RT/\bar{V}$, we see that:

$$\boxed{\beta = \frac{\bar{N}}{RT} = \frac{1}{kT},} \tag{5–37}$$

where k is called the BOLTZMANN CONSTANT and equals $R/\bar{N}$, or 1.38054×10^{-16} erg/molecule $\cdot$ °K.

The fact that β is proportional to the inverse of the ideal-gas absolute temperature is now seen to be consistent with its other attributes, shown in Eqs. (5–31), (5–32), and (5–35); namely, that:

a) The parameter β is equal in systems that are in thermal equilibrium with each other.

b) Expected uncertainty or entropy approaches zero or a constant as $\beta \to \infty$ and as $T \to 0$, which we recognize as the *Third Law of Thermodynamics* (Section 3).

Exercise 5–6. Some of the more obvious properties of the expectation operator, $\langle f(x) \rangle \equiv \sum_i \mathscr{P}_i [f(x_i)]$, are

$$\langle k \rangle = k \qquad \text{and} \qquad \langle k \cdot f(x) \rangle = k \cdot \langle f(x) \rangle,$$

where $k = $ constant. Use these and Eqs. (5–5), (5–7), and (5–8) to show that

$$S = k\,(\psi + \beta U). \tag{5–38}$$

Exercise 5–7. Use Eq. (5–38) to prove that

$$-\frac{\psi}{\beta} = A,$$

(5–39)

which is the *Helmholtz free energy*.

5–9
Negative temperatures

The foregoing section considered only positive values of β, but nothing in our discussion limits β, a statistical parameter, to positive values. More advanced treatments demonstrate that, when allowed energy levels are limited ("bounded"), negative values of β can exist.

Studies of lasers and of magnetic phenomena in certain perfect* crystals have yielded results explainable by postulating negative β's. Equation (5–34) indicates that when β is negative, the higher energy states are more probable than the lower. Negative temperatures are "hotter" than positive temperatures because spontaneous heat flow occurs from the negative to the positive temperatures.

5–10
Probability and partition function for single-particle quantum states

Figure 5–6 represents a set of hypothetical nondegenerate quantum states available to a certain microscopic particle. These single-particle states and their corresponding energies could have been determined from quantum-mechanical calculations or from spectroscopic data. The probability of each state is, however, not obtainable from these sources. If the expected energy of the *particle* is $\langle \epsilon \rangle$, then we can find the probability $\mathscr{P}_i$ of the particle being in single-particle quantum state i

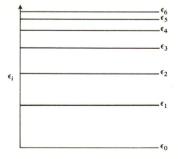

Fig. 5–6. Allowed single-particle energy states of a hypothetical system.

* N. F. Ramsay, *Phys. Rev.* **103**, 20 (1956). See also C. E. Hecht, *J. Chem. Educ.* **44**, 125 (1957).

by maximizing

$$S = -k \sum_i \mathscr{P}_i \ln \mathscr{P}_i \qquad (5\text{--}40)$$

subject to the constraints:

$$\sum_i \mathscr{P}_i = 1 \qquad \text{(The particle must be in one of its possible states),} \qquad (5\text{--}41)$$

$$\sum_i \mathscr{P}_i \, \epsilon_i = \langle \epsilon \rangle \qquad \text{(The particle has an expected energy).} \qquad (5\text{--}42)$$

We do this by means of the procedure demonstrated in Eqs. (5–9) through (5–14), with the result that

$$\mathscr{P}_i = e^{-\lambda - \beta \epsilon_i}, \qquad (5\text{--}43)$$

where λ and β are the Lagrangian multipliers. Proceeding as before, we find that

$$e^\lambda = \sum_i e^{-\beta \epsilon_i} \equiv \mathfrak{z} \equiv \text{single-particle partition function,} \qquad (5\text{--}44)$$

$$\lambda = \ln \sum_i e^{-\beta \epsilon_i} = \ln \mathfrak{z}, \qquad (5\text{--}45)$$

$$\left(\frac{\partial \lambda}{\partial \beta} \right) = -\langle \epsilon \rangle. \qquad (5\text{--}46)$$

Equation (5–43) makes a statement about the inverse relationship between $\mathscr{P}_i$ and ϵ_i of single-particle states similar to the statement that Eq. (5–14) makes about probability and energy of N-particle states.

Exercise 5–8. Derive Eqs. (5–43) through (4–46), starting from Eqs. (4–50), (5–41), and (5–42).

5–11
N-particle partition function from single-particle partition functions

Quantum mechanics deals with the energetics of individual microscopic particles, whereas *statistical mechanics* or *statistical thermodynamics* deals with the energetics of large assemblies of such particles, such as exist in macroscopic systems. We are interested in the energetics of macroscopic (multiparticle) systems. However, our knowledge of allowed energy states is usually limited to the allowed energy states of single particles. This is so because that knowledge is gained from quantum-mechanical considerations. In order to apply the information about single-particle energy states to multiparticle systems, we must be able to evaluate multiparticle partition functions (Z) from single-particle partition functions ($\mathfrak{z}$). This is readily done in systems in which individual particles do not interact with each other. Ideal-gas systems behave in this manner. Effectively each particle has its own energy, and there are negligible attractive or repulsive forces between particles. Particle interaction is limited to elastic collision without dissipation of energy.

Figure 5–7 shows the same set of quantum states as is shown in Fig. 5–6. However, three particles are now distributed among these states, and the horizontal width of each state represents its single-particle probability.

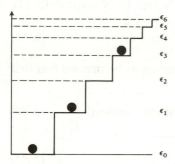

Fig. 5–7. A possible 3-particle quantum state of the system diagrammed in Fig. 5–6.

Assume that, as in Fig. 5–7, a system consists of three noninteracting *identical* particles. If the particles are momentarily in the single-particle states i, j, and k, the system is said to be in the three-particle state i, j, k, which we abbreviate as three-particle state l (where l stands for the combination of quantum numbers i, j, k). The energy of state l is

$$\mathscr{E}_l = \epsilon_i + \epsilon_j + \epsilon_k.$$

Now, if the particles are independent, the probability of finding the first particle in i, the second in j, and the third in k equals $\mathscr{P}_i \times \mathscr{P}_j \times \mathscr{P}_k$.

State l, however, consists of three *indistinguishable* particles occupying states i, j, k. Therefore all the *permutations* of i, j, k are also state l. Hence

$$\mathscr{P}_l = \mathscr{P}_i\mathscr{P}_j\mathscr{P}_k + \mathscr{P}_i\mathscr{P}_k\mathscr{P}_j + \mathscr{P}_k\mathscr{P}_i\mathscr{P}_j + \mathscr{P}_k\mathscr{P}_j\mathscr{P}_i + \mathscr{P}_j\mathscr{P}_i\mathscr{P}_k + \mathscr{P}_j\mathscr{P}_k\mathscr{P}_i \quad (5\text{–}47)^*$$

or

$$\mathscr{P}_l = 6(\mathscr{P}_i\mathscr{P}_j\mathscr{P}_k) = 3!(\mathscr{P}_i\mathscr{P}_j\mathscr{P}_k). \quad (5\text{–}48)$$

---

* For example, the probability of rolling a 3 and then a 4 and then a 5 on three consecutive rolls of a single die is $\mathscr{P}_3 \times \mathscr{P}_4 \times \mathscr{P}_5 = \frac{1}{6} \times \frac{1}{6} \times \frac{1}{6} = \frac{1}{216}$. However, the probability of rolling a 3 *and* a 4 *and* a 5 in any sequence is

$$\mathscr{P}_3\mathscr{P}_4\mathscr{P}_5 + \mathscr{P}_3\mathscr{P}_5\mathscr{P}_4 + \mathscr{P}_5\mathscr{P}_3\mathscr{P}_4 + \mathscr{P}_5\mathscr{P}_4\mathscr{P}_3 + \mathscr{P}_4\mathscr{P}_5\mathscr{P}_3 + \mathscr{P}_4\mathscr{P}_3\mathscr{P}_5 = \frac{3!}{216} = \frac{1}{36}.$$

When we substitute Eqs. (5–43) and (5–44) into Eq. (5–48), we obtain

$$\mathcal{P}_l = 3! \left(\frac{e^{-\beta \epsilon_i}}{e^{\lambda}}\right) \left(\frac{e^{-\beta \epsilon_j}}{e^{\lambda}}\right) \left(\frac{e^{-\beta \epsilon_k}}{e^{\lambda}}\right)$$

or

$$\mathcal{P}_l = 3! \frac{e^{-\beta(\epsilon_i + \epsilon_j + \epsilon_k)}}{(e^{\lambda})^3},$$

$$\mathcal{P}_l = 3! \frac{e^{-\beta \epsilon_l}}{(\sum e^{-\beta \epsilon_i})^3}. \tag{5–49}$$

For an N-particle state, Eq. (5–49) becomes

$$\mathcal{P}_l = N! \frac{e^{-\beta \mathcal{E}_l}}{(\sum_i e^{-\beta \epsilon_i})^N}. \tag{5–50}*$$

When we compare Eq. (5–50) with (5–17), it follows that the relationship of an N-particle partition function to a single-particle partition function is:

$$\sum_l e^{-\beta \mathcal{E}_l} = \frac{(\sum_i e^{-\beta \epsilon_i})^N}{N!}$$

or

$$Z_{N\text{-particle}} = \frac{(\mathfrak{Z}_{\text{single particle}})^N}{N!}. \tag{5–51}$$

Exercise 5–9. *Energy state* $\mathcal{E}_l$ in the above three-particle system may be degenerate; that is, it may contain a number of equal-energy three-particle *quantum* states. Assume that $\mathcal{E}_l = 6$ units and that the allowed single-particle states have nondegenerate energy levels of $\epsilon_i = 0, 1, 2, 3, 4, 5, 6, \ldots$. What is the degeneracy w_l of three-particle energy state $\mathcal{E}_l$?

5–12
Reversible work

The energy of a quantum state in an ideal-gas system is a function of the volume (see Eq. 5–1). Now if the volume is changed by an amount dV, then the energy $\mathcal{E}_l$

---

* The number of ways of ordering or permuting N objects all of which are different is $N!$. However, if some of the objects are the same, that is, if $N = n_i + n_j + n_k + \cdots$, where n_i is the number of objects of the ith kind, etc., then the number of permutations is $N!/n_i! \cdot n_j! \cdot \cdots$. To be perfectly general and to allow for more than one particle in each particle quantum state, this term should replace $N!$ in Eq. (5–50). In ideal-gas and other systems to which the above analysis applies, the number of particle *states* is far greater than the number of particles, so that there is small likelihood of multiple occupancy of any one particle state, and the error inherent in Eqs. (5–50) and (5–51) is negligible. In deriving Eqs. (5–50) and (5–49), we assumed single-particle states i, j, and k to be different from each other.

of multiparticle state, l, will change by

$$d\mathscr{E}_l = -P_l\,dV. \tag{5-52}$$

Therefore

$$P_l = -\frac{\partial \mathscr{E}_l}{\partial V}, \tag{5-53}$$

where P_l is the pressure of the system when it is in state $\mathscr{E}_l$.

The expected value of the pressure is, from Eqs. (5–6) and (5–53),

$$\langle P \rangle \equiv \sum_l \mathscr{P}_l P_l = -\sum_l \mathscr{P}_l \frac{\partial \mathscr{E}_l}{\partial V}. \tag{5-53a}$$

When we substitute Eq. (5–14) for $\mathscr{P}_l$:

$$\langle P \rangle = -\sum_l e^{-\psi - \beta \mathscr{E}_l} \frac{\partial \mathscr{E}_l}{\partial V}$$

$$= -e^{-\psi} \sum_l e^{-\beta \mathscr{E}_l} \frac{\partial \mathscr{E}_l}{\partial V}. \tag{5-54}$$

Now, from elementary calculus, we know that

$$d(e^{-\beta \mathscr{E}_l})_\beta = -\beta e^{-\beta \mathscr{E}_l}\,d\mathscr{E}_l,$$

which transforms Eq. (5–54) into:

$$\langle P \rangle = \frac{e^{-\psi}}{\beta}\left[\frac{\partial}{\partial V}\left(\sum_l e^{-\beta \mathscr{E}_l}\right)\right]_\beta. \tag{5-55}$$

If we now recall Eq. (5–16),

$$\sum_l e^{-\beta \mathscr{E}_l} = e^\psi,$$

Eq. (5–55) becomes

$$\langle P \rangle = \frac{1}{\beta}\left[\frac{\partial(e^\psi)}{\partial V}\right]_\beta \frac{1}{e^\psi}$$

$$= \frac{1}{\beta}\left[\frac{\partial(\ln e^\psi)}{\partial V}\right]_\beta$$

or

$$\langle P \rangle = \frac{1}{\beta}\left(\frac{\partial \psi}{\partial V}\right)_\beta. \tag{5-56}$$

The *expected pressure* is the pressure exerted by a system of particles characterized by a stable distribution of allowed multiparticle quantum states, such as that given by Eq. (5–14). It is the pressure exerted by a system in *equilibrium*. The work effect produced when a system acts on its surroundings with a pressure equal to the expected pressure is called a REVERSIBLE WORK EFFECT, W_{rev}. Thus

$$dW_{rev} = \langle P \rangle \, dV. \tag{5-57}$$

5–13
Work and heat

When we differentiate:

$$U = \sum_l \mathscr{P}_l \mathscr{E}_l, \tag{5-8}$$

we obtain

$$dU = \sum_l \mathscr{E}_l \, d\mathscr{P}_l + \sum_l \mathscr{P}_l \, d\mathscr{E}_l, \tag{5-58}$$

which states that a change in the accumulated internal energy may be accomplished in *two* ways: by changing the $\mathscr{P}_l$ at constant energy, and/or by changing the magnitudes of the energy states at constant $\mathscr{P}_l$.

    This is analogous to the First Law of classical thermodynamics, which also presents *two* ways of changing a system's energy:

$$dU = dQ - dW. \tag{from (2-15)}$$

We shall therefore let

$$dQ_{rev} \equiv \sum_l \mathscr{E}_l \, d\mathscr{P}_l \tag{5-59}$$

and

$$-dW_{rev} \equiv \sum_l \mathscr{P}_l \, d\mathscr{E}_l. \tag{5-60}$$

    Figure 5–8a shows the system of Fig. 5–7 after it has received reversible work. The number of particles at each level has not changed, but the magnitude of each

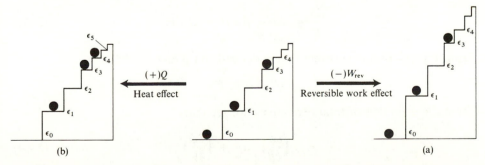

Fig. 5–8. The system diagrammed in Fig. 5–7 after it has received (a) reversible work, and (b) heat.

level has been increased as if the steps had been stretched upward. Figure 5–8b shows the same system after it has absorbed heat. The distribution of particles has shifted upward, whereas the magnitude of each level remains the same.

Equation (5–60) combines Eqs. (5–57) and (5–53a). It says that the effect of compressing a gas reversibly is to increase the magnitude of the allowed energy states without affecting their probabilities.

Equation (5–59) states that the effect of adding heat is to change the expected internal energy by changing the probabilities of all energy states without changing the external coordinates of the system. (The $\mathscr{E}_l$'s remain constant.) See Fig. 5–8.

5–14
Entropy

Let us now substitute the foregoing relations into the expected uncertainty equation, Eq. (5–5), and watch for familiar behavior:

$$S = -k \sum_l \mathscr{P}_l \ln \mathscr{P}_l.$$

Substituting Eq. (5–13) for $\ln \mathscr{P}_l$, we obtain

$$S = -k \sum_l \mathscr{P}_l(-\psi - \beta \mathscr{E}_l) = k\left(\sum_l \mathscr{P}_l \psi + \sum_l \mathscr{P}_l \beta \mathscr{E}_l\right). \tag{5–61}$$

Because ψ is a constant, we know that

$$\sum_l \mathscr{P}_l \psi = \psi \sum_l \mathscr{P}_l = \psi. \tag{5–62}$$

Also, substituting Eq. (5–4) into Eq. (5–61), we obtain

$$\boxed{S = k\psi + k\beta U.} \tag{5–38}$$

Taking the total derivative of (Eq. 5–38),

$$\frac{dS}{k} = d\psi + \beta\, dU + U\, d\beta. \tag{5–63}$$

From Eq. (5–16) it is apparent that ψ depends on β and V,* or

$$\psi = \psi(\beta, V).$$

From the definition of total derivatives, we see that

$$d\psi = +\left(\frac{\partial \psi}{\partial \beta}\right)_V d\beta + \left(\frac{\partial \psi}{\partial V}\right)_\beta dV. \tag{5–64}$$

---

* The statement, "ψ is a function of β and V" is abbreviated $\psi = \psi(\beta, V)$.

Substituting Eqs. (5–21) and (5–56) into Eq. (5–64) yields

$$d\psi = -U \, d\beta + \beta \langle P \rangle \, dV. \tag{5–65}$$

Substituting Eq. (5–65) into Eq. (5–63), we obtain

$$\frac{dS}{k} = \beta(dU + \langle P \rangle \, dV). \tag{5–66}$$

In the absence of external energy accumulations, $dU = dQ - dW$. Therefore

$$\frac{dS}{k} = \beta(dQ - dW + \langle P \rangle \, dV).$$

Remembering that $\beta = 1/kT$, we obtain

$$dS = \frac{dQ}{T} + \frac{1}{T} (\langle P \rangle \, dV - dW). \tag{5–67}$$

For a reversible process, $dW = dW_{rev} = \langle P \rangle \, dV$ (recall Eq. 5–57). Therefore

$$dS = \frac{dQ_{rev}}{T}, \tag{5–68}$$

which is the classical thermodynamic definition of *entropy* (Eq. 3–13).

5–15
The meaning of entropy

With the derivation of Eq. (5–68), we have achieved a major purpose of this section We see that:

The macroscopic thermodynamic property *entropy* is proportional to the maximum expected value of the ln of the probability of quantum states; Eq. (5–5).

Hence entropy is a measure of our *ignorance* about the precise microscopic quantum state in which a piece of macroscopic matter resides. The greater the number of states available to the system, the greater our *un*certainty and the greater the entropy. The more precise our information, the greater our certainty and the smaller the entropy.

5–16
The Clausius inequality

The term $(\langle P \rangle \, dV - dW)$, in Eq. (5–67), represents the difference between the reversible and the actual work done by or on a system. Since the *reversible work is the maximum work done by* (or the minimum work done on) *a system*, this term is *always positive* for real (irreversible) processes and zero for reversible processes.

Hence a conclusion which follows from Eq. (5–67) is that

$$dS \geqslant \frac{dQ}{T} . \tag{5–69}$$

Equation (5–69) is known as the *Clausius inequality*, and states that the change in entropy is always greater than or equal to dQ/T, where dQ refers to the *actual* heat effect. (See Section 3–34.)

5–17
Increase in entropy when heat is added to a system

We may further appreciate the reason why entropy increases when heat is added to a system if we think about an analogy. Consider the system shown in Fig. 5–9a, consisting of two steps on which two elastic balls are bouncing. Each step represents an allowed single-particle state. Thus there are three possible two-particle states of the system: both balls on the first step; both on the second; one on each. If one step is as likely as the other, we can assign probabilities to each state, and evaluate the entropy:

$$S = -k \sum_{l=1}^{3} \mathscr{P}_l \ln \mathscr{P}_l .$$

Now suppose that a third step is made accessible to the system (part b of Fig. 5–9). Then the entire system is capable of being in six states, and entropy has increased considerably.

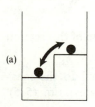

(a)

"Quantum" states of a 2-particle system in which particles may be in 2 states

l	$\mathscr{P}_l$	$\mathscr{P}_l \ln \mathscr{P}_l$
1,1	$\frac{1}{4}$	$-.3465$
2,2	$\frac{1}{4}$	$-.3465$
1,2	$\frac{1}{2}$	$-.3465$
		$-S_a/k = -1.0395$

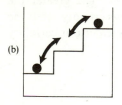

(b)

"Quantum" states of a 2-particle system in which particles may be in 3 states

l	$\mathscr{P}_l$	$\mathscr{P}_l \ln \mathscr{P}_l$
1,1	$\frac{1}{9}$	$-.2439$
2,2	$\frac{1}{9}$	$-.2439$
3,3	$\frac{1}{9}$	$-.2439$
1,2	$\frac{2}{9}$	$-.3340$
1,3	$\frac{2}{9}$	$-.3340$
2,3	$\frac{2}{9}$	$-.3340$
		$-S_b/k = -1.7337$

$\Delta S_{b-a} = 0.6942k$ entropy units

Fig. 5–9. Effect on entropy of increasing the number of allowed states.

A heat effect—or the addition of heat energy to a system—makes higher allowed energy levels accessible to the system's particles. The addition of heat is therefore analogous to the addition of the step (the new energy level) in Fig. 5–9b. The resulting increase in the possible number of configurations is reflected in the increase in the system's entropy. (Note again that Fig. 5–4 shows that the probability of a particle reaching a high energy level increases as β decreases or as temperature increases.)

Exercise 5–10. Calculate ΔS for the process which moves the system of Fig. 5–9 from state 5–9b:

i) To a state having four (4) steps.
ii) To a state in which the vertical distance (energy gap) between the steps is doubled, and everything else about the system remains the same. What might the process that effects this change of state be called?
iii) To a state having 3 particles, but in all other respects identical to state 5–9b.

5–18
Maxwell's distribution

By combining Eqs. (5–43) and (5–1), we may write the probability of a monatomic ideal-gas particle having a quantum number i (where i refers to the x-component of the particle's energy) as

$$\mathscr{P}_i = e^{-\lambda - \beta(h^2/8mV^{2/3})i^2}. \tag{5–70}$$

It also follows, from Eqs. (5–41) and (5–45), that

$$\lambda = \ln \sum_{i=0}^{\infty} e^{-\beta(h^2/8mV^{2/3})i^2}. \tag{5–71}$$

Equation (5–71) can be expressed in closed form (i.e., with elimination of the summation) by replacing the summation sign ($\sum$) by an integral sign. This is possible because the exponent in Eq. (5–71) is a very small number, so that

$$\sum_{i=0}^{\infty} e^{-\beta(h^2/8mV^{2/3})i^2} = \int_0^{\infty} e^{-\beta(h^2/8mV^{2/3})i^2}\, di. \tag{5–72}$$

The definite integral in Eq. (5–72) may be found in a table of integrals, so that

$$\lambda = \ln \left(\frac{V^{2/3}}{h^2} \frac{2\pi m}{\beta} \right)^{1/2},$$

which, when substituted back into Eq. (5–70), yields

$$\mathscr{P}_i = \left[\frac{h}{V^{1/3}} \left(\frac{\beta}{2\pi m} \right)^{1/2} \right] e^{-\beta \epsilon_i}, \tag{5–73}$$

where

$$\epsilon_i = \frac{h^2}{8mV^{2/3}} \cdot i^2 = \tfrac{1}{2}mu_i^2. \tag{5–1}$$

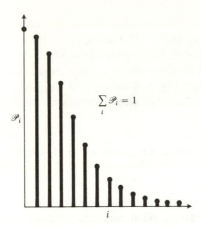

Fig. 5–10

For helium atoms at 300°K and 1 atm, the value of the terms in Eq. (5–73) are:

$$V = 4.087 \times 10^{-20} \text{ cm}^3/\text{molecule}$$

$$h = 6.6256 \times 10^{-27} \text{ erg-sec}$$

$$m = 6.645 \times 10^{-22} \text{ g/molecule}$$

$$\beta = 1/kT = 1/k(300°K)$$

$$k = 1.38054 \times 10^{-16} \text{ erg/molecule} \cdot °K$$

$$i = 0, 1, 2, 3, \ldots$$

The values of $\mathscr{P}_i$ calculated for these conditions are plotted against i in Fig. 5–10, and illustrate the conclusion of Eq. (5–34) that $\mathscr{P}_j/\mathscr{P}_i < 1$, when $j > i$.

Figure 5–11 plots energy contribution of each allowed single-particle state, that is, $\mathscr{P}_i\epsilon_i$, as a function of the quantum number of the particle. It follows from Eq. (5–42) that the sum of the ordinates in Fig. 5–11, $\sum_i \mathscr{P}_i\epsilon_i$, equals the internal energy per particle, $\overline{U}/\overline{N}$.

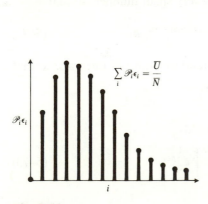

Fig. 5–11

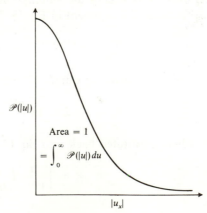

Fig. 5–12

As it appears in Eq. (5–73), $\mathscr{P}_i$ is a *discrete function*, which has values only at integral values of i. The continuous curve drawn through the discrete $\mathscr{P}_i$ values (Fig. 5–12) is represented by the equation

$$\mathscr{P}(|u|) = \left(\frac{2m}{\pi kT}\right)^{1/2} e^{-mu^2/2kT}, \tag{5–74}$$

where $\mathscr{P}(|u|)$ is the *probability density* of a particle having an x-component of absolute velocity between $|u|$ and $|u + du|$. The probability of a particle having this velocity is $\mathscr{P}(|u|)\, du$. Equation (5–74) is called the *Maxwell Distribution* after its discoverer,* J. C. Maxwell. It follows that the area under the curve in Fig. 5–12 is:

$$\int_0^\infty \mathscr{P}(|u|)\, du = 1. \tag{5–74a}$$

Exercise 5–11. Plot $\mathscr{P}(|u|)$, $\epsilon_u = mu^2/2$, and $\mathscr{P}(|u|)\cdot\epsilon_u$ versus u for helium at 300°K and 1 atm pressure. Repeat for 1500°K, and plot on the same graphs as previously, but in different color. Compare and comment.

5–19
Other distributions: open systems

The distribution of Eq. (5–14) does not apply to all systems. In particular, it does not apply to *open* systems in which material diffuses through the boundaries so that the number of particles in the system is uncertain.

The energy of an allowed state, $\mathscr{E}_i$, in such systems depends on the *amount of matter* (number of particles) in the state in addition to the quantum numbers and external coordinates which sufficed to describe the closed system. Thus

$$\mathscr{E}_i = \mathscr{E}(i, V, n_{Ai}, n_{Bi}, n_{Ci}, \cdots, n_m). \tag{5–75}$$

The energy of a state i depends on *composition i* as well as external coordinates. A meaningful description of an open system must therefore contain information on the *expected composition* in addition to the expected energy (that is, information on both $\langle n_A \rangle$, $\langle n_B \rangle$, $\cdots$, $\langle n_m \rangle$, and U).

We can obtain the unbiased distribution of allowed states in an open system by maximizing the entropy S (expected uncertainty) as before, but subject to an

---

* The Maxwell distribution (Eq. 5–74) is a form of the more general *Gaussian distribution* given by the equation $Y = e^{-x^2}$, where

$$Y = \mathscr{P}(u)\left(\frac{2\pi kT}{m}\right)^{1/2} \quad \text{and} \quad x = \left(\frac{mu^2}{2kT}\right)^{1/2};$$

u is no longer absolute, but may have plus and minus values. The equation plots as the symmetrical bell curve familiar to students of probability and statistics. Figure 5–11 is half a bell curve, with all the ordinates doubled so that Eq. (5–74a) is satisfied.

additional constraint imposed by the uncertainty in the number of particles in the system at any instant.* Thus we maximize

$$S = -k \sum_i \mathscr{P}_i \ln \mathscr{P}_i \tag{5–5}$$

subject to

$$\sum_i \mathscr{P}_i = 1 \qquad \text{(gives rise to Lagrangian multiplier } \Omega\text{),} \tag{5–7}$$

$$\sum_i \mathscr{P}_i \mathscr{E}_i = U \qquad \text{(gives rise to Lagrangian multiplier } \beta\text{),} \tag{5–8}$$

$$\sum_i \mathscr{P}_i n_{mi} = \langle n_m \rangle \qquad \text{(gives rise to an } \textit{additional} \text{ multiplier } \alpha_m\text{).} \tag{5–76}$$

Each component of composition (A, B, C, ...) provides an additional equation of constraint similar to Eq. (5–76), which gives rise to a corresponding Lagrangian multiplier α_A, α_B, etc.

The solution for $\mathscr{P}_i$ proceeds as on pages 196–197. The result is similar to Eq. (5–14), except for the addition of an α_m term to the exponent:

$$\mathscr{P}_i = e^{-\Omega - \beta \mathscr{E}_i - \alpha_m n_{mi}}. \tag{5–77}$$

Equation (5–77) contains only one α-exponent, and therefore applies to a pure material. If the system contained M components, then the exponent $\alpha_m n_{mi}$ would have been replaced by $\sum_{J=A}^{M} \alpha_J n_{Ji}$ terms.

By the procedures used previously, we arrive at the open-system analogs to Eqs. (5–18), (5–21), and (5–56):

$$\Omega = \ln \sum_i e^{-\beta \mathscr{E}_i - \alpha n_i}, \tag{5–78}$$

$$\left(\frac{\partial \Omega}{\partial \beta} \right)_{V,n} = -U, \tag{5–79}$$

$$\left(\frac{\partial \Omega}{\partial V} \right)_{\beta,n} = \beta \langle P \rangle, \tag{5–80}$$

---

* The number of quantum states in an open system is enormously greater than that in a closed system. In a closed system of N particles, the number of quantum states equals the number of "combinations" that can be obtained from the quantum indices of N particles. The $\langle N \rangle$-particle open system allows all these N-particle combinations, but, in addition, *all other* combinations of *all numbers* of particles are also possible, because the particle count fluctuates about $\langle N \rangle$, and states may momentarily contain 0, 1, 2, 3, ..., etc., particles.

$$\boxed{\left(\frac{\partial \Omega}{\partial \alpha}\right)_{V,\beta} = -\langle n \rangle.}$$
(5–81)

Exercise 5–12. Derive Eqs. (5–77) through (5–81) from Eqs. (5–5), (5–7), (5–8), and (5–76). Also see Exercise 5–5.

5–20
The nature of α

The parameter α behaves somewhat like β in that it has the same value in two systems after the systems *interact* so as to *exchange* matter as well as energy.

For example, let two open systems A and B (as in Fig. 5–5a), containing only particles A and particles B, respectively, be brought together (as in Fig. 5–5b), *but with the heat-conducting wall removed.* The systems may now exchange both particles and energies. We can calculate the distribution functions for the final state of each system by maximizing uncertainty,

$$S_{AB} = -k \sum_i \mathscr{P}_{Ai} \ln \mathscr{P}_{Ai} - k \sum_i \mathscr{P}_{Bi} \ln \mathscr{P}_{Bi},$$

subject to four restraints:

$$1 = \sum \mathscr{P}_{Ai},$$

$$1 = \sum \mathscr{P}_{Bi},$$

$$U_{AB} = U_{A\,\text{initial}} + U_{B\,\text{initial}} = \sum_i \mathscr{P}_{Ai} \mathscr{E}_{Ai} + \sum_i \mathscr{P}_{Bi} \mathscr{E}_{Bi},$$

$$\langle n_{AB} \rangle = \langle n_A \rangle + \langle n_B \rangle = \sum_i \mathscr{P}_{Ai} n_{Ai} + \sum_i \mathscr{P}_{Bi} n_{Bi}.$$

The four restraints give rise to only four Lagrangian multipliers, Ω_A, Ω_B, β, and α. Therefore

$$\mathscr{P}_{Ai} = e^{-\Omega_A - \beta \mathscr{E}_{Ai} - \alpha n_{Ai}},$$

$$\mathscr{P}_{Bi} = e^{-\Omega_B - \beta \mathscr{E}_{Bi} - \alpha n_{Bi}}.$$

Thus *systems that are in mass as well as thermal equilibrium* (that is, systems that share a stable statistical distribution of energy and matter) *have equal values of α and of β.*

Also two systems that have different α's, when they are allowed to exchange matter, will assume the same values. *In interacting systems, a difference in α results in mass transfer*, just as a difference in β results in heat transfer.

For an open system containing 1 mol* of pure material, we can combine Eqs. (5–77), (5–78), (5–75), (5–79), (5–80), and (5–81) with Eq. (5–5) to yield the open-system analog of Eq. (5–66):

$$\overline{S} = k\beta\overline{U} + k\alpha\overline{N} + k\beta\langle P\rangle\overline{V}. \tag{5–82}$$

Substituting $1/kT$ for β and $R/\overline{N}$ for k, we obtain

$$\overline{S} = \overline{U}/T + R\alpha + \langle P\rangle\overline{V}/T$$

or, rearranging and dropping the $\langle\ \rangle$, we have

$$RT\alpha = T\overline{S} - (\overline{U} + P\overline{V}) \equiv -\overline{G} \tag{5–83}$$

recalling the definition of free energy, Eq. (3–22). Therefore

$$\boxed{\alpha = -\frac{\overline{G}}{RT} = -\frac{\mu}{RT}.} \tag{5–84}$$

The statistical parameter α, for a pure substance, is proportional to the molar GIBBS FREE ENERGY divided by the temperature. For a pure substance, the molar free energy is called CHEMICAL POTENTIAL, and has the symbol μ.

For 1 mol of a homogeneous *mixture*, Eq. (5–82) becomes:

$$\overline{S} = k\beta\overline{U} + k\sum_{J=A}^{M}\alpha_J\langle n_J\rangle + k\beta P\overline{V}, \tag{5–85}$$

which, when we substitute Eqs. (3–22) and (5–37), rearranges to

$$\overline{G}_{\text{mixture}} = \overline{U} - T\overline{S} + P\overline{V} = -RT\sum_{J=A}^{M}\frac{\alpha_J\langle n_J\rangle}{\overline{N}}$$

or

$$\overline{G} = -RT\sum_{J}^{M}\alpha_J x_J \tag{5–86}$$

where x_J = mole fraction of component J. Substituting Eq. (5–84) into Eq. (5–86)

---

* $\langle$Number of particles in system$\rangle$ = 6.02×10^{23} = $\overline{N}$ = Avogadro's constant.

and expanding, we obtain

$$\bar{G} = \mu_A x_A + \mu_B x_B + \mu_C x_C + \cdots$$

or

$$\boxed{\bar{G} = \sum_{J=A}^{M} \mu_J x_J,}$$ (5–87)

where μ_J = the CHEMICAL POTENTIAL for component J in the mixture. It is also called the PARTIAL MOLAR FREE ENERGY.

Exercise 5–13. Prove that C_V, the specific heat at constant volume, which is defined as

$$C_V \equiv \left(\frac{\partial \bar{U}}{\partial T}\right)_V,$$

is also

$$C_V = \left(\frac{\partial^2 \Omega}{\partial \beta^2}\right) k\beta^2.$$ (5–88)

5–21
Bose-Einstein and Fermi-Dirac distributions

When the energy of an open-system multiparticle state, $\mathscr{E}_l$, can be assumed to be the sum of the individual particle energies, ϵ_i (no interaction energy),

$$\mathscr{E}_l = \sum_i n_i \epsilon_i,$$

then the logarithm of the partition function, Eq. (5–78), can be shown to be

$$\Omega_{BE} = -\sum_i \ln\left(1 - e^{-\beta\epsilon_i - \alpha}\right).$$ (5–89)

If, in addition, the system is so constrained that *no two particles may be in the same energy state* (the Pauli exclusion principle), then Eq. (5–78) is further transformed to:

$$\Omega_{FD} = \sum_i \ln\left(1 + e^{-\beta\epsilon_i - \alpha}\right).$$ (5–90)

Systems to which Eq. (5–89) applies are said to follow *Bose-Einstein statistics*. The particles in these systems are called *bosons*.

Systems to which Eq. (5–90) applies are said to follow *Fermi-Dirac statistics*. The particles in these systems are called *fermions*. Electrons and protons are fermions, whereas ideal-gas atoms in open systems are bosons, and so are protons and deuterons.

When α is large or when β is small (high temperature), the Bose-Einstein and Fermi-Dirac statistics produce the same distribution as the Boltzmann distribution, Eq. (5–74).

5–22
Other approaches to statistical thermodynamics

These last sections describe the salient ideas behind alternative methods that have been used to obtain the distribution function, stressing their similarities rather than their differences. Figure 5–13 sketches a somewhat whimsical view of the problem of achieving the distribution and partition functions.

5–23
Statement of basic problem

All approaches to the problem start from the following common ground:

1. Recognition that every macroscopic system has a fantastically detailed microscopic structure, and that the existence of this microstructure makes possible an astronomically large number of different arrangements of the microscopic elements (quantum states) which are completely consistent with the macroscopic system's properties.

2. A realization that there is no way of knowing which arrangement or state actually represents the system, and therefore, in determining the system's properties, we must consider *all* (or a representative portion of) the system's possible microstates.

The basic problem of statistical thermodynamics is therefore assigning, to each possible microstate, a weight (a probability) which reflects its contribution to the properties of the macroscopic system.

It is in the rationalization of the averaging technique—that is, in the derivation of the distribution function which assigns weight or probability to each microstate —that a variety of approaches are used. All approaches arrive at essentially the same distribution function, which is: For a closed fixed-volume system in equilibrium with a thermostated heat bath, the probability of the lth multiparticle microstate is equal to:

$$\mathscr{P}_l = \frac{e^{-\beta \mathscr{E}_l}}{Z},\qquad\qquad (5\text{–}17)$$

where $\mathscr{E}_l$ is the energy of state l and β and Z are constants of the equilibrium system.
The sum of all the probabilities $= 1$, and therefore

$$\sum_l \mathscr{P}_l = 1 = \sum_l \frac{e^{-\beta \mathscr{E}_l}}{Z},$$

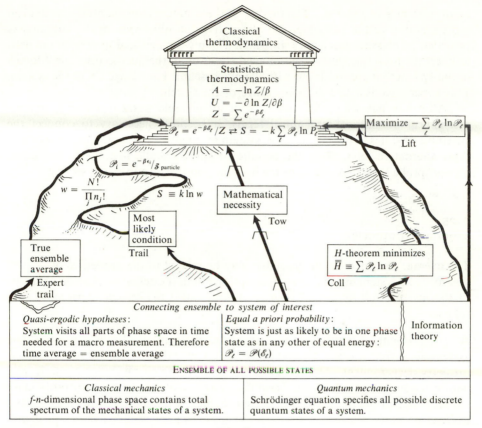

Fig. 5–13

where Z is the *partition function* or sum over states, β is always shown to be $1/kT$, and where k is the Boltzmann constant. The entropy of the system is equal to

$$S = -k \sum_l \mathscr{P}_l \ln \mathscr{P}_l \tag{5-5}$$

or

$$S = k \ln w. \tag{5-5a}$$

From these equations the expressions of classical thermodynamics are obtained in terms of ψ or $\ln Z$, as in Eqs. (5–21), (5–38), and (5–39).

5–24
Ensemble of states

At the base of all approaches to the distribution function is the concept of an ensemble of all possible microscopic states of a macroscopic system. In quantum mechanics, the Schrödinger equation specifies the possible discrete microscopic or quantum states of a system. The totality of these states is the quantum-mechanical representation of the ensemble. An alternative and older view of the ensemble is provided by classical mechanics, in which a many-dimensional hyperspace is used to chart the total spectrum of mechanical states of all the microscopic constituents of the system that are consistent with the macroscopic knowledge about the system. This hyperspace is called the *phase space* of the system.

After one has set up the ensemble of all possible states in either quantum-mechanical or classical-mechanical terms, it becomes necessary to connect the ensemble of states to the macroscopic system of interest. The connection is made in three alternative ways, using the ergodic or quasi-ergodic hypothesis, the equal *a priori* probability hypothesis, or the information-theory approach (see Fig. 5–13).

5–25
Ergodic hypothesis

The average properties of an ensemble may be related to the properties of a given macroscopic system by making an assumption about the actual *mechanical* behavior of the macroscopic system. The assumption, called the *ergodic hypothesis*, is as follows:

A measurement of a property (for example, pressure) made on a macroscopic system is a *time-average* property measurement rather than an instantaneous property measurement. The measurement time is long on a microscopic scale, and within the measurement time interval the system visits (or comes arbitrarily close to) all points in the phase space of the ensemble. It therefore follows that a time-average property of a macrosystem is the same as an ensemble average property.

The validity of the ergodic hypothesis is questionable, particularly because one can imagine systems for which the hypothesis does not hold: For example, an ideal gas in a rigid parallel-walled container whose particles are so arranged as to move perpendicular to the parallel faces of the container, and in such a manner that no collision occurs between the particles. This system would not visit all regions of phase space, that is, it would not go through all configurations of its particles' positions and velocities consistent with the total energy of the system.

5–26
Equal *a priori* probabilities

Another method of connecting the ensemble of states to the macrosystem of interest is to assign equal statistical weight or probability to all microstates of the ensemble that have equal energies. This is a reasonable assumption because, if we know only the energy of the system, we have no basis for choosing one microstate over any other microstate having the same energy. The system has an equal likelihood of being in all equal-energy microstates. Therefore its *average* property is the average over all the equally likely states.

A corollary of this approach is that the probability of a system being in a given microstate is a function of the energy of that state only. That is,

$$\mathcal{P}_{\text{state } l} = \mathcal{P} \text{ (energy of state } l). \tag{5–91}$$

The third way of connecting an ensemble of states to the macrosystem of interest is the information-theory or maximum-entropy approach used earlier in this section, which implicitly agrees with the equal-probability assumption, although it does not make the assumption explicitly.

5–27
True ensemble average

We now turn to the trails that ascend from the ensemble "base camp" to the distribution function (Fig. 5–13).

Given that the average properties of an ensemble are the same as the macroscopic properties of a system, a system property, M, is found by integrating over phase space:

$$M_{\text{system}} = \int_{\text{ensemble}} \rho(\mathbf{p}, \mathbf{q}, t) \cdot N(\mathbf{p}, \mathbf{q}) \, d\mathbf{p} \, d\mathbf{q},$$

where $\mathbf{p}$ and $\mathbf{q}$ are the generalized ensemble coordinates of phase space, t is time, ρ is a *density function*, which gives the probability of finding a state point in any unit volume of phase space, and $N(\mathbf{p}, \mathbf{q})$ is the value of the property of interest at coordinates $\mathbf{p}, \mathbf{q}$. A mathematical theorem, due to Liouville, is then used to show that the density function is independent of time,

$$\frac{d\rho}{dt} = 0,$$

when ρ is a constant, or a function of the energy of the entire system. (When this condition prevails, the ensemble is said to be in *statistical equilibrium*.) A suitable density function is $\rho = e^{-\lambda - \beta \mathscr{E} l}$, which leads to the conclusion that the probability of a system being in a given microstate is proportional to exponential $(-\beta \mathscr{E}_l)$.

This is the route taken by the professional statistical mechanician. It requires considerable mathematical sophistication.

5–28
Most likely condition

An alternative way of getting to the distribution function is via the most-likely-condition route. It is supposed to be a shortcut because it attempts to evaluate the average property of an ensemble of states by covering—not *all* states in the ensemble

$$n_3 = 1$$
$$n_1 = 3.$$
$$\overline{\Sigma n_i = 4}$$

$$\sum n_i \epsilon_i = 3 \cdot 1 + 1 \cdot 3 = 6$$

$$w_a = \frac{4!}{3! \times 1!} = 4$$

$$n_3 = 1$$
$$n_2 = 1$$
$$n_1 = 1$$
$$n_0 = 1$$
$$\overline{4}$$

$$\sum n_i \epsilon_i = 1 + 2 + 3 = 6$$

$$w_b = \frac{4!}{1! \times 1! \times 1! \times 1!} = 24$$

Condition (a)   Condition (b)

Fig. 5–14. Two conditions of a 4-particle, 6-energy-unit system. Condition (b) is more likely to occur than condition (a) because $w_b > w_a$.

—but only the *most likely* states, as represented by the most likely condition. The *condition* of a system is the *set of occupancy numbers* (n_i) which designate the number of microscopic particles in each of the energy levels accessible to a system's particles. For example, Fig. 5–14(a) shows a system which has only four particles. The condition of that system is given by the set of occupancy numbers (n_i): $n_1 = 3; n_3 = 1$. The sum of the n_i is equal to the total number of particles in the system, in this case 4; and the energy of the system is equal to

$$\mathscr{E}_l = \sum_i n_i \epsilon_i = (3 \times 1) + (1 \times 3) = 6 \text{ energy units.}$$

Now three 1-energy-unit particles and one 3-energy-unit particle can be permuted in $4!/(3! \times 1!) = 4$ ways.

The general rule for the number of permutations of N total objects, where N is equal to $\sum_i n_i$, is

$$w = N!/\prod_i n_i!. \tag{5–91a}$$

Quite clearly, the most likely condition of the system shown is that set of n_i's with energy of 6 units which produces the maximum number of permutations, that is, the condition, $n_1 = n_2 = n_3 = n_0 = 1$ (Fig. 5–14b). It can be shown that, as the number of particles becomes very large, the likelihood of any condition other than the most likely condition becomes small. Therefore the ensemble as a whole can be described with reasonable accuracy in terms of its most likely condition, and the set of n_i's that correspond to that most likely condition is simply found by maximizing the number of permutations, w (Eq. 5–91a), taking into account the fact that $\sum_i n_i = N$ and $\sum n_i \epsilon_i = \mathscr{E}_l$. This technique, if followed carefully, eventually leads to an expression for the partition function of a multi-particle system in terms of the allowed energy levels of its constituent particles (Eq. 5–50). Some rather questionable rationalizations* are used to arrive at this final result, and these offset the effort saved by averaging over the most likely condition rather than over the entire ensemble. In this approach,

$$S = k \ln w. \qquad (5\text{–}5a)$$

5–29
Mathematical necessity

When we use the equal *a priori* probability assumption, the probability of a state *l* is a function of its energy, $\mathscr{E}_l$, only:

$$\mathscr{P}_l = f(\mathscr{E}_l). \qquad (5\text{–}92)$$

If we have two systems at equilibrium within a thermostated bath, whose size is such that fluctuations of the energy of one system have no effect on the energy of the bath or on the energy of the other system, then we can state that

$$\mathscr{P}_i = f(\mathscr{E}_i), \qquad (5\text{–}93)$$

$$\mathscr{P}_j = f(\mathscr{E}_j), \qquad (5\text{–}94)$$

where $\mathscr{E}_i$ represents an allowed energy state of the first system and $\mathscr{E}_j$ represents an allowed energy state of the second system. Now, when we consider both systems together, the probability of the first system being at $\mathscr{E}_i$ and the second system at $\mathscr{E}_j$ must be

$$\mathscr{P}_{i \text{ and } j} = f(\mathscr{E}_i + \mathscr{E}_j) = \mathscr{P}_i \times \mathscr{P}_j. \qquad (5\text{–}95)$$

Therefore

$$f(\mathscr{E}_i + \mathscr{E}_j) = f(\mathscr{E}_i) \times f(\mathscr{E}_j). \qquad (5\text{–}96)$$

The only function satisfying Eq. (5–95) is an exponential:

$$\mathscr{P}_i = f(\mathscr{E}_i) = \frac{1}{Z} e^{-\beta \mathscr{E}_i} \qquad (5\text{–}17)$$

Therefore the exponential distribution function is a necessary consequence of the

---

* "Most likely condition" derivations of Eq. (5–50) use *Stirling's approximation* ($\ln n_i! = n_i \ln n_i - n_i$), which is valid only for large values of n_i, whereas n_i may be quite small. Also the particles are assumed to be distinguishable (?).

assumption of equal *a priori* probability for equal energy states. A mathematical consequence of Eq. (5–17) and the classical definition of entropy is that S can be shown to be equal to

$$S = -k \sum \mathscr{P}_i \ln \mathscr{P}_i. \tag{5-5}$$

Note that here we avoid the Lagrangian maximization computation.

5–30
Information-theory approach

The Information-Theory approach—although it uses mathematical forms that are exactly the same as those established in the older statistical-thermodynamic literature—has a somewhat different philosophical or logical orientation. It states that, given the ensemble of states, statistical thermodynamics is not a physical theory whose validity depends either on the truth of additional basic assumptions, such as ergodic behavior or equal probability, or on experimental verification. Statistical thermodynamics is instead a form of statistical inference, a technique for making the best estimates on the basis of incomplete information about the ensemble of states. If experimental verification is not obtained, this is not a shortcoming of statistical thermodynamics, but of the information supplied.

The relationship

$$S = -k \sum_i \mathscr{P}_i \ln \mathscr{P}_i \tag{5-5}$$

occupies the primal position in this approach. The equation, which is the basic equation of Shannon's *mathematical theory of information*, is identified with thermodynamic entropy.

It is the contention of the information theorists that maximizing $\sum \mathscr{P}_i \ln \mathscr{P}_i$, subject to constraints, is the logical, fundamental starting point for producing the least-biased distribution of probabilities (a distribution that is maximally noncommittal with regard to missing information). There is no need to postulate ergodicity or equiprobability or, for that matter, any other kind of behavior characteristic of the ensemble of states, because evaluating $\mathscr{P}_i$ is a problem in guessing (statistical inference) and not a problem in physics.

A similar technique, using a different rationale, was suggested by Pauli, who derived the distribution functions by minimizing the Boltzmann $\bar{\bar{H}}$-function,

$$\bar{\bar{H}} \equiv \sum \mathscr{P}_i \ln \mathscr{P}_i,$$

subject to the same constraints. (This technique is discussed in detail by Tolman.)

The information-theory approach is about as simple as the previously discussed mathematical-necessity approach. One accepts, as an axiom, that maximizing S subject to the known properties of a system produces a minimally biased set of $\mathscr{P}_i$'s. The mathematics of maximization are reasonably straightforward. The trouble with the axiom is that it does not relate to common experience, whereas other thermodynamic axioms do, and as a result have more or less intuitive acceptability. One way of making the axiom more acceptable is to demonstrate qualitatively that maximizing $-\sum_i \mathscr{P}_i \ln \mathscr{P}_i$ (or minimizing $+\sum_i \mathscr{P}_i \ln \mathscr{P}_i$) is a "smoothing" operation, which tends to lower the center of gravity (or minimize the moment) of a plot of $\mathscr{P}_i$ versus i.

As a qualitative example, assume that we have a system which is capable of existing in a great number of possible states, and that we are asked to arbitrarily assign probabilities to each of these states. The states can be ordered in a sequence, and indexed by an integral subscript i. Assume that all we know about this system is that it must be in some state ($\sum \mathscr{P}_i = 1$) (Fig. 5–15). Curve b is a plot of an arbitrarily assigned distribution for this system, which is constrained only by the fact that the sum of the ordinates equals unity.

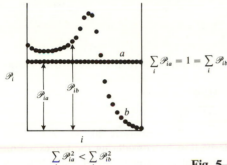

$$\sum_i \mathscr{P}_{ia}^2 < \sum_i \mathscr{P}_{ib}^2$$

Fig. 5–15. Unbiased (a) and biased (b) distributions.

This is *not* an unbiased distribution, because we have put maxima and minima in the distribution; that is, we have given some states more weight than others, without information that would justify our doing so. We can represent the relative smoothness of the arbitrary curve in Fig. 5–15 by the mathematical index

$$\sum \mathscr{P}_i \times \mathscr{P}_i, \tag{5–97}$$

which evaluates the *moment* of the distribution about the horizontal axis. The moment increases as the magnitude of the singularities or extrema in the system increases, and conversely, the moment decreases as the center of gravity of the distribution drops, that is, as the curve becomes more uniformly smooth. In fact, it is a straightforward exercise in the calculus of variation to show that the *minimum* moment corresponds to a constant value of $\mathscr{P}_i$. If, in the smoothness index (Eq. 5–97), we replace one $\mathscr{P}_i$ with a monotonic function of $\mathscr{P}_i$, that is, $\ln \mathscr{P}_i$, we should expect similar behavior. In other words, the effect of maximizing $\sum_i -\mathscr{P}_i \ln \mathscr{P}_i$ is to smooth out our distribution. (Curve a is the maximization result obtained with one restraint: $\sum_i \mathscr{P}_i = 1$.)

References

Fowler, R. H. and E. A. Guggenheim, *Statistical Thermodynamics*, Cambridge: Cambridge University Press, 1939

Tolman, R. C., *Statistical Mechanics*, Oxford: Oxford University Press, 1938

Pauling, L. and E. B. Wilson, *Introduction to Quantum Mechanics*, New York: McGraw-Hill, 1935

MacDonald, D. K. C., *Introductory Statistical Mechanics for Physicists*, New York: John Wiley, 1963

Tribus, M., *Thermostatics and Thermodynamics*, New York: Van Nostrand, 1961

Andrews, F. C., *Equilibrium Statistical Mechanics*, New York: John Wiley, 1963

Denbigh, K., *The Principles of Chemical Equilibrium*, Cambridge: Cambridge University Press, 1961

Problems

1. Derive the *barometric formula* for the variation of atmospheric pressure P with elevation h above sea level,

$$\ln \frac{P}{P_0} = - \frac{Mgh}{RT},$$

where P = atmospheric pressure at elevation h, P_0 = pressure at sea level, and M = molecular weight of air. Perform the derivation by considering the forces acting on a differential column of air of height dh, located at an elevation h such that the pressure on the lower end of the column is P. Then show that this leads directly to the Boltzmann factor:

$$\frac{\mathscr{P}}{\mathscr{P}_0} = \frac{n}{n_0} = e^{-\beta(E - E_0)\,\text{Grav.}}$$

where
$\quad \mathscr{P}$ = probability of a particle having energy E,
$\quad E = mgh$ = gravitational potential energy,
$\quad n$ = number of particles at energy E,
subscript 0 = sea level.

2. *Kinetic-Theory Derivation of Ideal-Gas Law*

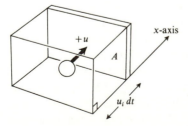

    1. *Q*: Consider a hard spherical elastic particle having a velocity whose x-component has the *absolute* value u_i. What is the probability of selecting such a particle?

A: $\mathscr{P}(|u_i|) = \mathscr{P}(+u_i) + \mathscr{P}(-u_i) = \mathscr{P}_i$.

2. Q: If the gas as a whole has no net velocity, what is the $\mathscr{P}(+u_i)$?

   A: $\mathscr{P}(+u_i) = \mathscr{P}(-u_i) = \frac{1}{2}\mathscr{P}_i$.

3. Q: What is the change in momentum of the particle on elastic collision with wall A, which is perpendicular to x-axis?

   A: $\Delta(mu_i) =$ final momentum $-$ initial momentum
   $= (-mu_i) \quad - (+mu_i) = -2mu_i$.

4. Q: How far will a particle with velocity u_i travel in time dt?

   A: _____ .

5. Q: How many i-particles will strike wall A in time dt?

   A: All such particles whose perpendicular distance from the wall is less than or equal to $u_i\, dt$. That is, all the particles whose velocity is u_i within the volume $Au_i\, dt$.

6. Q: What is the total number of particles in volume $Au_i\, dt$?

   A: $(N/V)\, Au_i\, dt$, where $N =$ total number of particles, and $V =$ total volume.

7. Q: How many of these particles have a velocity $+u_i$?

   A: _____ .

8. Q: How much momentum do these particles transfer to the wall in time dt?

   A: _____ .

9. Q: What is the magnitude of the force on the wall resulting from i-particle collisions?

   A: $F_i = \dfrac{d(mu_i)}{dt} =$ _____ .

10. Q: What is the pressure exerted by the i-particles?

    A: $P_i =$ _____ .

11. Q: Part (10) concerns the pressure of i-particles only. What is the total pressure on the wall?

    A: $P =$ sum of the pressures exerted by particles having all possible velocities $(u_0, u_1, u_2, u_3, \ldots, u_\infty)$. Thus $P =$ _____ .

12. Q: Since the kinetic energy of a particle is $\epsilon_i =$ _____ , how may part (11) be altered to contain this fact?

    A: $P =$ _____ .

13. Q: Can part (12) be written in terms of expectation values?

    A: Yes, because $\langle f(x) \rangle \equiv$ _____ .
    Therefore $P =$ _____ .

14. Q: Given that the $\langle \epsilon \rangle$ of a hard-sphere gas particle moving in a *rectangular coordinate* system is
    $$\langle \epsilon \rangle = \langle \epsilon_x \rangle + \langle \epsilon_y \rangle + \langle \epsilon_z \rangle = \tfrac{3}{2}\beta.$$
    What are $\langle \epsilon_x \rangle$ and P in terms of β?

    A: $\langle \epsilon_x \rangle =$ _____ .
    $P =$ _____ .
    This is Eq. _____ in the text.

15. Q: By analogy with the classical form of the ideal-gas law, what conclusions can be drawn about β?

A: _____.

_____.

3. Write out a representative part of the PARTITION-FUNCTION SERIES for a system having only two allowed particle-energy levels (ϵ_1 and/or ϵ_2), and in which:

a) The system is closed and contains only 2 noninteracting particles.
b) The system is open ($\langle N \rangle = 2$), and contains noninteracting particles.
c) Similar to (b), but particles obey the Pauli exclusion principle.

4. The only thing known about a certain system is that it may be in any of w states. Find the distribution function $\mathscr{P}_i$ for the system by maximizing

$$S = -k \sum_{i=1}^{w} \mathscr{P}_i \ln \mathscr{P}_i.$$

5. Repeat Problem 4 by minimizing $k\sum(\mathscr{P}_i)^2$. Does the distribution function differ from that found in Problem 4?

6. Suppose that, instead of using Eq. (5–5), we had decided to use maximum $-\langle \mathscr{P} \rangle$ as our selection criterion for a minimally based set of $\mathscr{P}_i$'s; that is, suppose that we had decided to maximize

$$\text{``}S\text{''} = -k\sum \mathscr{P}_i(\mathscr{P}_i).$$

Show that this new criterion for choosing $\mathscr{P}_i$ (when insufficient information or equations are given to make the $\mathscr{P}_i$ determinate) leads to the conclusions:

$$\text{``}S\text{''} = -k \sum_{i=1}^{w} \mathscr{P}_i^2$$

$$\mathscr{P}_i = -\text{``}\psi\text{''} - \text{``}\beta\text{''}\mathscr{E}_i$$

$$\text{``}\psi\text{''} = -\frac{1 + \beta \sum_{i}^{w} \mathscr{E}_i}{w}$$

$$\partial\text{``}\psi\text{''}/\partial\text{``}\beta\text{''} = -\bar{E} = \text{mean energy of states,}$$

$$\partial\text{``}\psi\text{''}/\partial V = -\text{``}\beta\text{''}\bar{P}; \qquad \bar{P} = \text{mean pressure of states.}$$

7. An *extensive property* is one whose magnitude is dependent on the mass or size of a system. If the entropy of a mass of material is S entropy units, the entropy of two such masses taken together is $2(S)$ units. Simultaneously, if the probability of an energy state in one single mass is $\mathscr{P}_i$ and in the other is $\mathscr{P}_j$, the probability of a state in the twofold system is

$$\mathscr{P}_{\text{twofold}} = \mathscr{P}_{ij} = (\mathscr{P}_i \cdot \mathscr{P}_j).$$

With the above in mind, show that S as defined by Eq. (5–5) is extensive; that is, that

$$S_{\text{twofold}} = S_i + S_j = -k \sum \mathscr{P}_i \ln \mathscr{P}_i - k \sum \mathscr{P}_j \ln \mathscr{P}_j = -k \sum \mathscr{P}_{ij} \ln \mathscr{P}_{ij},$$

whereas the "S" defined in Problem 6 is not extensive. (Can you extend the proof to an N-fold increase?)

8. Consider two systems, X and Y, each identical to that shown in Fig. 5–9a, except that the particles in system X have exactly twice the mass of the particles in system Y. Enumerate all the allowed 4-particle states that result from mixing systems X and Y, and calculate the ΔS for the mixing process.

9. In a mole of noble gas at a temperature of 300°K, the atoms each weigh 4.14×10^{-22} g (mass). What is the probability of finding a particle whose velocity is 1.414×10^{-4} cm/sec? What are the units of $mu^2/2kT$? of $m/2\pi kt$?

10. Suppose that you are given 2 boxes. One contains a gold and a silver coin; the other contains two gold coins. You select a box and withdraw a gold coin from it. What is the probability that the other coin in the box is gold?

11. Starting from Eq. (5–19a), show that

$$\left(\frac{\partial^2 \psi}{\partial \beta^2}\right) = \sum_{l} \mathscr{P}_l (\mathscr{E}_l)^2 - \left(\sum \mathscr{P}_i \mathscr{E}_i\right)^2 = \langle U^2 \rangle - \langle U \rangle^2 = -\left(\frac{\partial U}{\partial \beta}\right)_V.$$

12. Using the definition of expectation value Eq. (5–6), and the result of Problem (11) show that:

$$\langle (\mathscr{E}_l - \langle U \rangle)^2 \rangle = \left(\frac{\partial^2 \psi}{\partial \beta^2}\right) = -\left(\frac{\partial U}{\partial \beta}\right)_V.$$

Does the (expected) energy of a macroscopic system increase or decrease with β?

13. The term $\langle (\mathscr{E}_l - \langle U \rangle)^2 \rangle$ is the expected (average) value of the squared deviation of all allowed energies from the expected energy. In statistics this is called the *variance*, and is given the symbol $\sigma^2(\mathscr{E})$. From the previous two problems, you should be able to relate $\sigma^2(\mathscr{E})$ to C_V. Do so, and then comment on the macroscopic significance of C_V.

14. The allowed energy states for a certain kind of particle have magnitudes given by the expression

$$\epsilon_i = i(i + 1) \text{ energy units}, \qquad i = 0, 1, 2, 3, \ldots$$

These energy states are nondegenerate.

a) Compute the single-particle partition function for a closed system under conditions that $\beta = 1$ (energy unit)$^{-1}$; $\beta = 0.2$; $\beta = 5$.
b) Plot $\mathscr{P}_i$ versus i, and $\mathscr{P}_i$ versus ϵ_i, for each of the above β.

15. Compute the multiparticle partition function of a system of 4 particles whose allowed particle energies are given in Problem 14, at $\beta = 1$ (energy unit)$^{-1}$. Plot $\mathscr{P}_l$ versus $\mathscr{E}_l$ (energy of a 4-particle system).

16. Repeat Problem 15, with particle energy states being degenerate, and the degeneracy of energy state i being $w = 2i$. Remember that the partition function is the sum over *all* states, including each degenerate state.

17. Find the minimum cost of manufacturing an open-topped cardboard box (rectangular) whose volume is to be 10 liters, given that the cost of cardboard is Y cents/meter^2 and the cost of tape is Z cents/meter. The box is formed by folding flat cardboard stock cut into suitably dimensioned cruciform shapes. The cost of the cardboard depends on the surface area of the box only. All corners of the box are to be taped. Therefore the cost of the box C_B is

$$C_B = YA + Z \times 4h,$$

where A = surface area of box, h = height of taped corners, Y = \$.20/meter^2, and Z = \$.05/meter.

18. Extending the method of Problem 2, show that the number of molecules of ideal gas that strike a unit area of container wall in unit time is:

a) Wall collision rate $= \dfrac{\langle u_x \rangle N}{V}$

b) Suppose that $\langle u_x \rangle = \displaystyle\int_0^\infty \mathscr{P}(u) u \, du.$

Show that $\langle u_x \rangle = \left(\dfrac{kt}{2m\pi} \right)^{1/2},$

and that the wall collision rate $= P/(2\pi m k T)^{1/2}.$

c) The molecules stick to the wall after they strike it. Show that the force they exert per unit area of wall is $N/2V\beta.$

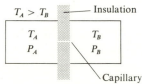

$T_A > T_B$ —— Insulation

T_A P_A T_B P_B

Capillary

19. A very small capillary connects two rigid tanks A and B, which are maintained at $T_A > T_B$, respectively, and which both contain helium. If pressure in each tank is to remain constant, the number of atoms passing from A to B must equal the number of atoms passing from B to A. The number of atoms passing in either direction is the same as the number impinging on the capillary opening in each tank. Show that $P_A > P_B.$ Use the results of Problem 18 in your considerations.

20. Imagine the system of Fig. 5–9a to be surrounded by a particle-permeable wall (i.e., the system is open), so that particles will occasionally leak into or out of the system. The expected number of particles in the system is two (that is, $\langle n \rangle = 2$). However, the particle count ranges from 0 to 4 ($\mathscr{P}_{n=0} = \mathscr{P}_1 = \mathscr{P}_2 = \mathscr{P}_3 = \mathscr{P}_4$). Enumerate the possible n-particle states of the system and the entropy of the system when open.

21. Assume that the steps in Fig. 5–9a represent two possible spin states of an electron: *spin up* and *spin down*. Each spin state is equally accessible to any particle. Given that individual particles are indistinguishable, how many different multiparticle spin quantum states are possible in a system of 2 particles? 10 particles? 10^{23} particles?

22. The typical roulette wheel has 38 numbered slots. There is equal likelihood that a roulette ball may fall into any one of these slots (assuming an honest wheel). Suppose that the croupier gets bored with the same old game, and decides to use two balls instead of one. When he adds the second ball, how many new states are created:

a) If the second ball is identical to the first, and the croupier calls out a number for each ball?

b) Similar to (a), but the croupier calls out the sum of the balls only?

c) If the second ball is red and the first is white, and the final state is a red number and a white number?

d) What probability is associated with each of the above states?

e) What is the expectation value of the probability for each kind of final state?

23. The *Maxwell Demon Bottle** is a device used to illustrate elementary concepts of

---

* See M. V. Sussman, *J. Chem. Educ.* **43**, February 1966, page 105.

statistical thermodynamics. Essentially it consists of a sealed vertical tube in which 10 loosely fitting balls (5 black, 5 white) may be arranged in a variety of black and white vertical arrays (see figure).

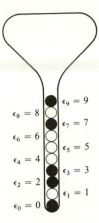

a) How many different arrays (quantum states) can the 10 balls form?
b) An energy may be assigned to each array by assuming that the black balls are unit mass particles and that the white balls are massless spacers, and assigning magnitudes of energy 0 through 9 to each of the 10 vertical ball positions. The energy of an array is then

$$\mathscr{E}_l = \sum_{i=0}^{9} n_i \epsilon_i,$$

where n_i is the number of black balls (0 or 1) having energy ϵ_i. For example, in the array pictured, $\mathscr{E}_l = 21$ energy units.

Construct a table showing the possible energy states of the system of 5 black and 5 white balls, and the degeneracy of each energy state.

24. In the system in Problem 23, assume that white spacing balls may be added or removed from the array, while the number of black spheres remains fixed (closed system).

a) Plot the number of arrays (quantum states) as a function of the number of spacing balls for 0 to 5 spacing balls.
b) Assuming that all quantum states associated with a given number of spacers are equally likely, plot the entropy versus the number of white spacers for 0 to 5 spacers (express your answer in terms of k).
c) Plot $U (= \langle \mathscr{E} \rangle)$ versus number of spacers between 0 and 5.

You may wish to use a computer to solve this problem, in which case it would be interesting to extend the computation to 20 spacer spheres.

25. In the system described in Problems 23 and 24, the energy may be changed by (a) adding or removing white spacer balls or (b) changing the elevation of the entire system. (For example, if the entire bottle is raised by two sphere levels, each energy level ϵ_i will be augmented by 2 units, and the energy of the system of 5 black spheres will increase by $5 \times 2 = 10$ units.) Which of the above energy-changing effects, (a) or (b), produces the smaller entropy change in the system? Comment on the above effects in light of Eqs. (5–58), (5–69), and (5–60).

26. The absolute entropy of 1 g-mol of hydrogen is 31.211 cal/g-mol $\cdot$ °K at 298.2°K and 1 atm. Assuming that all the quantum states of the mole of gas are equally likely, how many quantum states would the system have? How does this number compare with the number of molecules (particles) in the system?

27. The hydrogen gas in Problem 26 is heated to 400°K at 1 atm. Calculate the change in entropy and the change in the number of possible quantum states (again assuming equal probability for each quantum state of the system).

28. The stairway shown in the figure vibrates vertically as shown by the double-headed arrow. A large number of particles are resting on each stair tread, and are thrown by the vibration to an average height h_t above each tread. The bouncing particles tend to move from tread to tread. Eventually the number of particles directly over a tread becomes relatively stable; i.e., an equilibrium distribution of particles on tread levels is established, which depends on the frequency of the stair vibration, t.

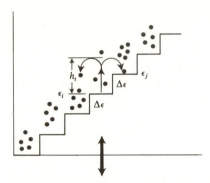

Assume that the probability of a particle moving from a level i to the adjacent higher level j is a function of the mean height of bounce, h_t, and the dimensions of the stair riser, which function is

$$\frac{\mathscr{P}_j}{\mathscr{P}_i} = \frac{h_t - \Delta\epsilon}{h_t} = 1 - \frac{\Delta\epsilon}{h_t},$$

where $\Delta\epsilon$ is the height of the stair riser (a constant), and $\Delta\epsilon < h_t$.

a) Show that the distribution function for the particles is given by

$$\mathscr{P}_i = \frac{c^{i-1}}{\sum_{i=1}^{n} c^{i-1}} = c^{i-1}(1 - c),$$

where $c = 1 - \Delta\epsilon/h_t$, and $n = $ total number of steps, a large number.

b) Plot $\mathscr{P}_i$ as a function of i, for $i = 0$ through 20 at $h_t = \Delta\epsilon$; $5\Delta\epsilon$; $20\Delta\epsilon$.

c) What value of h_t is analogous to absolute-zero temperature?

d) Show that the entropy of the particles is

$$S = +k \frac{\ln (1 - c)}{1 - c}\left[\left(\frac{1}{1 - c}\right) + \ln c(c + 2c^2 + 3c^3 + \cdots)\right].$$

(See M. V. Sussman, "Visualizing Statistical Thermodynamics," *Amer. J. Phys.* **34**, 12, 1966, page 1143.)

JAMES CLERK MAXWELL, 1831–1879

Section 6
RELATIONS AMONG THERMODYNAMIC PROPERTIES

So far we have considered five fundamental and three derived thermodynamic properties:

Temperature	T
Pressure	P
Volume	V
Internal energy	U
Entropy	S
Enthalpy	H
Free energy	G
Helmholtz function	A

This section will demonstrate the general manner in which such properties depend on each other.

6–1
Combined first and second law

If a system experiences a differential change in accumulated energy, then, because of the First Law and the fact that energy E is a state property, the change can *always* be expressed as

$$dE = dQ_{rev} - \sum_i dW_{rev_i} \qquad (6\text{--}1a)$$

if the change causes the system to move from one equilibrium state to another.†
The term $\sum_i dW_{rev_i}$ is used to allow for systems which can exchange more than one kind of reversible work with their surroundings. For example, a system in a gravitational and electric field, if its molecules had dipole moments, might exchange not only expansion work $(P\,dV)$, but also polarization work $(\mathscr{V}\,dp)$, elevation work $(-m_{system}\,g\,dX_{system})$, and acceleration work $(-m_{system}\,u\,du)$. In which case

$$\sum_i dW_{rev_i} = P\,dV - \mathscr{V}\,dp - mg\,dx - mu\,du, \qquad (6\text{--}1b)$$

where $\mathscr{V}$ is the intensity of the electric field and p is the dipole moment of the system's material. For the general case,

$$\sum_i dW_{rev_i} = P\,dV + \sum_i F_i\,dX_i - mg\,dx - mu\,du, \qquad (6\text{--}1c)$$

where $F_i\,dX_i$ represents any independent reversible kind of work other than $P\,dV$, $mg\,dx$, and $mu\,du$, that the system is capable of giving and receiving.

---

† Eq. (6–1a) is valid for dE even if the actual process changing E is *ir*reversible. dQ_{rev} and $\sum_i dW_{rev_i}$ are then the heat and work effects along a reversible process path that connects the actual initial and final equilibrium states.

Each property on the right-hand side of Eq. (6–1c) is a property of the system and not of the surroundings. Thus $mg\, dx$ refers to changes in the elevation of the system and not to masses that may have been elevated in the surroundings. When reversible $m_{\text{system}}\, g\, dx_{\text{system}}$ work is done by the system, the system loses an equal amount of its accumulated external (potential) energy.

When we substitute

$$dE = dE_{\text{ext}} + dU \qquad\qquad \text{(from Eq. 2–3)}$$

and

$$dQ_{\text{rev}} = T\, dS \qquad\qquad (3\text{–}13)$$

and Eq. (6–1c) into Eq. (6–1a), we obtain

$$dE_{\text{ext}} + dU = T\, dS - \left(P\, dV + \sum_i F_i\, dX_i - mg\, dx - mu\, du\right),$$

and, because the work of elevation and acceleration are equal to dE_{ext},[*]

$$\boxed{\; dU = T\, dS - P\, dV - \sum_i F_i\, dX_i. \;} \qquad\qquad (6\text{–}1\text{d})$$

Equation (6–1d) combines the First and Second Laws of Thermodynamics. It is called the GIBBS EQUATION.

6–2
The state principle

Equation (6–1d) says that a change in U depends on changes in the system's entropy, volume, and X_i's. This idea may be abbreviated as

$$U = U(S, V, X_i) \qquad\qquad (6\text{–}1\text{e})$$

which is read as, "U is a function of S, V, and the X_i's." The number of variables that U depends on is therefore two, plus the number of X_i's. Equation (6–1e) therefore permits the following generalization:

The thermodynamic state of a system is a function of as many independent properties of the system as there are independent reversible means of changing the accumulated internal energy of the system; i.e., as there are differential terms on the right-hand side of Eq. (6–1d).

This generalization is called the STATE PRINCIPLE.

In systems whose work can be expressed only in terms of its PV properties,[†]

---

[*] In the absence of relativity effects. Also see Eq. (3–23a).

[†] We are here confining our consideration to systems with constant mass and chemical composition. Reversible means for changing U include changes in mass and chemical composition; we shall discuss these changes in Sections 9 and 10.

there are *no* X_i's, and U is changed only by reversible heat and $P\,dV$ work. Therefore

$$U = U(2 \text{ properties})$$

or

$$\boxed{dU = T\,dS - P\,dV.}$$ (6–2)

6–3
Relations among thermodynamic properties

The thermodynamic state of a system of known mass that does *P dV work only* is uniquely defined when any two of the properties listed at the beginning of this section are specified. This means that the value of all the system's properties are fixed once the values of any two properties are fixed. For example, if the T and P of a system are specified, then V, U, S, H, G, and A take on corresponding fixed values. This in turn implies that there must be a function, $V = V(T, P)$, relating V to temperature and pressure; and another, $S = S(T, P)$, relating S to T and P; and so forth for H, G, and A. Furthermore, since *any* two properties fix all the properties of a $P\,dV$ work system, mathematical functions must exist which relate each property to any two of the others.*

Thermodynamic problems usually require us to determine the magnitude of changes in energy functions (U and H) and work functions (A and G) from changes in those properties of the system which are easily measured. The principal measurable properties are *temperature*, *pressure*, and *volume*. We shall therefore now develop relationships for U, H, G, and A in terms of T, P, and V. We shall also describe general techniques for generating relationships among any set of thermodynamic properties.

6–4
Differential property relationships

The differential of the accumulated internal energy in a fixed-composition, $P\,dV$-work system is

$$dU = T\,dS - P\,dV,$$ (6–2)

which is an expression involving P, V, T, and S. All the derived thermodynamic properties may also be expressed in terms of P, V, T, and S. Thus

$$dH = dU + d(PV) \qquad \text{(from Eq. 2–16)}$$

$$= dU + P\,dV + V\,dP.$$ (6–3)

---

* How many functions of the 8 listed properties can there be for a $P\,dV$ work system?

Substituting Eq. (6–2) in Eq. (6–3), we obtain

$$dH = T\,dS + V\,dP. \tag{6–4}$$

From Eq. (3–34), the equation defining the Helmholtz function A, we obtain

$$dA = dU - d(TS) = dU - T\,dS - S\,dT. \tag{6–5}$$

Substituting Eq. (6–2) in Eq. (6–5),

$$dA = -S\,dT - P\,dV. \tag{6–6}$$

From the Gibbs free energy equation (Eq. 3–22) and from Eq. (6–4),

$$dG = -S\,dT + V\,dP. \tag{6–7}$$

We have—in Eqs. (6–2), (6–4), (6–6), and (6–7)—expressed dU, dH, dA, and dG in terms of P, V, T, and S. We shall now try to eliminate S, because there are no convenient ways of measuring S (that is, you can't buy an S-meter the way you might buy a T-meter).

6–5
The Maxwell relations

We learned in Section 2 that thermodynamic properties have exact differentials. If a property M is a function of x and y,

$$M = M(x, y) \tag{6–7a}$$

then a differential change in M, dM, is the sum of the amount that M changes in the interval dx, with y held constant, plus the amount that M changes in the interval dy, with x held constant (see Fig. 6–1), or

$$dM = \left(\frac{\partial M}{\partial x}\right)_y dx + \left(\frac{\partial M}{\partial y}\right)_x dy. \tag{6–8}$$

The terms $(\partial M/\partial x)_y$ and $(\partial M/\partial y)_x$ are called the *partial derivatives* of M and dM is called the *total differential*.*

Equation (6–8) may be written

$$dM = B\,dx + C\,dy, \tag{6–9}$$

where B and C represent $(\partial M/\partial x)_y$ and $(\partial M/\partial y)_x$, respectively.

---

* If M is a function of many variables, $M = M(x, y, z, \ldots)$, then the total differential, dM, is the sum of the partial derivatives with respect to each variable:

$$dM = \left(\frac{\partial M}{dx}\right)_{y,z\ldots} dx + \left(\frac{\partial M}{\partial y}\right)_{x,z\ldots} dy + \left(\frac{\partial M}{\partial z}\right)_{x,y\ldots} dz + \cdots. \tag{6–8a}$$

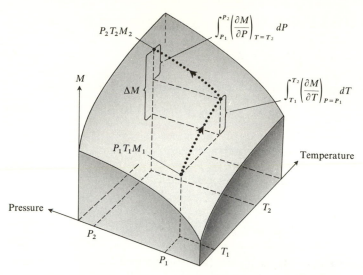

Fig. 6–1. If $M = M(P, T)$, then an MPT surface must exist.

Now Eqs. (6–2), (6–4), (6–6), and (6–7) are total differentials, and have the same form as Eq. (6–9). By comparison with Eqs. (6–7a) and (6–8), Eq. (6–2) may be written as

$$dU = \left(\frac{\partial U}{\partial S}\right)_V dS + \left(\frac{\partial U}{\partial V}\right)_S dV,$$

from which it follows that

$$T = \left(\frac{\partial U}{\partial S}\right)_V \quad \text{and} \quad P = \left(\frac{\partial U}{\partial V}\right)_S.$$

In a like manner, from Eq. (6–4) and (6–2), we obtain

$$T = \left(\frac{\partial H}{\partial S}\right)_P = \left(\frac{\partial U}{\partial S}\right)_V, \tag{6–10}$$

and from Eqs. (6–2) and (6–6),

$$P = -\left(\frac{\partial U}{\partial V}\right)_S = -\left(\frac{\partial A}{\partial V}\right)_T, \tag{6–11}$$

and from Eqs. (6–4) and (6–7),

$$V = \left(\frac{\partial H}{\partial P}\right)_S = \left(\frac{\partial G}{\partial P}\right)_T, \tag{6–12}$$

and from Eqs. (6–6) and (6–7),

$$S = -\left(\frac{\partial A}{\partial T}\right)_V = -\left(\frac{\partial G}{\partial T}\right)_P. \tag{6–13}$$

A characteristic of Eq. (6–9) and of *exact differentials* in general is that

$$\left(\frac{\partial B}{\partial y}\right)_x = \left(\frac{\partial C}{\partial x}\right)_y \tag{6–14}$$

or

$$\frac{\partial^2 M}{\partial x\, \partial y} = \frac{\partial^2 M}{\partial y\, \partial x}. \tag{6–14a}$$

When we apply Eq. (6–14)* to Eqs. (6–2), (6–4), (6–6), and (6–7), we obtain four differential equations in P, V, T and S, known as the *Maxwell Relations*:
From Eq. (6–2),

$$\boxed{-\left(\frac{\partial T}{\partial V}\right)_S = \left(\frac{\partial P}{\partial S}\right)_V.} \tag{6–15}$$

From Eq. (6–4),

$$\boxed{\left(\frac{\partial T}{\partial P}\right)_S = \left(\frac{\partial V}{\partial S}\right)_P.} \tag{6–16}$$

From Eq. (6–6),

$$\boxed{\left(\frac{\partial S}{\partial V}\right)_T = \left(\frac{\partial P}{\partial T}\right)_V.} \tag{6–17}$$

From Eq. (6–7),

$$\boxed{-\left(\frac{\partial S}{\partial P}\right)_T = \left(\frac{\partial V}{\partial T}\right)_P.} \tag{6–18}$$

Exercise 6–1. Consider a system consisting of a rubber band reversibly elongated in a vacuum. The only work done by this system is $dW_{\text{rev}} = -\tau\, dL$ (rather than $+ P\, dV$), where τ is tension and L is length. Inserting this fact in Eq. (6–1a) produces the analog of Eq. (6–2) for a rubber band:

$$dU = T\, dS + \tau\, dL. \tag{6–18a}$$

Write the rubber-band analog of the Eq. (6–15) Maxwell Relation.

---

* This equation is called the *Euler criterion for integrability or exactness*.

6–6
The Clapeyron equation

Equation (6–17) is the basis of an important physico-chemical relationship called the *Clapeyron equation*. It enables us to estimate the change in pressure that accompanies a change in temperature in a system in which two phases are in equilibrium with each other.

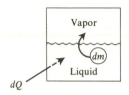

Fig. 6–2

In systems having two phases in equilibrium (for example, water in equilibrium with steam at 100°C and 1 atm), the movement of a differential amount of material, dm, from one phase to the other (Fig. 6–2) involves an entropy change:

$$dS = \frac{dH}{T} = \frac{\Delta \bar{H}_{vap}\, dm}{T}$$

and a volume change

$$dV = (\bar{V}_{vap} - \bar{V}_{liq})\, dm.$$

Combining the above with Eq. (6–17), we obtain

$$\left(\frac{\partial S}{\partial V}\right)_T = \boxed{\frac{\Delta \bar{H}_{vap}}{T(\bar{V}_{vap} - \bar{V}_{liq})}} = \left(\frac{\partial P}{\partial T}\right)_V \qquad (6\text{–}19)$$

where $\bar{V}_{vap}$ and $\bar{V}_{liq}$ are the specific volumes of the final and initial phases and $\Delta \bar{H}_{vap}$ is the enthalpy of vaporization or of phase transformation/unit mass of material (i.e., vaporization, fusion, sublimation, etc.) The boxed part of Eq. (6–19) is called the CLAPEYRON EQUATION.

Example 6–1. The enthalpy of vaporization of water at 212°F and 1 atm is 970.3 Btu/lb$_m$ and the specific volume of the vapor and liquid phases are 26.80 ft^3/lb$_m$ and 0.01672 ft^3/lb$_m$, respectively.

Estimate the pressure inside a boiler containing liquid water and steam, given that the boiler thermometer reads 220°F.

Answer: Rearranging Eq. (6–19), we find that

$$dP = \frac{\Delta \bar{H}_V}{\Delta \bar{V}} \cdot \frac{dT}{T}.$$

The values of $\Delta \bar{H}_V$ and $\Delta \bar{V}$ are substantially constant over small temperature intervals. Therefore we may integrate the previous equation to obtain

$$P_2 - P_1 = \frac{\Delta \bar{H}_V}{\Delta \bar{V}} \ln \frac{T_2}{T_1}.$$

We are given that

$$\Delta \bar{H}_V = 970.3 \text{ Btu/lb}_m$$

and

$$\Delta \bar{V} = 26.80 - 0.02 = 26.78 \text{ ft}^3/\text{lb}_m.$$

Therefore

$$P_2 - P_1 = \frac{970.3 \text{ Btu/lb}_m}{26.78 \text{ ft}^3/\text{lb}_m} \times \frac{778 \text{ ft-lb}_f}{\text{Btu}} \times \ln \frac{680}{672},$$

$$P_2 - P_1 = 338 \text{ lb}_f/\text{ft}^2, \qquad P_2 = 17.0 \text{ lb}_f/\text{in}^2.$$

The handbook value for vapor pressure (see Appendix II) at 220°F is 17.19 lb$_f$/in^2.

Example 6-2. Liquid oxygen boils at 90.15°K. At 100°K its vapor pressure is 2.50 atm. Estimate the ΔH_{vap} of liquid oxygen.

Answer: In Example 6-1, the specific volume of the liquid was negligible compared with the volume of the vapor with which it was in equilibrium. (This is generally true when the temperature is more than 50°C below the critical.) Not having any other information on the specific volume of liquid oxygen, we shall consider it negligible, and also assume that the vapor behaves like an ideal gas.

Equation (6–19) becomes

$$\frac{dP}{dT} = \frac{\Delta \bar{H}_V}{T \bar{V}_{vap}} = \frac{\Delta \bar{H}_V}{TRT/P} = \frac{\Delta \bar{H}_V P}{RT^2}$$

or

$$\frac{dP}{P} = \frac{\Delta \bar{H}_V}{R} \cdot \frac{dT}{T^2}, \qquad\qquad (6\text{–}19\text{a})$$

which can be integrated if ΔH_V is assumed constant:

$$\ln \frac{P_2}{P_1} = \frac{\Delta \bar{H}_V}{R} \left(\frac{1}{T_1} - \frac{1}{T_2} \right). \qquad\qquad (6\text{–}19\text{b})$$

Equation (6–19b) is called the *Clapeyron-Clausius equation* because Clausius suggested this modification of Eq. (6–19). Since we know two vapor pressures and their corresponding temperatures, we can use Eq. (6–19b) to estimate $\Delta \bar{H}_V$:

$$\ln \frac{2.50}{1.00} = \frac{\Delta \bar{H}_V}{1.98 \text{ cal/mol} \cdot °\text{K}} \left(\frac{1}{90.15°\text{K}} - \frac{1}{100°\text{K}} \right)$$

$$\Delta \bar{H}_V = 1658 \frac{\text{cal}}{\text{g-mol}}.$$

The measured enthalpy of vaporization at 90.15°K is 1630 cal/g-mol (value from *Chemical Engineers' Handbook*, fourth edition, New York: McGraw-Hill).

6–7
Jacobians

To simplify the method of finding relations among properties, we shall adopt a procedure for handling partial derivatives developed by Shaw,* which uses a mathematical notation called JACOBIANS.

Jacobians are determinants of partial derivatives. We will simply state their relevant properties. Expanded presentations of their use and derivations of their properties can be found in the references cited.†

In Jacobian notation, a partial derivative is written as:

$$\left(\frac{\partial M}{\partial N}\right)_z = \frac{[M, Z]}{[N, Z]}. \tag{6–20}$$

The notation $[M, Z]$ may be verbalized as *Jacobian MZ*. The relevant properties of the Jacobian are:

$$[M, N] = -[N, M], \tag{6–21}$$

$$[M, M] = 0. \tag{6–22}$$

An exact differential, $dY = a\,dM + b\,dN$, appears in Jacobian form as:

$$[Y, X] = a[M, X] + b[N, X]. \tag{6–23}$$

For example, Eq. (6–2) may be written as

$$[U, X] = T[S, X] - P[V, X], \tag{6–24}$$

where X is any thermodynamic property other than the given ones. Jacobians may be combined algebraically within the limits of the restrictions given.

Example 6–3. Prove that the Maxwell relation, Eq. (6–17), is equivalent to

$$\boxed{[T, S] = [P, V].} \tag{6–24a}$$

Proof: We write Eq. (6–17) in Jacobian form. From Eq. (6–20),

$$\frac{[S, T]}{[V, T]} = \frac{[P, V]}{[T, V]}.$$

But from Eq. (6–21),

$$[V, T] = -[T, V].$$

---

* A. N. Shaw, *Phil. Trans. Roy. Soc. A*, **334**, 299–328 (1935).

† M. Tribus, *Thermostatics and Thermodynamics*, Princeton, N.J.: Van Nostrand, 1961, page 246. See also H. Margenau and G. M. Murphy, *The Mathematics of Physics and Chemistry*, Princeton, N.J.: Van Nostrand, 1956, page 18.

Therefore the denominators cancel, and

$$-[S, T] = [P, V].$$

Again recalling Eq. (6–21), we obtain

$$[T, S] = [P, V]$$

Q.E.D.*

Exercise 6–2

a) Prove that *all* the Maxwell Relations—Eqs. (6–15), (6–16), (6–17), and (6–18)—are equivalent to Eq. (6–24a).

b) From Eq. (6–18a) in Exercise 6–1, show that $[T, S] = -[\tau, L]$, and derive the rubber-band analogs of Eqs. (6–16), (6–17), and (6–18).

From the above exercise it should be apparent that all the Maxwell Relations, when written as ratios of Jacobians, contain $\pm[T, S]$ or $\pm[P, V]$ in the numerator, and that the Jacobians in the denominator on both sides of the equals sign are identical except for sign, and contain one property appearing in the numerator. Thus remembering Eq. (6–24a) enables us to generate all the Maxwell Relations. In much of our work we shall find that it is much easier to use Eq. (6–24a) than the Maxwell Relations.

6–8
Specific heats and expansion coefficients

In addition to P, V, and T, we can measure certain other intensive properties of matter and use them to establish values of derived thermodynamic properties. The more important are the following.

Specific heat at constant P:

$$C_P \equiv \left(\frac{\partial Q_{rev}}{\partial T}\right)_P = T\left(\frac{\partial \bar{S}}{\partial T}\right)_P = T\frac{[\bar{S}, P]}{[T, P]}. \qquad (6\text{--}25)$$

Specific heat at constant V:

$$C_V \equiv \left(\frac{\partial Q_{rev}}{\partial T}\right)_V = T\left(\frac{\partial \bar{S}}{\partial T}\right)_V = T\frac{[\bar{S}, V]}{[T, V]}. \qquad (6\text{--}26)$$

Ratio of specific heats:

$$\gamma \equiv \frac{C_P}{C_V} = \frac{[\bar{S}, P]}{[T, P]} \cdot \frac{[T, V]}{[\bar{S}, V]}. \qquad (6\text{--}27)$$

---

* $[M, N]$ is equivalent to $\oint M\, dN$ (see preceding references). Therefore Eq. (6–24a) may be written as $\oint T\, dS - \oint P\, dV = 0$, which is Eq. (2–38).

Coefficient of thermal expansion:

$$\beta_T \equiv \frac{1}{V}\left(\frac{\partial V}{\partial T}\right)_P = \frac{1}{V}\frac{[V, P]}{[T, P]}. \tag{6–28}$$

Coefficient of compressibility:

$$\kappa_T \equiv -\frac{1}{V}\left(\frac{\partial V}{\partial P}\right)_T = -\frac{1}{V}\frac{[V, T]}{[P, T]}. \tag{6–29}$$

6–9
Thermodynamic properties as functions of measurable properties

We now have the tools to express most derivatives in terms of measurable properties. The steps to follow are:

a) Express the derivative in terms of Jacobians.
b) Eliminate energy and derived properties (U, H, G, A) by expressing them in terms of P, V, T, and S, using Eqs. (6–2), (6–4), (6–6), and (6–7) in the Jacobian form given by Eq. (6–23).
c) Express all the entropy Jacobians in terms of C_P or C_V by substituting Eqs. (6–25) or (6–26), or in terms of $[P, V]$ by substituting Eq. (6–24a). (The partial derivatives of G and A at constant P or V will retain an entropy term outside the derivative which is not eliminated by the above procedure.)

Example 6–4. Express the following in terms of $P, \overline{V}, T$, and measurable properties:

$$\left(\frac{\partial \overline{G}}{\partial P}\right)_T, \qquad \left(\frac{\partial \overline{H}}{\partial P}\right)_T, \qquad \left(\frac{\partial \overline{H}}{\partial T}\right)_P.$$

Answer: Using Jacobian notation and Eqs. (6–23) and (6–7),

$$\left(\frac{\partial \overline{G}}{\partial P}\right)_T = \frac{[\overline{G}, T]}{[P, T]} = \frac{-\overline{S}[T, T] + \overline{V}[P, T]}{[P, T]}.$$

Substituting Eq. (6–22) in the above, we obtain

$$\left(\frac{\partial \overline{G}}{\partial P}\right)_T = \overline{V}. \qquad \text{(Answer)} \tag{6–12}$$

Similarly, using Eq. (6–23) and Eq. (6–4), we have

$$\left(\frac{\partial \overline{H}}{\partial P}\right)_T = \frac{[\overline{H}, T]}{[P, T]} = \frac{T[\overline{S}, T] + \overline{V}[P, T]}{[P, T]} = T\frac{[\overline{S}, T]}{[P, T]} + \overline{V}.$$

Recalling Eqs. (6–24a) and (6–28),

$$\left(\frac{\partial \overline{H}}{\partial P}\right)_T = -T\left(\frac{\partial \overline{V}}{\partial T}\right)_P + \overline{V} = \overline{V}(1 - \beta_T T). \qquad \text{(Answer)}$$

Again using Eq. (6–23) and Eq. (6–4), we obtain

$$\left(\frac{\partial \bar{H}}{\partial T}\right)_P = \frac{[\bar{H}, P]}{[T, P]} = \frac{T[\bar{S}, P] + \bar{V}[P, P]}{[T, P]}$$

$$= T\frac{[\bar{S}, P]}{[T, P]} = C_P. \qquad \text{(Answer)}$$

Exercise 6–3. Prove the following equations.

$$\left(\frac{\partial \bar{H}}{\partial T}\right)_V = C_V + \bar{V}\left(\frac{\partial P}{\partial T}\right)_V,$$

$$\left(\frac{\partial \bar{H}}{\partial V}\right)_T = -\frac{1}{\kappa_T} + T\left(\frac{\partial P}{\partial T}\right)_V.$$

Example 6–5. How does the temperature of any substance change with pressure during an isentropic compression? Evaluate the rate of change of temperature in nitrogen at 1 atm and 0°C.

Answer: The question, in effect, asks us to evaluate $(\partial T/\partial P)_S$. Using Jacobians and Eq. (6–24a), we write

$$\frac{[T, \bar{S}]}{[P, \bar{S}]} = \frac{[P, \bar{V}]}{[P, \bar{S}]} = \frac{[\bar{V}, P]}{[\bar{S}, P]} = \frac{[\bar{V}, P]}{[T, P]} \cdot \frac{[T, P]}{[\bar{S}, P]},$$

and from Eqs. (6–25) and (6–28), we have

$$\left(\frac{\partial T}{\partial P}\right)_S = \frac{\bar{V}\beta_T T}{C_P}. \qquad (6\text{–}29a)$$

Equation (6–29a) states that the temperature *usually increases* during an isentropic compression, because $\bar{V}$, T, and C_P are *always positive* and β_T, the coefficient of thermal expansion, is usually positive. (However, rubber and some other polymeric substances have negative expansion coefficients.)

For an ideal gas: $\bar{V} = RT/P$,

$$\beta_T = \frac{1}{\bar{V}}\left(\frac{\partial \bar{V}}{\partial T}\right)_P = \frac{1}{\bar{V}}\frac{R}{P} = \frac{1}{T}.$$

Therefore, substituting back into Eq. (6–29a), we get

$$\left(\frac{\partial T}{\partial P}\right)_S = \frac{RT}{PC_P} = \frac{\bar{V}}{C_P}. \qquad \text{(Ideal gas only)} \qquad (6\text{–}29b)$$

For 1 mol of nitrogen at 0°C and 1 atm pressure, $\bar{V} = 22.4$ liters/mol, and $C_P = 6.96$ cal/g-mol·°K.

Therefore

$$\left(\frac{\partial T}{\partial P}\right)_S = \frac{22.4 \text{ l/g-mol}}{6.96 \text{ cal/g-mol} \cdot {}^\circ\text{K}} \times \frac{\text{cal}}{41.3 \text{ l-atm}} = 0.078 \frac{{}^\circ\text{K}}{\text{atm}}.$$

Exercise 6–4. Integrate Eq. (6–29b) to obtain Eq. (2–36).

Exercise 6–5. Compare the isentropic rate of change of temperature with pressure of an ideal gas with the isentropic rate of change of temperature with pressure for water and lead at 0°C and 1 atm.

Water: $\beta_T = 0.207/{}^\circ\text{K}$

Lead: $\beta_T = 0.84/{}^\circ\text{K}$; $\rho = 11.4 \text{ g/cm}^3$; C_P (see Table 2–3).

Exercise 6–6. Prove that

$$\left(\frac{\partial P}{\partial T}\right)_V = \frac{\beta_T}{\kappa_T}. \tag{6–29c}$$

Exercise 6–7. Prove that, for an ideal gas,

$$\left(\frac{\partial P}{\partial V}\right)_S = -\frac{C_P}{C_V} \cdot \frac{P}{V},$$

and therefore, for an isentropic process,

$$P_1 V_1^{C_P/C_V} = P_2 V_2^{C_P/C_V}. \tag{2–35}$$

Exercise 6–8. Using the obvious Jacobian property,

$$\left[\frac{PV}{PV}\right] \cdot \left[\frac{VT}{VT}\right] \cdot \left[\frac{TP}{TP}\right] = 1,$$

show that

$$-\left(\frac{\partial P}{\partial V}\right)_T = \left(\frac{\partial P}{\partial T}\right)_V \left(\frac{\partial T}{\partial V}\right)_P.$$

6–10
Evaluation of changes in thermodynamic properties

Our concern with the thermodynamic state of a system is usually motivated by a need to determine (a) the magnitudes of the system's energy interactions with its surroundings (work or heat effects); (b) the position of the system relative to an equilibrium state. To make these determinations, we shall need to calculate the *changes* in the system's

Entropy, Internal Energy, Enthalpy, Gibbs Free Energy,
and/or Helmholtz Free Energy

from changes in the system's P, V, and/or T.

In general terms, our problem is to calculate the change in any thermodynamic property, M, when the system goes from P_1, T_1, which characterizes state (1), to P_2, T_2, which characterizes state (2).

We now know that in a fixed-composition, $P\,dV$-work system,

$$M = M(P, T).$$

Therefore

$$dM = \left(\frac{\partial M}{\partial P}\right)_T dP + \left(\frac{\partial M}{\partial T}\right)_P dT \tag{6–30}$$

and

$$\Delta M = \int_{P_1}^{P_2} \left(\frac{\partial M}{\partial P}\right)_{T_1} dP + \int_{T_1}^{T_2} \left(\frac{\partial M}{\partial T}\right)_{P_2} dT. \tag{6–31}$$
$$\underset{(T_1=\text{constant})}{} \qquad\qquad \underset{(P_2=\text{constant})}{}$$

Equation (6–31) says that the total change in the magnitude of M consists of two kinds of changes; one resulting from a temperature change and the other from a pressure change. And each partial change can be separately evaluated and summed to find the total change. Figure 6–1 shows the operation schematically.

Evaluations of the integrals in Eq. (6–31) present two problems.

First: $(\partial M/\partial P)_T$ and $(\partial M/\partial T)_T$ must be expressed in terms of measurable properties. This is accomplished via Eqs. (6–2), (6–4), (6–6), and (6–7) and the techniques of Jacobian substitution discussed in Section 6–9.

Example 6–6. Let us replace M in Eq. (6–31) with $\bar{S}$, representing entropy. Write the expression for $\bar{S}$ accompanying a change from state (1) to state (2) in terms of measurable properties.

Answer: From Eq. (6–31), we obtain

$$\Delta \bar{S} = \int_{P_1 T_1}^{P_2 T_1} \left(\frac{\partial \bar{S}}{\partial P}\right)_T dP + \int_{T_1 P_2}^{T_2 P_2} \left(\frac{\partial \bar{S}}{\partial T}\right)_P dT.$$

But

$$\left(\frac{\partial \bar{S}}{\partial P}\right)_T = -\left(\frac{\partial \bar{V}}{\partial T}\right)_P \tag{3–18) or (6–24a}$$

and

$$\left(\frac{\partial \bar{S}}{\partial T}\right)_P = \frac{C_P}{T}. \tag{6–25}$$

Therefore

$$\boxed{\Delta \bar{S} = -\int_{P_1}^{P_2} \left(\frac{\partial \bar{V}}{\partial T}\right)_P dP + \int_{T_1}^{T_2} C_P \frac{dT}{T}.} \tag{6–32}$$

Second: In order to carry out the integration, the partial *derivative* of the measurable properties—such as $(\partial V/\partial T)_P$ in Eq. (6–32), Example 6–6—must be expressed in terms of P or T. This requires additional information in the form of an equation or relationship between P, V, and T, called an EQUATION OF STATE. The simplest equation of state is the familiar Ideal-Gas Law,

$$P\overline{V} = RT.$$

Equations of state are *addenda* to classical thermodynamics. They are not derivable from the premises or "laws" of classical thermodynamics. (However, certain equations of state are derivable via methods of statistical thermodynamics.)

Example 6–7. Derive a general expression for the change in internal energy for a system that moves from state P_1, T_1 to state P_2, T_2 and then apply this expression to an ideal gas.

Answer: Using Eq. (6–31), we get

$$\Delta\overline{U} = \int_{P_1}^{P_2} \left(\frac{\partial \overline{U}}{\partial P}\right)_T dP + \int_{T_1}^{T_2} \left(\frac{\partial \overline{U}}{\partial T}\right)_P dT . \qquad (6\text{–}31\text{a})$$

First, using Jacobian notation and Eq. (6–24), we eliminate U from the partial derivatives:

$$\left(\frac{\partial \overline{U}}{\partial P}\right)_T = \frac{[\overline{U}, T]}{[P, T]} = \frac{T[\overline{S}, T] - P[\overline{V}, T]}{[P, T]}$$

$$= T\frac{[\overline{S}, T]}{[P, T]} - P\frac{[\overline{V}, T]}{[P, T]}.$$

Substituting Eq. (6–24a) in the above, and reversing the order of $[P, T]$ as in Eq. (6–21), we obtain

$$\frac{[\overline{U}, T]}{[P, T]} = -T\frac{[\overline{V}, P]}{[T, P]} - P\frac{[\overline{V}, T]}{[P, T]}$$

or

$$\left(\frac{\partial \overline{U}}{\partial P}\right)_T = -T\left(\frac{\partial \overline{V}}{\partial T}\right)_P - P\left(\frac{\partial \overline{V}}{\partial P}\right)_T . \qquad (6\text{–}31\text{b})$$

Similarly,

$$\left(\frac{\partial \overline{U}}{\partial T}\right)_P = \frac{[\overline{U}, P]}{[T, P]} = T\frac{[\overline{S}, P]}{[T, P]} - P\frac{[\overline{V}, P]}{[T, P]} .$$

When we substitute Eq. (6–25) in the above,

$$\left(\frac{\partial \overline{U}}{\partial T}\right)_P = C_P - P\left(\frac{\partial \overline{V}}{\partial T}\right)_P \qquad (6\text{–}31\text{c})$$

and on substituting Eqs. (6–31b) and (6–31c) in Eq. (6–31a), we obtain a general expression for $\Delta \overline{U}$,

$$\Delta \overline{U} = \int_{P_1}^{P_2} - T \left(\frac{\partial \overline{V}}{\partial T} \right)_P dP - \int_{P_1}^{P_2} P \left(\frac{\partial \overline{V}}{\partial P} \right)_T dP$$

$$+ \int_{T_1}^{T_2} C_P \, dT - \int_{T_1}^{T_2} P \left(\frac{\partial \overline{V}}{\partial T} \right)_P dT . \qquad (6\text{–}31\text{d})$$

Second, to evaluate the remaining derivatives, we use an equation of state. For 1 mol of ideal gas, $P \overline{V} = RT$ and $\overline{V} = RT/P$. Therefore

$$\left(\frac{\partial \overline{V}}{\partial T} \right)_P = \frac{R}{P} = \frac{\overline{V}}{T} \quad \text{and} \quad \left(\frac{\partial \overline{V}}{\partial P} \right)_T = - \frac{RT}{P^2} .$$

Substituting these back into Eq. (6–31d), we obtain

$$\Delta \overline{U} = \int_{P_1}^{P_2} - TR \frac{dP}{P} + RT \frac{dP}{P} + \int_{T_1}^{T_2} C_P \, dT - R \, dT.$$

Using Eq. (2–24), we have

$$\Delta \overline{U}_{\text{ideal}} = \int_{T_1}^{T_2} C_V \, dT. \qquad \text{(Answer)}$$

We see that, for an ideal gas, $(\partial \overline{U}/\partial P)_T = 0$. Therefore $\Delta \overline{U}$ depends only on the change in temperature.

Exercise 6–9. Prove that for any material,

$$\left(\frac{\partial \overline{U}}{\partial T} \right)_P = C_P - P \overline{V} \beta_T \quad \text{and} \quad \left(\frac{\partial U}{\partial P} \right)_T = P V \kappa_T - T V \beta_T.$$

Real gases require equations of state considerably more complex than the ideal-gas law. Evaluation of the derivatives in Eq. (6–31) may then involve difficulties in computation which warrant the use of computers, or require extensive graphical integration. This is generally true of equations of state that are non-analytical (the P, V, T relationship is expressed as a graph, for example), or contain numerous constants. Section 7 presents equations of state for various real gases and describes how to use them in evaluating thermodynamic properties.

Problems

1. Using only the saturation pressure data for steam, available in the steam table, prepare a graph of pressure versus temperature and estimate the $\Delta \overline{H}_{\text{vap}}$ for liquid water at 70°F, 212°F, and 600°F. Compare your estimated values with values given in the steam table.

2. What is the minimum working pressure for a tank which stores liquid CO_2 at room temperature (70°F)? ΔH_{vap} of CO_2 at 0°F $= 120.1$ Btu/lb and P_{CO_2} at 0°F $= 305.8$ psia.

3. The blade of an ice skate is $\frac{3}{32}$ in. thick. It contacts the ice on 6 in. of its length and supports a 160-lb skater. What is the freezing point of water directly beneath the skate blade? (See a chemistry handbook for the enthalpy of fusion of ice.)

4. For engineers designing refrigeration and gas-liquefaction equipment, the variation of the temperature of a gas with pressure in a free expansion or a throttling process is an important coefficient. Derive general expressions for this coefficient for an adiabatic free expansion and for a throttling process. One of these,

$$\left(\frac{\partial T}{\partial P}\right)_H,$$

is called the *Joule-Thomson coefficient*. Show that for an ideal gas it is zero.

5. Derive general expressions for

$$\left(\frac{\partial U}{\partial S}\right)_T, \quad \left(\frac{\partial U}{\partial V}\right)_T, \quad \left(\frac{\partial U}{\partial H}\right)_T,$$

and evaluate these for an ideal gas.

6. Using Jacobians, show that

a) $\left(\dfrac{\partial V}{\partial T}\right)_P dP = -\left(\dfrac{\partial P}{\partial T}\right)_V dV$ at constant T

b) $\left(\dfrac{\partial P}{\partial T}\right)_V \left(\dfrac{\partial V}{\partial P}\right)_T \left(\dfrac{\partial T}{\partial V}\right)_P = -1$

c) $\left(\dfrac{\partial P}{\partial V}\right)_T \left(\dfrac{\partial T}{\partial P}\right)_V = -\left(\dfrac{\partial T}{\partial V}\right)_P$

7. A certain system shows a decrease in entropy when its volume increases (at constant temperature).

a) How will the system's temperature respond to a small adiabatic decrease in volume?
b) What will happen to the system's volume as its temperature increases at constant pressure?

8. In the course of interplanetary exploration of an asteroid circling Quasar S–37, space-probe data indicated that materials found on that remote body have a positive isothermal modulus of compression. Discuss the implications of this finding.

9. The density of liquid water is a maximum at 4°C. What is the effect of pressure on the entropy of liquid water at this temperature? at temperatures above and below 4°C?

10. Prove that an elastic band that does only $(-\tau\, dL)$ work and that follows the equation of state $\tau = KLT$—where $\tau = $ tension, $L = $ length, $T = $ temperature, and K is a constant—is like an ideal gas in that $U = U\,(T$ only). Also find:

a) The effect on the tension of a temperature increase at constant L.

b) The effect on temperature of adiabatic reversible stretching.

c) An expression for dU in terms of τ, T, L, and S.

11. Show that if $\bar{U} = \bar{U}(P, \bar{V})$, then, for an ideal gas,

$$\left(\frac{\partial \bar{U}}{\partial P}\right)_V dP + \left(\frac{\partial \bar{U}}{\partial \bar{V}}\right)_P d\bar{V} = C_V\, dT$$

12. Using data given in Exercise 6–5 and the steam table, and an equation derived in Example 6–4, compare the change in enthalpy of water going from 193.2°F and 300 psia to 417.3°F and 300 psia, with the change in enthalpy of water going from 193.2°F and 10 psia to 417.3°F and 300 psia. See Exercise 4–4.

13. Compare $(\partial V/\partial P)_T$ and $(\partial V/\partial P)_S$. Which is larger?

14. Write expressions for the slopes of lines of constant pressure and constant temperature on an $\bar{H}$-$\bar{S}$ plane,

a) for an ideal gas,

b) for a liquid which is incompressible and has a constant specific heat.

15. Is C_P always larger than C_V? How do C_τ and C_L compare in a system such as that described in Exercise 6–1?

16. Consider the relations derived in this section and devise experiments to measure (a) C_P and (b) $\Delta H_{\text{vaporization}}$, without using a calorimeter (i.e., without measuring Q effects).

17. Write a general expression for the ΔH that accompanies a change from equilibrium state (1) at P_1, T_1 to equilibrium state (2) at P_2, T_2, for a system whose only reversible work mode is expansion work ($P\, dV$), in terms of the measurable properties of the system.

18. Repeat Problem 17 for ΔG.

19. Repeat Problem 17 for ΔA.

20. The *departure function* of a property is the difference between the magnitude of that property in an ideal-gas system and the magnitude of the property at the same state of pressure and temperature in a real-material system. Thus, for a property $M = M(P, T)$, the departure function for M is

$$M_{\text{departure}} = \Delta M_{\text{ideal}} - \Delta M.$$

Show that the departure function for enthalpy is

$$\Delta \bar{H}_{\text{ideal}} - \Delta \bar{H} = \int_{P_1}^{P_2} \left[T \left(\frac{\partial \bar{V}}{\partial T}\right)_P - \bar{V} \right] dP.$$

21. Extend Problem 20 by deriving general expressions for the departure functions of entropy and internal energy.

22. An additional useful property of Jacobians is that:

$$[A, B]\, [C, X] + [B, C]\, [A, X] + [C, A]\, [B, X] = 0.$$

Therefore

$$[P, V]\, [T, S] + [V, T]\, [P, S] + [T, P]\, [V, S] = 0.$$

Use this relation to prove that

a) $\left(\dfrac{\partial \overline{V}}{\partial T}\right)_S = \dfrac{1}{1-k} \left(\dfrac{\partial \overline{V}}{\partial T}\right)_P$

b) $C_P - C_V = T \left(\dfrac{\partial P}{\partial T}\right)_V \left(\dfrac{\partial \overline{V}}{\partial T}\right)_P$

c) From part (a), derive Eq. (2–37).

23. a) Show that

$$\left(\frac{\partial \overline{H}}{\partial P}\right)_T = -T^2 \left(\frac{\partial (\overline{V}/T)}{\partial T}\right)_P.$$

b) The following data are available on n-heptane at 100 psia. Use them to estimate the specific volume of n-heptane vapor at 224.5°F and 100 psia and the slope of the vapor-pressure curve, dP/dT.

Boiling point of n-heptane at 100 psia	224.5°F
$\Delta \overline{H}_{\text{vaporization}}$ at 100 psia	122.4 Btu/lb
Specific volume of saturated liquid	0.03045 ft³/lb
Specific volume of vapor at 460°F and 100 psia	1.2975 ft³/lb

T, °F	$-(\partial \overline{H}/\partial P)_T$, Btu/lb · psia
460	0.0592
430	0.0618
400	0.0647
370	0.0688
340	0.0749
310	0.0836
280	0.0957
250	0.1126

JOHANNES DIDERIK VAN DER WAALS, 1837–1923

Section 7
EQUATIONS OF STATE

7–1
Deviation from ideal-gas behavior: compressibility factor

In order to apply thermodynamics to engineering problems, we must recognize the fact that real gases and vapors obey the Ideal-Gas Law (Eq. 2–21) only approximately and over very limited ranges of temperatures and pressures. Like so many ideals, it is honored largely in the breach. For real gases, it is more generally true that:

$$\frac{P\overline{V}}{RT} < 1. \tag{7–1}$$

(At sufficiently high temperatures, the direction of the inequality sign reverses. The temperature at which reversal occurs is called the BOYLE TEMPERATURE.) The ratio $P\overline{V}/RT$ is called the COMPRESSIBILITY or *compressibility factor*.

Exercise 7–1. Table 7–1 gives empirical values of the molar volume of various real gases under specified conditions. Use the ideal-gas law to calculate volumes, and compare the calculated to the actual values. Also calculate the compressibility factors, $P\overline{V}/RT$. Enter these values in Table 7–1.

A concern with the thermodynamic properties of real materials is therefore a necessary and characteristic part of applied thermodynamics. Considerable effort has been—and is being—expended to develop graphical or analytical relationships

Table 7–1

PVT values for real gases*

Gas	Temperature, °F	Pressure, psia	Volume, ft^3/lb · mol	Ideal volume	$P\overline{V}/RT$
Steam	212	14.7	482		
	500	14.7	698		
	1400	14.7	1356		
	1400	2000	9.63		
	1000	3000	4.45		
Air	32	14.7	359		
	500	14.7	701		
	500	1470	7.29		
	32	1470	3.566		
Ammonia	0	5	980		
	32	14.7	357		
	70	14	402		
	70	120	42.7		
	270	300	24.2		

* Data from J. H. Perry, *Chemical Engineers' Handbook*, fourth edition, New York: McGraw-Hill, 1963.

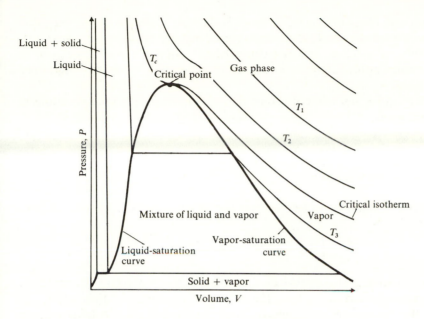

Fig. 7–1. Pressure–volume behavior of a real material.

between pressure, volume, and temperature of real materials. Such relationships are called EQUATIONS OF STATE. An equation of state for a real gas simulates all or part of the complex behavior illustrated in Fig. 7–1, which shows the relation of pressure to volume at selected temperatures for a typical real gas, carbon dioxide.

7–2
Real gases

The behavior of an ideal gas is postulated on the *absence* of any interaction between gas molecules, other than that of elastic collision. Molecules of a real gas approach such independent behavior when they are far apart and are moving at a high average speed; that is, when the pressure is low and the temperature is high (region to the right of T_1 in Fig. 7–1). As pressure increases and temperature drops, real gas molecules "feel" each other's attractive-force fields. This causes deviation from ideality (isotherms T_2, T_c, and T_3) and eventually results in the liquefaction or solidification of the gas.

Equations which attempt to describe the behavior of a real gas will obviously be more complex than $P\overline{V} = RT$. Two criteria have been used to select such equations:

1) At low pressures, the equation must reduce to the ideal-gas law.

2) At the *critical point* (see Fig. 7–1), the equation must show a horizontal inflection point.

The CRITICAL POINT of a pure substance is the state identified by the *maximum temperature* (called the CRITICAL TEMPERATURE), and the *minimum pressure* (CRITICAL PRESSURE) at which a substance can exist as a liquid. A pure gas *cannot* be liquefied at temperatures above its critical temperature (T_c), no matter what pressure is imposed on it.

The following sections describe some of the more frequently used equations of state for real gases.

7–3
The van der Waals equation

The following equation is called the *van der Waals equation of state*, after its originator:

$$\left(P + \frac{a}{\overline{V}^2}\right) (\overline{V} - b) = RT. \tag{7–2}$$

Equation (7–2) attempts to improve on the ideal-gas law by reducing the volume in which the gas molecules cavort by b, the volume of the molecules themselves, and increasing the pressure by $a/\overline{V}^2$, which allows for the fact that molecules attract each other to an extent inversely proportional to the square of the molar volume. The intermolecular attraction is considered an internal pressure which adds to the actual pressure.

Table 7–2 presents values of the van der Waals constants for common gases. The values of the constants may be estimated from the critical properties by applying

Table 7–2

Van der Waals constants*

Gas	a	b
Air	1.33×10^6	36.6
NH_3	4.19×10^6	37.3
H_2O (vapor)	5.48×10^6	30.6
CO_2	3.60×10^6	42.8
CH_4	2.25×10^6	42.8
H_2	0.245×10^6	26.6
N_2	1.347×10^6	38.6
O_2	1.36×10^6	31.9

$$a = \text{atm} \cdot \frac{\text{cm}^6}{(\text{g-mol})^2}, \qquad b = \frac{\text{cm}^3}{\text{g-mol}}$$

* Data are from *International Critical Tables*, New York: McGraw-Hill, 1926.

the mathematical criterion for a horizontal inflection point,

$$\left(\frac{\partial P}{\partial \overline{V}}\right)_T = 0 = \frac{\partial^2 P}{\partial \overline{V}^2}, \tag{7-3}$$

to the equation of state when P, $\overline{V}$, and T equal P_c, $\overline{V}_c$, and T_c. (The subscript c refers to the *critical state*.) It follows that:

$$a = 3P_c\overline{V}_c^2, \tag{7-4}$$

$$b = \frac{\overline{V}_c}{3}. \tag{7-5}$$

The van der Waals equation yields a qualitative picture of the behavior of a real gas in that it shows a critical point and a vapor–liquid envelope. It is not accurate, however, for gases at high densities.

Exercise 7–2. Solve Eq. (7–2) for P and derive Eqs. (7–4) and (7–5) from the critical-point inflection conditions, Eq. (7–3). At the critical point $P = P_c$, $\overline{V} = \overline{V}_c$ and $T = T_c$.

7–4
Other two-constant equations of state

$$P = \frac{RT}{\overline{V} - b} e^{-a/RT\overline{V}}, \tag{7-6}$$

$$P = \frac{RT}{\overline{V} - b} - \frac{a}{T^{1/2}(\overline{V}^2 + \overline{V}b)}, \tag{7-6a}$$

$$P = \frac{RT}{\overline{V} - b} - \frac{a}{T\overline{V}^2}. \tag{7-7}$$

Equations (7–6), (7–6a), and (7–7) are known as the *Dieterici, Redlich-Kwong,* and *Berthelot equations of state,* respectively. They represent the behavior of a real gas over limited ranges of conditions.

The Dieterici equation (7–6) can be used for nonpolar gases in the critical region.*

Each of these equations may be written as compressibility equations, in which form their correction to ideal-gas behavior is emphasized.

---

* Beattie and Stockmayer, *Rept. Prog. Phys.* **7**, 194 (1940).

From Eq. (7–2), we can proceed to

$$\frac{P\overline{V}}{RT} = 1 + \frac{Pb}{RT} - \frac{a}{RT\overline{V}} + \frac{ab}{RT\overline{V}^2}. \tag{8–7}$$

From Eq. (7–7), we get

$$\frac{P\overline{V}}{RT} = 1 + \frac{Pb}{RT} - \frac{a}{RT^2\overline{V}} + \frac{ab}{RT^2\overline{V}^2}. \tag{7–9}$$

From Eq. (7–6), replacing the exponential with a series, we obtain

$$\frac{P\overline{V}}{RT} = 1 + \frac{Pb}{RT} - \frac{a}{RT\overline{V}} + \frac{a^2}{2(RT\overline{V})^2} - \frac{a^3}{6(RT\overline{V})^3} + \cdots. \tag{7–10}$$

Exercise 7–3. Derive Eqs. (7–8), (7–9), and (7–10) from Eqs. (7–2), (7–7), and (7–6), respectively.

7–5
Benedict-Webb-Rubin equation*

$$P = \frac{RT}{\overline{V}} + \frac{1}{\overline{V}^2}\left(B_0 RT - A_0 - \frac{C_0}{T^2}\right) + \frac{1}{\overline{V}^3}(bRT - a)$$

$$+ \frac{a\alpha}{\overline{V}^6} + \frac{c}{\overline{V}^3 T^2}\left(1 + \frac{\gamma}{\overline{V}^2}\right)e^{-\gamma/\overline{V}^2}. \tag{7–11}$$

Equation (7–11) contains 8 empirical constants: A_0, B_0, C_0, a, b, c, γ, and α. In view of the number of constants, it is not surprising that this equation offers a good description of the behavior of a real gas.† Techniques for applying this equation to mixtures of real gases have also been developed [1]. Tabulations of the constants for common gases appear in the references cited below [1, 2, 3, 4].

Although cumbersome in appearance, Eq. (7–11) is self-consistent and analytic. It may therefore be placed on a computer. It has been used in this manner for a large project involving the compilation of thermodynamic data, sponsored by the American Petroleum Institute [5].

---

* M. Benedict, G. Webb, and L. C. Rubin, *J. Chem. Phys.* **8**, 334 (1940).
1. M. Benedict and G. Webb, *Chem. Eng. Prog.* **47**, 419 (1951).
2. E. I. Organick and W. R. Studhalter, *Chem. Eng. Prog.* **44**, 847 (1948).
3. H. H. Stotler and M. Benedict, *Chem. Eng. Prog.*, Symp. No. 49, **6** 25 (1953).
4. L. N. Canjar *et al.*, *Ind. Eng. Chem.* **47**, 1028 (1955).
† The mathematician Gauss is supposed to have said, "I can describe an elephant with four arbitrary constants and with a fifth I will have it wagging its tail."

7–6
Virial equations of state

If the interactions of gas molecules are considered first two at a time, then three at a time, then four at a time, etc., using the methods of statistical mechanics, an equation can be derived for the departure from ideality called the VIRIAL EQUATION OF STATE [6]:

$$\frac{P\overline{V}}{RT} = 1 + \frac{\mathbf{B}}{\overline{V}} + \frac{\mathbf{C}}{\overline{V}^2} + \cdots, \qquad (7\text{–}12)$$

where the constants **B** and **C** are called the *second and third virial coefficients.* Electronically computed values of **B** and **C** (based on a Lennard-Jones potential between particles) are available in the literature.* It can also be shown that when $\overline{V} \rightarrow \infty$, $\mathbf{B} = \overline{V} - RT/P$.†

The virial equation lends theoretical validity to the previous empirical equations, which have similar forms.

7–7
The generalized equation of state

A useful way to relate the thermodynamic properties of real gases depends on the LAW OF CORRESPONDING STATES, proposed by van der Waals in 1873. This law suggests that *all materials obey a single universal $P\overline{V}T$ relation* when their pressures, volumes, and temperatures are compared at *corresponding states.* Corresponding states are usually taken to be states with equal REDUCED PROPERTIES. A reduced property is the ratio of the absolute value of the property to its value at the *critical temperature and pressure* of the material. As previously noted, T_c, the critical temperature, is the maximum temperature at which a given material can exist in a liquid phase, regardless of the pressure applied to it. P_c, the critical pressure, is the minimum pressure required to liquefy a gas at its critical temperature. $\overline{V}_c$ is the specific volume at T_c and P_c, and has the same value in both the liquid and gas phases. Table 7–3 lists critical constants of common gases.

---

5. L. N. Canjar and F. D. Rossini, "The Work of the API Research Project 44 on *P–V–T* Properties," presented at IUPAC Joint Conference on Thermodynamics and Transport Properties of Fluids, London, July 10, 1957.

6. K. Onnes, *Commun. Phys. Lab.*, Leiden, No. 71 (1901).

* J. O. Hirschfelder, C. F. Curtis, and R. B. Bird, *The Molecular Theory of Gases and Liquids*, New York: John Wiley, 1954.

† J. J. Martin, "Thermodynamics," *Chem. Eng. Symp. Series*, No. 44, **59** (1963).

Table 7-3

Critical properties of common gases*

Substance	T_c, °K	P_c, atm	$\overline{V}_c$, cm^3/g-mol	Z_c
Air	132.41	37.25	92.35	
Argon, A	150.72	47.99	75	0.291
Helium, He	5.19	2.26	58	0.308
Carbon monoxide, CO	132.91	34.529	93	0.294
Hydrogen, H$_2$	33.24	12.797	65	0.304
Nitrogen, N$_2$	126.2	33.54	90	0.291
Oxygen, O$_2$	154.78	50.14	74	0.292
Carbon dioxide, CO$_2$	304.20	72.90	94	0.275
Sulfur dioxide, SO$_2$	430.7	77.8	122	0.269
Water, H$_2$O	647.27	218.167	56	0.230
Ammonia, NH$_3$	405.5	111.5	72.4	0.243
Acetylene, C$_2$H$_2$	309.5	61.6	113	0.274
Ethane, C$_2$H$_6$	305.48	48.20	148	0.285
Ethylene, C$_2$H$_4$	283.06	50.50	124	0.270
n-butane, C$_4$H$_{10}$	425.17	37.47	255	0.274
Methane, CH$_4$	190.7	45.8	99	0.290
Propane, C$_3$H$_8$	370.01	42.1	200	0.277

* Modified from a compilation by J. Lay, *Thermodynamics*, Columbus, Ohio: Charles E. Merrill Books, 1963.

Thus REDUCED TEMPERATURE is defined as

$$T_r \equiv \frac{T}{T_c},$$

REDUCED PRESSURE is defined as

$$P_r \equiv \frac{P}{P_c},$$

and REDUCED VOLUME is defined as

$$V_r \equiv \frac{V}{V_c}.$$

The term

$$z \equiv \frac{P\overline{V}}{RT} \tag{7-13}$$

is called the COMPRESSIBILITY FACTOR. It obviously equals 1.00 for an ideal gas.

The difference between z and 1.00 is a measure of the deviation from ideality of a real gas or vapor.

The Law of Corresponding States does not tell us what the universal $P\overline{V}T$ relation is, but it does say that *all gases have the same compressibility factor, z, at the same reduced conditions*. In effect, it states that z depends only on T_r and P_r, and is *independent* of molecular species; that is,

$$z = z(T_r, P_r). \tag{7–14}$$

It necessarily follows that all gases must show the same compressibility factor at their critical conditions ($T_r = P_r = 1$). Therefore

$$z_c = \frac{P_c \overline{V}_c}{RT_c} = \text{Universal constant.} \tag{7–15}$$

The great utility of this law is that it implies that if z were determined for *one gas* over a broad range of states, the data, if expressed in terms of reduced temperatures and pressures, would suffice for *all gases*. Such a compilation of data is referred to as a GENERALIZED EQUATION OF STATE.

The validity of the Law of Corresponding States is demonstrated by Fig. 7–2, which plots the experimentally determined compressibility factor, z, against reduced pressure, P_r, at various values of reduced temperature T_r for a number of different gases.

At a given value of T_r, all the gases have compressibilities which fall close to a single curve. Therefore, given the temperature and pressure of any real gas and its critical properties (Table 7–3), we may estimate its volume by reading a generalized z-value from Fig. 7–2 and then calculating $\overline{V}$ from Eq. (7–13).

Exercise 7–4

  i) What are the units of T_r, $\overline{V}_r$, P_r, T_c, $\overline{V}_c$, and P_c?
 ii) Compare the actual temperatures, pressures, and volumes of O_2, methane, and water vapor when all are in the *same* corresponding state of $P_r = 1.1$, $T_r = 1.2$.
iii) Compare reduced properties of these gases when all are at 1 atm and 212°F.

7–8
Improved generalized correlation

A closer examination of Fig. 7–2 reveals that all z's do not fall precisely on the constant T_r lines. In addition z_c Eq. (7–15) is not constant for all gases, but varies between 0.20 and 0.30 (Table 7–3).

Lydersen, Greenkorn, and Hougen* improved on these shortcomings of the

---

* O. A. Hougen, K. M. Watson, and R. A. Ragatz, *Chemical Process Principles, Part Two: Thermodynamics*, New York: John Wiley, 1959.

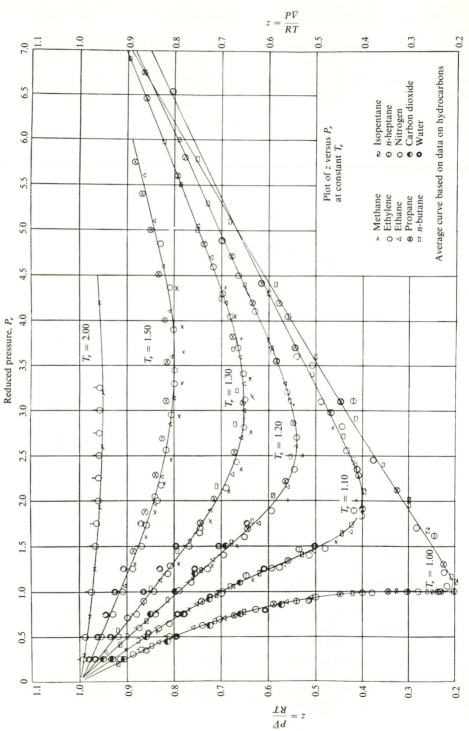

$$z = \frac{P\bar{V}}{RT}$$

Plot of z versus P_r at constant T_r

× Methane   ⧉ Isopentane
○ Ethylene   ⊖ n-heptane
△ Ethane   ○ Nitrogen
⊗ Propane   ⊕ Carbon dioxide
□ n-butane   ● Water

Average curve based on data on hydrocarbons

Reduced pressure, P_r

$T_r = 2.00$

$T_r = 1.50$

$T_r = 1.30$

$T_r = 1.20$

$T_r = 1.10$

$T_r = 1.00$

$$z = \frac{P\bar{V}}{RT}$$

Fig. 7–2. Compressibility factor versus reduced pressure for a number of gases. (From the paper, "Modified Law of Corresponding States," by Gouq-Jen Su, *Ind. Eng. Chem.* **38**, pages 803–806; reprinted by permission.)

simple generalized correlation by adding a third parameter, z_c, to Eq. (7–14). In effect:

$$z = z'(P_r, T_r, z_c). \qquad (7–16)$$

They prepared separate correlations for four different ranges of z_c (0.23, 0.25, 0.27, and 0.29) and demonstrated that gases with similar z_c's correlated well with each other. The correlations are available in extensive tabular form in the cited reference.

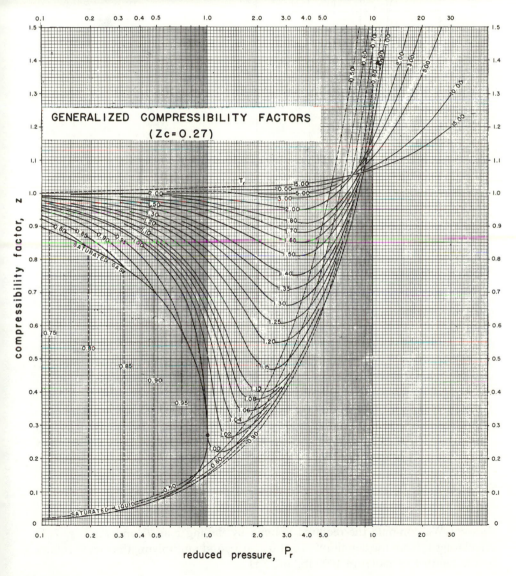

Figure 7–3

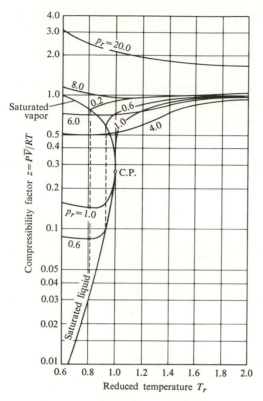

Fig. 7–4. Generalized compressibility factors for gases and liquids versus reduced temperature; $z_c = 0.27$. (From Hougen, Watson, and Ragatz, *Chemical Process Principles, Part Two: Thermodynamics*, second edition, New York: John Wiley, 1959, pages 619–635, by permission.)

Compressibility factors and other thermodynamic properties are tabulated as functions of T_r and P_r, for z_c equal to 0.27. (We can find properties at z_c other than 0.27 by using tabulated deviation terms, D_a and D_b, which are also functions of P_r and T_r. D_a is used when $z_c > 0.27$ and D_b is used when $z_c < 0.27$.) Figures 7–3 and 7–4 present the correlation of this compressibility factor graphically for pure gases and liquids when z_c equals 0.27. The reader should consult the book by Hougen, *et al.*, when z_c differs appreciably from 0.27 and precise computations are required.

Example 7–1. Estimate the volume required to contain 1000 lb_m of ethylene at 150°F and 1000 psig (gage pressure).

Answer: We compute the required volume by finding z at T_r and P_r (Fig. 7–3), calculating $\overline{V}$ from the definition of z (Eq. 7–13), and multiplying the molar volume by the number of moles in 1000 lb_m of C_2H_4.

For ethylene:

$$T_c = 283.1°K = 509°R,$$

$$P_c = 50.50 \text{ atm} = 743 \text{ lb}_f/\text{in}^2 \text{ (Table 7–3),}$$

$$T_r = \frac{T}{T_c} = \frac{610}{509} = 1.197,$$

$$P_r = \frac{P}{P_c} = \frac{1015}{743} = 1.37,$$

$$z = 0.71 = \frac{P\bar{V}}{RT}.$$

From this we obtain

$$\bar{V} = \frac{RTz}{P} = 10.73 \frac{\text{ft}^3 \cdot \text{lb}_f}{\text{lb-mol} \cdot °R \text{ in}^2} \times \frac{610°R}{1015 \text{ lb}_f/\text{in}^2} \times 0.71 = 4.58 \frac{\text{ft}^3}{\text{lb-mol}},$$

$$V = 4.58 \frac{\text{ft}^3}{\text{lb-mol}} \times \frac{1000}{28} \text{ lb-mol} = 163.8 \text{ ft}^3. \quad \text{(Answer)}$$

Example 7–2. A 10-cubic-foot cylinder, having a safe working pressure of 500 psig, contains 1 lb-mol of propane. What is the maximum operating temperature for the tank?

Answer: Here we are not given temperature data. A trial-and-error calculation will therefore be required to find a T that satisfies both Fig. 7–3 and Eq. (7–13). We proceed as follows.

$$C_3H_8: \quad T_c = 370°K = 666°R, \quad \text{(Table 7–3)}$$

$$P_c = 42.1 \text{ atm} = 619 \text{ lb}_f/\text{in}^2,$$

$$P_r = \frac{515}{619} = 0.833,$$

$$T = 666°R \times T_r,$$

$$z = \frac{515 \times 10}{10.73 \, T \, (°R)} = \frac{479}{T \, (°R)}. \quad \text{(Eq. 7–13)}$$

First trial: Let $T_r = 1.2$. Then

$$z = 0.83 \quad \text{(Fig. 7–3),}$$

$$T = 666 \times 1.2 = 799°R,$$

$$z = 0.60 \quad \text{(Eq. 7–13).}$$

Second trial: Use the last value of z to choose an improved value of T_r at $P_r =$ 0.833. Let $T_r = 1.00$ (Fig. 7–3). Then

$$z = 0.6 \qquad \text{(Fig. 7–3)},$$

$$T = 666 \times 1.0 = 666°\text{R},$$

$$z = 0.72 \qquad \text{(Eq. 7–13)}.$$

Third trial: Let $T_r = 1.06$ (Fig. 7–3). Then

$$z = 0.72 \qquad \text{(Fig. 7–3)},$$

$$T = 706°\text{R},$$

$$z = 0.68 \qquad \text{(Eq. 7–13)}.$$

Fourth trial: Let $T_r = 1.04$. Then

$$z = 0.68,$$

$$T = 692°\text{R},$$

$$z = 0.693.$$

Fifth trial: Let $T_r = 1.042$. Then

$$z = 0.693$$

$$T = 694°\text{R} \qquad \text{(Answer)}$$

$$z = 0.692 \qquad \text{(Check)}$$

Exercise 7–5. Calculate the pressure in a 2-cubic-foot cylinder holding one pound of methane at 500°R.

7–9
Thermodynamic properties from equations of state

We shall now demonstrate how Eq. (6–31) may be operated on and combined with different equations of state to express the *changes in enthalpy and entropy* that occur when a gas goes from an initial reference state P_0, T_0, to any final state P, T. (The *reference state* P_0, T_0, is one at which the pressure is so low that a real gas behaves ideally.) The final expressions obtained will depend on the equation of state selected.

Enthalpy Change, Any Gas: We begin by replacing M by $\bar{H}$ in Eq. (6–31), to obtain:

$$\Delta \bar{H} = \int_{P_0}^{P} \left(\frac{\partial \bar{H}}{\partial P}\right)_T dP + \int_{T_0}^{T} \left(\frac{\partial \bar{H}}{\partial T}\right)_P dT. \qquad (7\text{–}17)$$

But

$$\left(\frac{\partial \overline{H}}{\partial T}\right)_P = C_P. \tag{2–18}$$

Now using Jacobian notation and Eqs. (6–4) and (6–23), we get

$$\left(\frac{\partial \overline{H}}{\partial P}\right)_T = \frac{[\overline{H}, T]}{[P, T]} = \frac{T[\overline{S}, T] + \overline{V}[P, T]}{[P, T]}$$

or

$$\left(\frac{\partial \overline{H}}{\partial P}\right)_T = T\frac{[\overline{S}, T]}{[P, T]} + \overline{V}. \tag{7–18}$$

Substituting Eq. (6–24a) into Eq. (7–18), we may write

$$\left(\frac{\partial \overline{H}}{\partial P}\right)_T = \overline{V} - T\left(\frac{\partial \overline{V}}{\partial T}\right)_P. \tag{7–19}$$

We now insert Eqs. (2–18) and (7–19) into Eq. (7–17) and obtain:

$$\Delta \overline{H} = \int_{P_0}^{P} \left[\overline{V} - T\left(\frac{\partial \overline{V}}{\partial T}\right)_P\right] dP + \int_{T_0}^{T} C_P \, dT. \tag{7–20}$$

Equation (7–20) is a *general expression for enthalpy change* and applies to any thermodynamic system that does only expansion work.

Entropy Change, Any Gas: We derived the *general equation for entropy change* in Example 6–6:

$$\Delta \overline{S} = \int_{P_0}^{P} -\left(\frac{\partial \overline{V}}{\partial T}\right)_P dP + \int_{T_0}^{T} C_P \frac{dT}{T}. \tag{6–32}$$

7–10
Effect of different equations of state on property computation

In this section and Sections 7–11 and 7–12 we shall evaluate Eqs. (7–20) and (6–32) by three different equations of state: the Ideal-Gas Law (Eq. 2–26); van der Waals' equation (Eq. 7–2), and the Generalized Equation of State (Fig. 7–3), in order to demonstrate that the computed magnitude of a property change depends on the equation of state that is used.

Ideal gas: $P\overline{V} = RT.$

Enthalpy change (per mole of gas): The equation of state is used to determine

$(\partial \overline{V}/\partial T)_P$ in Eq. (7–20):

$$\overline{V} = \frac{RT}{P}.$$

Therefore

$$\left(\frac{\partial \overline{V}}{\partial T}\right)_P = \frac{R}{P} = \frac{\overline{V}}{T}. \tag{7–21}$$

On substituting Eq. (7–21) into Eq. (7–20), we obtain

$$\Delta \overline{H} = \int_{P_0}^{P} \left[\overline{V} - T\left(\frac{\overline{V}}{T}\right)\right] dP + \int_{T_0}^{T} C_P \, dT$$

or

$$\Delta \overline{H}_{\text{ideal}} = \int_{T_0}^{T} C_P \, dT \equiv \overline{H}^*, \tag{7–22}$$

which indicates that *the enthalpy of an ideal gas depends only on the temperature and is independent of pressure* (see Eq. 2–24a). The *asterisk* (*) is used to indicate ideal-gas properties. The state (P_0, T_0) is one in which all gases behave ideally. Therefore $\overline{H}^*$ is the enthalpy of the ideal gas at P and T, measured with reference to a state P_0, T_0 in which all gases are ideal.

Exercise 7–6. Calculate $\Delta \overline{H}$ for methane heated at 1 atm from 100°C to 300°C. Use C_P values from Table 2–3. Assume ideal-gas behavior.

Entropy change: When we substitute Eq. (7–21) into the general equation for entropy Eq. (6–32), we have

$$\Delta \overline{S} = -\int_{P_0}^{P} Rd \ln P + \int_{T_0}^{T} C_P \frac{dT}{T} \tag{7–23}$$

or

$$\Delta \overline{S}_{\text{ideal}} = R \ln \frac{P_0}{P} + \int_{T_0}^{T} C_P \frac{dT}{T} \equiv \overline{S}^*. \tag{7–24}$$

Again the asterisk (*) indicates an ideal-gas property. The entropy of an ideal gas depends on *both* temperature and pressure.

Exercise 7–7. Write an expression for $\Delta \overline{G}$ for an ideal gas corresponding to a change of state from P_0, T_0 to P, T_0.

Van der Waals gas: If we solve Eq. (7–2) for P, we obtain:

$$P = \frac{RT}{\overline{V} - b} - \frac{a}{\overline{V}^2}.$$ (7–24a)

Enthalpy change. The contribution of the change in temperature to the enthalpy change is given by the second integral on the right-hand side of Eq. (7–20), and is the same as for the ideal gas, Eq. (7–22).

We now use Eq. (7–24a) to express the enthalpy change resulting from a change in pressure [the first integral on the right-hand side of Eq. (7–20)]:

$$\int_{P_0}^{P} \left(\frac{\partial \overline{H}}{\partial P}\right)_T dP = \int_{P_0}^{P} \left[\overline{V} - T \left(\frac{\partial \overline{V}}{\partial T}\right)_P\right] dP.$$ (7–25)

It is convenient to change Eq. (7–25) to a form containing P and $(\partial P/\partial T)$ because Eq. (7–24a) is *explicit* in pressure, i.e., it has the form $P = P(V, T)$. We therefore recall the elementary rule of differentiation:

$$\overline{V} dP = d(P\overline{V}) - P\, d\overline{V},$$ (7–26)

which transforms the $V\, dP$ term on the right-hand side of Eq. (7–25).

Using Jacobians, it is evident that:

$$\left[\frac{P, \overline{V}}{P, \overline{V}}\right] \times \left[\frac{\overline{V}, T}{\overline{V}, T}\right] \times \left[\frac{T, P}{T, P}\right] = 1,$$ (7–27)

which may be rearranged to yield

$$\left[\frac{\overline{V}, P}{T, P}\right] = - \left[\frac{P, \overline{V}}{T, \overline{V}}\right] \times \left[\frac{\overline{V}, T}{P, T}\right],$$

from which it follows that, for an isothermal process (see Exercise 6–8),

$$\left(\frac{\partial \overline{V}}{\partial T}\right)_P dP = - \left(\frac{\partial P}{\partial T}\right)_{\overline{V}} d\overline{V}.$$ (7–28)

Substituting Eqs. (7–28) and (7–26) in Eq. (7–25), we obtain

$$\Delta \overline{H}_T = \int_{P_0}^{P} d(P\overline{V}) + \int_{\overline{V}_0}^{\overline{V}} \left[T \left(\frac{\partial P}{\partial T}\right)_{\overline{V}} - P\right] d\overline{V}.$$ (7–29)

By differentiating Eq. (7–24a), we obtain

$$\left(\frac{\partial P}{\partial T}\right)_{\overline{V}} = \frac{R}{\overline{V} - b}.$$ (7–30)

Replacing P with Eq. (7–24a) and substituting Eq. (7–30) in Eq. (7–29), we get

$$\Delta \bar{H}_T = \int_{V_0}^{V} d(P\bar{V}) + \frac{a}{\bar{V}^2} \, d\bar{V} = [P\bar{V} - P_0\bar{V}_0] + \left[-\frac{a}{\bar{V}} + \frac{a}{\bar{V}_0} \right].$$
$$(7\text{–}31)$$

If the reference state $(P_0, \bar{V}_0, T)$ is such that the gas behaves ideally, then P_0 must be very low $(P_0 \to 0)$ and the corresponding V_0 very large $(V_0 \to \infty)$. Also $P_0\bar{V}_0 = RT$. Therefore Eq. (7–29) integrates to

$$\Delta \bar{H}_T = P\bar{V} - RT - \frac{a}{\bar{V}}. \qquad (7\text{–}32)$$

When we substitute Eqs. (7–32) and (7–22) back into Eq. (7–20), we obtain the total change in enthalpy for a van der Waals gas:

$$\Delta \bar{H} = P\bar{V} - RT - \frac{a}{\bar{V}} + \bar{H}^*,$$

or (dropping the Δ sign),

$$(\bar{H}^* - \bar{H})_{\text{van der Waals}} = RT - P\bar{V} + \frac{a}{\bar{V}}. \qquad (7\text{–}33)$$

Equation (7–33) is called an ENTHALPY DEPARTURE function, since it shows the amount by which the enthalpy of a van der Waals gas departs from the enthalpy of an ideal gas. Note that the enthalpy of a van der Waals gas depends on *pressure*. The departure function (Eq. 7–33) is simply the effect of pressure on enthalpy.

Entropy change (van der Waals gas): On substituting Eq. (7–28) and (7–30) into Eq. (6–32), we obtain

$$\Delta \bar{S} = \int_{V_0}^{V} \frac{R \, d\bar{V}}{\bar{V} - b} + \int_{T_0}^{T} C_P \, d\ln T.$$

Integrating the first term on the right between $\bar{V}_0$ and $\bar{V}$,

$$\Delta \bar{S}_T = R \ln \frac{\bar{V} - b}{\bar{V}_0 - b}.$$

The total change in entropy is:

$$\Delta \bar{S} = R \ln \frac{\bar{V} - b}{\bar{V}_0 - b} + \int_{T_0}^{T} \frac{C_P}{T} \, dT. \qquad (7\text{–}34)$$

When we combine Eqs. (7–24) and (7–34), we find that the ENTROPY DEPARTURE of a van der Waals gas from the entropy of an ideal gas at the *same pressure and temperature* is:

$$(\bar{S}^* - \bar{S})_{\text{van der Waals}} = R \ln \frac{\bar{V}^*}{\bar{V} - b}, \qquad (7\text{--}35)$$

where $\bar{V}^* = P_0 V_0/P$, the volume which an ideal gas would have at the final state. $\bar{S}$ is the same as $\Delta \bar{S}$ in Eq. (7–34).

Exercise 7–8. Derive Eq. (7–35) from Eqs. (7–34), and (7–24).

Example 7–3. Compute $\Delta \bar{H}$ and $\Delta \bar{S}$ for 1 g-mol of methane (CH_4) initially at 1 atm and 300°K and finally at 30 atm and 200°K. The C_P of $CH_4 = 5.34 + 11.5 \times 10^{-3}T$ cal/g-mol · °K.* Assume (a) ideal-gas behavior, then (b) van der Waals gas behavior.

a) *Ideal gas:*

$$\Delta \bar{H} = \bar{H}^* = \int_{300°}^{200°} (5.34 + 0.0115T)\, dT \qquad (7\text{--}22)$$

$$= 5.34\,(200 - 300) + \frac{0.0115}{2}\,(200^2 - 300^2)$$

$$= -821\, \frac{\text{cal}}{\text{g-mol}}. \qquad \text{(Answer)}$$

$$\Delta \bar{S} = \bar{S}^* = R \ln \frac{1}{30} + \int_{300°}^{200°} (5.34 + 0.0115T)\, \frac{dT}{T}$$

$$= -1.98 \times 3.40 + 5.34 \ln \frac{200}{300} + 0.0115\,(200 - 300)$$

$$= -10.06\, \frac{\text{cal}}{\text{g-mol} \cdot °\text{K}}. \qquad \text{(Answer)}$$

b) *Van der Waals gas:* To evaluate the departure functions—Eqs. (7–33) and (7–35)—we must first find $\bar{V}$ from Eq. (7–2), at 30 atm and 200°K. The van der Waals constants are taken from Table 7–2:

$$a = 2.25 \times 10^6\, \frac{\text{atm} \cdot \text{cm}^6}{(\text{g-mol})^2},$$

$$b = 42.8\ \text{cm}^3/\text{g-mol}.$$

---

* K. Kelley, *U.S. Bureau of Mines Bulletin* **371** (1934).

We then find $\overline{V}$ by trial (iteration) using the van der Waals equation in the form of Eq. (7–8). The gas is assumed to behave ideally in the initial state. For the final state,

$$\overline{V}^* = 547 \text{ cm}^3/\text{g-mol} \qquad \text{(ideal gas)}$$

$$\overline{V} = 435 \text{ cm}^3/\text{g-mol} \qquad \text{(van der Waals gas)}$$

From Eq. (7–33), we obtain

$$(\overline{H}^* - \overline{H})_{\text{van der Waals}} = 1.98 \frac{\text{cal}}{\text{g-mol} \cdot {}^\circ\text{K}} \times 200{}^\circ\text{K} - 30 \text{ atm} \times 435 \frac{\text{cm}^3}{\text{g-mol}}$$

$$\times \frac{0.0242 \text{ cal}}{\text{atm} \cdot \text{cm}} + \frac{2.25 \times 10^6 \text{ atm} \cdot \text{cm}^6}{435 \text{ cm}^3 \text{ g-mol}} \times \frac{0.0242 \text{ cal}}{\text{atm} \cdot \text{cm}^3}$$

$$= (396 - 316 + 125.1) \text{ cal/g-mol}$$

$$= +205 \text{ cal/g-mol}. \qquad \text{(A sizable departure!)}$$

Therefore

$$\Delta\overline{H}_{\text{van der Waals}} = -821 - 205 = -1026 \text{ cal/g-mol}. \qquad \text{(Answer)}$$

From Eq. (7–35), we obtain

$$(\overline{S}^* - \overline{S})_{\text{van der Waals}} = 1.98 \frac{\text{cal}}{\text{g-mol}{}^\circ\text{K}} \ln \frac{\overline{V}^*}{435 - 43} = +0.66,$$

$$\overline{V}^* = RT/P = 547.$$

Therefore

$$\Delta\overline{S}_{\text{van der Waals}} = -10.06 - 0.66 = -10.72 \text{ cal/g-mol}{}^\circ\text{K}. \qquad \text{(Answer)}$$

Exercise 7–9. From Eq. (7–2), determine the $\overline{V}$ of methane (CH_4) at 300°K and 1 atm. Then evaluate the van der Waals enthalpy and entropy departures for this state of methane. How large an error in properties of a van der Waals gas is introduced by assuming, as we did in Example 7–3, that this state is ideal?

7–11
Departure functions

Figure 7–5 illustrates the general manner in which departure functions are used to calculate a change in the value of a thermodynamic property. The departure function for the initial state permits one to climb from the real-gas surface to the ideal-gas surface. The process is then carried out on the ideal-gas surface, and the departure function for the final state is then used to regain the real-gas surface. If

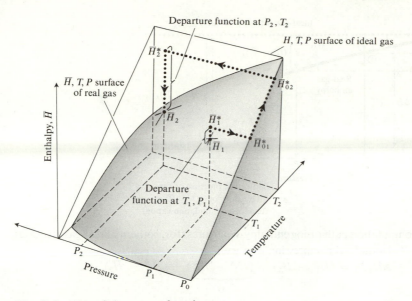

Fig. 7–5. Use of departure functions:

$$\bar{H}_2 - \bar{H}_1 = (\bar{H}_1^* - \bar{H}_1) + \underbrace{(\bar{H}_{01}^* - \bar{H}_1^*)}_{0} + (\bar{H}_{02}^* - \bar{H}_{01}^*) + \underbrace{(\bar{H}_2^* - \bar{H}_{02}^*)}_{0} - (\bar{H}_2^* - \bar{H}_2).$$

departure functions $(\bar{H}^* - \bar{H})$ are available at states (1) and (2), then the difference in enthalpy is

$$\Delta\bar{H}_{1-2} = (\bar{H}_1^* - \bar{H}_1) + (\bar{H}_{01}^* - \bar{H}_1^*) + (\bar{H}_{02}^* - \bar{H}_{01}^*) + (\bar{H}_2^* - \bar{H}_{02}^*) - (\bar{H}_2^* - \bar{H}_2)$$

$$(7\text{--}35a)$$

(Fig. 7–5 shows the meaning of each term). [Usually we know the specific heat as a function of temperature only for the ideal state of real gases, that is, $C_{P_0}(T)$. The computational path in Fig. 7–5 must therefore move the system from its initial pressure to its ideal-state pressure, P_0, T_1, and then from T_1 to T_2, instead of going directly from T_1 to T_2 at P_1.]

For the enthalpy property, $\bar{H}_{02}^* = \bar{H}_2^*$ and $\bar{H}_{01}^* = \bar{H}_1^*$, and Eq. (7–35a) becomes

$$\Delta\bar{H}_{1-2} = (\bar{H}_1^* - \bar{H}_1) + (\bar{H}_2^* - \bar{H}_1^*) - (\bar{H}_2^* - \bar{H}_2)$$

$$= (\bar{H}_1^* - \bar{H}_1) + \int_{T_1}^{T_2} C_{P_0}\, dT - (\bar{H}_2^* - \bar{H}_2)$$

$$= (\bar{H}\ \text{Departure at 1}) + (\Delta\bar{H}\ \text{Ideal gas}) - (\bar{H}\ \text{Departure at 2}).$$

$$(7\text{--}35b)$$

The ideal-gas surface and the real-gas surface become tangent at P_0, the very low reference-state pressure. The shape of the real-gas surface depends on the equation of state used to describe the real gas.

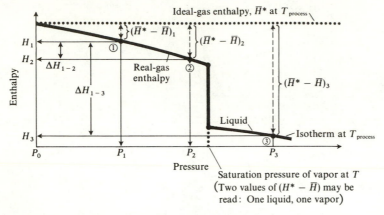

Fig. 7–6. Computation path (using enthalpy departures) for isothermal real-gas processes.
Process $1 \rightarrow 2$: Isothermal compression of a gas:
$$\Delta \bar{H}_{1-2} = (\bar{H}_1^* - H_1) + \underbrace{(\bar{H}_2^* - H_1^*)}_{0} - (\bar{H}_2^* - \bar{H}_2).$$
Process $1 \rightarrow 3$: Condensation of a gas to a liquid by isothermal compression:
$$\Delta \bar{H}_{1-3} = (\bar{H}_1^* - H_1) = (\bar{H}_3^* - H_3).$$

Figure 7–6 shows the computation path for the departure function for the change in enthalpy which takes place during an isothermal process involving a change in phase. Note that the properties of a condensed (*liquid*) gas can also be considered as departures from ideal-gas behavior.

The calculation of a change in entropy has the same form as Eq. (7–35a):

$$\Delta \bar{S}_{1-2} = (\bar{S}_1^* - \bar{S}_1) + (\bar{S}_{01}^* - \bar{S}_1^*) + (\bar{S}_{02}^* - \bar{S}_{01}^*) + (\bar{S}_2^* - \bar{S}_{02}^*) - (\bar{S}_2^* - \bar{S}_2)$$
$$(7\text{–}35c)$$

Since the S of an ideal gas depends on pressure, we must evaluate all the ideal entropy terms $(\bar{S}_i^* - \bar{S}_{0i}^*)$.

$$\bar{S}_{01}^* - \bar{S}_1^* = R \ln \frac{P_1}{P_0}$$

$$\bar{S}_{02}^* - \bar{S}_{01}^* = \int_{T_1}^{T_2} C_{P_0} \frac{dT}{T}$$

$$\bar{S}_2^* - \bar{S}_{02}^* = R \ln \frac{P_0}{P_2}$$

$$\Delta \bar{S}_{1-2} = (\bar{S}_1^* - \bar{S}_1) - (\bar{S}_2^* - \bar{S}_2) + R \ln \frac{P_1}{P_2} + \int_{T_1}^{T_2} C_{P_0} \frac{dT}{T} \,.$$
$$(7\text{–}35d)$$

Therefore

$$\Delta \bar{S}_{1-2} = (\bar{S} \text{ Departure at } 1) - (\bar{S} \text{ Departure at } 2) + (\Delta \bar{S} \text{ Ideal gas}). \quad (7\text{–}35\text{e})$$

In either calculation the net property change is the ideal-gas property change corrected by the difference between the initial and final departure functions.

7–12
Generalized equation of state: departure functions

Let us now demonstrate how departure functions are computed, using the generalized equation of state.

$$\frac{P\bar{V}}{RT} = z = z'(P_r, T_r, z_c). \quad (7\text{–}16)$$

Generalized enthalpy departure function: As noted previously, the enthalpy departure function is concerned with the effect of pressure on enthalpy:

$$\left(\frac{\partial \bar{H}}{\partial P}\right)_T = \bar{V} - T\left(\frac{\partial \bar{V}}{\partial T}\right)_P. \quad (7\text{–}19)$$

When

$$\bar{V} = z\frac{RT}{P}, \quad \text{(from 7–13)}$$

then, by differentiating,

$$\left(\frac{\partial \bar{V}}{\partial T}\right)_P = \frac{Rz}{P} + \frac{RT}{P}\left(\frac{\partial z}{\partial T}\right)_P. \quad (7\text{–}36)$$

Substituting Eqs. (7–13) and (7–36) in Eq. (7–19), we have

$$\left(\frac{\partial \bar{H}}{\partial P}\right)_T = -\frac{RT^2}{P}\left(\frac{\partial z}{\partial T}\right)_P. \quad (7\text{–}37)$$

In terms of reduced properties (note that $T = T_c \times T_r$; $dT = T_c\, dT_r$; $P = P_c \times P_r$; $dP = P_c\, dP_r$), Eq. (7–37) becomes

$$\left(\frac{\partial \bar{H}}{\partial P_r}\right)_{T_r} = -T_c\frac{RT_r^2}{P_r}\left(\frac{\partial z}{\partial T_r}\right). \quad (7\text{–}38)$$

Integrating at constant T_r, we obtain

$$\frac{\Delta \bar{H}_T}{T_c} = -RT_r^2 \int_{P_{0r}}^{P_r} \left(\frac{\partial z}{\partial T_r}\right)_{P_r} \frac{dP_r}{P_r} \quad (7\text{–}39)$$

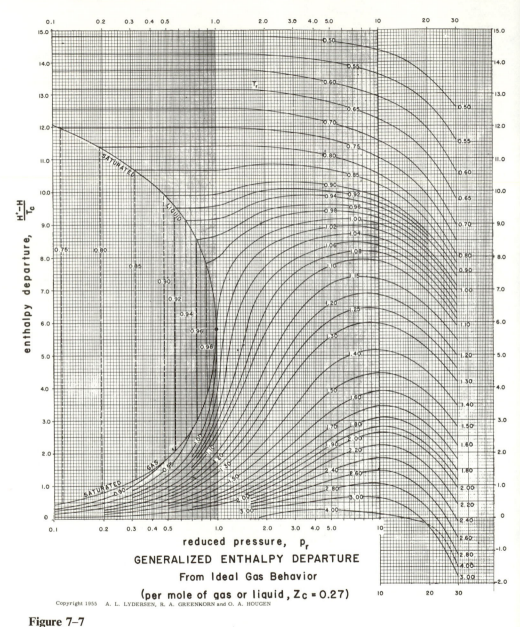

Figure 7–7

and expressing Eq. (7–39) as an enthalpy-departure function, we get

$$\left(\frac{\overline{H}^* - \overline{H}}{T_c}\right) = +RT_r^2 \int_{P_{0r}}^{P_r} \left(\frac{\partial z}{\partial T_r}\right)_{P_r} d \ln P_r. \qquad (7\text{–}40)$$

The right-hand side of Eq. (7–40) must be integrated numerically or graph-ically because $z = z(P_r, T_r, z_c)$ is a graphical function (Figs. 7–3 and 7–4). A brief outline of the procedure used is as follows.

Graphs or tables of z versus T_r, at fixed values of P_r, are prepared. Then $(\partial z/\partial T_r)_{P_r}$ values are measured graphically or obtained by numerical differentiation at fixed values of T_r. These are then plotted against $\ln P_r$ and the enthalpy departure obtained by graphical or numerical integration of the area under the $(\partial z/\partial T_r)_{P_r}$ versus $\ln P_r$ curve.

Lydersen, Greenkorn, and Hougen have performed these computations, with the results shown in Fig. 7–7 for $z_c = 0.27$. The figure shows departure from ideal-gas behavior of both *gas and liquid* phases. For more precise data on enthalpy departure, consult the original reference.

Example 7–4. Use the Lydersen *et al.*, generalized correlation (Fig. 7–7) to find the ΔH of 1 g-mol of methane (CH_4) moving through the process described in Example 7–3, given that: $T_c = 191.1°K$, $P_c = 45.8$ atm, and $\bar{V}_c = 99$ cm³/g-mol. The initial conditions are $P = 1$ atm and $T = 300°K$, and the final conditions are $P = 30$ atm and $T = 200°K$.

The reduced conditions are

$$T_r = \frac{200}{191.1} = 1.047 \quad \text{and} \quad P_r = \frac{30}{45.8} = 0.655.$$

From Fig. 7–7, we find the enthalpy departure,

$$\left(\frac{\bar{H}^* - \bar{H}}{T_c}\right) = 1.58 \text{ cal/g-mol} \cdot °K,$$

$$(\bar{H}^* - \bar{H}) = 1.58 \times 191.1° = 302 \text{ cal/g-mol}$$

(which is more than the van der Waals departure). Using the value of $\bar{H}^*$ from Example 7–3, we have

$$\Delta \bar{H} = -821 - 302 = -1123 \frac{\text{cal}}{\text{g-mol}}.$$

Also, from Fig. 7–3, we have $z = 0.78$, and therefore:

$$\bar{V} = \frac{zRT}{P} = \frac{0.78 \times 82.06 \times 200°}{30} = 427 \text{ cm}^3/\text{g-mol}.$$

(Compare this result with the van der Waals volume in Example 7–3.)

Exercise 7–10. Using the generalized equation of state enthalpy departures, calculate the enthalpy of methane (CH_4) at 50 atm and 300°K relative to its enthalpy at 1 atm and 300°K.

Exercise 7–11. Find $\Delta \bar{H}$ for methane which is heated from 300°K to 600°K at a constant pressure of 50 atm. Use the generalized equation of state enthalpy departures.

Exercise 7–12. From Fig. 7–7, determine the $\Delta \bar{H}_{vap}$ of CO_2 at 243.4°K.

Generalized entropy departure function: We can obtain an expression for the entropy departure of a real gas from the entropy of an ideal gas at the same temperature and pressure by subtracting Eq. (6–32) from Eq. (7–23):

$$(\bar{S}^* - \bar{S}) = \int_{P_0}^{P} \left[\left(\frac{\partial \bar{V}}{\partial T} \right)_P - \frac{R}{P} \right] dP. \tag{7–41}$$

Substituting Eq. (7–36) in Eq. (7–41), we obtain

$$(\bar{S}^* - \bar{S}) = -R \int_{P_0}^{P} \left[\frac{1 - z}{P} - \frac{T}{P} \left(\frac{\partial z}{\partial T} \right)_P \right] dP. \tag{7–42}$$

Replacing T and P with reduced properties ($T = T_r T_c$ and $P = P_r P_c$),

$$(\bar{S}^* - \bar{S}) = -\int_{P_{0r}}^{P_r} R \frac{1 - z}{P_r} dP_r + RT_r \left(\frac{\partial z}{\partial T_r} \right) d \ln P_r. \tag{7–43}$$

Recalling Eq. (7–40), we have

$$(\bar{S}^* - \bar{S}) = -R \int_{P_{0r}}^{P_r} (1 - z) d \ln P_r + \frac{1}{T_r} \left(\frac{\bar{H}^* - \bar{H}}{T_c} \right). \tag{7–44}$$

Figure 7–8 shows the Lydersen, *et al.*, solutions to this equation for $z_c = 0.27$. (The original reference contains information for values of z_c other than 0.27.)

The term $\int (1 - z) d \ln P_r$ is related to the *fugacity coefficient*, a term which will be described in Section 8.

Example 7–5. Find ΔS of 1 g-mol of methane for the process described in Example 7–4, by using the Lydersen, *et al.*, entropy-departure functions in Fig. 7–8.

Answer: From Example 7–4, for the final state, we have

$$T_r = 1.047, \qquad P_r = 0.655,$$

assuming that methane behaves ideally in the initial state. From Fig. 7–8, we find that

$$\bar{S}^* - \bar{S} = 1.2.$$

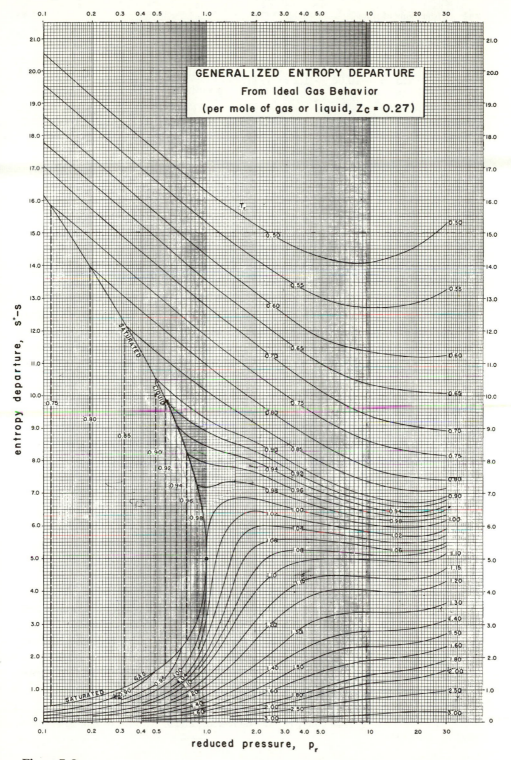

GENERALIZED ENTROPY DEPARTURE

From Ideal Gas Behavior

(per mole of gas or liquid, $Z_c = 0.27$)

entropy departure, $s^* - s$

reduced pressure, p_r

Figure 7–8

Using $\bar{S}^*$ from Example 7–3, we obtain

$$\Delta \bar{S} = -10.06 - 1.2 = -11.26 \frac{\text{cal}}{\text{g-mol} \cdot {}^\circ \text{K}} . \quad \text{(Answer)}$$

Summary: When 1 g-mol of CH_4 goes from 1 atm and 300°K to 30 atm and 200°K:

	Ideal	van der Waals	Generalized	Chemical literature*
$\Delta \bar{H}$	−821	−1026	−1123	−946
$\Delta \bar{S}$	−10.06	−10.72	−11.26	−10.24
$\bar{V}$	547	435	427	439

* Mathews and Hurd, *Trans. Amer. Inst. Chem. Eng.* **42**, 55 (1946).

The table illustrates the variation of property magnitudes that may occur when you use different equations of state. The values of the generalized compressibility factor are usually more accurate than the values of the generalized departure functions, which are obtained by graphically integrating measured differences. We would expect that methane, with a $z_c = 0.29$, would show deviations from experimental values, because the generalized equation of state correlations used (Figs. 7–3, 7–7, and 7–8) are based on $z_c = 0.27$. Correction factors for $z_c = 0.29$ can be found in the previous Lydersen, *et al.*, reference.

An analytical form of the generalized equation of state has been developed by J. O. Hirschfelder, R. J. Buehler, H. A. McGee, Jr., and J. R. Sutton.† With this equation it is possible to use a digital computer to calculate the generalized thermodynamic functions.

7–13
The need for experimental data

The various real-gas equations of state are not a substitute for experimental measurements of simultaneous values of P, $\bar{V}$, and T. Experimental measurements are needed to determine the empirical constants of the real-gas equations of state, and the equations of state in turn make it possible for us to interpolate and extrapolate experimental data.

Once the equation of state has been established, it is then possible to compute other thermodynamic properties as outlined here.

† J. O. Hirschfelder, R. J. Buehler, H. A. McGee, Jr., and J. R. Sutton, *Ind. Eng. Chem.* **50**, 375, 386 (1958).

7–14
Presentation of thermodynamic data

The practicing engineer is usually spared the task of computing the thermodynamic properties of common pure substances because the computations have been made and the results recorded in the literature. Thermodynamic properties may be recorded either graphically or in tabular form.

The best-known and most widely used tabulation is the *steam table*, which compiles the thermodynamic properties of steam. The importance of this compilation is reflected by the fact that it is kept under the surveillance of an international commission whose duty it is to extend and improve it. Abridged steam tables appear in Appendix II.

Diagrams of thermodynamic properties are not as accurate as experimental tabulations, but for some applications they offer the advantage of conciseness and convenience. The diagrams plot pressure, temperature, specific volume, enthalpy, and entropy. Steam diagrams also contain quality lines. (Recall that *quality* is the number of pounds of dry steam per pound of dry steam plus suspended liquid water.)

Diagrams with enthalpy–entropy coordinates are called MOLLIER DIAGRAMS after their originator, Richard Mollier. Figure 7–9 is the Mollier diagram for steam. Vertical displacements on the diagram represent isentropic processes; whereas horizontal displacements represent constant-enthalpy processes.

Figure 7–10 is the pressure–enthalpy diagram for Freon-12.* This type of plot is commonly used for refrigerants and is convenient when it comes to analyzing refrigeration cycles.

Figures 7–11, 7–12, and 7–13 are simplified enthalpy–entropy, temperature–entropy, and pressure–entropy diagrams which show how properties other than the coordinate properties appear in each type of diagram. Note that constant P and T lines coincide in the vapor–liquid regions.

An extensive discussion of practical techniques for charting and correlating thermodynamic properties of hydrocarbons may be found in *Applied Hydrocarbon Thermodynamics*, by W. C. Edmister (Gulf Publishing Co., Houston, 1961). Also, R. C. Reid and T. K. Sherwood, in their book, *Properties of Gases and Liquids* (McGraw-Hill, New York, 1958) critically review existing methods of correlating and estimating physical properties.

7–15
Thermodynamic properties of liquids and solids

The thermodynamic properties of liquids and solids are more difficult to predict or correlate than those of gases. The reason is that no comprehensive theories of the liquid and solid states are available, which in turn reflects the fact that the liquid and solid states are more complex than the gaseous state.

---

* Freon-12 is a Dupont Company trade name for a halogenated hydrocarbon refrigerant.

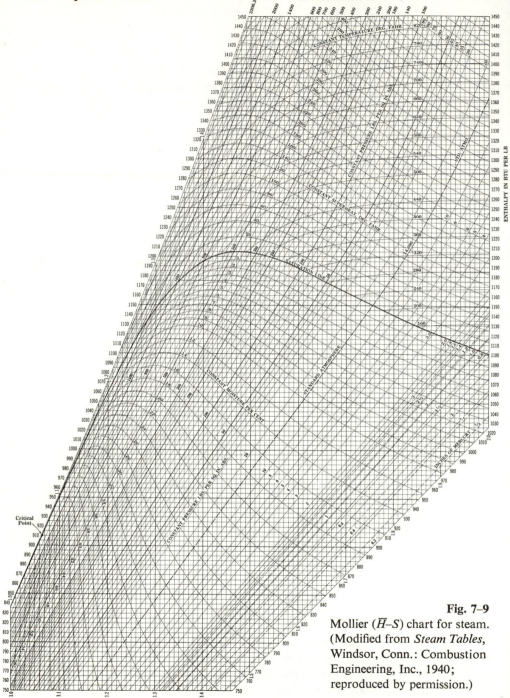

Fig. 7–9
Mollier ($\overline{H}$–S) chart for steam.
(Modified from *Steam Tables*,
Windsor, Conn.: Combustion
Engineering, Inc., 1940;
reproduced by permission.)

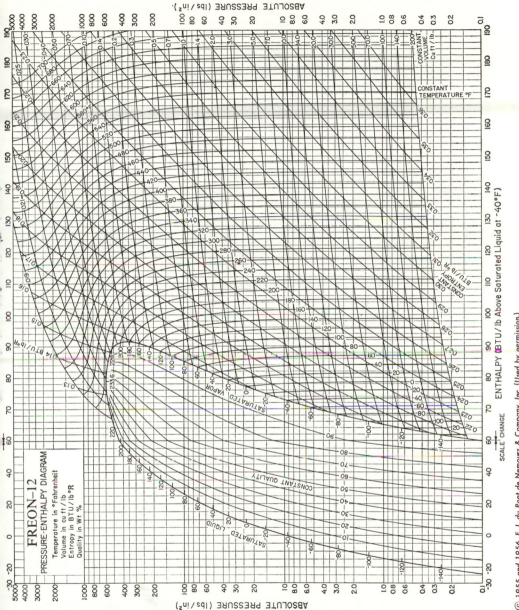

Fig. 7–10. Pressure–enthalpy diagram for Freon-11 (dichlorodifluoromethane). Temperature in °F, volume in ft³/lb, and entropy in Btu/lb · °R.

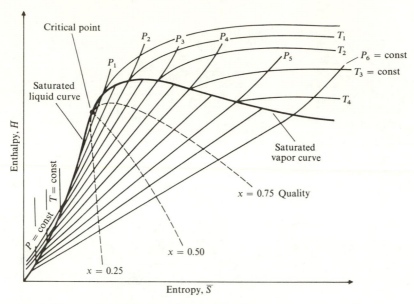

Fig. 7–11. Simplified Mollier diagram (enthalpy versus entropy).

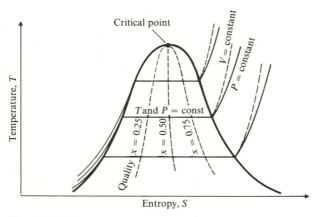

Fig. 7–12. Temperature–entropy diagram.

Compilations of experimental and computed thermodynamic data for liquids and solids appear in:

International Critical Tables, McGraw-Hill, New York, 1926–1933 (seven volumes)

ASHRAE Guide and Data Book, Amer. Soc. Heating, Refrig., and Air Cond. Eng., New York, 1967

J. H. Perry (editor), *Chemical Engineers' Handbook*, fourth edition, McGraw-Hill, New York, 1963

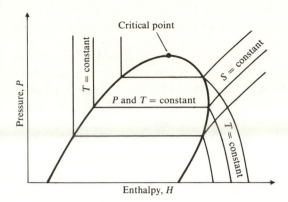

Fig. 7–13. Pressure–enthalpy diagram.

N. A. Lange, *Handbook of Chemistry*, tenth edition, McGraw-Hill, New York, 1961

"Selected Values of Properties of Hydrocarbons and Related Compounds," Amer. Petrol. Inst. Research Project 44, Thermodynamics Research Center, Texas A. and M. University, College Station, Texas

Handbook of Chemistry and Physics, Chemical Rubber Company, forty-third edition, Cleveland, Ohio, 1961–1962

Data Bank, National Bureau of Standards, Washington, D.C.

Methods are available for estimating the properties of liquids, and in the absence of experimental data, these methods are useful. The previously cited generalized property charts of Lydersen, Greenkorn, and Watson apply to the liquid phase at and near saturation conditions; in addition liquid densities have been correlated as functions of P_r, T_r, and z_c. Other methods of estimating thermodynamic properties of liquids are described in the following published material:

J. M. Brown, University of Wisconsin Department of Chemical Engineering, Special Project Report, Madison, Wis., June 1953

B. C. Sakaides and J. C. Coates, *J. Amer. Inst. Chem. Eng.* **288** (1956)

R. R. Wenner, *Thermochemical Calculations*, McGraw-Hill, New York (1941)

W. C. Edmister, *ibid.*

R. C. Reid and T. K. Sherwood, *ibid.*

J. F. Fallon and K. M. Watson, *Nat. Petrol. News*, tech. section, June 7, 1944

O. A. Hougen, K. M. Watson, and R. A. Ragatz, *Chemical Process Principles*, Part I, pages 266–283, second edition, John Wiley, New York (1938)

"Physical Property Estimation System," Amer. Inst. Chem. Eng., New York (computerized system of estimating physical properties of compounds and mixtures).

The Clapeyron equation (Eqs. 6–19 and 6–19b) and Trouton's Rule (Eq. 3–20a) have already been cited as being useful in estimating vapor pressures and enthalpies of phase change.

For simple crystalline solids, C_V can be predicted by the methods of statistical thermodynamics. An approximate empirical relationship called *Kopp's Law* may be used to estimate the molar specific heats of solids which are near room temperature. This law states that the molar specific heat of a solid equals the sum of the atomic specific heats of all its elements. The specific heat of all atoms equals 6.2 cal/g-mol · °K, except for the lighter atoms, whose specific heats are taken to be: carbon, 1.8; hydrogen, 2.3; boron, 2.7; silicon, 3.8; oxygen, 4.0; fluorine, 5.0; phosphorus, 5.4; sulfur, 5.4.

Exercise 7–13

i) Normal hexane (C_6H_{14}, molecular weight 86.2) boils at 68.7°C, at which temperature $\Delta \bar{H}_{vap} = 79.3$ cal/g. How well does hexane obey Trouton's Rule?

ii) What $\Delta \bar{H}_{vap}$ does Trouton's Rule predict for water? How does this compare with the $\Delta \bar{H}_{vap}$ value given in the steam tables?

iii) Compare the Kopp's-Rule estimate for the C_P of ordinary salt (NaCl, molecular weight = 58.5) with the experimental value of 0.207 cal/g · °K.

Exercise 7–14. A *throttling calorimeter* is a device for measuring the quality of wet saturated steam. Steam samples drawn from a steam line are throttled adiabatically in the calorimeter, from line pressure to atmospheric pressure. The temperature of the throttled steam then fixes its thermodynamic state, and given the characteristic of adiabatic throttling processes (Example 2–23), it also fixes the state of the steam in the line if the pressure in the line is known.

A throttling calorimeter on a 500-psia steam line indicates an exhaust temperature of 270°F. What is the quality and temperature of the steam in the line? Use Fig. 7–9 to answer these questions.

Exercise 7–15. Draw the refrigeration cycle of Fig. 4–10 on Fig. 7–10, given that point (1) is saturated Freon-12 vapor (dry) at 0°F; point (2) is at 160 psia; and point (3) is saturated liquid.

What is the evaporator pressure and the condenser temperature?

Estimate the ratio of liquid to vapor at point (4).

Problems

1. Compute the change in free energy of ammonia, NH_3, which takes place when NH_3 goes from 1 to 1000 psia at 800°F, assuming that it follows the van der Waals equation of state. Repeat the computation, using the Berthelot equation of state.

2. Determine $\Delta \bar{G}$ and $\Delta \bar{U}$ for the ammonia in the above problem, using the charts of the generalized departure functions.

3. Derive the equations for the departure functions for internal energy and free energy for the generalized equation of state.

4. One g-mol of sulfur dioxide, initially at 72°C and 23.3 atm, goes through a process which increases its pressure and temperature to 373°C and 156 atm. Calculate the change in V, S, H, U, and G for the process, using the generalized equation of state. Use 273°K and 1 atm as an ideal reference state for the S and G properties.

How much work is needed for this process? Is the process reversible or irreversible?

5. In a certain thermodynamic state, the slope of an isobar in a Mollier diagram is found to be 800 units of slope. Find the temperature. Use the methods of Section 6 to relate slope to temperature.

6. Show that the van der Waals equations of state may be written as:

$$\left(z + \frac{1}{z} \frac{27}{64} \frac{P_r}{T_r^2} \right) \left(1 - \frac{1}{8z} \frac{P_r}{T_r} \right) = 1.$$

Note that, in this equation, z appears as a function of P_r and T_r only. Comment.

7. Write a computer program to prepare a table of simultaneous values of z, P_r, and T_r which would satisfy the van der Waals equation as given above. Plot z as a function of P_r at various values of T_r, and compare with Figs. 7–2 and 7–3.

8. Derive the entropy departure function for a gas that follows the Redlich-Kwong equation of state:

$$P = \frac{RT}{\overline{V} - b} - \frac{a}{T^{1/2}(\overline{V}^2 + \overline{V}b)},$$

where a and b are constants.

9. Estimate the value of the constants a and b in the Redlich-Kwong equation for methane, using data on critical properties.

10. The vapor pressure of many liquids is well represented by the equation log $P = A - (B/T)$. The boiling point of liquid oxygen at 1 atm is -182.8°C and that of liquid nitrogen is -118.6°C. Using the data in Table 7–3 and assuming that the principle of corresponding states applies to the equations for vapor pressure, so that log $P_r = A_r - (B_r/T_r)$, determine the constants A_{oxygen}, B_{oxygen}, $A_{nitrogen}$, and $B_{nitrogen}$ for the original equation.

11. Experimental measurements of the coefficient of thermal expansion β_T (Eq. 6–8) and coefficient of compressibility κ_T (Eq. 6–9) of a certain gas yield the following equations:

$$\beta_T = \frac{1}{\overline{V}} \left[\frac{R}{P} + bf(T) \right], \qquad \kappa_T = -\frac{1}{\overline{V}} \left[\frac{RT}{P^2} \right],$$

where b is a constant and $f(T)$ is a function of temperature only.

a) Show that the equation of state $P\overline{V} = RT + (bP/T^2)$ satisfies the above experimental coefficients and that $f(T)$ in the equation for β_T is $-2/T^3$.

b) Are there other equations of state that will satisfy the coefficients?

12. Given that, for a certain gas,

$$\overline{U} = a + bP\overline{V} + fP,$$

where a, b, and f are known constants, write an expression for the work needed to compress this gas isentropically from initial state P_1, $\overline{V}_1$, to final state P_2, $\overline{V}_2$, using only known quantities.

13. a) Given the following data, and using charts of generalized departure functions, estimate the change in enthalpy accompanying a rise in the temperature of methyl amine (CH_3NH_2) from 450°K to 700°K at 80 atm pressure:

$$T_c = 430°K \qquad P_c = 73.1 \text{ atm,} \qquad \overline{V}_c = 0.207 \text{ liters/g-mol,} \qquad z_c = 0.294$$

$$C_P \text{ (vapor)} = 2.996 + 3.61 \times 10^{-2}T - 1.64 \times 10^{-5}T^2 + 2.95 \times 10^{-9}T^3.$$

b) Evaluate the entropy change of CH_3NH_2, given that the initial state is as above, but the final state is 450°K and 30 atm.

14. The properties of a certain gas may be approximated by the following equation:

$$P(\overline{V} - b) = RT,$$

where

$$R = 0.37 \text{ Btu/lb} \cdot °R, \qquad b = 0.22 \text{ ft}^3/\text{lb,} \qquad C_P = 1.33 \text{ Btu/lb} \cdot °R,$$

and

$$S = 40.00 \text{ Btu/lb} \cdot °R \text{ at } 40°F, 10 \text{ lb/in}^2.$$

a) Calculate $\Delta\overline{V}$, $\Delta\overline{H}$, $\Delta\overline{S}$, $\Delta\overline{G}$, $\Delta\overline{A}$, $\overline{Q}_R$ and $\overline{W}_R$ for the isothermal compression of the gas from 10 to 100 lb/in² at a constant temperature of 40°F.

b) Calculate the same quantities as in part (a) for the adiabatic compression of the gas from 10 to 100 lb/in² at a constant temperature of 40°F.

15. a) Compare the values of critical constants P_c, $\overline{V}_c$, T_c in terms of a, b obtained from

    i) the van der Waals equation of state (Eq. 7–2),
    ii) the Dieterici equation of state (Eq. 7–6).

b) Show that the reduced Dieterici equation of state is

$$P_r = \frac{T_r}{2\overline{V}_r - 1} e^{(2 - 2/T_r\overline{V}_r)}.$$

16. Construct a temperature–entropy diagram for water, using property magnitudes computed from the chart of generalized departure functions (Fig. 7–8). Draw the diagram so that it shows the vapor–liquid saturation curve; the critical point; four isobars at 25, 40, 100, and 218.2 atm; and a 50% quality line. Let the isobars extend through the vapor–liquid region, and let the diagram cover the temperature range from 250° to 1000° Celsius. The reference state at which the entropy is arbitrarily taken to be zero is: saturated liquid water at 1 atm and 100°C. The $\Delta\overline{H}_{vap}$ of water at the reference state is 9713 cal/g-mol. Data on the vapor pressure of water may be taken if needed from Appendix II. In the diagram, use units of cal/g-mol · °C and atmospheres.

Discussion of Computation: Refer to Eqs. (7–35d) and (7–35e).

To construct the diagram, you must repeatedly evaluate ΔS_{1-2} for a sufficient number of states to make it possible to draw smooth curves for the saturation curve and the isobars.

    State (1) is always the reference state (liquid water at 100°C), whereas state (2) moves from point to point along the saturation curve and the isobars. It will prove convenient to do the computation in the tabular form suggested by Tables A and B.

S departure at 1: State (1) is saturated liquid at 1 atm; $P_r = 0.005$; $T_r = 0.422$. At this low pressure the gas phase behaves like an ideal gas: $(S^* - S_{vap})_1 = 0$. However, the liquid phase, by virtue of its being liquid, departs from ideal-gas behavior by an amount equal to $\Delta \bar{S}_{vap}$. Therefore

$$(\bar{S}_1^* - \bar{S}_1) = \frac{9713}{373} = 26.2 \text{ cal/g-mol} \cdot {}^\circ\text{C}.$$

S departure at 2: Along the saturation curve, vapor and liquid phases coexist at the same P and T. Separate departure functions must be determined for each phase. In the table below, the values marked T_{sat} are the saturation temperatures corresponding to P_2.

Table A

Saturation Curve

P_2, atm	25	40	100	160	200	218.2
P_r		0.183				1.00
$(S^* - S_{liq})_2$		14.17†				5.00
$(S^* - S_{vap})_2$		0.69†				5.00
T_{sat}		524°K				647.30

†Values read from intersection of P_r with saturation curve in Fig. 7–8.

Table B

Isobars ($\bar{S}_2^* - \bar{S}_2$)

Temp. state (2)			$P_2 =$	25	40	100	218.2 atm
°C	°K	T_r	$P_r =$	0.115	0.183	0.458	1.000
250	523	0.809			14.2 (liq)		
300					0.56 (vap)		
374.3					0.50 (vap)		
500					0.17 (vap)		
600							
800					Negligible		
1000					Negligible		

$$\Delta S \text{ ideal gas:} \qquad \Delta S_{ideal} = -R \ln P_2 + \int_{273^\circ K}^{T_2} (C_{P_0}/T) \, dT.$$

Evaluate ΔS_{ideal} for all the (2) states and combine these with corresponding departure functions, as in Eq. (7–35e).

Construct the diagram on 10 in. × 16 in. graph paper. The equation for C_{P_0} is to be taken from Appendix I.

17. Construct an entropy–enthalpy (Mollier) diagram for propane using data computed from the charts of generalized departure functions (Figs. 7–7 and 7–8). The diagram should contain the vapor–liquid saturation curve with critical point; also isobars at 10, 20, 40, 42.1, and 60 atm; and isotherms at 40°, 80°, T_c, and 130°C. Use the C_{P_0} equation from Appendix I. The reference state for zero entropy and enthalpy is: Saturated liquid at 0°C.

Data on ΔH_{vap} and the vapor pressure of propane may be found in the *Chemical Engineers' Handbook*, edited by J. H. Perry, or the *ASHRAE Guide and Data Book*.

18. Use generalized departure functions to:

a) Compute the final temperature, the change in enthalpy, and the change in entropy for the throttling of ethylene from 200 atm and 300°K to 10 atm.

b) Repeat part (a) for an adiabatic reversible compression process.

19. Estimate the specific volume of steam in cubic feet per pound at 1000°F and 1000 pounds per square inch, by

a) Edmister's method [*Ind. Eng. Chem.* **30**, 352 (1938)],

b) Pitzer's method [*J. Am. Chem. Soc.* **77**, 3433, (1955)].

Compare the values with those taken from the steam table.

GILBERT NEWTON LEWIS, 1875–1946

Section 8
FUGACITY AND ACTIVITY

The ease with which the Ideal-Gas Law lends itself to the calculation of thermo-dynamic properties serves as the rationale for introducing two additional thermodynamic functions called FUGACITY and ACTIVITY. These functions may be thought of as the *idealized pressure* and the *idealized concentration* of materials that are not ideal gases. They are particularly useful in calculating the free energy and the equilibrium in multicomponent systems of real materials.

8–1
Fugacity

The change in free energy of an isothermal process in an expansion-work system is

$$\Delta \bar{G}_T = \int_{P_1}^{P_2} \bar{V} \, dP. \qquad \text{(from (3–29a) or (6–7))}$$

Integration of Eq. (3–29a) requires a correlation between V and P, and is quite simple if the Ideal-Gas Law applies:

$$\Delta \bar{G}_{T, \text{ ideal}} = \int_{P_1}^{P_2} \bar{V} \, dP = \int_{P_1}^{P_2} \frac{RT}{P} \, dP = RT \ln \frac{P_2}{P_1}. \qquad (8\text{–}1)$$

Nonideal gases require more complex computations. For example, the iso-thermal free-energy change of 1 mol of a van der Waals gas is

$$\Delta \bar{G}_{T, \text{ van der Waals}} = \int_{V_1}^{V_2} \left(\frac{2a}{\bar{V}^2} - \frac{RT\bar{V}}{(\bar{V} - b)^2} \right) d\bar{V}. \qquad (8\text{–}2)$$

Exercise 8–1
a) Derive Eq. (8–2).
b) Express, in a form analogous to Eq. (8–2), the change in free energy for the isothermal compression of a Benedict-Webb-Rubin gas (Eq. 7–11) from $\bar{V}_1$ to $\bar{V}_2$.

G. N. Lewis* suggested that a function, which he called FUGACITY, f, be defined in a manner that would make it possible to express the change in isothermal free energy for *any real gas* in the same form as Eq. (8–1). Let us therefore define fugacity by the following equation:

$$\boxed{\Delta \bar{G}_T \equiv RT \ln \frac{f_2}{f_1}.} \qquad (8\text{–}3)$$

---

* G. N. Lewis, *Proc. Am. Acad.* **37**, 49 (1901).

Combining Eqs. (3–29a) and (8–3), we obtain

$$\Delta \bar{G}_T = RT \int_{f_1}^{f_2} d \ln f = \int_{P_1}^{P_2} \bar{V} \, dP. \tag{8–4}$$

For an ideal gas,

$$\Delta \bar{G}_T = RT \int d \ln P = RT \int d \ln f$$

or

$$f_{\text{ideal}} = P_{\text{ideal}}. \tag{8–5}$$

It follows that the fugacity of a real gas approaches the pressure when the real gas approaches an ideal state (that is, $P \to 0$). Therefore, for any gas,

$$\boxed{\lim_{P \to 0} \frac{f}{P} = 1.} \tag{8–6}$$

If the change in free energy is evaluated *with reference to an ideal state, P_0*, then Eq. (8–4) becomes

$$\Delta \bar{G}_T = RT \ln \frac{f}{P_0}. \tag{8–7}$$

8–2
Fugacities computed from the generalized equation of state

If

$$\bar{V} = \frac{zRT}{P}, \tag{7–13}$$

then

$$\bar{V} \, dP = zRT \frac{dP}{P}. \tag{8–8}$$

Combining Eqs. (8–8) and (8–4),

$$d \ln f = \frac{z}{P} \, dP. \tag{8–9}$$

On integrating Eq. (8–9) between a reference *ideal* state, P_0, and P, we obtain

$$\ln \frac{f}{f_0} = \int_{P_0}^{P} \frac{z}{P} \, dP. \tag{8–10}$$

The right-hand integral in Eq. (8–10) cannot be evaluated as it stands because

$z/P \to \infty$ at the lower integration limit ($P_0 \to 0$). To overcome this difficulty, add a plus dP/P and a minus dP/P to the integral and rearrange:

$$\ln \frac{f}{f_0} = \int_{P_0}^{P} \frac{(z-1)}{P}\, dP + \ln P - \ln P_0 \tag{8–11}$$

or

$$\ln \frac{f}{P} = \int_{P_0}^{P} \frac{(z-1)}{P}\, dP + \ln \frac{f_0}{P_0}. \tag{8–12}$$

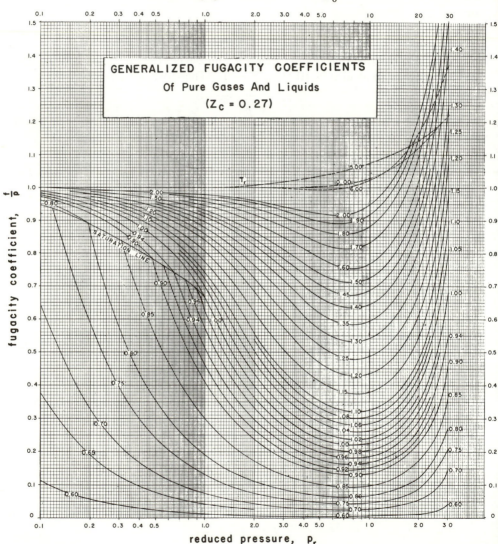

Figure 8–1

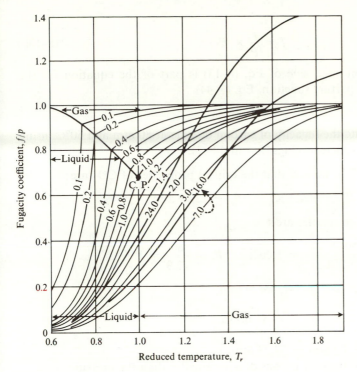

Fugacity coefficient, f/P

Reduced temperature, T_r

Fig. 8–2. Fugacity coefficients of gases and liquids; $z_c = 0.27.$*

Equation (8–6) applies to the right-hand term in Eq. (8–12). Therefore

$$\ln \frac{f}{P} = - \int_{P_0}^{P} \frac{(1-z)}{P}\, dP,$$

which may be written in reduced coordinates as

$$\ln \frac{f}{P} = - \int_{P_{0r}}^{P_r} \frac{(1-z)}{P_r}\, dP_r. \tag{8–13}$$

Equation (8–13) may be evaluated by graphical integration, using the data in Fig. 7–3. At the lower integration limit, as P approaches zero, $(1-z)/P$ approaches the initial slope of the Fig. 7–3 compressibility isotherm.

The integration has been performed by Lydersen, Greenkorn, and Hougen,* whose results are shown in Figs. 8–1 and 8–2 as graphs of f/P versus P_r and T_r.

The ratio f/P is called the FUGACITY RATIO or the FUGACITY COEFFICIENT. The fugacity of a real gas or vapor is the product of its absolute pressure and its fugacity

---

* O. A. Hougen, K. M. Watson, and R. A. Ragatz, *ibid.*

ratio:

$$f = \frac{f}{P} \times P. \tag{8–13a}$$

Note that the right-hand side of Eq. (8–13) is part of the equation for the generalized entropy departure function, Eq. (7–44).

Example 8–1. Compute the fugacity of carbon dioxide at 80 atm and 50°C, using generalized fugacity coefficients (Figs. 8–1 and 8–2).

Answer: From Table 7–3, we find that the critical properties of CO_2 are:

$$T_c = 304.2°K; \qquad P_c = 72.90 \text{ atm}.$$

Therefore the reduced properties are:

$$T_r = \frac{50 + 273}{304.2} = 1.062, \qquad P_r = \frac{80}{72.9} = 1.098.$$

Using these coordinates in Fig. 8–1, we obtain

$$\frac{f}{P} = 0.716.$$

Therefore $f = 0.716 \times 80 \text{ atm} = 57.28 \text{ atm}$.

In this example the fugacity is considerably smaller than the pressure.

Exercise 8–2. Using Eq. (8–7), determine the free energy of the gas in Example 8–1 relative to its ideal state (assumed to be at 1 atm and 50°C). Check your result by means of the generalized departure functions and Eq. (3–22). This check is a test of the internal consistency of Fig. 8–1 and Figs. 7–7 and 7–8.

8–3
Fugacity of liquids and solids

In Section 3, we showed that the movement of dm moles of material from one equilibrium phase into another causes no change in free energy (Example 3–6). This fact enables us to determine the fugacities of pure liquids and solids from the fugacities of their equilibrium vapors.

Given that ΔG for an equilibrium phase change is zero,

$$\Delta \bar{G}_{T,P} = 0 = RT \ln \frac{f_{\text{vapor}}}{f_{\text{liquid}}}, \tag{8–14}$$

it follows that

$$\boxed{f_{\text{vapor}} = f_{\text{liquid}}} \qquad \text{(at equilibrium)}. \tag{8–15}$$

Equation (8–15) may be extended to any number and variety of phases in equilibrium with each other:

$$f_1 = f_2 = f_3 = f_i, \tag{8–16}$$

where f_1, f_2, f_3, and f_i are fugacities of any one material in equilibrium phases 1, 2, 3, and i, respectively. Thus *the fugacity of a solid is equal to the fugacity of the vapor and/or liquid in equilibrium with the solid.** At low pressure, the vapor behaves as an ideal gas, and fugacities of liquids and solids are equal to their equilibrium vapor pressures:

$$f_i = P_i = \text{the vapor pressure of substance } i.$$

Exercise 8–3. Compute the fugacity of ice at $-10°C$, and of liquid water at $100°C$ and at $300°C$, and corresponding saturation pressures. Use appropriate handbooks for vapor pressure of ice. Consult the steam table in Appendix II for the vapor pressure of liquid water.

8–4
The effect of pressure on fugacity of liquids and solids

Solids and liquids are frequently considered at total pressures which are very much higher than their saturation vapor pressures. Does the increased pressure affect their fugacities? It does to the extent that it increases their vapor pressures, or more precisely to the extent that it increases the fugacity of their equilibrium vapors. (This is the *Poynting effect*, described in Section 3 in terms of ideal-gas pressures.)

Exercise 8–4. From suitable handbooks, find the vapor pressures of water, mercury, benzene, and copper at temperatures at or near room temperature. Do these materials normally occur at total pressures equal to their vapor pressures?

The magnitude of the pressure effect on fugacities of liquids or solids is small except at very high pressures. It may be evaluated from Eq. (8–4) because the isothermal compression of a solid or liquid produces a change in free energy ($\int \overline{V} \, dP$) which is (by definition of fugacity) related to the ratio of the fugacities in the low- and high-pressure state.

Thus, rearranging Eq. (8–4), we obtain

$$\int_{f_1}^{f_2} d \ln f = \frac{1}{RT} \int_{P_1}^{P_2} \overline{V} \, dP, \tag{8–17}$$

where f_1 is the fugacity at the saturation *vapor pressure* P_1 (corresponding to T);

---

* We here assume that the boundaries between equilibrium phases (interfaces) are flat and of constant dimension.

f_2 is the new fugacity at a *total pressure* P_2, higher than the equilibrium vapor pressure; and $\overline{V}$ is the molar volume of the liquid. Liquids and solids are relatively incompressible, so that we may treat $\overline{V}$, or $\overline{V}_{average}$, as a constant and integrate Eq. (8–17) to

$$\ln \frac{f_2}{f_1} = \frac{\overline{V}}{RT}(P_2 - P_1). \tag{8–18}$$

Compare Eq. (8–18) with Eq. (3–33).

Example 8–2. Calculate the fugacity of liquid water at 25°C and at its equilibrium vapor pressure. Compare with its fugacity at 25°C and 1 atm and at 25°C and 100 atm.

Answer: The vapor pressure of H_2O at 25°C = 23.76 mm Hg* = 0.0313 atm, and the critical and reduced properties are:

$$P_c = 218.2 \text{ atm (Table 7–3)}, \qquad T_c = 647.3°K, \qquad T_r = \frac{298}{647} = 0.461.$$

a)   At equilibrium vapor pressure,

$$P_r = \frac{0.0313}{218.2} = 1.434 \times 10^{-4}.$$

At this low value of P_r, the vapor is in an ideal state. Therefore, from Eq. (8–5),

$$f_{liq} \text{ (at } P_{H_2O}) = P_{H_2O} \text{ (at 25°C)} = 0.0313 \text{ atm},$$

which is the fugacity of liquid water and water vapor at 25°C and 0.0313 atm.

b)   If the total pressure on the system is increased to 1 atm:

$$\ln \frac{f_{1 \text{ atm}}}{0.0313} = \frac{\overline{V}}{RT}(1{-}0.0313),$$

and, assuming $\overline{V}_{liq}$ to be independent of pressure,

$$\overline{V}_{25°C} = 1 \text{ cm}^3/\text{g} = 18 \text{ cm}^3/\text{g-mol}.$$

Thus we obtain

$$\ln \frac{f_1}{0.0313} = \frac{18 \text{ cm}^3}{(82.06 \text{ cm}^3/\text{g-mol} \cdot °K) \cdot \text{atm} \times 298°K} (0.969 \text{ atm}) = 0.000713.$$

Therefore, when we remember that $\ln (1 + x) = x$, when x is small,

$$\frac{f_1}{0.0313} = 1.0007.$$

---

* *Chemical Engineers' Handbook*, edited by J. H. Perry, *ibid.*

This means that the fugacity of liquid water at 1 atm is essentially the same as the fugacity of liquid water at its equilibrium vapor pressure.

c) At 100 atm total pressure:

$$\ln \frac{f}{0.0313} = \frac{18}{82.06 \times 298} (99.969) = 0.07355 \quad \text{and} \quad \frac{f}{0.0313} = 1.077.$$

Therefore

$$f_{100} = 0.0337 \text{ atm.} \quad \text{(Answer)}$$

Note that, although the total pressure has increased 3000-fold, the fugacity of the liquid water has increased less than 8%.

Exercise 8–5

i)   Calculate the fugacity of water at 25°C and 1000 atm.
ii) Calculate the fugacity of liquid water in the absence of water vapor, at 300°C and 1000 atm. Compare with Exercise 8–3.
iii) Calculate the fugacity of water vapor in an air space at 300°C and 1000 atm that is in equilibrium with liquid water at 300°C and 1000 atm assuming that no air dissolves in the liquid water.

8–5
Activity

The ACTIVITY a is defined as a ratio of fugacities,

$$a \equiv \frac{f}{f_0}, \tag{8–19}$$

where f is the fugacity of the state under consideration and f_0 refers to an ideal or reference state. Substituting Eq. (8–19) into Eq. (8–3), we obtain

$$\Delta \bar{G}_T = RT \ln a, \tag{8–20}$$

where the change in free energy is measured between the state under consideration and the standard reference state (subscript 0).

Exercise 8–6. What is the activity of any material in its reference state? What is the activity of an ideal gas at 0.8 atm relative to a 1-atm reference state?

8-6
Activity of liquids and solids

For incompressible solids or liquids, Eq. (8-4) becomes

$$\Delta \bar{G}_T = \bar{V}(P - P_0). \tag{8-21}$$

Combining Eq. (8-20) with Eq. (8-21), we obtain

$$\ln a = \frac{\bar{V}}{RT}(P - P_0), \tag{8-22}$$

where P_0 is the pressure at an ideal or reference state. Compare Eq. (8-22) with Eq. (8-18).

Equation (8-22) enables us to calculate solid or liquid activities (relative to a reference state) *without* recourse to data on vapor pressure, if $\bar{V}$ remains reasonably constant. For most solids and liquids $\bar{V}$ is reasonably constant up to pressures of several hundred atmospheres.

Example 8-3. Calculate the activity of pure iron (density $\rho = 7.8$ g/cm^3; molecular weight $= 55.85$ g/g-mol) at 1000 psia and 20°C and at 10,000 psia and 20°C, relative to a reference state at 1 atm.

Answer: When we combine Eq. (8-20) with Eq. (8-4), we obtain

$$RT \ln a = \int_{\text{reference}}^{P} \bar{V} \, dP. \tag{8-23}$$

Assuming $\bar{V}$ constant, we may write

$$\ln a = \frac{\bar{V}}{RT}(P - P_{\text{reference}})$$

and knowing that

$$\bar{V}_{25°C} = \frac{\text{cm}^3}{7.8 \text{ g}} \times \frac{55.85 \text{ g}}{\text{g-mol}},$$

we may then calculate the activity as

$$\ln a = \frac{\text{cm}^3 \times 55.85 \text{ g/g-mol} \times \Delta P}{7.8 \text{ g} \times \dfrac{82.06 \text{ cm}^3 \cdot \text{atm}}{\text{g-mol°K}} \times 298°\text{K}}$$

$$= 2.98 \times \frac{10^{-4}}{\text{atm}} \times \Delta P.$$

a) At 1000 psia (= 68.1 atm), we have

$$\ln a = 2.98 \times 10^{-4} \times 68.1 = 0.02026,$$

$$a = 1.02. \qquad \text{(Answer)}$$

b) At 10,000 psia (= 681 atm), we obtain

$$\ln a = 2.98 \times 10^{-4} \times 681 = 0.2026,$$

$$a = 1.224. \qquad \text{(Answer)}$$

Exercise 8–7. Compare the activity of liquid water at 25°C and its equilibrium vapor pressure; at 25°C and 1 atm; and at 25°C and 100 atm, using 25°C and 1 atm as a reference state. See Example 8–2.

Exercise 8–8. What is the free energy of iron at 25°C and 10,000 psia relative to its free energy at a reference state of 1 atm and 25°C?

Activities are most often used for expressing effective concentration of *liquids* or *solids*. As seen from the illustrative examples, activities of *pure* liquids and solids are *unity* or very close thereto, except at extremely high pressures.

8–7
General dependence of fugacity and activity on pressure and temperature

From Eq. (6–7), for a mole of pure substance,

$$d\bar{G} = \bar{V}\, dP - \bar{S}\, dT. \qquad (6\text{–}7)$$

Dependence of Fugacity on Pressure

Combining Eq. (6–7) and the definition of fugacity (Eq. 8–3) at constant temperature (or taking the derivative of Eq. (8–4), we obtain

$$(d\bar{G})_T = RT\, d \ln f = \bar{V}\, dP. \qquad (8\text{–}24)$$

Now, dividing by dP, at constant T, we obtain an equation for the dependence of fugacity on pressure:

$$\boxed{RT \left(\frac{\partial \ln f}{\partial P} \right)_T = \bar{V}} = \left(\frac{\partial \bar{G}}{\partial P} \right)_T. \qquad (8\text{–}25)$$

Equation (8–25) also follows quite directly from Eqs. (6–12) and (8–4).

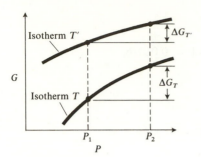

Fig. 8–3. The magnitude of an isothermal change in free energy, ΔG_T, depends on T.

Dependence of Fugacity on Temperature

Equation (8–3) can be differentiated, at constant pressure, with respect to temperature, to obtain an expression

$$\left(\frac{\partial \Delta \bar{G}_T}{\partial T}\right)_P = R \ln \frac{f_2}{f_1} + RT \left(\frac{\partial \ln f_2}{\partial T}\right)_P - RT \left(\frac{\partial \ln f_1}{\partial T}\right)_P \qquad (8\text{–}26)$$

for the manner in which $\Delta \bar{G}_T$ depends on the *temperature T of the isothermal process* that moves the system from state (1) at P_1 and T to state (2) at P_2 and T (Fig. 8–3). If state (1) is a reference state in which the system's vapor behaves ideally, that is, if $f_1 = f_0 = P_0$, then, dropping the subscript for state (2), $\Delta \bar{G}_T = (\bar{G} - \bar{G}_0)_T$, and Eq. (8–26) becomes

$$\left(\frac{\partial \Delta \bar{G}_T}{\partial T}\right)_P = \left(\frac{\partial \bar{G}}{\partial T}\right)_P - \left(\frac{\partial \bar{G}_0}{\partial T}\right)_P = R \ln \frac{f}{f_0} + RT \left(\frac{\partial \ln f}{\partial T}\right)_P. \qquad (8\text{–}27)$$

The last term in Eq. (8–26) is zero now, because

$$\left(\frac{\partial \ln f_0}{\partial T}\right)_P = \left(\frac{\partial \ln P_0}{\partial T}\right)_P = 0.$$

Substituting Eqs. (6–13) and (8–3) in (8–27), we get

$$\bar{S}_0 - \bar{S} = \frac{1}{T}(\bar{G} - \bar{G}_0) + RT \frac{(\partial \ln f)}{(\partial T)_P}, \qquad (8\text{–}28)$$

and using the definition of free energy (Eq. 3–22c) in the above,

$$\boxed{\left(\frac{\partial \ln f}{\partial T}\right)_P = \frac{\bar{H}_0 - \bar{H}}{RT^2}.} \qquad (8\text{–}29)$$

Note that $\bar{H}_0 - \bar{H}$ is the departure function for the enthalpy (the difference in the enthalpy of an ideal gas and a real material at the same temperature T).

We can obtain the analogous equations for activity by substituting the defining equation for activity, Eq. (8–19), in Eqs. (8–25) and (8–29):

$$RT \left(\frac{\partial \ln a}{\partial P}\right)_T = \left(\frac{\partial \bar{G}}{\partial P}\right)_T, \qquad (8\text{–}25a)$$

$$\left(\frac{\partial \ln a}{\partial T}\right)_P = \frac{\bar{H}_0 - \bar{H}}{RT^2}. \qquad (8\text{–}29a)$$

Exercise 8–9. Is Eq. (8–29) consistent with the behavior of a gas in an ideal state?

Problems

1. Steam is fed to a reversible isothermal turbine, in which it expands from 50 to 5 atm at 300°C. Estimate the maximum work delivered, using:
a) generalized fugacity coefficients,
b) van der Waals equation of state,
c) steam table.

2. Using generalized fugacity coefficients, compare the work output of the above isothermal turbine when it is supplied with (a) nitrogen (molecular weight 28), and (b) ammonia (molecular weight 17), at equal mass-flow rates and at the previous conditions. Repeat the comparison on the basis of equal volumetric feed rates at the turbine inlet.

3. Helium, water, and a small gold bar are contained in a piston-cylinder machine initially at 20°C and 1 atm. The contents of the cylinder are compressed isothermally to 2000 atm. Assuming mutual insolubility of the contents of the cylinder (except water vapor in He), compute the activities of the contents of the cylinder with reference to 1 atm.

4. Calculate the fugacities of steam, air, and ammonia at the conditions given in Table 7–1. Compare the magnitudes of the fugacity coefficients with those of the compressibility factors.

5. Liquid propane is being stored at 100°F and 100 psia. Estimate the fugacity of the liquid. The vapor pressure of liquid propane at 100°F is 190 psia, and its specific volume is 0.033 ft^3/lb_m.

6. Calculate the fugacity of benzene at 80.1°C and 100 atm. The following data are available. Benzene, C_6H_6: molecular weight, 78.11; normal boiling point, 80.1°C; density at boiling point, 0.801 g/cm^3; T_c, 288.5°C; and P_c, 47.7 atm. Repeat the calculation for benzene at 100 atm and 300°C.

7. From the following data for hydrogen at 144°R, calculate the fugacity of hydrogen at 144°R and 100 atm.

P, atm	H, Btu/lb	S, Btu/lb · °R
100	509	6.54
0.1	583	13.77

8. The fugacity coefficient f/P for propane at 100°C is higher at 1 atm than at 20 atm. Calculate $\Delta \bar{G}$ for 1 mol of propane compressed from 1 to 20 atm at 100°C. What is the minimum isothermal work required to compress the propane?

9. The volume of hydrogen at 28°C can be approximated within 1% by: $\bar{V} = RT\,(1/P) + 0.00064$, for pressure up to 1500 atm. Calculate the fugacity of hydrogen at 1200 atm and the maximum work obtainable by expanding the H_2 isothermally from 1200 atm to 1 atm.

10. Prove that the fugacity of a gas following the van der Waals equation of state, Eq. (7–2), is given by:

$$\ln f = \ln \left(P + \frac{a}{\bar{V}^2} \right) + \frac{2a}{\bar{V}} - \frac{RTb}{\bar{V} - b} .$$

11. Derive an expression for the fugacity or fugacity coefficient of a gas which follows the Redlich-Kwong equation of state (see Problem 8 at the end of Chapter 7).

12. Derive an expression in terms of P, $\bar{V}$, and T for the fugacity of a nonideal gas whose equation of state is

$$P\bar{V} = RT + bP^2\bar{V}.$$

13. The density ρ of solid magnesium at 25°C is 1.745 g/cm³. Calculate the activity of magnesium at 3000 psia and 25°C, relative to its activity at a reference state at 1 atm and 25°C. Assume that the magnesium is incompressible over this pressure range.

14. Determine the fugacity of isobutane at 150 psia and 190°F,

a) assuming that isobutane follows the ideal-gas law,
b) using generalized fugacity coefficients, Figs. 8–1 and 8–2,
c) using actual P-V-T data at 190°F [B. H. Sage, *et al.*, *Ind. Eng. Chem.* **30**, 673 (1938)].

Pressure, psia	Specific volume, ft³/lb$_m$
10	11.854
14.7	8.023
20	5.860
30	3.861
40	2.861
50	2.260
50	1.860
80	1.360
100	1.060
125	0.8183
150	0.6557
175	0.5391
200	0.4504

The critical properties of isobutane are: $P_c = 36.0$ atm, $T_c = 408°$K.

JOSIAH WILLARD GIBBS, 1839–1903

Section 9
THERMODYNAMICS OF MIXING AND
COMPOSITION CHANGE

Up to this point our thermodynamic analysis has been largely limited to systems of pure materials or of fixed composition. We shall now extend the analysis to systems *whose composition can change*. Let us first examine the thermodynamics of simple solutions and of mixing processes. The generalizations we derive will then permit us to treat phase equilibrium, separation processes, and chemical-reaction processes.

We begin by considering *solutions*. Solutions are uniform mixtures of materials which can form a single homogeneous phase over a range of compositions. Some common examples are mixtures of gases (such as air), salt dissolved in water, gasoline, glass, etc. Solutions can form homogeneous phases that are liquid, solid, or gaseous.

9-1
Ideal solutions

A mixture is said to be an IDEAL SOLUTION if it is *homogeneous* and if the *forces between differing molecules* of the mixture *are the same as the forces between like molecules*. A molecule in an ideal solution therefore interacts with its surroundings exactly as it would in the pure state, and contributes its $\overline{V}$, $\overline{U}$, and $\overline{H}$ to the properties of the mixture in proportion to its numbers (or its mole fraction) in the mixture.

Therefore the *molar volume of an ideal solution* is

$$\overline{V} = x_A\overline{V}_A + x_B\overline{V}_B + x_C\overline{V}_C + \cdots = \sum_i x_i\overline{V}_i, \qquad \text{(Ideal solution only)}$$

(9-1)

where subscripts A, B, C, and i refer to components of the solution, and

x_i = mole fraction of component i = $n_i/\sum_i n_i$,

$\overline{V}_i$ = molar volume of pure i at the T and P of the mixture,

$\overline{V}$ = volume of 1 mol of the mixture.

Similarly, the *molar enthalpy of an ideal solution* is

$$\overline{H} = x_A\overline{H}_A + x_B\overline{H}_B + x_C\overline{H}_C + \cdots = \sum_i x_i\overline{H}_i, \qquad \text{(Ideal solution only)}$$

(9-2)

where $\overline{H}_i$ is the molar enthalpy of pure i at the T and P of the solution.

The (accumulated) *internal energy of 1 mol of an ideal solution* is

$$\overline{U} = x_A\overline{U}_A + x_B\overline{U}_B + x_C\overline{U}_C + \cdots = \sum_i x_i\overline{U}_i, \qquad \text{(Ideal solution only)}$$

(9-3)

where $\overline{U}_i$ is the molar accumulated internal energy of pure component i at the T and P of the solution.

In an ideal solution which is a mixture of gases, each component gas interacts (collides) with the container walls in proportion to its mole fraction in the mixture, and therefore contributes to the total pressure in proportion to its mole fraction. Thus

$$\mathbb{P} = y_A \mathbb{P} + y_B \mathbb{P} + y_C \mathbb{P} + \cdots = p_A + p_B + p_C + \cdots = \sum_i p_i,$$

where

(9–4)

$$\boxed{p_i \equiv y_i \mathbb{P},}$$

(9–5)

y_i is the mole fraction of component i in the gas solution,* and $\mathbb{P}$ is the total pressure of a mixture of gases. p_i is called the PARTIAL PRESSURE of component i, and is defined by Eq. (9–5).

Note that ideal solutions need not consist of ideal gases, but may be mixtures of real gases or real liquids or real solids. The requisite for solution ideality is that intermolecular interaction be the same for both like and unlike molecules. The requirement for gas ideality is much more stringent: For a gas to be ideal, there must be *no* intermolecular interaction except for instantaneous elastic collision.

Equations (9–1) through (9–3) can be looked upon as the equations that define an ideal solution. Equations (9–4) and (9–5) apply only to gas solutions.

9–2
Ideal solutions of ideal gases: Dalton's and Amagat's laws

If the components of an ideal solution are all ideal gases, then each pure component exerts a pressure, $p_{\text{pure } i}$, which depends on the temperature and total volume, V, of the mixture and on the number of moles of that component, n_i in the mixture:

$$p_{\text{pure } i} = \frac{n_i RT}{V} = \frac{n_i RT}{(\sum_i n_i)\overline{V}} = y_i \frac{RT}{\overline{V}}.$$

Therefore

$$p_{\text{pure } i} = y_i \mathbb{P} \equiv p_i. \qquad \text{(Ideal gas only)}$$

(9–6)

Equation (9–6) states that the *partial* pressure of an ideal gas which is a component of an ideal solution of gases equals the pressure that the pure component would exert if it alone occupied the entire volume of the solution. The total pressure of the solution of ideal gas equals the sum of the pressures of the pure components:

$$\mathbb{P} = \sum_i p_{\text{pure } i}.$$

(9–6a)

This concept, which is known as *Dalton's law*, is illustrated in Fig. 9–1.

---

* Both x_i and y_i will be used to designate mole fraction (or number of moles of component i per mole of mixture). When y_i is used, it refers to compositions in a *gas* phase. x_i will be used to designate liquid-phase composition or, in the general case, the composition of a phase of unspecified form.

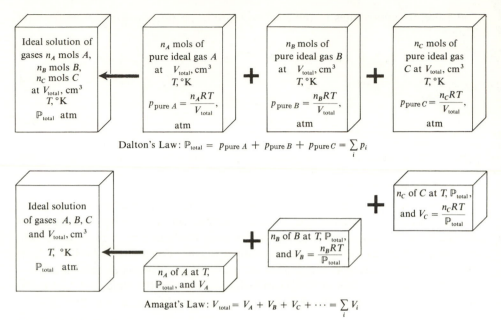

Fig. 9–1. Ideal solutions of ideal gases.

If a pure ideal-gas component, i, of an ideal solution or mixture of gases were held at the total pressure, $\mathbb{P}$, it would occupy a volume

$$V_i = n_i \frac{RT}{\mathbb{P}} = n_i \overline{V}_{\text{mixture}} = n_i \frac{V_{\text{mixture}}}{\sum_i n_i},$$

or

$$V_i = y_i V_{\text{mixture}}. \tag{9–7}$$

It follows that

$$V_{\text{mixture}} = \sum V_i. \tag{9–8}$$

Equation (9–8), which states that the *partial volumes*, V_i, of an ideal-gas mixture are additive, is known as *Amagat's law* (Fig. 9–1).

Example 9–1. Given a cubic foot of gas solution (air) in an ideal-gas state, at 70°F and 1 atm, determine: (i) the partial pressure, (ii) the partial volume, (iii) the volume percent, (iv) the weight percent, of each component gas, and (v) the average molecular weight (weight in pounds of one pound-mole), and (vi) the density. The chemical composition of air is as follows.

Air component	Mole % (= mole fraction × 100)
N_2	78.03
O_2	20.99
Argon	0.95
CO_2	0.03
	100.00

Answer: The problem is best solved in tabular form, as follows.

Component	Mole %	p_i, atm	V_i, ft^3	Molecular weight	y_i (MW)$_i$
N_2	78.03	0.7803	0.7803	28.02	21.87
O_2	20.99	0.2099	0.2099	32.00	6.72
Ar	0.95	0.0095	0.0095	39.94	0.38
CO_2	0.03	0.0003	0.0003	44.01	0.01
	100.00	1.0000	1.0000		28.98

i) Partial pressures are defined by Eq. (9–5). Hence

$$p_{N_2} = \mathbb{P}y_{N_2} = 0.7803 \times 1 \text{ atm} = 0.7803 \text{ atm}.$$

The column headed p_i records all the partial pressures.

ii) The V_i are found by applying Eq. (9–7).

iii) The volume percent is $V_i/V_{\text{mixture}} \times 100$. It is identical to the mole percent, because each mole of any ideal gas occupies the same volume at the same T and P.

iv) Weight percent is found by multiplying each pure component mole fraction by its molecular weight, summing the component weights, and dividing the individual component weight by the total weight. That is,

$$\text{Weight \% of component } i = \frac{y_i \times (\text{molecular weight of } i)}{\sum_i y_i \times (\text{molecular weight of } i)} \times 100,$$

Weight % of N_2

$$= \frac{(0.7803)(28)}{(0.7803)(28) + (0.2099)(32) + (0.0095)(39.94) + (0.0003)(44)} \times 100$$

$$= 75.5.$$

v) The average molecular weight of a gas mixture is the weight of one mole of the mixture, which is $\sum_i y_i \times (\text{molecular weight of } i) = 28.98$.

vi) Density ρ is mass per unit volume. Therefore

$$\rho = \left(\frac{n}{V}\right) \times \text{(average molecular weight)} = \left(\frac{P}{RT}\right) \times \text{(average molecular weight)}$$

$$= \frac{1\ \text{atm} \times 14.7\ \dfrac{\text{lb}_f/\text{in}^2}{\text{atm}} \times 28.98\ \dfrac{\text{lb}_m}{\text{lb-mol}}}{10.73\ \dfrac{(\text{lb}_f/\text{in}^2)(\text{ft}^3)}{\text{lb-mol} \cdot \,^{\circ}\text{R}} \times 530\,^{\circ}\text{R}} = 0.0746\ \frac{\text{lb}_m}{\text{ft}^3}\ .$$

Exercise 9–1. Calculate the partial pressure, partial volume, average molecular weight, and total weight of 10,000 ft^3 of flue gas (ideal) measured at 1 atm and 250°F and having the following composition.

	Mole percent
CO_2	16
N_2	76
CO	2
O_2	2
H_2O (vapor)	4

9–3
Ideal solutions of liquids: Raoult's Law

The partial pressure of components in a vapor phase that is in equilibrium with an ideal *liquid* solution (Fig. 9–2) is given by RAOULT'S LAW, in terms of the vapor pressure of the pure component and its mole fraction, x_i, in the liquid phase:

$$p_i = P_i x_i, \tag{9–9}$$

where p_i = partial pressure of *vapor* component i in equilibrium with a *liquid* solution containing x_i mole fraction of liquid i,

and P_i = the *vapor pressure* of pure liquid i at the temperature and pressure of the solution.*

---

* It is important to be clear on the distinction between the terms *partial pressure* (p_i) and *vapor pressure* (P_i). Partial pressure refers to that part of the total pressure of a *gas mixture* contributed by a particular gas or vapor. Vapor pressure is the equilibrium pressure exerted by vapors emanating from a liquid or solid, and depends strongly on temperature. Raoult's Law relates *vapor-phase composition* (or partial pressure) to *liquid-phase composition* of ideal solutions in equilibrium. This law states that a component's vapor-phase partial pressure equals its liquid-phase vapor pressure times its liquid-phase mole fraction.

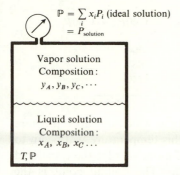

$$\mathbb{P} = \sum_i x_i P_i \ \text{(ideal solution)}$$
$$= P_{\text{solution}}$$

Vapor solution
Composition:
$y_A, y_B, y_C, \cdots$

Liquid solution
Composition:
$x_A, x_B, x_C \ldots$
$T, \mathbb{P}$

Fig. 9–2. Volatile liquid solution and its equilibrium vapor solution. Usually vapor composition differs from liquid composition ($x_A \neq y_A$). Liquid and vapor are at the same temperature and pressure. Composition of each phase does not change so long as temperature and total pressure remain constant.

Combining Eq. (9–5) with Eq. (9–9), we obtain

$$p_i = x_i P_i = y_i \mathbb{P}. \tag{9–10}$$

We may rearrange Eq. (9–10) to

$$\frac{y_i}{x_i} = \frac{P_i}{\mathbb{P}}, \tag{9–10a}$$

which shows that the *ratio* of the mole fraction of i in the vapor and liquid phases is a *constant* (at constant temperature) equal to the ratio of the vapor pressure of pure i (P_i) to the total pressure ($\mathbb{P}$).

The *vapor pressure* of an ideal solution is the sum of its component partial pressures, as given by Eq. (9–9). Thus

$$P_{\text{solution}} = \sum_i p_i = \sum_i x_i P_i. \tag{9–11}$$

Example 9–2. Benzene and toluene mix to form an ideal solution. Calculate the equilibrium vapor pressure of a solution of 40 mole % benzene and 60 mole % toluene at the following temperatures: 75°C, 100°C, 125°C. Calculate the boiling (or bubble) point of this solution at 1 atm total pressure, and the composition of the vapor phase at the boiling point.

Answer: The vapor pressure of an ideal solution is the sum of the partial pressures of its components (Eq. 9–11). The vapor pressures of the pure components benzene and toluene at the temperatures in question are listed below. (Pressures are given in mm of Hg.)

Temperature	$P_{benzene}$	$P_{toluene}$	$P_{solution} = 0.40P_b + 0.60P_t$
75°C	650	240	404
100°C	1350	540	864
125°C	2700	1100	1740

The bubble point of a solution is the temperature at which its vapor pressure equals atmospheric pressure. A solution heated to this temperature will begin to boil. To determine the boiling point of the given solution, we plot the calculated vapor pressures of the solution versus temperature. The boiling point is the temperature at which the curve intersects the 760 mm Hg line.

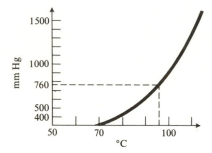

From the diagram, we see that the boiling point of the solution is 96°C. Checking at 96°C, we find that

$$p_b = 0.4 \times P_b = 0.4 \times 1180 = 472$$

$$p_t = 0.6 \times P_t = 0.6 \times 480 = 288$$

$$P_{solution} = 760 \text{ mm Hg}$$

The composition of the vapor is

$$y_b = \frac{p_b}{\mathbb{P}} = \frac{472}{760} = 0.62,$$

$$y_t = 1 - y_b = 1 - 0.62 = 0.38.$$

The vapor is richer in benzene than the liquid.

9–4
Entropy of mixing ideal solutions: statistical approach

The preparation of an ideal solution is accompanied by *no* anomolous changes in volume, by no heat effects, and by no chemical reactions. The volume and energy properties of the solution are simply the sum of the properties of its pure components. However, this simple additivity on mixing does not apply to entropy, and the entropy-related properties, G and A.   Intuitively we associate mixing with confusion, and thermodynamics bears out our intuition, in that mixing—even of ideal solutions—is usually accompanied by an increase* in entropy.

In Section 5 we showed that entropy (uncertainty) is a function of the number of states of a system.   When two different pure systems are mixed, the entropy increases beyond the sum of the pure system entropies because the mixed system (solution) contains new, two-component states which did not exist prior to mixing. The increase in entropy that attends mixing depends on the *number of new 2-component states* formed.   The number of 2-component states created by mixing N_A A states with N_B B states equals the number of permutations, w, of N_A plus N_B things:

$$w_N^{N_A+N_B} = \frac{(N_A + N_B)!}{N_A! \, N_B!} .$$
(5–91a)

Therefore from Eq. (5–5a), we obtain

$$\Delta S_{\text{mixing}} = k \ln \frac{(N_A + N_B)!}{N_A! \, N_B!}$$
(9–12)

(since each of the permutations is equally likely).

Exercise 9–2. A system consists of 10 balls, some of which are black and some of which are white.   What ratio of black to white balls produces the greatest number of permutations?   What is the maximum number of permutations?   What is the entropy of this system (assuming that all permutations are equally likely)?   What is the entropy of a system of 3 black and 7 white balls?

When N is very large (as it is when it refers to the number of molecules in a macroscopic piece of matter), the logarithm of N factorial may be approximated by

$$\ln N! = N \ln N - N.$$
(9–13)

This is known as *Stirling's Approximation*.   Thus, if $N = N_A + N_B$, Eq. (9–12) becomes

$$\frac{\Delta S_{\text{mixing}}}{k} = N \ln N - N - (N_A \ln N_A - N_A + N_B \ln N_B - N_B).$$

We divide by N, let $x_A = N_A/N$, and $N = N^{x_A} N^{x_B}$ (with $x_A = 1 - x_B$).

Therefore

$$\frac{\Delta S_{\text{mixing}}}{Nk} = \ln N - x_A \ln N_A - x_B \ln N_B$$

$$= -x_A \ln x_A - x_B \ln x_B$$
(9–14)

---

* In certain nonideal solutions, reaction or ionization effects may offset the confusion of mixing by reducing the number or freedom of particles in the mixed system, so that entropy decreases.

or, letting $N/\overline{N} = n =$ number of moles and $R = \overline{N}k$ ($\overline{N} =$ Avogadro's number),

$$\frac{(\Delta S_{\text{mixing}})_T}{nR} = -\sum_i x_i \ln x_i. \qquad \text{(Ideal solutions only)} \qquad (9\text{–}15)$$

9–5
Entropy of mixing ideal solutions: macroscopic approach

We can also find the entropy of mixing from macroscopic thermodynamics by devising a reversible mixing process and evaluating $\int dQ_{\text{rev}}/T$ for that process. Figure 9–3 shows a device (a *gedanken apparat*) which is, in principle, capable of continuously mixing two pure gases reversibly (and isothermally). Pure gas A, at point (1), at a total pressure $\mathbb{P}$, expands isothermally through a reversible turbine to pressure p_A. It then flows through a semipermeable membrane* into the gas mixture chamber, where its partial pressure is p_A and the total chamber pressure is $\mathbb{P}$.

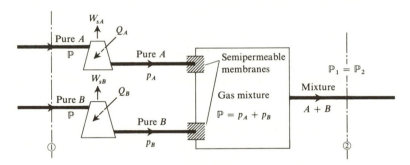

Fig. 9–3. Reversible isothermal process for mixing gases.

Similarly, pure gas B flows through another reversible isothermal turbine, expanding from $\mathbb{P}$ to p_B, and entering the mixture chamber through a semipermeable membrane which permits only B to pass. The gas mixture at total pressure $\mathbb{P} = p_A + p_B$ leaves the device at point (2). Heat is added to both turbines to maintain isothermal conditions. (One could reverse the direction of flow by making an infinitesimal increase in the total pressure of the gas mixture. The gas mixture

---

* A semipermeable membrane is a membrane which allows the free passage of one kind of gas but not of another. It is a kind of molecular filter. For example, palladium is almost completely permeable to H_2 but it is impermeable to most other gases. A given pure gas moves through a corresponding semipermeable membrane when its partial pressure downstream of the membrane is less than its partial pressure upstream, even though the total downstream pressure may be greater than the total upstream pressure. Although perfect semipermeable membranes do not exist in nature, there appears to be no objection to devising "thought" experiments based on their existence.

would then be *separated* reversibly into pure components by flowing back through the semipermeable membranes. The pure components at p_A and p_B would be compressed to $\mathbb{P}$ by passing backward through the turbines, which would now function as compressors.)

Applying the First Law (Eq. 2–45) to the flowing gas streams, we have

$$\bar{Q}_A = \bar{W}_{sA} + \Delta\bar{H}_A, \qquad \bar{Q}_B = \bar{W}_{sB} + \Delta\bar{H}_B.$$

If the gases are ideal, then $\Delta\bar{H} = 0$ (isothermal process) and $\bar{Q} = \bar{W}_s$. The process in the turbine is an isothermal reversible expansion and applies to each mole of gas expanded. If the mixture contains n_A moles of A and n_B moles of B, then

$$Q_A = W_{sA} = n_A\, RT \ln \frac{\mathbb{P}}{p_A},$$

$$Q_B = W_{sB} = n_B\, RT \ln \frac{\mathbb{P}}{p_B}.$$

Therefore the entropy changes of gases A and B are:

$$\Delta S_A + \Delta S_B = \frac{Q_A}{T} + \frac{Q_B}{T} = R\left(n_A \ln \frac{\mathbb{P}}{p_A} + n_B \ln \frac{\mathbb{P}}{p_B}\right).$$

Consequently the entropy of mixing 1 mol of gas mixture from pure components is

$$\frac{\Delta S_A + \Delta S_B}{n_A + n_B} = \frac{\Delta S_{\text{mixing}}}{n} = R\left(x_A \ln \frac{1}{x_A} + x_B \ln \frac{1}{x_B}\right),$$

where

$$x_A = \text{mole fraction of component } A \text{ in the mixture}$$

$$= n_A/(n_A + n_B) = p_A/\mathbb{P} \qquad \text{(Eq. 9–5)}.$$

For any number of ideal-gas components,

$$\frac{\Delta S_{\text{mixing}}}{nR} = -\sum_i x_i \ln x_i,$$

which is Eq. (9–15).

9–6
Entropy, free energy, and Helmholtz function of mixing ideal gases

Consider the following irreversible mixing of ideal gases.

A container holding 1 g-mol of argon at 25°C and 1 atm is connected to a similar container holding 1 g-mol of helium at the *same temperature* and *pressure*. The gases diffuse from one container into the other and mix freely.

At the final equilibrium state, the gases will have mixed completely to form an ideal solution of ideal gases. Each container will hold a uniform mixture of 50 mole % of argon and 50 mole % of helium, at 25°C and 1 atm total pressure. What changes in equilibrium thermodynamic properties result from this mixing?

The thermodynamic properties in the final state are:

Temperature: 25°C, $\Delta T = 0$,

Total pressure: 1 atm, $\Delta \mathbb{P} = 0$,

Partial pressure of helium $= 0.5$ atm,

Partial pressure of argon $= 0.5$ atm,

$$\Delta U_{mixing} = C_V \, \Delta T = 0,$$

$$\Delta H_{mixing} = \Delta U + \Delta(\mathbb{P}V) = 0 + 0 \qquad (\mathbb{P}V \text{ is constant at constant } T),$$

$$\Delta S_{mixing} = \Delta \bar{S}_{Ar} + \Delta \bar{S}_{He} \neq 0.$$

$\Delta \bar{S}_{Ar}$ is the entropy of mixing 1 mol of argon with 1 mol of any other nonreacting gas, both of which are at 25°C and 1 atm. The mixing described is highly irreversible. However, the entropy change (a state property) is the same as for the reversible mixing process shown in Fig. 9–3, which is the same as for the *reversible isothermal expansion* of argon from 1 atm to the final partial pressure, $\frac{1}{2}$ atm. Therefore, using Eq. (3–16), we obtain

$$\Delta S_{Ar} = (1) \, R \ln \frac{1}{0.5} = 1.378 \, \frac{cal}{g\text{-mol} \cdot °K} \, .$$

Similarly

$$\Delta S_{He} = (1) \, R \ln \frac{1}{0.5} = 1.378 \, \frac{cal}{g\text{-mol} \cdot °K}$$

and

$$\Delta S_{mixing} = 1.378 + 1.378 = 2.756.$$

Therefore

$$\Delta G_{mixing} = \Delta H_{mixing} - T \, \Delta S_{mixing}$$

$$= 0 - (298)(2.756) = -820 \, \frac{cal}{g\text{-mol}} \, . \qquad (9\text{-}16)$$

An alternative way of finding ΔG_{mixing} is

$$\Delta G_{mixing} = \Delta G_{Ar} + \Delta G_{He}. \qquad (9\text{-}17)$$

ΔG_{He} is the change in free energy corresponding to an isothermal expansion of helium from *1 atm* to the *partial pressure of the mixture*. Therefore

$$\Delta G_{He} = \int_{1.0}^{0.5} \bar{V} \, dP = RT \ln 0.5 = -410 \, \frac{cal}{g\text{-mol}} \, .$$

Similarly,

$$\Delta G_{Ar} = RT \ln 0.5 = -410 \, \frac{cal}{\text{g-mol}}.$$

Therefore

$$\Delta G_{mixing} = (-410) + (-410) = -820 \, \frac{cal}{\text{g-mol}}.$$

Exercise 9–3. Determine ΔA for the above mixing process.

The changes in entropy and free energy that accompany the isothermal mixing of pure ideal gases are the same as the changes that occur during isothermal expansion from the *initial pressure* of the pure component gas to its *partial pressure* in the gas mixture.

9–7
Free energy of mixing: ideal liquid solutions

The free energy of mixing of liquid or solid solutions is equal to the free energy of mixing of the vapor mixtures that are in equilibrium with these solutions. The reason for this is that there is no change in free energy when one moves matter from one equilibrium phase to another. For example, the change in free energy for an isothermal process wherein 1 mol of liquid benzene and 1 mol of liquid toluene are mixed to form 2 mols of liquid solution is the same as the change in free energy that accompanies the expansion of 1 mol of benzene vapor and 1 mol of toluene vapor from their respective *vapor pressures* at the process temperature to their respective *partial* pressures in the vapor that is in equilibrium with the liquid solution.

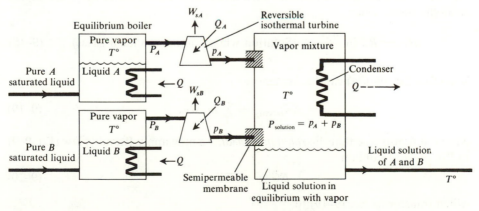

Fig. 9–4. Reversible isothermal flow process for mixing liquids.

Figure 9–4 shows a *gedanken apparat* for mixing liquids reversibly and isothermally in a flow process. This device is like the apparatus used for reversible mixing of gases, except that vaporizing and condensing chambers have been added. Pure (saturated) liquids (*A* and *B*) enter on the left at equilibrium vapor pressures corresponding to the process temperature. Each is separately and reversibly vaporized, and the pure vapors are then isothermally expanded in reversible engines down to their respective mixture partial pressures. Each pure vapor then passes freely through a semipermeable membrane into the mixing chamber. There cooling coils condense the vapor mixture into a saturated liquid solution, which exits on the right. The *only* free energy changes occur in the reversible isothermal expansion engines,

$$\Delta G = \Delta G_A + \Delta G_B = n_A \, RT \ln \frac{p_A}{P_A} + n_B \, RT \ln \frac{p_B}{P_B}, \qquad (9\text{–}17\text{a})$$

and the sum of these changes must therefore be the ΔG_{mixing}. (Vapors are assumed to be in an ideal-gas state.) Note that the free energy changes are equal to (minus) the reversible shaft work produced by the expansion engines (see Eq. 3–31b).

To reverse the mixing process, one can reverse the flow and interchange the condenser and vaporizer coils. The solution then moves through the apparatus from right to left. The liquid is then vaporized, and the vapors are separated into low-pressure pure components by the semipermeable membranes. Each pure vapor is then compressed reversibly and isothermally to the vapor pressure of the pure liquid, and then condensed to pure liquid. The work supplied to the compressors during the separation process is the same amount as that delivered during the mixing process. Initial and final temperatures are equal.

Let us now generalize the ideas behind Eq. (9–17a).

The change in free energy $\Delta \bar{G}_i$ of 1 mol of a liquid *i* which takes place when the liquid goes from its *pure state* to the *composition* x_i in a solution at constant temperature may be computed from the ratio of partial pressure to vapor pressure, if the vapors in equilibrium with the pure liquid and the final solution are in the ideal-gas state. Thus

$$\Delta \bar{G}_i = RT \ln \frac{p_i}{P_i}. \qquad \text{(Solution whose vapors are ideal gases)} \qquad (9\text{–}18)$$

Therefore Eq. (9–17a) can be written as

$$\Delta G_{\text{mixing}} = \sum_i n_i \, \Delta \bar{G}_i. \qquad (9\text{–}19)$$

Now if it is also true that the *solution is ideal*, we may apply Raoult's law (Eq. 9–9) to Eq. (9–18) to obtain

$$\Delta \bar{G}_i = RT \ln x_i, \qquad (9\text{–}20)$$

which transforms Eq. (9–19) to

$$\Delta G_{\text{mixing}} = \sum_i n_i \, (RT \ln x_i) \qquad (9\text{–}21)$$

or, dividing by $n = \sum_i n_i$,

$$\boxed{\Delta \bar{G}_{\text{mixing}} = RT \sum_i x_i \ln x_i.} \quad \text{(Ideal solution)} \quad (9\text{--}21a)$$

Example 9–3. Calculate the free energy of 1 mol of a solution of 40 mol % benzene and 60 mol % toluene at its equilibrium vapor pressure and 96°C, relative to its pure components at 96°C and their equilibrium vapor pressures.

Answer: The G of the solution relative to its pure components at the same temperature is the G loss on mixing, or ΔG_{mixing},

$$P_{\text{benzene}} \text{ at } 96°C = 1180 \text{ mm Hg,}$$

$$P_{\text{toluene}} \qquad\quad = \quad 480 \text{ mm Hg.}$$

The composition of vapor in equilibrium with the solution is found by Raoult's law (Eq. 9–9):

$$p_b = 0.40 \times 1180 = 472 \text{ mm Hg,}$$

$$p_t = 0.60 \times \quad 480 = 288 \text{ mm Hg.}$$

Therefore: $\Delta G_{\text{mixing}} = \Delta G_b + \Delta G_t$

$$= n_b RT \ln \frac{p_b}{P_b} + n_t RT \ln \frac{p_t}{P_t}$$

$$= 1.987 \frac{\text{cal}}{\text{g-mol} \cdot °K} \times 369°K (0.4 \ln 0.4 + 0.6 \ln 0.6)$$

$$= 1.987 \times 369 (-0.3667 - 0.306) = -493 \frac{\text{cal}}{\text{g-mol}}.$$

The previously discussed process of mixing liquids dealt with liquids at their equilibrium vapor pressures. If the pressure on the solution is higher than the equilibrium vapor pressure (suppose that the benzene-toluene solution of Example 9–3 were at 1 atm), then the computation path should, in addition, evaluate the ΔG for altering the pressure on the pure liquid components and the solution from the given total pressure to their respective equilibrium vapor pressures. We can evaluate this ΔG for a change in pressure of a liquid by using Eq. (8–21). Its value is negligible for moderate pressure changes. Figure 9–5 shows the modified process required to find ΔG_{mixing} for solutions at elevated pressures.*

---

* Although the reversible mixing processes in Figs. 9–4 and 9–5 are shown as continuous-flow processes, a reversible batch process could also be used. To do this, one simply replaces the turbines with piston-cylinder machines, which contain the pure components that enter the process. Reversible vaporization and expansion are performed sequentially in each cylinder, and vapors then pass through the semipermeable membranes to the mixing chamber.

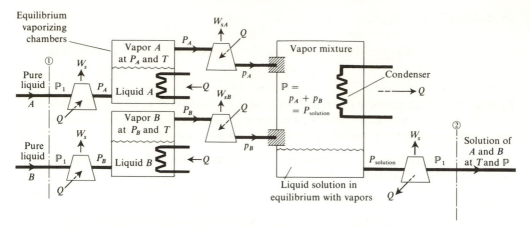

Fig. 9–5. Reversible isothermal flow process for preparing a liquid solution at any T and P from pure liquids at the same T and P.

Exercise 9–4. Compute the free energy of the solution in Example 9–3 at 96°C and 1000 atm, and at 96°C and 1 atm, relative to its pure components at 96°C and 1 atm. How much work could be done on the surroundings if the solution were mixed reversibly from its pure components?

9–8
Composition change and chemical work

An important fact is implicit in Figs. 9–3, 9–4, and 9–5, and in the results of the preceding analysis of the free energy of mixing. It is that:

A system capable of a COMPOSITION CHANGE *can interchange reversible work with its surroundings without changing its total pressure, temperature, or volume. The work interchange equals minus the change in free energy associated with the composition change.*

We shall call such composition-dependent work CHEMICAL WORK, and give it the symbol W_c. When n moles of ideal solution are made isothermally from pure components, the reversible chemical work that can be obtained from the process is the same as the shaft work produced by the turbines in Fig. 9–5, or

$$W_c = -\Delta G_{\text{mixing entire solution}} = -\sum_i \Delta \bar{G}_i n_i, \tag{9–22}$$

where $\Delta \bar{G}_i$ is given by Eq. (9–18). $\Delta \bar{G}_i$ is an *intensive* property, since it does not depend on the magnitude of n_i (although it does depend on x_i). It therefore follows that for the formation of a differential amount of solution,

$$dW_c = -\sum_i \Delta \bar{G}_i \, dn_i. \qquad \text{(Solution whose vapors are ideal gases)} \tag{9–22a}$$

In order to apply Eqs. (9–22) and (9–22a) to all solutions, we replace $\Delta \bar{G}_i$ with μ_i, which is called the CHEMICAL POTENTIAL. μ_i is the rate at which the free energy of an *entire* solution changes per mole of i added at composition x_i. Hence Eq. (9–22a) is transformed into Eq. (9–22b),

$$dW_c = - \sum_i \mu_i \, dn_i, \qquad (9\text{–}22\text{b})$$

which is the general form of the differential of chemical work, and which applies to *real* solutions as well as to ideal solutions.

The differential of the chemical work, dW_c, is a form of $F_i \, dX_i$ or dW_{other} (refer to Eqs. (6–1d), (3–23a), and (3–27); also footnote in Section 6–2). The intensive property μ_i is the generalized force, and the extensive property change dn_i is the generalized displacement. Each independent component i gives rise to a $\mu_i \, dn_i$ reversible work term in the summation which appears on the right-hand side of Eq. (9–22b). All these terms must be included in the energy-conservation expression, Eq. (6–1d). As a consequence, any thermodynamic property of a variable-composition $P \, dV$-work system is a function of as many variables as there are independent components, plus two. (Reread Section 6–2.) For example, the free energy,

$$G = G[(\text{number of components} + 2) \text{ variables}]. \qquad (9\text{–}23)$$

The most convenient additional variables are the number of moles of each component, n_i, in the system. Therefore Eq. (9–23) may be written

$$G = G(T, P, n_A, n_B, \ldots, n_i \ldots). \qquad (9\text{–}24)$$

Similar expressions can be written for other thermodynamic properties.

Exercise 9–5. Using Eq. (9–24), write the total differential of G (Eq. 6–8a). Compare your result with Eqs. (6–7), (6–12), (6–13), (3–27), and (3–28). Show that

$$\left(\frac{\partial G}{\partial n_i} \right)_{T,P,n_j} = \mu_i. \qquad (9\text{–}24\text{a})$$

Exercise 9–6. Which thermodynamic properties of an ideal solution relative to its pure components cannot be computed with only temperature and pressure information? What additional information is required? Evaluate these properties for the ideal-gas solution in Example 9–1 relative to a pure material whose reference state is at 70°F and 1 atm.

9–9
Minimum work of separation

The theoretical minimum work needed to separate a mixture is the same as (minus) the free energy of the separation process. The actual energy expended in a separation is frequently many times greater than the theoretical minimum.

Example 9–4. What is the minimum work required to separate 1 g-mol of pure water from a large reservoir of sea water (3.5% dissolved salts) at 25°C? The vapor pressure of sea water at 25°C $= 23.24$ mm Hg. The vapor pressure of pure water at 25°C $= 23.76$ mm Hg. (The vapor pressure of the dissolved salts is negligible.)

Answer

$$\text{Minimum work} = -\Delta G = RT \ln \frac{23.24}{23.76} = -13.61 \text{ cal/g-mol.}$$

The energy used in distilling 1 g-mol of fresh water from sea water in a simple distillation apparatus without heat recovery is:

$$C_P \int_{25}^{100} dT + \Delta \overline{H}_{\text{vap, 100°C}} = 135 + 9713 = 9848 \text{ cal/g-mol,}$$

which is 725 times greater than the minimum work.

9–10
Ideal solutions of real materials

From a practical point of view, it is more important to be able to estimate the properties of *solutions* than of pure materials because it is conceivable that experimental determinations of thermodynamic properties will eventually be made for all important pure materials, whereas it will never be possible to measure the properties of all possible mixtures of these materials. The unfortunate fact is that there is no completely satisfactory method of estimating the properties of real solutions from the properties of pure components.

Where empirical thermodynamic data are unavailable, the properties of real solutions are frequently estimated by assuming that real materials form ideal solutions. The simple ideal-solution and ideal-gas relationships of Eqs. (9–5), (9–9), (9–10), and (9–11) are used, with fugacity and activity terms substituted for pressure and concentration terms. Thermodynamic properties computed by such techniques are approximations which may be adequate for design purposes; in addition, when compared with experimental data, they can serve to identify nonideal behavior.

9–11
Compressibility and fugacity of real gas solutions: pseudo-critical properties

The compressibility (Eq. 7–13) and fugacity coefficient of a solution of real gases may be estimated from Figs. 7–3 and 8–1 by using the PSEUDO-CRITICAL TEMPERATURE AND PRESSURE to calculate the reduced temperature and pressure of the gas mixture (Eq. 7–13). Pseudo-critical conditions are the *molar average critical con-*

ditions of a solution:

Pseudo-critical temperature $\equiv T_{pc} = y_A T_{cA} + y_B T_{cB} + y_C T_{cC} + \cdots$ (9–25)

or $\qquad\qquad\qquad\qquad\qquad\qquad = \sum_i y_i T_{ci},$ (9–26)

Pseudo-critical pressure $\equiv P_{pc} = \sum_i y_i P_{ci},$ (9–27)

where T_{ci} and P_{ci} are the *critical temperature and pressure of component i*, and y_i is the *mole fraction of i* in the gas solution.

Pseudo-critical properties bear no definite relation to the actual critical properties of a gas mixture. Nevertheless they yield surprisingly accurate estimates of compressibility and fugacity, particularly for hydrocarbon mixtures.

9–12
Ideal solution of real gases

The fugacity of component *i* of an ideal solution of real gases is found by using an equation similar to Eq. (9–5), with fugacities replacing pressures,

$$f_i = y_i f_{i\mathbb{P}} \qquad\qquad (9\text{–}28)$$

where f_i = fugacity of gas *i* in the *solution* of real gases.

$\qquad f_{i\mathbb{P}}$ = fugacity of *pure* gas *i* at the *total pressure* and temperature of the solution.

Example 9–5. Calculate the molar volume and fugacity of the flue-gas mixture (treated as a real gas) in Exercise 9–1, at 100 atm pressure and 300°K. Also compute the fugacity of each component, using Eq. (9–28).

Answer. Properties of the flue-gas mixture are obtained by computing pseudo-critical conditions for the mixture and using these in Figs. 7–3 and 8–1. The computation is conveniently carried out in tabular form.

	Mole %	T_c, °K	P_c, atm	$y_i T_{ci}$	$y_i P_{ci}$, atm	T_r	P_r	$f_{i\mathbb{P}}/\mathbb{P}$ (Fig. 8–1)	$f_{i\mathbb{P}}$	$f_i = y_i f_{i\mathbb{P}}$, atm
CO_2	16	304.2	72.9	48.6	11.67	0.98	1.37	0.489	48.9	7.82
N_2	76	126.2	33.5	95.8	25.45	2.38	2.98	0.950	95.0	71.3
CO	2	132.9	34.5	2.66	0.69	2.26	2.90	0.950	95.0	1.90
O_2	2	155.8	50.1	3.1	1.00	1.92	2.00	0.955	95.5	1.91
H_2O	4	647.3	218.2	25.9	8.73	0.46	0.46	—	—	0.085
				176.1	47.54					

$$T_{pc} = \sum y_i T_{ci} = 176.1°K, \qquad P_{pc} = \sum y_i P_{ci} = 47.5 \text{ atm}$$

Reduced T of the mixture $= 300/176.1 = 1.72$.

Reduced P of the mixture $= 100/47.5 = 2.11$.

Using these reduced conditions in Fig. 7–3, we find that $z = 0.916 =$ compressibility factor of the flue-gas mixture. Therefore

$$\overline{V} = \frac{zRT}{P} = \frac{0.916 \times 0.08206 \ l\text{-atm} \times 300°K}{\text{g-mol} \times 100 \text{ atm}} = 0.226 \text{ liters/g-mol of flue gas.}$$

From Fig. 8–1:

$$\frac{f}{P} = 0.918; \qquad f_{\text{mixture}} = 91.8 \text{ atm.} \qquad \text{(Answer)}$$

Computations for component fugacities appear on the right-hand side of the table. Water is liquid at the conditions specified. Therefore $f_{\text{H}_2\text{O}}$ is equal to the vapor pressure of water at 300°K.

Exercise 9–7. Calculate the pressure required to contain 100 lb-mol of a mixture of 50 mole % of propane and 50 mole % of n-butane in a volume of 3000 ft³ at 1000°F.

9–13
Solutions of real liquids: Henry's Law

Equilibrium compositions of vapor and liquid phases of real solutions are often related over *part* of their composition range by *Henry's Law*:

$$\boxed{p_i = K'x_i,} \qquad\qquad (9\text{–}29)$$

where p_i = partial pressure of i in the *vapor* phase,

x_i = mole fraction of i in the *liquid* phase,

K' = Henry's Law constant.

Equation (9–29) holds for *all* solutions as x_i *approaches zero*.

Note that Raoult's Law (Eq. 9–9) is a special case of Henry's Law; that is, when $K' = P_i$, Henry's Law is the same as Raoult's Law. Raoult's Law may be applied to real solutions as x_i approaches *unity*. In fact, for binary solutions, Raoult's Law generally applies to one liquid component (the major component or solvent) over the concentration range in which Henry's Law applies to the other (the minor component or solute).

9–14
Ideal solutions of real liquids

Fugacities of components of a *liquid* solution may be computed by using an equation similar to Raoult's Law (Eq. 9–9):

$$f_i = x_i f_{Pi}, \tag{9–30}$$

where f_i = fugacity of component i in liquid solution at concentration x_i,

    f_{Pi} = fugacity of *pure* liquid i at its *vapor pressure* at solution T and P,

    x_i = mole fraction of i in the liquid solution.

    Implicit in both Eqs. (9–28) and (9–30) is the assumption that there is no molar volume change when the components are mixed. That is, real liquids or gases may mix with no energetic interaction and therefore may form *ideal* solutions.

9–15
K factors: composition of vapor and liquid phases in equilibrium

When a liquid and vapor phase are in equilibrium (both phases being ideal solutions), we may combine Eqs. (9–28) and (9–30) to obtain

$$f_i = x_i f_{Pi} = y_i f_{iP}$$

or

$$\frac{y_i}{x_i} = \frac{f_{Pi}}{f_{iP}} = K. \tag{9–31}$$

This K is called a *vaporization equilibrium constant*, and gives the ratio of mole fractions in the gas and liquid phases. K depends on both temperature and pressure.

Exercise 9–8. What is K for an ideal solution of liquids whose vapors are ideal gases? How does this latter K vary with the composition of the solution? How should the K defined by Eq. (9–31) depend on composition? Distinguish between f_{Pi} and f_{iP}.

    Values of K for light hydrocarbon mixtures and mixtures of similar inorganic gases based on the Benedict-Webb-Rubin equation of state (Eq. 7–11) are available and have been widely used in the design of chemical processing equipment.*

---

* M. Benedict, G. B. Webb, and L. C. Rubin, *Chem. Eng. Progr.* 47, 419, 449, 571, 609 (1951); F. C. Schiller and L. N. Canjar, *Chem. Eng. Progr.*, Symp. Ser. 7, 49, 67 (1953); C. L. DePriester, *Chem. Eng. Progr.*, Symp. Ser. 7, 49, 1 (1953); and W. C. Edmister and C. L. Ruby, *Chem. Eng. Progr.* 51, 95-F (1955).

9–16
Real liquid solutions: activity coefficients

A relatively convenient way of treating real solutions of liquids (particularly at low pressures, at which vapor-phase behavior is near ideal) is to correct for deviation from ideal-solution behavior by using an *activity coefficient*.

For example, for a liquid solution whose vapors are ideal gases, the isothermal change in free energy per mole of liquid component i as i goes from the pure to the mixed state x_i is

$$\Delta \bar{G}_i = RT \ln \frac{p_i}{P_i} . \tag{9–18}$$

If the solution is also ideal and follows Eq. (9–9), then

$$\Delta \bar{G}_i = RT \ln x_i. \tag{9–20}$$

If we now recall the definition of activity Eq. (8–19) and apply it to an ideal vapor-phase component i, we obtain

$$a_i \equiv \frac{f_i}{f_0} = \frac{p_i}{P_{\text{reference}}} . \tag{9–32}$$

Applying Eq. (9–32) to Eq. (9–18), we have

$$\Delta \bar{G}_i = RT \ln a_i. \tag{9–33}$$

From Eqs. (9–33) and (9–20), it follows that

$$a_i = x_i \quad \begin{array}{l} \text{(Ideal solution of liquid} \\ \text{whose vapors are ideal gases)} \end{array} \tag{9–34}$$

where the reference state for a_i is the *pure liquid* component i at the temperature and pressure of the liquid solution.

The *nonideality* of real solutions of liquids is corrected for by introducing the ACTIVITY COEFFICIENT γ_i in Eq. (9–34). Thus

$$a_i = \gamma_i x_i \quad \text{(Real solutions)} \tag{9–35}$$

and using Eq. (9–35) in Eq. (9–33),

$$\hat{G}_i = RT \ln \gamma_i x_i \tag{9–35a}$$

$$= RT \ln x_i + RT \ln \gamma_i,$$

or

$$\hat{G}_i = \bar{G}_i^* + \bar{G}_i^{\text{ex}}, \tag{9–36}$$

where $\bar{G}_i^* = $ the change in free energy per mole of i when i goes from the pure state to the concentration x_i in an *ideal* solution $= \Delta \bar{G}_i$ as in Eq. (9–20),

$\bar{G}_i^{ex}$ = the *excess* free energy per mole of i when i goes from *ideal* solution to *real* solution at composition x_i,

$\hat{G}_i$ = the free energy of a mole of i when in a *real* solution at concentration x_i, relative to G of a mole of pure i.

The immediate importance of Eq. (9–36) is that it suggests that the thermodynamic properties of *real* solutions can be expressed as sums of ideal and excess properties.

Equation (9–35) may also be combined with Eqs. (9–18) and (9–33) to correct for deviations from Raoult's Law. Thus

$$p_i = (\gamma_i x_i)P_i. \tag{9-37}$$

9–17
Partial molar properties

In the paragraphs preceding Eq. (9–24), we said that, if we wish to uniquely specify the state of a system capable of expansion work and of changing its composition, we must describe it in terms of composition variables in addition to its pressure and temperature (or any other two thermodynamic properties). Equation (9–24) may be generalized* as

$$M = M(Y, Z, n_A, n_B, n_C, \ldots), \tag{9-38}$$

where M might represent H, V, G, A, S or other extensive property, and where Y and Z are any two thermodynamic properties other than the compositions.

If we let Y be the temperature T, and Z the pressure P, Eq. (9–38) then states that property M depends on the pressure, temperature, and number of moles of components A, B, C, $\ldots$, etc., in the system. It therefore follows that the total differential of M is

$$dM = \left(\frac{\partial M}{\partial P}\right)_{T, n_A, n_B, n_C, \ldots} dP + \left(\frac{\partial M}{\partial T}\right)_{P, n_A, n_B, n_C, \ldots} dT + \left(\frac{\partial M}{\partial n_A}\right)_{P, T, n_B, n_C, \ldots} dn_A$$

$$+ \left(\frac{\partial M}{\partial n_B}\right)_{P, T, n_A, n_C, \ldots} dn_B + \cdots. \tag{9-39}$$

When P and T are constant,

$$dM_{P, T} = \sum_i \left(\frac{\partial M}{\partial n_i}\right)_{P, T, n_A, n_B, \ldots, n_j} dn_i, \tag{9-40}$$

---

* The generalization is limited to systems that do expansion and chemical work only. If more kinds of reversible work can be performed, then additional variables must be included in the Eq. (9–38) parentheses.

which may be written as

$$dM_{P,T} = \sum_i \widehat{M}_i \, dn_i, \tag{9–41}$$

where

$$\boxed{\widehat{M}_i \equiv \left(\frac{\partial M}{\partial n_i}\right)_{P,T,n_A,n_B,\ldots,n_j}.} \tag{9–41a}$$

The $\widehat{M}_i$ in Eq. (9–41) is an abbreviation for $(\partial M/\partial n_i)_{P,T,n_A,n_B,\ldots,n_j}$, and is called a PARTIAL MOLAR PROPERTY† of component i. The subscript on the partial derivative means that pressure, temperature, and all components other than i are constant. The designation "partial molar property" applies only to partial derivatives with respect to n_i at *constant P and T*.

A partial molar property is the *rate* at which an extensive property, M, of the *entire* solution changes with the number of moles of component i in the solution when temperature, pressure, and extent of all other components remain fixed. Put another way, it is the slope of an M_{solution} versus n_i plot made at constant T, P, $n_A, n_B, \ldots, n_h, n_j, n_k, \ldots$ (see Fig. 9–6).

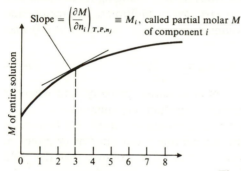

Slope $= \left(\dfrac{\partial M}{\partial n_i}\right)_{T,P,n_j} \equiv M_i$, called partial molar M of component i

M of entire solution

n_i, number of moles of i in solution
(T, P, and all other components held constant)

Fig. 9–6. A partial molar property is the slope of an M_{solution} versus n_i curve.

The partial molar property is *intensive* and its value generally depends on the *composition* of the solution. Equation (9–41) is homogeneous in extensive differentials and may be integrated at a given temperature, pressure, and composition to

$$\boxed{M_{P,T,x_i} = \sum_i \widehat{M}_i n_i,} \tag{9–42}$$

---

† The $\widehat{M}_i$ in Eq. (9–41) is called a "partial mol*al* property" in many textbooks. I consider the term mol*al* confusing, in part because of its use in chemistry to designate volumetric concentration (moles/1000 cm^3 of liquid solution). I have therefore chosen (following K. Denbigh, *The Principles of Chemical Equilibrium*, Cambridge: Cambridge University Press, 1966) to use the term "partial mol*ar* property" for the property of a mole of component material in solution, corresponding to the use of "molar property" for the property of a mole of material in pure form.

where M_{P, T, x_i} is the M value of a *solution*, at P, T, and composition x_i, relative to an arbitrary preassigned reference state.* If the reference state is taken to be the pure components at the temperature and pressure of the solution, then (Eq. 9–42) expresses the ΔM *of mixing* of the solution.

Equation (9–42) expresses the extensive property, M, of a solution as the sum of intensive property contributions from each of its components. It follows that all the extensive properties of a solution may be expressed as sums of component contributions. Thus

$$V_{P, T, x_i} = \sum_i \widehat{V}_i n_i, \tag{9–43}$$

$$U = \sum_i \widehat{U}_i n_i, \tag{9–44}$$

$$H = \sum_i \widehat{H}_i n_i, \tag{9–45}$$

$$S = \sum_i \widehat{S}_i n_i, \tag{9–46}$$

$$A = \sum_i \widehat{A}_i n_i, \tag{9–47}$$

$$G = \sum_i \widehat{G}_i n_i. \tag{9–48}$$

For *ideal solutions*, the volume and energy *partial-molar* properties are the same as the *pure*-material *molar* properties. Thus

$$
\begin{aligned}
\overline{V}_i &= \widehat{V}_i, \\
\overline{U}_i &= \widehat{U}_i, \qquad \text{(Ideal solutions only)} \\
\overline{H}_i &= \widehat{H}_i.
\end{aligned}
\tag{9–49}
$$

The partial molar entropy and free energy, however, are not the same as the molar entropy and free energy because, even in ideal solutions, entropy increases when components are mixed.

Equations (9–43), (9–44), and (9–45), when applied to a mole of ideal solution, are identical to Eqs. (9–1), (9–2), and (9–3). If, as is usual, we take the reference state for the ideal-solution properties to be pure components at the solution temperature and pressure, then Eqs. (9–43), (9–44), (9–45), and (9–49) all become zero, in keeping with the defining fact that ideal solutions show no volume or energy effects on mixing. In the limiting case of a solution consisting of one pure component, the partial molar property is also the molar property.

Exercise 9–9. Which of the following is the correct statement?

a) The partial molar entropy, $\widehat{S}_i$, of material i in an ideal solution is the same as the change in entropy of a mole of pure i moving from the pure state to the solution state.

---

* The result of this integration is characteristic of homogeneous functions. See Appendix III and Section 9–23.

b) The partial molar entropy, $\widehat{S}_i$, of material i in an ideal solution is the same as the molar entropy, $\bar{S}_i$, of pure i.

Repeat the question, replacing S with G, U, H, and A.

It must be emphasized that, in general, a partial molar property is *not* the same as a molar property:

$$\widehat{M}_i \neq \bar{M}_i. \qquad \text{(For real solutions)} \qquad (9\text{–}50)$$

Examples of where they differ and the significance of these differences are discussed below.

9–18
Partial molar volume

Let us replace M with V in Eq. (9–40) and discuss partial molar properties in terms of volume. The expression

$$\left(\frac{\partial V}{\partial n_A}\right)_{P,\,T,\,n_B,\,\ldots}$$

refers to the rate at which the *total* volume of a solution changes per mole of component A added, all other components and intensive properties remaining constant. For example, take 1 mol of hexane liquid in a large graduated vessel at 1 atm and 25°C. Add to it small measured amounts of liquid heptane and repeat the additions many times, recording the total volume after each addition. The results, when plotted, form a straight line, as in Fig. 9–7, because hexane and heptane liquid form an ideal solution, whose volume is equal to the sum of the volumes of the pure components used to make up the solution. The slope of the line is

$$\left(\frac{\partial V}{\partial n_{\text{hep}}}\right)_{P,\,T,\,n_{\text{hex}}} \qquad \text{or} \qquad \widehat{V}_{\text{hep}},$$

the *partial molar volume* of heptane, which (if we recall the characteristics of ideal solutions) must equal the *molar volume* of pure heptane, $\bar{V}_{\text{hep}}$.

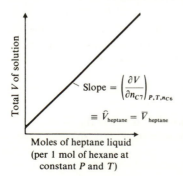

Moles of heptane liquid
(per 1 mol of hexane at
constant P and T)

Fig. 9–7. Partial molar volume of an *ideal* solution.

Why, then, is there a need for the additional qualification "partial"? To answer, let us repeat the above volume-mixing experiment with ethyl alcohol and water. The results appear as in Fig. 9–8, and do not form a straight line. The total volume of the solution is less than the sum of the volumes of its pure components. The slope,

$$\left(\frac{\partial V}{\partial n_{H_2O}}\right)_{P, T, n_{alc}},$$

is not constant, but depends on composition. Here $\hat{V}_{H_2O} \neq \bar{V}_{H_2O}$, except when n_{H_2O} is very large.

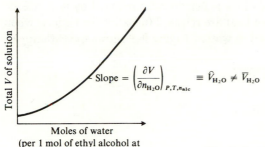

Fig. 9–8. Partial molar volume of a *real* solution.

Hexane and heptane resemble each other closely and form an ideal solution. Water and alcohol interact on mixing, so that a solution made from them is not ideal. The former situation is analogous to mixing black billiard balls with white billiard balls, while the latter may be compared to mixing billiard balls with marbles. The marbles occupy some of the interstices between the billiard balls, and hence the volume of the mixture is less than the sum of the volumes of the pure components.

Partial molar volumes reflect the behavior of materials in a solution. Only in ideal solutions are partial molar volumes the same as the molar volumes of the pure materials.

Exercise 9–10. Suppose that 12-gauge bird shot ($\bar{V} = 0.003$ ft^3/lb, diameter = 0.040 in.) is being poured into a silo which is one-fourth full of 2-inch cannonballs. What is $\hat{V}_{shot}$ at the beginning of the operation, and what is it when the silo is three-fourths full?

Exercise 9–11. Figures 9–7 and 9–8 are based on absolute volumes. Redraw these figures, using as the reference state the volume of the pure components at the temperature and pressure of the solution. That is, plot ΔV of mixing (total volume of solution minus volume of the components in their pure state) as the ordinate.

Do the values of the partial molar volumes change as the reference state is changed? Evaluate $\hat{V}_{heptane}$ at 50 mole % composition. Use the following data (based on Table 3–129, J. H. Perry, *ibid.*) for computing the volumes of the ethanol solutions.

$\dfrac{\text{cm}^3 \text{ solution}}{\text{mole alcohol}}$	58.7	63.0	68.6	76.2	86.6	101.2	123.4	161.4	248.2	469.8
$\dfrac{\text{g-mol } H_2O}{\text{g-mol alcohol}}$	0	0.284	0.639	1.096	1.703	2.56	3.82	5.94	10.24	23.04

9–19
Partial molar enthalpies of a nonideal binary solution

The dissolution of solids in liquid solutions is usually accompanied by heat effects. For example, considerable amounts of heat are released (negative heat effect) when sodium hydroxide (NaOH) is dissolved in water. Figure 9–9 shows the enthalpy of

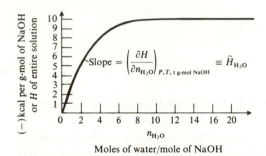

$$\text{Slope} = \left(\frac{\partial H}{\partial n_{H_2O}}\right)_{P,T,\, 1\text{ g-mol NaOH}} \equiv \hat{H}_{H_2O}$$

n_{H_2O}

Moles of water/mole of NaOH

Fig. 9–9. Enthalpy of solution of NaOH and H_2O. Reference state: pure water and pure sodium hydroxide at 25°C and 1 atm.

a NaOH solution (per mole of NaOH) at 1 atm and 25°C as a function of the number of moles of water added to the solution. The curve is obtained by measuring and plotting the total heat effect accompanying successive small additions of pure water to 1 mol of pure NaOH at constant temperature and pressure.* This isothermal, isobaric heat effect corresponds to the enthalpy change for the process

$$1\,NaOH_{(s)} + (n_i)H_2O_{(l)} \rightarrow (1 + n_i)NaOH_{\text{solution}}, \qquad \text{at 25°C and 1 atm.}$$

It is also the enthalpy of the entire solution when the enthalpy reference state is taken as being the *pure* components at 25°C and 1 atm.

The partial molar enthalpy of *water in* NaOH *solution* is the slope of the curve in Fig. 9–9, and clearly depends on composition:

$$\left(\frac{\partial H}{\partial n_{H_2O}}\right)_{P,\,T,\,n_{NaOH}=1} = \hat{H}_{H_2O} = \text{function of } n_{H_2O}.$$

---

* Data such as Fig. 9–9 are also called *integral heat of solution* curves, since they give the total or integrated heat effect obtained when one mixes a given solution from pure ingredients. See Fig. 2–31.

Exercise 9–12. Why does $\hat{H}_{H_2O}$ approach zero as n_{H_2O} gets large in Fig. 9–9?

From Eq. (9–45), it is apparent that

$$H_{solution} = \hat{H}_{H_2O} \cdot n_{H_2O} + \hat{H}_{NaOH} \cdot n_{NaOH}, \tag{9–51}$$

from which, by using Fig. 9–9, we can compute $\hat{H}_{NaOH}$ corresponding to a given composition and $\hat{H}_{H_2O}$ (slope).

Exercise 9–13. Compute $\hat{H}_{NaOH}$ for a 5-mol % solution of NaOH in 95% water at 25°C and 1 atm. (Use Figs. 9–9 or 2–31 to solve this problem. You must first, however, compute how many g-mols of water need be added to 1 g-mol of NaOH to form a 5-mol % solution of NaOH.)

9–20
Graphical computation of partial molar properties

Partial molar properties of *binary* solutions can be found graphically as follows. Divide Eq. (9–51) by the total number of moles, so that it becomes

$$\frac{H_{solution}}{n_{H_2O} + n_{NaOH}} = \bar{H}_{solution} = \hat{H}_{H_2O} \cdot x_{H_2O} + \hat{H}_{NaOH} \cdot x_{NaOH}. \tag{9–52}$$

If you now plot the data in Fig. 9–9 as $\bar{H}_{solution}$ versus the mole fraction of H_2O (x_{H_2O}), they will appear as in Fig. 9–10. A tangent to this latter curve has a slope equal to $(b - a)$ and an intercept (a). Hence the equation for the tangent line is

$$\bar{H}_{solution} = (b - a)x_{H_2O} + a$$
$$= bx_{H_2O} + a(1 - x_{H_2O}) = bx_{H_2O} + ax_{NaOH}.$$

But $\bar{H}_{solution}$ is also given by Eq. (9–52). Therefore

$$b = \hat{H}_{H_2O}, \qquad a = \hat{H}_{NaOH}.$$

Exercise 9–14. Prepare a plot of volume per mole of solution versus mole fraction of water, x, for a solution of ethyl alcohol and water, using the data of Exercise 9–11, and using the pure components as the reference state. From this plot, evaluate $\hat{V}_{alcohol}$ and $\hat{V}_{H_2O}$ at 25 mole % and 75 mole % alcohol.

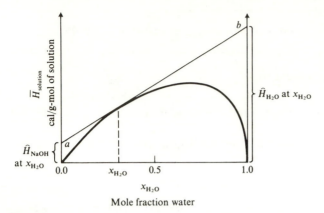

Fig. 9–10. Graphical computation of partial molar enthalpies (NaOH–H_2O solution).

9–21
The energy-conservation equation for chemical work systems

The generalized energy-conservation equation (6–1d), when applied to a system that does $P\,dV$ work and whose composition can change (i.e., the system can also do chemical work), takes the form

$$dU = T\,dS - P\,dV + \sum \mu_i\,dn_i. \qquad (9\text{–}53)$$

Equation (9–53) applies to both open or closed systems moving between one equilibrium state and another. An open system experiences a composition change if the matter entering the system is not identical in mass and composition to the matter leaving it. A closed system experiences a composition change only if the following two conditions prevail: (a) It contains more than one independent chemical species which are kept apart by *inner constraints*. The constraints may take the form of a partition (as in the system on the left-hand side of Fig. 2–35a), or the absence of a spark or catalyst (as in an unreacted mixture of hydrogen and oxygen), or the diffusion barriers in an electrochemical battery on open circuit. (b) The constraints separating the species are removed. That is, the partition is removed, or the reaction barrier is overcome by introducing a spark or catalyst, or the battery circuit is closed. A composition change will then follow, and a new equilibrium state will result. The magnitude of the $\sum \mu_i\,dn_i$ change will be the $dG_{T,P}$ for the reaction or mixing process.

When mixing or reaction-inhibiting constraints are absent from a *closed-equilibrium* system, there can be no composition changes with time, and $\sum \mu_i\,dn_i = 0$.

9–22
Differential relations in chemical work systems

We can clarify the nature of the $\mu_i \, dn_i$ terms by realizing that Eq. (9–53) implies that

$$U = U(S, V, n_A, n_B, \ldots, n_i, \ldots). \tag{9–54}$$

As a consequence, the total differential of U (Eq. 6–8a) is

$$dU = \left(\frac{\partial U}{\partial S}\right)_{V,n_i} dS + \left(\frac{\partial U}{\partial V}\right)_{S,n_i} dV + \sum_i \left(\frac{\partial U}{\partial n_i}\right)_{S,V,n_j} dn_i. \tag{9–55}$$

When we compare Eq. (9–53) with Eq. (9–55), it follows that

$$\left(\frac{\partial U}{\partial S}\right)_{V,n_i} = T, \tag{6–10}$$

$$\left(\frac{\partial U}{\partial V}\right)_{S,n_i} = -P, \tag{6–11}$$

$$\left(\frac{\partial U}{\partial n_i}\right)_{S,V,n_j} = \mu_i. \tag{9–56}*$$

Equation (9–56) says that μ_i is the rate at which the internal energy of the entire system changes per mole of component i added at constant S, V, and constant quantity of all other components.

Earlier in this section we expressed the chemical potential in terms of the system's free energy (Eq. 9–24a). We now have an alternative expression:

$$\mu_i = \left(\frac{\partial G}{\partial n_i}\right)_{P,T,n_j} = \left(\frac{\partial U}{\partial n_i}\right)_{S,V,n_j}. \tag{9–57}$$

Note that the constant parameters differ on each partial derivative in the preceding equation. The expression $(\partial U/\partial n_i)_{S,V,n_j}$ is *not* the partial molar internal energy because partial molar properties are defined as partial derivatives at constant P and T. On the other hand, $(\partial G/\partial n_i)_{P,T,n_j}$ is $\widehat{G}_i$, the *partial molar free energy*.

The chemical potential may also be expressed in terms of partial derivatives of H and A. From the definition of enthalpy,

$$H \equiv U + PV, \tag{2–16}$$

it follows that

$$dH = dU + P \, dV + V \, dP. \tag{9–58}$$

---

* The subscript n_j on the partial derivative with respect to n_i means that all components other than i are held constant.

Substituting Eq. (9–53) for dU in Eq. (9–58),

$$dH = T \, dS + V \, dP + \sum_i \mu_i \, dn_i, \qquad (9\text{–}59)$$

from which it follows that

$$\left(\frac{\partial H}{\partial n_i}\right)_{S,P,n_j} = \mu_i. \qquad (9\text{–}60)$$

By similar means, we can show that

$$dA = -P \, dV - S \, dT + \sum_i \mu_i \, dn_i, \qquad (9\text{–}61)$$

and therefore

$$\left(\frac{\partial A}{\partial n_i}\right)_{V,T,n_j} = \mu_i.$$

Also

$$dG = V \, dP - S \, dT + \sum_i \mu_i \, dn_i, \qquad (9\text{–}62)$$

and therefore

$$\left(\frac{\partial G}{\partial n_i}\right)_{P,T,n_j} = \mu_i.$$

Summarizing, we obtain

$$\mu_i = \left(\frac{\partial G}{\partial n_i}\right)_{P,T,n_j} = \left(\frac{\partial U}{\partial n_i}\right)_{V,T,n_j} = \left(\frac{\partial H}{\partial n_i}\right)_{S,P,n_j} = \left(\frac{\partial A}{\partial n_i}\right)_{V,T,ni}. \qquad (9\text{–}63)$$

Maxwell relations for chemical potential are derived by applying the Euler criterion (Eq. 6–14) to Eqs. (9–53), (9–58), (9–61), and (9–62). For example, from Eq. (9–53):

$$\left(\frac{\partial \mu_i}{\partial V}\right)_{S,n_j,n_i} = -\left(\frac{\partial P}{\partial n_i}\right)_{S,V,n_j}, \qquad (9\text{–}64)$$

$$\left(\frac{\partial \mu_i}{\partial S}\right)_{V,n_j,n_i} = \left(\frac{\partial T}{\partial n_i}\right)_{S,V,n_j}. \qquad (9\text{–}64a)$$

Exercise 9-15

i) Show that:

$$\left(\frac{\partial \mu_i}{\partial T}\right)_{P,n_j n_i} = -\left(\frac{\partial S}{\partial n_i}\right)_{P,T,n_j} = -\widehat{S}_i, \qquad (9\text{-}65)$$

$$\left(\frac{\partial \mu_i}{\partial P}\right)_{T,n_j n_i} = \left(\frac{\partial V}{\partial n_i}\right)_{P,T,n_j} = \widehat{V}_i. \qquad (9\text{-}65a)$$

ii) Derive Eqs. (9-61), (9-62), and (9-63) from the definitions of A and G.

9-23
Gibbs-Duhem equation

Imagine a process in which a differential quantity of matter capable of expansion work and composition change is augmented by the addition of identical differential quantities of matter until it grows into a system of finite size. The internal energy accumulated in each differential piece of matter is

$$dU = T\, dS - P\, dV + \sum_i \mu_i\, dn_i. \qquad (9\text{-}53)$$

The internal energy of the final finite system will be the sum total of the differential energies of all pieces added to the system. Because dU, dS, dV, and dn_i in Eq. (9-53) are extensive properties, they are the only terms that are augmented as the size of the system increases, and it follows that the energy of the finite system is

$$U = TS - PV + \sum_i \mu_i\, n_i. \qquad (9\text{-}66)^*$$

Now if Eq. (9-66) is substituted in the defining equations for H, G, and A— Eqs. (2-16), (3-22), and (3-34)—then

$$H = TS - \sum_i \mu_i n_i, \qquad (9\text{-}67)$$

$$G = \sum_i \mu_i n_i, \qquad (9\text{-}68)$$

$$A = -PV + \sum_i \mu_i\, n_i. \qquad (9\text{-}69)$$

---

* See Appendix III for an alternative derivation.

The total derivative of Eq. (9–66) is

$$dU = T \, dS \qquad\qquad - P \, dV \qquad\qquad + \sum_i \mu_i \, dn_i$$

$$+ S \, dT \qquad\qquad - V \, dP \qquad\qquad + \sum_i n_i \, d\mu_i, \qquad (9\text{–}70)$$

and when we compare Eq. (9–70) with Eq. (9–53), it follows that

$$\boxed{S \, dT - V \, dP + \sum_i n_i \, d\mu_i = 0,} \qquad (9\text{–}71)$$

which is known as the GIBBS-DUHEM EQUATION.

Exercise 9–16. Perform the substitutions which yield Eqs. (9–67), (9–68), and (9–69).

We may also obtain these equations by integrating Eqs. (9–53), (9–61), and (9–62) with respect to extensive differentials only, i.e., by carrying out the accretion process used to obtain Eq. (9–66). At the completion of the accretion process, differentials of *intensive* properties (dT and dP) remain as differentials and have not increased in value. They may therefore be dropped from the final summation.

Exercise 9–17.

i) Show that the Gibbs-Duhem equation (Eq. 9–71) may be obtained from Eqs. (9–67), (9–68), and (9–69) by comparing the total derivatives of these equations with the total derivatives given in Eqs. (9–59), (9–61), and (9–62).

ii) Write a generalized version of the Gibbs-Duhem equation for a system that may do $F_i \, dX_i$ reversible work, in addition to expansion and chemical work. F_i is an intensive and X_i an extensive property.

The Gibbs-Duhem equation provides a limiting relationship among the *intensive* variables of any reactive or open system.

At constant temperature and pressure, Eq. (9–71) reduces to

$$\boxed{\left(\sum_i n_i \, d\mu_i \right)_{T,P} = 0.} \qquad (9\text{–}72)$$

Let us apply this equation to a binary liquid solution which is in equilibrium with vapors above it. The liquid phase, taken by itself, constitutes an open system because matter continuously enters and leaves its boundaries (the liquid surface). For a 2-component system, Eq. (9–72) is

$$n_1 d\mu_1 + n_2 d\mu_2 = 0. \qquad (9\text{–}73)$$

Now, from Eqs. (9–24a), (9–41a), and (8–3), we have

$$d\mu_i = d\widehat{G}_i = RT \, d \ln f_i. \qquad (9\text{–}74)$$

For solutions in equilibrium with an *ideal* vapor phase the f_i of a liquid-phase component equals the partial pressure of that component in the vapor phase.

Therefore

$$d\mu_i = RT \, d \ln p_i,$$

which transforms Eq. (9–73) into

$$n_1 d \ln p_1 + n_2 d \ln p_2 = 0$$

or, dividing by $(n_1 + n_2)$ and by $dx_1(= -dx_2)$, we obtain

$$x_1 \frac{d \ln p_1}{dx_1} - x_2 \frac{d \ln p_2}{dx_2} = 0. \qquad (9\text{–}75)$$

Exercise 9–18. In Section 9–13, we said that when one component of a binary (two-component) liquid solution follows Raoult's Law, the other follows Henry's Law. Show that this follows as a consequence of Eq. (9–75). That is, if

$$p_1 = P_1 x_1, \qquad \text{(Raoult's Law)}$$

then

$$p_2 = K_2 x_2. \qquad \text{(Henry's Law)}$$

For real binary solutions, Eq. (9–73) is combined with Eqs. (9–74) and (9–35a),

$$n_1 d \ln \gamma_1 + n_1 d \ln x_1 + n_2 d \ln \gamma_2 + n_2 d \ln x_2 = 0,$$

and, again dividing by $(n_1 + n_2)$ and $dx_1(= -dx_2)$, we obtain

$$x_1 \frac{d \ln \gamma_1}{dx_1} + x_1 \frac{d \ln x_1}{dx_1} - x_2 \frac{d \ln \gamma_2}{dx_2} - x_2 \frac{d \ln x_2}{dx_2} = 0.$$

But

$$x_1 \frac{d \ln x_1}{dx_1} = \frac{x_1 \, dx_1}{x_1 \, dx_1} = 1.$$

Therefore

$$x_1 \frac{d \ln \gamma_1}{dx_1} = x_2 \frac{d \ln \gamma_2}{dx_2}. \qquad (9\text{–}76)$$

If we know the activity coefficient of one component of a binary solution as a function of composition, then Eq. (9–76) enables us to determine the activity coefficient of the second component. Most often Eqs. (9–75) and (9–76) are used to examine the consistency of experimentally determined activity coefficients, or partial pressures, of binary solutions, since the slopes of the curves of $\ln p_i$ or $\ln \gamma_i$ versus composition must be related to each other as in these equations.

Exercise 9–18 demonstrated that Raoult's and Henry's laws are possible solutions of Eq. (9–75). Other solutions involving empirically determined constants have been proposed. Commonly used solutions are those proposed by Margules*

---

* M. Margules, *Sitzber. Akad. Wiss. Wien, Math. Naturw. Kl.* **104**, 1243 (1895).

and van Laar,* which are useful in estimating, from limited experimental data, the vapor pressures or activities of binary solutions over their entire composition range.

The *van Laar equation*, Eq. (9–77), contains two empirical constants $\mathbb{A}$ and $\mathbb{B}$:

$$\log \gamma_1 = \frac{\mathbb{A}x_2^2}{\left(\frac{\mathbb{A}}{\mathbb{B}} x_1 + x_2\right)^2}, \qquad \log \gamma_2 = \frac{\mathbb{B}x_1^2}{\left(x_1 + \frac{\mathbb{B}}{\mathbb{A}} x_2\right)^2}. \qquad (9\text{–}77)$$

These constants may be determined from one experimental measurement of liquid and vapor composition. Once $\mathbb{A}$ and $\mathbb{B}$ are known, one can compute the complete composition-activity coefficient curve.

Some binary solutions form azeotropes. *Azeotropes* are solutions, with a minimum or maximum boiling point, in which the compositions of the equilibrium liquid and vapor phases are identical. Therefore at the azeotropic composition, because $x_1 = y_1$ and $x_2 = y_2$, we may use Eqs. (9–37) and (9–5) to obtain:

$$\gamma_1 = \frac{y_1 \mathbb{P}}{x_1 P_1} = \frac{\mathbb{P}}{P_1}, \qquad \gamma_2 = \frac{\mathbb{P}}{P_2},$$

where the vapor pressures are taken at the boiling point of the azeotrope.

We then rearrange Eq. (9–77) to obtain

$$\mathbb{A} = \log \gamma_1 \left(1 + \frac{x_2 \log \gamma_2}{x_1 \log \gamma_1}\right)^2, \qquad \mathbb{B} = \log \gamma_2 \left(1 + \frac{x_1 \log \gamma_1}{x_2 \log \gamma_2}\right)^2.$$
$$(9\text{–}78)$$

Example 9–6.† Acetone and chloroform form an azeotrope at 66.6 mole % chloroform, which boils at 64.5°C, at 760 mm Hg. Estimate the constants of the van Laar equation and evaluate the activity coefficients at the azeotrope and at 20 mole % chloroform in the liquid phase. Vapor pressures (in mm Hg) of pure acetone and chloroform are as follows.

T, °C	Acetone, P_A	Chloroform, P_C
45	510.5	439
50	612.6	526
55		625.2
56.3	760	
60	860.6	739.6
60.9		760
70	1190	1019
80	1611	1403

---

* J. J. van Laar, *Z. Physik. Chem.* **72**, 723 (1910); **83**, 599 (1913).

† Based on an example in J. M. Smith and H. C. Van Ness, *Introduction to Chemical Engineering Thermodynamics*, New York: McGraw-Hill, 1959.

Answer. Assuming that the vapor phase is ideal, we write

$$p_A = y_A \mathbb{P} = \gamma_A x_A P_A, \qquad \gamma_A = \frac{y_A \mathbb{P}}{x_A P_A}. \tag{a}$$

But at the azeotrope, $x_A = y_A$. Therefore

$$\gamma_A = \mathbb{P}/P_A = \frac{760}{1000} = 0.760; \qquad \log \gamma_A = -0.1193,$$

$$\gamma_C = \mathbb{P}/P_C = \frac{760}{858} = 0.886; \qquad \log \gamma_C = -0.0525.$$

We obtain P_A and P_C at 64.5°C from the given data, by plotting that data as $\log P_i$ versus $1/T°K$. Then, from Eq. (9–78),

$$\mathbb{A} = -0.1193 \left(1 + \frac{0.666}{0.333} \cdot \frac{0.0525}{0.1193}\right)^2 = -0.4205,$$

$$\mathbb{B} = -0.0525 \left(1 + \frac{0.333}{0.666} \cdot \frac{0.1193}{0.0525}\right)^2 = -0.2396.$$

And from Eq. (9–77), we obtain

$$\log \gamma_A = \frac{-0.4205(x_C)^2}{\left(\frac{0.4205}{0.2396} x_A + x_C\right)^2}, \qquad \log \gamma_B = \frac{-0.2396(x_A)^2}{\left(x_A + (x_C)\frac{0.2396}{0.4205}\right)^2}. \tag{b}$$

At $x_C = 0.20$ and $x_A = 0.80$, the above equations yield

$$\gamma_A = 0.985, \qquad \gamma_C = 0.655.$$

The total vapor pressure of the liquid solution is

$$\mathbb{P} = P_{\text{solution}} = \gamma_A x_A P_A + \gamma_C x_C P_C. \tag{c}$$

We see that the vapor pressure of a solution (unlike that of a pure liquid) is not fixed by its temperature, but depends also on its composition. Similarly the boiling point of a solution (i.e., the temperature at which $P_{\text{solution}} = 760$ mm Hg) depends on its composition.

Figure 9–11 shows how the boiling-point and condensation-point temperatures of acetone–chloroform solutions vary with composition. The bubble-point line indicates the temperatures and compositions at which liquid solutions begin to boil. The dew-point line shows temperatures and compositions at which vapor begins to condense. A horizontal line drawn between the two connects equilibrium vapor and liquid compositions. The curves meet at the azeotrope and at the pure-component compositions. We can compute vapor compositions, y_A and y_C, from

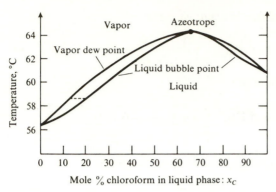

Fig. 9–11. Variation with composition of the boiling-point and condensation-point temperatures of acetone–chloroform solutions. Total pressure = 1 atm.

liquid compositions by combining equations (a) and (c) with the corresponding equation for γ_B to yield

$$y_A = \frac{1}{\dfrac{\gamma_C}{\gamma_A}\dfrac{x_C}{x_A}\dfrac{P_C}{P_A} + 1}. \tag{d}$$

If the vapor pressure ratio, P_C/P_A, is reasonably constant over the boiling range of the solution, then we can find y_A from (d) for all x_A's. However, if the P_C/P_A ratio changes appreciably, then a trial-and-error calculation is needed to determine the boiling temperature at each x_A, and the boiling temperature is then used to read P_C and P_A from the vapor-pressure plot. In this solution, P_C/P_A is constant.

Using (d), we obtain $y_C = 0.125$ when $x_C = 0.20$.

An experimental value was reported as $y_C = 0.126$ [Rasanoff and Easley, *J. Am. Chem. Soc.* **30**, 953 (1909)].

If the temperature of a binary system of vapor and liquid solutions (as in Fig. 9–11) is fixed, then its vapor pressure is a function of its composition. Similarly, if the total pressure of the system is fixed, then its temperature varies with its composition. Fixing both temperature and pressure fixes composition. Therefore it is impossible to vary the composition of a binary liquid–vapor system while maintaining both T and P constant, unless an inert gas is mixed with the vapors. Strictly speaking, Eqs. (9–73), (9–76), and (9–77) hold only for conditions of constant P and T. However, because the vapor pressure and the activity of a liquid depend only slightly on pressure [Poynting effect and Eq. (8–18)], the equations may be applied to systems in which the total pressure does vary slightly. The correct forms of the equation to be used under conditions of variable temperature and pressure have been given by Ibl and Dodge.*

---

* N. V. Ibl and B. F. Dodge, *Chem. Eng. Sci.* **2**, 120 (1953).

Exercise 9–19. Complete each of the following equations and show how they are interrelated.

a) $\sum_i \mu_i\, dn_i + \sum_i n_i\, d\mu_i =$

b) $\sum_i \mu_i n_i =$

c) $TS - VP - U =$

d) $TS - VP + \sum_i \mu_i n_i =$

e) $T\, dS - P\, dV + \sum_i \mu_i\, dn_i =$

f) $S\, dT - V\, dP - \sum_i \mu_i\, dn_i =$

g) $S\, dT - V\, dP + \sum_i n_i\, d\mu_i =$

h) $\left(\sum_i \mu_i\, dn_i \right)_{T,P} =$

i) $\left(\sum_i n_i\, d\mu_i \right)_{T,P} =$

9–24
Equilibrium in multiphase systems; constancy of μ_i

A *heterogeneous system* is one that consists of a number of phases. A *phase* is a *homogeneous, bounded* region of matter. At equilibrium, the intensive properties of a phase are constant throughout its extent.

For equilibrium to exist in a heterogeneous system, the temperature or pressure of each phase must be the same, so that no heat or bulk flow occurs. (We assume that there are no internal restraining walls or sharply curved interfaces.) In addition, the chemical potential of any component, μ_i, must be the same in all phases, so that there is no net diffusion of component i from one region of the system to another. The requirement for uniform μ_i's may be shown to be a consequence of the equilibrium criterion

$$dG_{T,P} = 0. \tag{3–42}$$

For Eq. (3–42) to hold in a chemical and expansion-work system, it follows from Eq. (3–43) that

$$\sum_i \mu_i\, dn_i = 0. \tag{9–79}$$

Let us apply this equation to a system having, for example, three phases. [The phases are designated by single prime ('), double prime ("), and triple prime (''').]

Equation (9–79) becomes

$$\sum_i \mu_i \, dn_i = \sum_i \mu_i' \, dn_i' + \sum_i \mu_i'' \, dn_i'' + \sum_i \mu_i''' \, dn_i''' = 0. \tag{9–80}$$

Taken together, the three phases constitute a heterogeneous *closed* system. Therefore the total number of moles of each component is constant (number subscripts refer to components 1, 2, 3, . . . , etc.), and

$$\sum_{\text{all phases}} dn_1 = 0 = \sum_{\text{all phases}} dn_2 = \sum_{\text{all phases}} dn_3 = \cdots.$$

Hence

$$dn_1' = -dn_1'' - dn_1''',$$
$$dn_2' = -dn_2'' - dn_2''', \tag{9–81}$$
$$dn_3' = -dn_3'' - dn_3''',$$
$$\vdots \qquad \vdots \qquad \vdots$$

Substituting Eq. (9–81) in Eq. (9–80) so as to eliminate the dn_i' terms, we obtain

$$\sum_i \mu_i \, dn_i = (\mu_1'' - \mu_1') \, dn_1'' + (\mu_1''' - \mu_1') \, dn_1'''$$
$$+ (\mu_2'' - \mu_2') \, dn_2'' + (\mu_2''' - \mu_2') \, dn_2''$$
$$+ (\mu_3'' - \mu_3') \, dn_3'' + (\mu_3''' - \mu_3') \, dn_3''' + \cdots = 0. \tag{9–82}$$

With the single-prime phase terms, dn_i', eliminated, all the remaining dn_i'' and dn_i''' terms are independent, and may have any value. Therefore, for $\sum \mu_i \, dn_i$ to always equal zero, it must be that

$$\mu_i'' = \mu_i' = \mu_i'''. \tag{9–83}$$

Therefore μ_i has the same value in all phases of a heterogeneous equilibrium system.

9–25
The phase rule

We now see that the Gibbs-Duhem equation, (9–71), is a relation among just those intensive properties which must be uniform throughout all phases of a heterogeneous equilibrium system. The equation implies that for every phase (considering each phase as a separate open system), there must be a function (ϕ),

$$\phi^{\text{phase}} (P, T, \mu_i) = 0. \tag{9–84}$$

This equation is the basis of the PHASE RULE, a rule with far-reaching consequences and applications, which was first formulated by J. Willard Gibbs.

The considerations leading to the Phase Rule are simply the following:

The number of variables in Eq. (9–84) is exactly equal to the number needed to specify the thermodynamic state of a system capable of expansion and chemical work. (See the State Principle in Section 6.) A function such as (9–84) exists for each phase in a heterogeneous system. These functions are the equations of state of each phase:

$$\text{phase (')} \qquad \phi'(P, T, \mu_1, \mu_2, \mu_3, \ldots) = 0,$$

$$\text{phase (")} \qquad \phi''(P, T, \mu_1, \mu_2, \mu_3, \ldots) = 0, \qquad (9\text{--}85)$$

$$\text{phase (''')} \qquad \phi'''(P, T, \mu_1, \mu_2, \mu_3, \ldots) = 0,$$

$$\vdots \qquad\qquad \vdots \qquad\qquad\qquad\qquad\qquad \vdots$$

where P, T and μ_i have the *same* value in each phase.

The number of variables in each function is equal to

$$C + 2,$$

the 2 referring to pressure and temperature, and the C being the number of *independent* components in the entire heterogeneous system. (An *independent component* is one whose amount can be changed without affecting the amount of other independent components in the system.)

Now, in a function like (9–84), with $C + 2$ variables, only $C + 1$ are *independent*. This means that if all but one of the variables are assigned arbitrary values, the value of the one remaining variable becomes fixed. Also, if *additional* functions relating any of the $C + 2$ variables exist, each additional independent function reduces the number of independent variables by one. For example, if $C + 2$ independent equations were available, they could all be solved simultaneously to determine all $C + 2$ variables and there would be *no* independent variable.

Now each phase has its own equation of state (Eq. 9–85), which is just such an independent function as will reduce the number of independent variables by one.

Therefore the number of independent intensive variables in a multiphase equilibrium system depends on the number of phases, as follows:

$$\boxed{\mathbf{F} = C + 2 - \phi,} \qquad\qquad (9\text{--}86)$$

where $\mathbf{F}$ = the number of independent intensive variables in a system (also called *degrees of freedom* or *variance*), and ϕ = number of phases.

Equation (9–86) is called the *Phase Rule*.

Example 9–7. Consider a system of pure water. (1) What is the maximum number of phases that can coexist in such a system? (2) What are the maximum degrees of freedom? (3) How many degrees of freedom are there in 2-phase regions?

Answer. (1) From Eq. (9–86), when $C = 1$ (pure water),

$$\mathbf{F} = (1 + 2) - \phi.$$

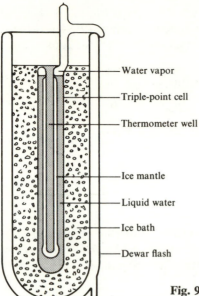

Fig. 9–12. Diagram of the National Bureau of Standards triple-point cell, in use in an ice bath within a Dewar flask.

The maximum number of phases occurs when $F = 0$, that is, the system is invariant. Thus

$$\phi_{max} = 3.$$

This condition at which ice, water vapor, and liquid water coexist is called the *triple point*. The temperature and pressure of the triple point are unique ($0.01°C$) and any attempt to vary either results in the disappearance of one of the three phases. In fact, the triple point of water is so unique that it is used as a reference standard for temperature calibrations. Figure 9–12 shows the triple-point cell designed by the National Bureau of Standards for use as a temperature standard.

2) Maximum degrees of freedom:

$$F \text{ is a maximum when } \phi \text{ is a minimum.}$$

Since ϕ can never be less than 1,

$$F = (1 + 2) - 1 = 2,$$

which is consistent with our statement that we can fix a state in a pure single work-mode system by fixing two of its intensive variables.

3) In the regions in which two phases (ice and liquid; liquid and vapor; vapor and ice) coexist in equilibrium:

$$F = (1 + 2) - 2 = 1.$$

Only *one* intensive variable is independent. Thus, in a two-phase, liquid–vapor system containing only one component, every value of pressure is associated with a unique value of temperature. For example, if the pressure of an isothermal steam–water system is increased, the vapor phase disappears. Conversely, if the pressure is decreased and the temperature remains fixed, the liquid phase disappears. When two phases are present in a pure system, fixing the pressure automatically fixes the temperature at a value given by the vapor pressure or saturation curve for the system.

It is often hard to determine C in reacting systems. C equals the least number of pure materials required to make up the entire system. It also equals the number of components whose quantity can be varied without changing the concentration of other components. Therefore each pure phase is considered a component.

Exercise 9–20. A solution of liquid benzene and toluene is contained in a piston-cylinder machine. The temperature of the liquid is raised and eventually vapor begins to form. The piston rises so as to keep the pressure of the system constant. Show that the phase rule requires that additional intensive variables begin to change as the temperature continues to rise, as soon as vapor appears. Which variables change and how are the changes calculated?

Exercise 9–21. Time: Saturday morning. A male suburban citizen (U.S.) is handed a shopping list by his wife, reading, "Buy: cookies, cheese, butter, milk, yogurt, meat for Sunday dinner, apples, oranges, carrots, peas, potatoes."

a) As he walks to the closet for his hat, he is told, "Get twice as many kilos of apples as oranges."

b) At the living-room threshold, he hears, "We're having Cub Scouts. Spend $5 on cookies and milk, and spend as much on butter as on cheese."

c) In the entrance hall, "Spend twice as much on meat as you do on fruit and vegetables."

d) At the front door, "Get the same weight of carrots as peas, and twice that weight of potatoes."

e) At the garage door, "Harry! Spend exactly $20."

f) As the car moves out of the driveway, "Harry! Forget the yogurt."

A comprehensive price list will meet Harry's eye at the entrance to the grocery store. How many degrees of freedom does Harry have initially, and after each stage from (a) through (f) (with regard to the number of kilos of items on his shopping list, of course)?

Problems

1. What is the free energy and the entropy of a 50–50 mole % solution of liquid *n*-hexane and liquid *n*-octane at 230°C and at its equilibrium pressure, relative to the pure components at 230°C?

2. A liquid solution containing 20 mole % heptane and 80 mole % hexane is at its equilibrium vapor pressure and at a temperature of 100°C.

a) Determine the free energy per mole of the mixture relative to its pure components at the same conditions.

b) What is the minimum isothermal work required to separate 1 mol of the mixture into its components?

3. Calculate the enthalpy, free energy, and entropy of mixing 5 g-mols CO_2, 2 g-mols C_3H_8, and 2.5 g-mols C_4H_{10}, each at 400°C and 20 atm, assuming that these real gases form an ideal gaseous solution.

4. A *gas chromatograph* is an instrument used to separate components of a gas mixture for purposes of analysis or purification. Essentially it consists of a long tube or column which acts as a differential delay line for gases flowing through the tube. It operates as follows: A small amount of the gas mixture to be separated into its individual components is introduced into a stream of inert (carrier) gas immediately upstream of the entrance of the chromatographic column. The inert gas flows into the column, carrying the gas mixture with it. The passage of the components of the mixture is, however, retarded by the column and each component moves through the column at its own characteristic flow rate. If the column is properly constructed, each component of the original mixture will leave the tube at a distinct time different from the time of the other components. The original mixture is therefore "unmixed." Does the gas chromatograph violate the second law of thermodynamics? Explain.

5. Compute the change in entropy of 1 milliliter of an equimolar binary mixture of ideal gases A and B, between the inlet and outlet of a gas chromatograph that uses helium as its carrier, given that gas A is swept out of the chromatographic tube by the helium 100 sec before gas B, and at the time of exit each component gas constitutes 20% of the exit-gas stream.

What is the maximum possible concentration of a component gas in the carrier gas at exit conditions?

6. By the pseudo-critical method, compute the molar volume of a mixture containing 30 mole % N_2 and 70 mole % CH_4 at 800 atm and 100°C. Compare this volume with that obtained from the ideal-gas law.

7. We need to produce a mixture of N_2 and CH_4 having a molar volume 0.05 liter/lb-mol at 300 atm and 60°C. Using the data in Table 7–3, calculate the composition of the mixture by each of the following methods:

a) Ideal-gas law and Amagat's Law.

b) Pseudo-critical method.

8. a) Assuming Raoult's Law and Dalton's Law, determine the composition of the vapor in equilibrium with liquid for the system butane and hexane at 100°C and 350 psia total pressure.

b) Compare the result of (a) with the composition obtained by using the fugacity method, Eq. (9–27).

9. A gas mixture containing 20 mole % methane, 40 mole % propane, and 40 mole % butane originally at 1 atm and 100°C is compressed isothermally to its dew point.

a) Find the composition of the equilibrium liquid condensed at the dew point.

b) Find the equilibrium pressure of the mixture.

c) If the compression is carried out reversibly, what is the minimum work required?

10. A gaseous mixture contains 40% methane and 60% nitrogen at 80°C and 25 atm pressure. Calculate $\widehat{H}, \widehat{G}, \widehat{S}, \widehat{V}$ for each of the components in this mixture for the following conditions:

a) The mixture behaves as a real gas.

b) The mixture behaves as an ideal gas.

(In both (a) and (b), the mixture forms an ideal solution.)

11. If 1 mol of H_2SO_4 (pure liquid) is diluted with n moles of water, the heat evolved is given by the relation:

$$Q = \frac{17.860n}{n + 1.7983} \text{ cal}$$

at 18°C and 1 atm for $n < 20$. What is the partial molar enthalpy of water and H_2SO_4—that is, $\widehat{H}_{water}$ and $\widehat{H}_{acid}$—for a solution containing 4 mol of acid and 6 mol of water? [Thompson, *Thermochemistry*, London: Longmans Green, page 75 (1908).]

12. For a mixture of CO, H_2, and CH_4 containing 3:1:4 g-mol of each, respectively, determine

$$\left(\frac{\partial \mu_{CH_4}}{\partial n_{CH_4}}\right)_{T,P,n_{CO},n_{H_2}}$$

at 300°K and 1 atm. Assume an ideal-gas and ideal-solution behavior.

13. The biological polymer *collagen* changes its dimensions reversibly when its chemical environment is altered. Thus the polymer shrinks in the presence of certain aqueous salt solutions and regains its length when washed free of salt. Fibers made from this polymer will do work on shrinking, and engines have been devised which take advantage of this fact.*

The accumulated internal energy of a system consisting of collagen fiber, absorbed salt, and water is therefore

$$U = U(S, \text{fiber length } l, n_{H_2O}, n_{salt}).$$

a) Write the total differential of U.

b) Show that for cyclic operation at constant T, the tension-contraction work done by the fiber is equal to minus the chemical work released. That is,

$$\oint \tau \, dL = - \oint \sum \mu_i \, dn_i, \quad \text{where } \tau = \text{tension.}$$

c) Show that

$$\left(\frac{\partial S}{\partial L}\right)_{\tau,n_i} = - \left(\frac{\partial \tau}{\partial T}\right)_{L,n_i}.$$

---

* M. V. Sussman and A. Katchalsky, "Mechano-Chemical Turbine," *Science*, **167**, 45 (1970).

d) Show that

$$d\tau = \frac{\Delta H_{contraction}}{\Delta L_{contraction}} d\ln T \qquad \text{(analog of the Clapeyron equation).}$$

14. The following data are experimental vapor pressures of HNO_3 solutions (K. Denbigh, *ibid.*). Check the consistency of these data by means of Eq. (9–75).

HNO_3, % by weight	P_{H_2O}, mm Hg	P_{HNO_3}, mm Hg
50	7.53	0.49
60	4.93	0.89
70	2.83	3.08
80	1.35	10.49
90	0.46	26.03
100	0.00	48.0

15. The partial pressures of the acetone vapor above solutions of acetone and water of various compositions at 60°C are as follows.

Acetone	0	3.3	11.7	31.8	55.4	73.6	100.0
P acetone, mm Hg	0	190	443	588	672	711	860

The vapor pressure of pure water at 60°C is 149.4 mm Hg. By means of the Gibbs–Duhem equation, compute the partial pressure of water above this solution at 60°C.

16. At 80°C the vapor pressure of pure water and of dioxane are 355.1 and 383 mm Hg, respectively. Measured total pressure above liquid mixtures of dioxane and water at 80°C are as follows:

Mole % water	10	20	30	40	60	70	80	90
Total pressure, mm Hg	476	526.5	566	571	575.5	569.5	550	501.5

a) By means of the Gibbs–Duhem equation, compute the partial pressures of dioxane and water above liquid solutions of various compositions at 80°C.

b) Using the result obtained in part (a), evaluate the Henry's-Law constant for dioxane in a dioxane-water system at 80°C.

c) Over what range of liquid composition would Henry's Law be satisfactory?

d) Make a plot of partial pressure of dioxane versus liquid composition, including lines for Henry's Law and Raoult's Law.

17. Ethyl alcohol and ethyl acetate form an azeotrope at 53.9 mole % ethyl acetate, which boils at 71.8°C at 76 mm Hg.

a) Determine the values of the van Laar constants A and B in Eq. (9–78), assuming these to be independent of pressure and temperature.

b) Determine the azeotrope composition and the total pressure if phase change occurs at 56.3°C. Use the following vapor-pressure data.

Temperature, °C	Ethyl acetate, mm Hg	Ethyl alcohol, mm Hg
56.3	360	298
71.8	636	587

18. How many degrees of freedom are there in the following systems?

a) A liquid solution of ethanol in water in equilibrium with its vapor.

b) A pure substance at its critical point.

c) A binary azeotrope.

d) Any mixture of CO, O_2, and CO_2 at room temperature.

JACOBUS HENRICUS VAN'T HOFF,
1852–1911

Section 10
CHEMICAL EQUILIBRIUM

A major achievement of thermodynamics is its ability to predict the conditions at which a chemical-reaction process is at equilibrium and the direction a chemical-reaction process will take. For example, if we should want to know whether material A can react with material B to form material C (where C is a compound combining the atoms in A and B, such as $H_2 + \frac{1}{2}O \rightarrow H_2O$), thermodynamic techniques tell us whether the tendency for $A + B$ to form C is greater than or less than the tendency for C to decompose to A and B. In addition, these techniques tell us the concentrations of reactant and product materials that can coexist without change, i.e., the equilibrium concentrations at a given temperature and pressure.

The *Gibbs free energy*, G, is the property used to make these predictions about reaction processes. *From the free-energy point of view, all chemical-reaction processes appear as combinations of mixing and separation (unmixing) processes.* Chemical-equilibrium thermodynamics is therefore an extension of the thermodynamics of mixing and separation processes described in Section 9.

10–1
The chemical-equilibrium state

Suppose that we have materials A and B which can react chemically to form products C and D, as shown in Eq. (10–1),

$$aA + bB \rightarrow cC + dD, \qquad (10-1)$$

where the lower-case letters (a, b) indicate the number of moles of each reactant material needed to produce the c and d moles of each of the product materials.

If a moles of A and b moles of B are mixed in a suitable vessel (a catalyst may be required to promote the reaction), they form products C and D. However, if, as in gaseous reactions, the components of the reaction are mutually soluble, the quantities of C and D formed are always less than c and d moles, simply because the reaction stops before the reactants (A and B) are completely consumed. This condition prevails because product molecules (C and D) accumulate to a concentration such that they meet to form reactant molecules (A and B) as fast as the reactant molecules are forming product. Composition then becomes time invariant, as are temperature and pressure, and the system—consisting of a mixture of reactants and products—is in a state of CHEMICAL EQUILIBRIUM.

As an example, sulfur dioxide reacts with oxygen in the presence of a platinum or vanadium catalyst to form sulfur trioxide:

$$SO_2 + \tfrac{1}{2}O_2 \rightarrow SO_3.$$

(This reaction is the basis of the *contact process* for manufacturing sulfuric acid.) If 1 mol of SO_2 is mixed with $\frac{1}{2}$ mol of O_2 in a 680°C reaction vessel containing catalyst, half the SO_2 is quickly converted to SO_3. The composition in the reaction vessel then becomes time invariant, because *a chemical-equilibrium state* has been reached. If SO_3 is now removed from the mixture, more unreacted SO_2 and O_2 will combine to compensate for the SO_3 removed. If SO_3 is added to the equilibrium mixture, it will in part decompose to SO_2 and O_2. Adding SO_2 and O_2 to (or removing them from) the equilibrium reaction mixture will similarly cause the reaction to move in the forward (or reverse) direction.

In the chemical-equilibrium state, mixtures of reactants and products coexist at fixed temperatures and pressures and in predictable proportions.

10–2
ΔG of equilibrium chemical-reaction process

Conversion of reactants to products or products to reactants, *at equilibrium concentration*, proceeds with *zero change in free energy*. To demonstrate this fact, we employ a van't Hoff equilibrium box (Fig. 10–1) which is the same *gedanken apparat* used to reversibly mix and/or separate vapors in Section 9, adapted to carrying out a chemical-reaction process (Eq. 10–1) reversibly. Although we shall describe gaseous reactions, the principles derived will later be shown to apply to reactions of materials in any state of aggregation.

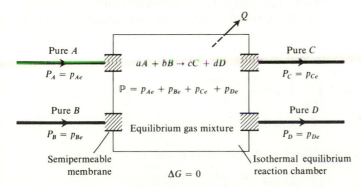

Fig. 10–1. Van't Hoff equilibrium box for carrying out a chemical-reaction process reversibly when reactants and products are pure and at their *equilibrium* conditions.

Pure gases A and B enter the box at a steady rate through separate conduits and semipermeable membranes in the proportions given by Eq. (10–1). That is, for every a moles of A entering, b moles of B enter. The box is maintained at a constant temperature and constant *total* pressure. Chemical equilibrium exists within the box, in which reactants and products—A, B, C, and D—are present in constant proportions. The equilibrium partial pressures of the components in the box are p_{Ae}, p_{Be}, p_{Ce}, and p_{De} (the subscript e's denote equilibrium). As more A and B enter the equilibrium chamber, they mix, react, and convert to C and D, which leave through their respective semipermeable membranes and conduits. If the reactants are fed to the box in the proportions of Eq. (10–1), then products leave in these proportions. The partial pressures within the box remain fixed at the equilibrium values.

Let us evaluate the ΔG for the isothermal process, wherein the pure gases in the conduits are each at a total pressure equal to the equilibrium partial pressure within the equilibrium chamber. That is,

$$P_A = p_{Ae}; \qquad P_B = p_{Be}; \qquad P_C = p_{Ce}; \qquad P_D = p_{De},$$

and a moles of A plus b moles of B are converted to cC and dD by passage through the equilibrium box. The overall process is

$$aA \begin{Bmatrix} \text{pure and} \\ \text{at } P_A \end{Bmatrix} + bB \begin{Bmatrix} \text{pure and} \\ \text{at } P_B \end{Bmatrix} \rightarrow cC \begin{Bmatrix} \text{pure and} \\ \text{at } P_C \end{Bmatrix} + dD \begin{Bmatrix} \text{pure and} \\ \text{at } P_D \end{Bmatrix},$$

$$(10\text{--}2)$$

all at constant temperature T.

To maintain a constant temperature in the box, a quantity of heat equal to the ΔH_R enters or leaves the system from a thermostated bath surrounding the apparatus.

The ΔG for the process is equal to:

i) ΔG of mixing pure aA at P_A with pure bB at P_B, plus
ii) ΔG of converting aA and bB to cC and dD at equilibrium conditions, plus
iii) ΔG of separating cC and dD from the equilibrium mixture, at P_C and P_D respectively.

For mixing the reactants, step (i), using Eq. (3–29a),

$$\Delta G_{(i)} = a \int_{P_A}^{P_{Ae}} \bar{V} \, dP + b \int_{P_B}^{P_{Be}} \bar{V} \, dP = 0 + 0,$$

because the initial pressures of A and B are the same as their partial pressures in the mixture.

Similarly, for step (iii), separating the products C and D, we obtain

$$\Delta G_{(iii)} = c \int_{P_{Ce}}^{P_C} \bar{V}_C \, dP + d \int_{P_{De}}^{P_D} \bar{V}_D \, dP = 0 + 0,$$

because the pure products leave at total pressures equal to their partial pressures in the mixture (Eq. 10–2).

To evaluate ΔG_R for the conversion step (ii), we recall the definition of the Gibbs free energy G (Eq. 3–22c). Hence, for an isothermal reaction process,

$$\Delta G_R = \Delta H_R - T \, \Delta S_R.$$

The conversion is reversible and isothermal. Therefore

$$\Delta S_R = \frac{Q_R}{T} = \frac{\Delta H_R}{T}.$$

Hence

$$\Delta G_{R\,(\text{equilibrium})} = \Delta H_R - T\left(\frac{\Delta H_R}{T}\right) = 0. \qquad (10\text{–}3)$$

In other words, *no* change in free energy is required to interconvert reactants and products that are in a state of chemical equilibrium.

10–3
ΔG of a chemical-reaction process at standard conditions

The chemical-reaction process,

$$aA \text{ (at 1 atm and } T_{\text{std}}) + bB \text{ (at 1 atm and } T_{\text{std}})$$

$$\rightarrow cC \text{ (at 1 atm and } T_{\text{std}}) + dD \text{ (at 1 atm and } T_{\text{std}}), \qquad (10\text{–}4)$$

which starts with gaseous reactants, each *pure* and at a pressure of *one* atmosphere and at a standard temperature of 298°K, and ends with *pure* gaseous products, each at the same 1-atm pressure and standard temperature conditions, may be carried out continuously and reversibly in the van't Hoff equilibrium box shown in Fig. 10–2. It differs from that in Fig. 10–1 in that *reversible isothermal engines* have been added to each reactant and product line to expand the pure reactant gases from the standard pressure to their respective equilibrium pressures and to compress the pure product gases from equilibrium pressure to standard pressure.

    The change in free energy for the process is called the STANDARD FREE ENERGY

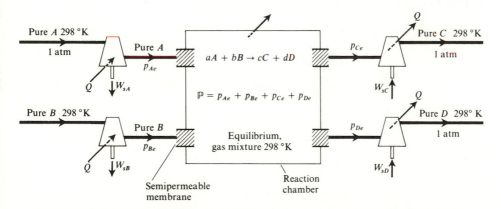

Fig. 10–2. Van't Hoff equilibrium box for carrying out a chemical-reaction process reversibly when the reactants and products are pure and at *standard* conditions.

OF REACTION, and equals

$$\Delta G_R^0 = a\,\Delta\bar{G}_A + b\,\Delta\bar{G}_B + c\,\Delta\bar{G}_C + d\,\Delta\bar{G}_D$$

$$= a\int_1^{P_{Ae}} \bar{V}_A\,dP + b\int_1^{P_{Be}} \bar{V}_B\,dP + c\int_{P_{Ce}}^1 \bar{V}_C\,dP + d\int_{P_{De}}^1 \bar{V}_D\,dP,$$

$$(10\text{-}5)$$

which is the sum of the ΔG's for isothermally expanding each reactant from its initial to its partial pressure in the equilibrium mixture and compressing each product from its equilibrium pressure to the pressure of the final pure state. In other words, it is the ΔG of mixing the reactants into—and separating the products from—the equilibrium reaction mixture. Applying the definition of fugacity, Eq. (8-4), we get

$$\Delta G_R^0 = aRT\ln\frac{f_{Ae}}{f_{A1}} + bRT\ln\frac{f_{Be}}{f_{B1}} + cRT\ln\frac{f_{C1}}{f_{Ce}} + dRT\ln\frac{f_{D1}}{f_{De}}$$

$$= -RT\ln\frac{(f_{Ce})^c(f_{De})^d}{(f_{Ae})^a(f_{Be})^b} + RT\ln\frac{(f_{C1})^c(f_{D1})^d}{(f_{A1})^a(f_{B1})^b},\qquad (10\text{-}6)$$

or letting the ratio of fugacities be represented by K and K_{initial},

$$\Delta G_R^0 = -RT\ln\frac{K}{K_{\text{initial}}}.\qquad (10\text{-}6a)$$

If the gases are ideal, then fugacity equals pressure and since the standard pressure is 1 atm, $K_{\text{initial}} = 1$. Equation (10-6) becomes

$$\boxed{\Delta G_R^0 = -RT\ln\frac{(p_{Ce})^c(p_{De})^d}{(p_{Ae})^a(p_{Be})^b},}\qquad (10\text{-}7)$$

which may also be written as

$$\boxed{\Delta G_R^0 = -RT\ln\frac{\Pi_c(p_{\text{products }e})^c}{\Pi_a(p_{\text{reactants }e})^a}}\qquad (10\text{-}8)$$

where $\Pi_c(p_{\text{products }e})^c$ means the mathematical product of all the product-component equilibrium partial pressures raised to the power of their stoichiometric coefficients. The denominator is the corresponding term for all reactant components.

10–4
Graphical representation of the ΔG_R^0 computation path

For a generalized gaseous reaction such as that shown by Eq. (10–9),

$$aA \rightarrow cC \text{ (initial and final materials, pure and at STP*),} \qquad (10\text{–}9)$$

the computation path for the free energy can be shown graphically in a manner that elucidates the computation technique (Fig. 10–3).

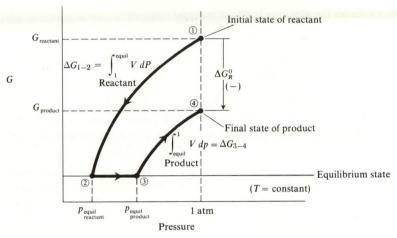

Fig. 10–3. Computation path for ΔG_R^0, the standard free energy of reaction.

Pure reactant(s) at an initial standard temperature (usually taken as 25°C or 298°K) and pressure of 1 atm (point 1 in Fig. 10–3) expands isothermally and reversibly to its equilibrium partial pressure (2), thereby dropping to the *minimum* free energy at which the system of A and C may coexist. The transformation of aA into cC in the equilibrium state occurs at constant free energy (2 to 3), and the free energy required to purify the product is that needed to compress it from its equilibrium partial pressure (3) to the standard pressure (4). The total change in free energy for the reaction process is therefore

$$\Delta G_R^0 = G_4 - G_1 = (G_2 - G_1) + (G_3 \overset{0}{\overset{\nearrow}{-} G_2}) + (G_4 - G_1)$$

$$= \underbrace{\int_1^{P_e} V \, dP}_{\text{Reactants}} + \underbrace{\int_{P_e}^1 V \, dP}_{\text{Products}}$$

$$= -RT \ln \frac{(f_{Ce})^c}{(f_{Ae})^a}$$

$$= -RT \ln \frac{(p_{Ce})^c}{(p_{Ae})^a}. \qquad \text{(Ideal gas only)}$$

---

* STP ≡ standard temperature and pressure = 298°K and 1 atm.

Note that in the Fig. 10–3 process, less isothermal shaft work is required to compress the product to the standard state (3 → 4) than is obtained in expanding the reactant from the standard to the equilibrium state (1 → 2). This means that this particular process goes from its initial to its final state with a net loss in free energy (ΔG_R^0 is negative). This process does work on its surroundings; therefore, it tends to occur spontaneously.

Exercise 10–1. Consider the ideal-gas reaction process:

$$B \to D \ (B \text{ initially at 1 atm and 25°C; final state of } D \text{ has same } T \text{ and } P).$$

In the equilibrium state for this system at 25°C and 1 atm total pressure, the partial pressure of B is found to be greater than that of D:

$$p_{Be} > p_{De} \ .$$

Draw the computation path for the process on a G–P plane (similar to Fig. 10–3). Label all points as in Fig. 10–3. What is the sign of ΔG_R^0? Does the process tend to occur spontaneously?

Exercise 10–2. Draw the computation path for finding ΔG of a chemical-reaction process in which pure A at STP forms product C as per Eq. (10–9) in an equilibrium reactor, and the *equilibrium mixture* is discharged *without* separation, as the final product. Write the equation for ΔG.

10–5
The equilibrium constant

The standard free energy of reaction, for a given set of pure reactant and product components, is a *constant* because it is the change in free energy for a process which moves a system between two fixed thermodynamic states, each with fixed, characteristic thermodynamic properties. (T, P, and composition are fixed for the initial and final state; therefore G is fixed for each state.)

$$\Delta G_R^0 = G_{products}^0 - G_{reactants}^0 = \text{constant.} \tag{10–10}$$

It therefore follows that the pressure or fugacity ratio in Eqs. (10–6), (10–7), or (10–8) is constant for a given reaction at a given T and P. This constant is called the EQUILIBRIUM CONSTANT. Thus

$$\boxed{\frac{(f_{Ce})^c (f_{De})^d}{(f_{Ae})^a (f_{Be})^b} = K \equiv \text{Equilibrium constant.}} \tag{10–11}$$

For *ideal-gas* reactions, Eq. (10–11) becomes

$$\boxed{\frac{(p_{Ce})^c (p_{De})^d}{(p_{Ae})^a (p_{Be})^b} = K_p.} \tag{10–12}$$

Although K is constant for a given reaction, the equilibrium fugacities or partial pressures are *not* constant and may take on any values which satisfy Eqs. (10–11) and (10–12). The ratio of partial pressures in Eq. (10–12) could have been written as in Eq. (10–8).

Note that the fugacity ratio K in Eq. (10–11) is always constant, whereas the partial pressure ratio K_p in Eq. (10–12) is constant only when the assumption of ideal-gas behavior is valid. From Eq. (10–11) or (10–12) and Eq. (10–6) or (10–7), it follows that

$$\Delta G^0_R = -RT \ln K. \tag{10–13}$$

Exercise 10–3. What are possible units of K in Eqs. (10–11) and (10–12)? Can you evaluate $\ln (5.2 \text{ atm})$? What term implicit in Eq. (10–13) makes evaluation of $\ln K$ possible when $a + b - c - d \neq 0$?

10–6
Tabulation of $\Delta G^0_{formation}$

The $\Delta G^0_{formation}$ of a chemical compound is the change in free energy for the process wherein a *compound* at standard temperature and pressure is formed from its *elements* at the same standard state. The standard temperature and pressure (STP) is usually taken as being 25°C and 1 atm. Table 10–1 contains values of ΔG^0_F of common chemical compounds, from a compilation by the U.S. Bureau of Standards.* The reference state for the table is the elements in their normal state of aggregation at 25°C and 1 atm.

10–7
Additivity of standard free energies: calculation of $\Delta G^0_{reaction}$

Standard free energies of formation, $\Delta G^0_{formation}$, of compounds can be combined arithmetically to obtain the standard free energy of reaction, $\Delta G^0_{reaction}$, in exactly the same way that standard enthalpies of formation are combined to obtain $\Delta H^0_{reaction}$:

$$\Delta G^0_{reaction} = G^0_{\text{final state}} - G^0_{\text{initial state}}$$

$$= \sum_{\text{Products}} \Delta G^0_{formation} - \sum_{\text{Reactants}} \Delta G^0_{formation}. \tag{10–14}$$

* From: D. D. Wagman, W. H. Beans, V. B. Barker, I. Halow, S. M. Bailey, and R. H. Schum, "Selected Values of Chemical Thermodynamic Properties," Washington, D.C.: *National Bureau of Standards Technical Notes TN*-270-3 and 4, 1968, 1969.

Table 10–1

Thermodynamic properties of selected compounds*

Compound	State	$\Delta G^0_{formation}$, kcal/g-mol	$S^0_{absolute}$, cal/°K · g-mol	
CO	g	− 32.8079	47.301	
CO_2	g	− 94.2598	51.061	
CH_4	g	− 12.140	44.50	
C_2H_2	g	50.000	47.997	
C_2H_4	g	16.282	52.45	Based on
C_2H_6	g	− 7.860	54.85	carbon
CH_3OH	l	− 39.73	30.3	as
$(COOH)_2$	c	− 166.8	28.7	graphite
C_2H_5OH	l	− 41.77	38.4	
$(CH_2OH)_2$	l	− 77.12	39.9	
CCl_4	l	− 16.4	51.25	
CS_2	l	15.2	36.10	
$(CN)_2$	g	70.81	57.86	
$CdCl_2$	c	− 81.88	28.3	
$CdSO_4$	c	− 195.99	32.8	
CdS	c	− 33.6	17.0	
CuO	c	− 30.4	10.4	
CuCl	c	28.4	21.9	
$CuSO_4$	c	− 158.2	27.1	
CaC_2	c	− 16.2	16.8	
CaO	c	− 144.4	9.5	
$Ca(OH)_2$	c	− 214.33	18.2	
$CaCl_2$	c	− 179.3	27.2	
$CaSO_4$	c	− 315.56	25.5	
$CaCO_3$	c	− 269.78	22.2	
Fe_2O_3	c	− 177.1	21.5	
Fe_3O_4	c	− 242.4	35.0	
FeS_2	c	− 39.84	12.7	
H_2O	g	− 54.6357	45.109	
H_2O	l	− 56.6902	16.716	
HF	g	− 64.7	41.47	
HCl	g	− 22.769	44.617	
HBr	g	− 12.72	47.437	
HI†	g	+ 0.31	49.314	
H_2S	g	− 7.892	49.15	
HNO_3	l		37.9	
HCOOH	g	− 80.24	60.0	
HCHO	g	− 26.3	52.26	
H_2CO_3	(aq)m = 1	− 149.00	45.7	
HCN	g	28.7	48.23	
KOH	(aq)m = 1	− 105.061	22.0	
$KMnO_4$	c	− 170.6	41.04	

Table 10-1 (*Continued*)

Compound	State	$\Delta G^0_{formation}$, kcal/g-mol	$S^0_{absolute}$, cal/°K·g-mol
KCl	c	−97.592	19.76
K$_2$SO$_4$	c	−314.62	42.0
KBr	c	−90.63	23.05
KI	c	−77.03	24.94
MnO$_2$	c	−111.4	12.7
MgCO$_3$	c	−246.0	15.7
NO	g	20.719	50.339
NO$_2$	g	12.390	57.47
N$_2$O	g	24.76	52.58
N$_2$O$_4$	g	23.491	72.73
NH$_3$	g	−3.976	46.01
NH$_4$Cl	c	−48.73	22.6
(NH$_4$)$_2$SO$_4$	c	−215.19	52.65
NaOH	(aq)m = 1	−100.184	11.9
NaCl	c	−91.785	17.30
Na$_2$SO$_3$	c	−239.5	34.9
Na$_2$SO$_4$	c	−302.78	35.73
NaNO$_3$	c	−87.45	27.8
Na$_2$CO$_3$	c	−250.4	32.5
SO$_2$	g	−71.79	59.40
SO$_3$	g	−88.52	61.24
SiO$_2$	(glass)	−190.9	11.2
ZnCl$_2$	c	−88.255	25.9
ZnS	c	−47.4	13.8
ZnSO$_4$	c	−208.31	29.8
ZnO	c	−76.05	10.5

* From D. D. Wagman, *ibid.*
† From solid I$_2$.

Example 10-1. What is the ΔG^0_R for the reaction process

$$CO_{(g)} + \tfrac{1}{2}O_{2\,(g)} \rightarrow CO_{2\,(g)} \qquad \text{at } 298°K, \text{ 1 atm?}$$

Answer: From Table 10-1:

$$CO_{(g)}: \qquad \Delta G^0_F = 32{,}808 \text{ cal/g-mol,}$$

$$O_{2\,(g)}: \qquad \Delta G^0_F = 0,$$

$$CO_{2\,(g)}: \qquad \Delta G^0_F = 94{,}260 \text{ cal/g-mol.}$$

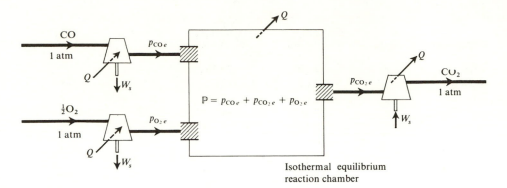

Fig. 10–4. The reversible production of pure CO_2 at STP from pure CO and O_2 at STP.

Therefore
$$\Delta G_R^0 = -94,260 - (-32,808) = -61,452 \text{ cal/g-mol.}$$

Example 10–2. The above chemical-reaction process is carried out isothermally and reversibly in a van't Hoff equilibrium box (Fig. 10–4) with the reactants and products pure and at standard temperature and pressure. Calculate, per mole of CO_2, the following values. (Assume ideal-gas behavior.)

a) Flow work
b) Shaft work performed by turbines
c) ΔH of process
d) ΔU of process
e) The heat load, Q, at each turbine and at the van't Hoff equilibrium box, to maintain isothermal conditions; and the net heat load.
f) ΔS of process
g) ΔG of process

Answer

a) Flow work $= (P\bar{V})_{CO_2} - (P\bar{V})_{CO} - \frac{1}{2}(P\bar{V})_{O_2} = -\frac{1}{2}P\bar{V} = -\frac{1}{2}RT$.
b) Shaft work performed by or on reversible isothermal turbines

$$= W_s = RT \ln \frac{P}{p_{CO\,e}} + \frac{1}{2}RT \ln \frac{P}{p_{O_2\,e}} + RT \ln \frac{p_{CO_2\,e}}{P}$$

$$= +RT \ln \frac{p_{CO_2\,e}}{p_{CO\,e}(p_{O_2\,e})^{1/2}} = -\Delta G_R^0 = 61,452 \text{ cal/g-mol.}$$

c) ΔH at each of the isothermal turbines $= 0$.
   ΔH for equilibrium box $= \Delta H_F^0 (CO_2) - \Delta H_F^0 (CO)$. Therefore, using data from Table 2–5, we find that

$$\Delta H = -94,052 + 26,416 = -67,636 \text{ cal/g-mol.}$$

d) $\Delta U = U^0_{CO_2} - \frac{1}{2}U^0_{O_2} - U^0_{CO} = \Delta H^0_R - \Delta(PV^0_R) = \Delta H^0_R + \frac{1}{2}RT$

$$= -67,636 + \frac{1}{2}\frac{1.99 \text{ cal}}{\text{g-mol} \cdot {}^\circ K} \times 298{}^\circ K = -67,339 \text{ cal/g-mol.}$$

e) For an isothermal ideal-gas expansion-flow process,

$$Q_{\text{turbines}} = W_s + \overset{0}{\cancel{\Delta H}} = -\Delta G^0_R = 61,452 \text{ cal/g-mol,}$$

$$Q_{\text{reaction}} = \Delta H^0_{\text{reaction}} = -67,636 \text{ cal/g-mol.}$$

Therefore

$$\text{Net } Q = -6,184 \text{ cal/g-mol.}$$

f) $\Delta S_{\text{turbines}} = \dfrac{Q_{\text{turbines}}}{T} = \dfrac{W_s}{T} = -\dfrac{\Delta G^0_R}{T}$

$\Delta S_{\text{reaction chamber}} = \dfrac{\Delta H^0_R}{T}$

$\Delta S_{\text{process}} = \Delta S_{\text{turbines}} + \Delta S_{\text{reaction chamber}} = \Delta S^0_R$

$$= \frac{\Delta H^0_R - \Delta G^0_R}{T} = -\frac{6184}{298} = 20.8 \text{ cal/g-mol} \cdot {}^\circ K$$

g) $\Delta G_{\text{process}} = \Delta G^0_R = -61,452 \text{ cal/g-mol}$

Exercise 10–4. Using data from Table 10–1, calculate ΔG^0_R and the equilibrium constant for the chemical-reaction process

$$NO + \tfrac{1}{2}O_2 \rightarrow NO_2 \qquad \text{at } 25{}^\circ C \text{ and } 1 \text{ atm.}$$

10–8
Absolute entropy and standard free energy of reaction

From the definition of G (Eq. 3–22c) and the fact that we are considering an isothermal process, it follows that

$$\Delta G^0_R = \Delta H^0_R - T\,\Delta S^0_R . \tag{10–15}$$

Therefore we can compute ΔG^0_R from data on standard enthalpy and entropy of reaction. We compute ΔH^0_R as described in Section 2, using the data of Table 2–5. We compute the STANDARD ENTROPY OF REACTION, ΔS^0_R, similarly, by subtracting the reactant entropies from the product entropies when all are at standard temperature and pressure:

$$\Delta S^0_R = \sum_{\text{Products}} S^0 - \sum_{\text{Reactants}} S^0. \tag{10–16}$$

However, the computation of ΔS_R^0 *differs* from the computation of ΔH_R^0 in two important respects:

1.   The tabulated values of S_{product}^0 and S_{reactant}^0 (Tables 10–1 and 10–2) are standard *absolute* entropies (see Eq. 3–21). These are the entropies of chemical substances at 25°C and 1 atm, relative to their entropies at 0°K (and not relative to the entropies of their elements at 298°K).

2.   It follows that both elements and compounds have entropies at standard temperature and pressure, and that the entropies of *elements* as well as those of compounds must be included in the summations in Eq. (10–16).

    The entropies of elements at 25°C and 1 atm are not arbitrarily set at zero (as is done with the enthalpies of elements in ΔH_R^0 compilations).   Therefore the

Table 10–2

Absolute entropies of selected elements at 298.16°K
(cal/°K · g-mol or g-atom)*

Element	State	S^0
Al	c	6.769
Ba	c	16.0
B	c	1.56
Br_2	g	58.639
Ca	c	9.95
C	gr	1.3609
C	diamond, c	0.5829
Cl_2	g	53.286
F_2	g	48.6
H_2	g	31.211
I_2	g	62.280
I	c	27.9
K	c	15.2
Mg	c	7.77
Mn	c	7.59
Ni	c	7.20
O_2	g	49.003
N_2	g	45.767
Na	c	12.2
P, white	c	10.6
S, rhombic	c	7.62
S, monoclinic	c	7.78
Ti	c	7.24
Zn	c	9.95

* From D. D. Wagman, *ibid.*

standard entropy of formation, ΔS_F^0, of a compound differs from the standard absolute entropy, $\Delta S_{compound}^0$, of that compound:

$$\Delta S_{formation}^0 = S_{compound}^0 - \sum S_{elements}^0. \qquad (10\text{-}17)$$

Values of absolute entropies are calculated either from spectroscopic or calorimetric data. Typical values are listed in Tables 10–1 and 10–2. When published or experimental data are lacking, we can estimate ΔS_R^0 by methods described in the literature.*

Example 10–3. Calculate the standard entropy of reaction for each of the following reactions occurring at 25°C and 1 atm and involving pure reactants and products. Use data from Tables 10–1 and 10–2.

a) $CO_{(g)} + \frac{1}{2}O_{2\,(g)} \rightarrow CO_{2\,(g)}$

b) $H_2O_{(liq)} \rightarrow H_2O_{(vap)}$

c) $Ag_{(s)} + \frac{1}{2}Cl_{2\,(g)} \rightarrow AgCl_{(s)}$

d) $C_{(graphite)} \rightarrow C_{(diamond)}$

Answer

a) $\Delta S_R^0 = S_{CO_2}^0 - S_{CO}^0 - \frac{1}{2}S_{O_2}^0$
    $= 51.061 - 47.301 - \frac{1}{2}49.003 = -20.742$ cal/g-mol · °K.

b) $\Delta S_R^0 = S_{vap}^0 - S_{liq}^0 = 45.109 - 16.716 = 28.393$ cal/g-mol · °K.

Note that this is *not* the entropy of vaporization, but the difference between absolute entropies of liquid and vapor at 25°C and 1 atm. That is, both the initial pure water (liquid) and the final pure water (vapor) are at 25°C and 1 atm. (The latter is a hypothetical or metastable state for H_2O vapor, which normally condenses to liquid at pressures above 0.46 lb$_f$/in^2 at 25°C.) The ΔS_R^0 differs from the entropy of vaporization at 25°C by the loss of entropy that would occur if water vapor were compressed to 1 atm from its vapor pressure at 25°C, without being condensed.

c) $\Delta S_R^0 = S_{AgCl}^0 - S_{Ag}^0 - \frac{1}{2}S_{Cl_2}^0$
    $= 22.97 - 10.206 - \frac{1}{2} \times 53.286 = -13.88$ cal/g-mol · °K.

d) $\Delta S_R^0 = S_{diamond}^0 - S_{graphite}^0 = 0.5829 - 1.3609 = -0.7780$ cal/g-mol.

---

* H. C. Weber and H. P. Meissner, *Thermodynamics for Chemical Engineers*, New York: John Wiley, 1959, pages 408–410. O. A. Hougen, K. M. Watson, and R. A. Ragatz, *Chemical Process Principles, Part II, Thermodynamics*, second edition, New York: John Wiley, 1964, pages 1011–1013.

Example 10-4. Calculate $\Delta G^0_{\text{reaction}}$ for the following reactions (at 298°K and 1 atm), using data on $\Delta G^0_{\text{formation}}$ and $\Delta H^0_{\text{formation}}$ and on standard absolute entropy.

a) $CO_{(g)} + \frac{1}{2}O_{2\,(g)} \rightarrow CO_{2\,(g)}$

b) $H_2O_{(liq)} \rightarrow H_2O_{(vap)}$

c) $N_{2\,(g)} + \frac{1}{2}O_{2\,(g)} \rightarrow N_2O_{(g)}$

Answer

a) From Table 10-1, we find that

$$\Delta G^0_{\text{reaction}} = \Delta G^0_{\text{formation}}\,(CO_2) - \Delta G^0_{\text{formation}}\,(CO) = -61,452 \text{ cal/g-mol}$$
$$\text{(Example 10-1),}$$

or

$$\Delta S^0_R = -20.742 \text{ cal/g-mol} \cdot °K \text{ (Example 10-3),}$$
$$\Delta H^0_R = -94,052 - (-26,416) = -67,636 \text{ cal/g-mol.}$$

Therefore

$$\Delta G^0_R = -67,636 - (-298 \times 20.742) = -61,455 \text{ cal/g-mol.}$$

(The differences between the two computed values of ΔG^0_R indicate a slight inconsistency in the data on ΔG^0_F, ΔH^0_F, and S^0.)

b) $\Delta G^0_R = \Delta G^0_F\,(H_2O, \text{vap}) - \Delta G^0_F\,(H_2O, \text{liq}) = -54,636 - (-56,690)$
$\qquad = 2054 \text{ cal/g-mol.}$

$\Delta G^0_R = \Delta H^0_R - T\,\Delta S^0_R$; we know ΔS^0_R from Example 10-3.

$\Delta H^0_R = \Delta H^0_F\,(H_2O, \text{vap}) - \Delta H^0_F\,(H_2O, \text{liq}) = -57,798 - (-68,317)$
$\qquad = 10,519 \text{ cal/g-mol.}$

$\Delta G^0_R = 10,519 - 298 \times 28.393 = 2058 \text{ cal/g-mol.}$

As was pointed out in Example 10-3, we are *not* dealing with an equilibrium vaporization, but with a process which converts liquid at STP to vapor at STP. (What would ΔG be for the equilibrium vaporization? Compute this value from $\Delta H - T\,\Delta S$.)

c) $\Delta G^0_R = 24,760 \text{ cal/g-mol.}$

$\Delta H^0_R = \Delta H^0_F\,(N_2O) = 19,490 \text{ cal/g-mol.}$

$\Delta S^0_R = 52.58 - (-45.767 - \frac{1}{2} \times 49.003) = -17.689 \text{ cal/g-mol} \cdot °K.$

$\Delta G^0_R = 19,490 - (-298 \times 17.689) = 24,761 \text{ cal/g-mol.}$

The positive value of ΔG^0_F for N_2O means that when the reaction is carried out reversibly in a van't Hoff equilibrium box, more work is supplied to the reaction system to compress N_2O from its equilibrium partial pressure to 1 atm than is obtained when the reactant elements are expanded to equilibrium pressures. There-fore 1 *g-mol* of *pure* N_2O at 1 atm and 298°K will not form spontaneously from N_2 and O_2. On the other hand, ΔG_R^0 for the decomposition reaction,

$$N_2O_{(g)} \rightarrow N_{2\,(g)} + \tfrac{1}{2}O_{2\,(g)},$$

is simply $-24{,}760$ cal/g-mol, and N_2O should tend to decompose spontaneously. However, the decomposition rate in the absence of a catalyst is so slow that N_2O appears to be a stable gas at 25°C.

10–9
Composition of chemical equilibrium mixtures

We can calculate the composition of chemical equilibrium mixtures if we know the total pressure in the equilibrium chamber, the stoichiometric reaction equation, and the composition of the reactants fed to the equilibrium chamber. We shall now demonstrate the calculation at conditions of standard total pressure and temper-ature. Later we shall show how the calculation can be extended to any conditions of temperature and pressure.

Example 10–5. Calculate the equilibrium composition in an isothermal, constant-pressure (1 atm) reaction chamber initially charged with pure N_2O_4 at STP. The N_2O_4 tends to decompose into nitric oxide as follows:

$$N_2O_4 \rightarrow 2NO_2.$$

Answer: Using data from Table 10–1, we find that

$$\Delta G_R^0 = 2(12{,}390) - 23{,}491 = +1289 \text{ cal/g-mol.}$$

Therefore, from Eq. (10–13),

$$\ln K = -\frac{1289}{1.987 \times 298} = -2.176; \qquad K = 0.1137.$$

Now, if the gases are ideal gases,

$$K_p = \left(\frac{p_{NO_2\,e}}{p_{N_2O_4\,e}}\right)^2 = \frac{(y_{NO_2}P)^2}{y_{N_2O_4}P} = \frac{(y_{NO_2})^2}{y_{N_2O_4}} \times 1 = \frac{(y_{NO_2})^2}{1 - y_{NO_2}} = 0.1137.$$

Hence

$$y_{NO_2} = 0.2852, \qquad y_{N_2O_4} = 0.7148,$$

and

$$p_{NO_2\,e} = 0.2852 \text{ atm}, \qquad p_{N_2O_4\,e} = 0.7148 \text{ atm.}$$

An important point to notice in Example 10–5 is that, although ΔG_R^0 was positive for N_2O_4 decomposition, some N_2O_4 *did* decompose. A positive value of ΔG_R^0 means that $K < 1$ and that the equilibrium mixture contains *less* product than reactant.

Example 10–6. What is the composition of the equilibrium state at 298°K for the oxidation of carbon monoxide to carbon dioxide in a STP reactor fed with 1 mol of CO and 1 mol (not the stoichiometric quantity) of O_2?

Answer

$$CO_{(g)} + \tfrac{1}{2}O_{2\,(g)} \rightarrow CO_{2\,(g)}.$$

Again, we use data from Table 10–1 to obtain

$$\Delta G_R^0 = -61{,}452 \text{ cal/g-mol.}$$

It therefore follows that

$$\ln K = \frac{-(-61{,}452) \text{ cal/g-mol}}{1.987 \text{ (cal/g-mol} \cdot {}^{\circ}\text{K)} \times 298{}^{\circ}\text{K}} = 103.7,$$

$$K = 1.18 \times 10^{45}.$$

Using Eqs. (10–12) and (9–9), we obtain

$$K = \frac{p_{CO_2\,e}}{(p_{CO\,e})(p_{O_2\,e})^{1/2}} = \frac{y_{CO_2}}{(y_{CO})(y_{O_2})^{1/2}\underbrace{P^{1/2}}_{=1}}.$$

It is clear from the immense value of K that the equilibrium mixture will contain almost 1 mol of CO_2, with a microscopic trace of CO and an amount of O_2 microscopically greater than $\tfrac{1}{2}$ mol. To calculate the amounts present, we proceed as follows.

$$\text{Let } x = \text{number of moles of } CO_2 \text{ formed}$$
$$1 - x = \text{number of moles of CO remaining}$$
$$1 - \tfrac{1}{2}x = \text{number of moles of } O_2 \text{ remaining}$$

$$2 - \tfrac{1}{2}x = \text{total number of moles in equilibrium chamber}$$

Therefore

$$y_{CO_2} = \frac{x}{2 - 0.5x} \;; \qquad y_{CO} = \frac{1 - x}{2 - 0.5x} \;; \qquad y_{O_2} = \frac{1 - 0.5x}{2 - 0.5x} \;;$$

and

$$K = \frac{x(2 - 0.5x)^{1/2}}{(1 - x)(1 - 0.5x)^{1/2}} = 1.18 \times 10^{45}. \qquad (10\text{–}18)$$

Hence

$$x = 1 - (1.5 \times 10^{-45})$$

and

$$y_{CO_2} = 0.666\ldots; \qquad y_{CO} = 10^{-45}; \qquad y_{O_2} = 0.333\ldots.$$

Equations of unusual order, such as Eq. (10–18), frequently occur in computations to determine equilibrium composition. They must be solved by trial or by iteration techniques—on a computer.

Exercise 10–5. How would the expression for K in Eq. (10–18) differ if $\frac{1}{2}$ mol of O_2 (rather than 1 mol of O_2) were used in the reactor feed of Example 10–6? Would the number of moles of CO_2 formed in the equilibrium chamber be larger or smaller?

Exercise 10–6. Derive an equation for

$$\Delta S_R^{0T} = \begin{array}{l} \text{the entropy of isothermal isobaric reaction} \\ \text{at 1 atm and } \textit{any} \text{ temperature } T. \end{array}$$

See Eq. (2–56) and Fig. 2–33.

10–10
Effect of pressure on $\Delta G_{\text{reaction}}$ and the equilibrium constant K

The previous sections considered reaction processes confined to 1 atm. Let us now consider isothermal processes at any pressure.

We can compute $\Delta G_{\text{reaction}}$ for pressures other than $P = 1$ atm via the path shown in Fig. 10–5, which applies to a general isothermal reaction process, wherein

$$r \text{ moles reactant} \to n \text{ moles product,}$$

and where the reactants are each initially at a pressure of P_R and the products are finally each at a pressure P_N. The path leads through an equilibrium state (b–e) at a total pressure which may be different from that of the equilibrium state (c–d) that existed when reactants and products were at 1 atm pressure. In going from b to e, reactants are converted to products without any change in free energy:

$$\Delta G_{\text{reaction}} = \Delta G_{R \to b} + \Delta G_{e \to N}$$

$$= \sum_{\text{Reactants}} \int_{P_R}^{p_e} V \, dP + \sum_{\text{Products}} \int_{p_e}^{P_N} V \, dP. \qquad (10\text{–}19)$$

Equation (10–19) is the same as Eq. (10–5), except for the integration limits.

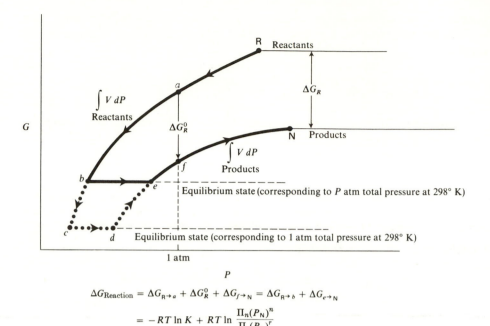

$$\Delta G_{\text{Reaction}} = \Delta G_{R \to a} + \Delta G_R^0 + \Delta G_{f \to N} = \Delta G_{R \to b} + \Delta G_{e \to N}$$

$$= -RT \ln K + RT \ln \frac{\Pi_n (P_N)^n}{\Pi_r (P_R)^r}$$

Fig. 10–5. Computation paths for evaluating isothermal $\Delta G_{\text{reaction}}$ at *any* pressure.

It follows that (for ideal gases)

$$\Delta G_{\text{reaction}} = -RT \ln \frac{\Pi_n (p_{\text{products }e})^n}{\Pi_r (p_{\text{reactants }e})^r} + RT \ln \frac{\Pi_n (P_N)^n}{\Pi_r (P_R)^r}$$

$$= -RT \ln K' + RT \ln \frac{\Pi_n (P_N)^n}{\Pi_r (P_R)^r}, \qquad (10\text{–}20)$$

where K' represents the ratio of the partial pressures in the equilibrium state corresponding to nonstandard pressure conditions of reactants and products.

To show that K' is the same as the standard K of Eqs. (10–12) and (10–13), we need only note from Fig. 10–5 that ΔG_R may also be evaluated over path *R-a-f-N*, or

$$\Delta G_R = \sum \int_{P_R}^{1\text{ atm}} V \, dP + \Delta G_R^0 + \sum \int_{1\text{ atm}}^{P_N} V \, dP$$

$$= RT \ln \frac{1}{\Pi_r (P_R)^r} - RT \ln K + RT \ln \frac{\Pi_n (P_N)^n}{1}. \qquad (10\text{–}21)$$

Therefore

$$\boxed{\Delta G_{\text{reaction}} = -RT \ln K + RT \ln \frac{\Pi_n (P_N)^n}{\Pi_r (P_R)^r},} \qquad (10\text{–}22)$$

which is identical to Eq. (10–20). Therefore K is the same as K', and *the equilibrium* constant is independent of the total pressure in the equilibrium chamber. The equilibrium composition and ΔG_R, however, are generally *not* independent of pressure.

When the reaction components are not ideal gases, the P's in the above equations are replaced by corresponding fugacities or activities.

Example 10–7

i)   Calculate the equilibrium composition in an isothermal, isobaric (10 atm) reaction chamber in which N_2O_4 is decomposing into NO_2 at 298°K:

$$N_2O_4 \rightarrow 2NO_2.$$

The chamber was initially charged with pure N_2O_4 at 10 atm pressure. Assume ideal-gas behavior.

ii)   Calculate K for the above reaction conditions.

iii) Calculate $\Delta G_{\text{reaction}}$ for the reaction which converts 1 g-mol of N_2O_4 completely to NO_2, each at 10 atm and 298°K.

iv) Calculate ΔG for the reaction converting 1 g-mol of pure N_2O_4 at 10 atm to the equilibrium mixture at 10 atm total pressure and 298°K. Why does this result differ from ΔG_R^0?

Answer

i) and (ii). The equilibrium constant at 298°K is independent of pressure and is a function of ΔG_R^0 only. It is the same as calculated in Example 10–5: $K = 0.1137$. Also from Example 10–5 we have

$$K = \frac{(y_{NO_2})^2}{y_{N_2O_4}}\,\mathbb{P} = \frac{(y_{NO_2})^2 \times (10)}{1 - y_{NO_2}}.$$

Therefore the equilibrium composition at 10 atm total pressure is

$$y_{NO_2} = 0.1011, \qquad y_{N_2O_4} = 0.8989.$$

Note that the higher total pressure has reduced the amount of N_2O_4 decomposed (see Example 10–5). Increasing pressure tends to reduce the extent of any equilibrium reaction in which the volume of product exceeds the volume of reactant, and to increase the extent of reactions in which the volume of product is less than the volume of reactant. In other words, an equilibrium reaction moves in a direction which tends to counteract the effect of a pressure change.

iii) $\Delta G_R = \Delta G_R^0 + RT \ln 10$
$\qquad = +1289 + (1.987 \times 298) \ln 10 = 2652 \text{ cal/g-mol.}$

iv) ΔG is the loss of free energy that accompanies the expansion of 1 g-mol of N_2O_4 to its equilibrium partial pressure. (See Exercise 10–2.)

$$\Delta G = \int_{10}^{P_{N_2O_4\,e}} \bar{V}\,dP = RT \ln \frac{0.8989 \times 10}{10} = -62.7 \text{ cal/g-mol.}$$

This process can be represented by a path similar to *N-f-e* in Fig. 10–5. The ΔG_R^0 for this reaction would follow a path analogous to *f-d-c-a* because ΔG_R^0 is > 0.

Exercise 10–7

i) Calculate the change in free energy accompanying the compression of 2 g-mol of NO_2 from its equilibrium partial pressure in the above reactor to a pressure of 10 atm.

ii) Use the result obtained above to find ΔG_R for the reaction:

$$N_2O_4 \rightarrow 2NO_2 \quad \text{at } 298°K \text{ and } 10 \text{ atm.}$$

Exercise 10–7a. Calculate ΔG for the process which starts with a stoichiometric *mixture* of CO and $\frac{1}{2}O_2$ at a total pressure of 1 atm and ends with pure CO_2 at 1 atm, all at 298°K:

$$[CO_{2\ (g)} + \tfrac{1}{2}O_{2\ (g)}] \text{ (mixture at 1 atm)} \rightarrow CO_{2\ (g)\ (at\ 1\ atm)}.$$

How does ΔG compare with ΔG_R^0?

10–11
Effect of temperature on the equilibrium constant

Equation (10–13), $\Delta G_R^0 = -RT \ln K$, refers to the change in free energy accompanying a reaction process at 298°K and 1 atm. The equation has exactly the same form if it is derived for an isothermal reaction process at *any* temperature other than 298°K*:

$$\Delta G_R^{0T} = -RT \ln K_T. \tag{10–23}$$

Superscript and subscript *T*'s have been appended to *G* and *K* to differentiate them from the *G* and *K* at standard temperature. The variation of K_T with temperature is found as follows: The partial derivative of Eq. (10–23) with respect to temperature is

$$\left(\frac{\partial(\Delta G_R^{0T})}{\partial T}\right)_P = -RT\left(\frac{\partial \ln K_T}{\partial T}\right)_P - R \ln K_T.$$

Using Eq. (10–23) to replace $(R \ln K_T)$, we get

$$\left(\frac{\partial \Delta G_R^{0T}}{\partial T}\right)_P = -RT\left(\frac{\partial \ln K_T}{\partial T}\right)_P + \frac{\Delta G_R^{0T}}{T}. \tag{10–24}$$

But Eqs. (6–13) and (3–22c) tell us that

$$\left(\frac{\partial \Delta G_R^{0T}}{\partial T}\right)_P = -\Delta S_R^{0T} = \frac{\Delta G_R^{0T} - \Delta H_R^{0T}}{T}. \tag{10–25}$$

---

* The effect of pressure is given by Eq. (10–22). Total pressure is taken as 1 atm in Eq. (10–23).

Combining Eqs. (10–24) and (10–25), we have

$$\left(\frac{\partial \ln K_T}{\partial T} \right)_P = \frac{\Delta H_R^{0T}}{RT^2}.$$ (10–26)

Equation (10–26) is called the *van't Hoff equation*. It indicates that the equilibrium constant decreases with temperature for an exothermic reaction (negative ΔH_R^{0T}) and increases with temperature for an endothermic reaction (positive ΔH_R^{0T}).

Integration of Eq. (10–26) requires that we express ΔH_R^{0T} as a function of temperature, as in Eq. (2–56). *If ΔH_R^{0T} does not change* much with temperature [that is, if in Eq. (2–56) the combined $\sum n \int C_P \, dT$ terms are small compared with ΔH_R^0], then

$$\Delta H_R^{0T} \approx \Delta H_R^0 = \text{constant}$$ (10–27)

and Eq. (10–26) may be integrated in the form

$$\int_{298}^{T} d \ln K = \frac{\Delta H_R^0}{R} \int_{298}^{T} \frac{dT}{T^2}.$$

Therefore

$$\ln \frac{K_T}{K} = - \frac{\Delta H_R^0}{R} \left(\frac{1}{T} - \frac{1}{298°K} \right)$$ (10–28)

or

$$\ln K_T = - \frac{\Delta H_R^0}{RT} + \text{constant}.$$ (10–29)

We can therefore plot $\ln K_T$ as a straight-line function of $1/T$ with a slope of $-\Delta H_R^0/R$, whenever Eq. (10–27) is valid. Figure 10–6 shows plots of $\ln K$ versus $1/T$ for selected gaseous reactions that may occur during combustion processes.

Exercise 10–8. From Fig. 10–6, compute the ΔH_R^0 for the reaction

$$CO + \tfrac{1}{2}O_2 \rightarrow CO_2,$$

and compare with the value computed from tables of $\Delta H_{\text{formation}}^0$.

If data on constant-pressure heat capacity are available for all reactants and products in the form of Eq. (2–20), then we can express ΔH_R^{0T} as

$$\Delta H_R^{0T} = \Delta H_R^0 + \alpha T + \beta T^2 + \gamma T^3,   \text{from (2–56)}$$

which, when substituted into Eq. (10–26) and integrated, yields

$$\ln K_T = - \frac{\Delta H_R^0}{RT} + \frac{\alpha \ln T}{R} + \frac{\beta T}{R} + \frac{\gamma T^2}{2R} + \text{integration constant}.$$

(10–30)

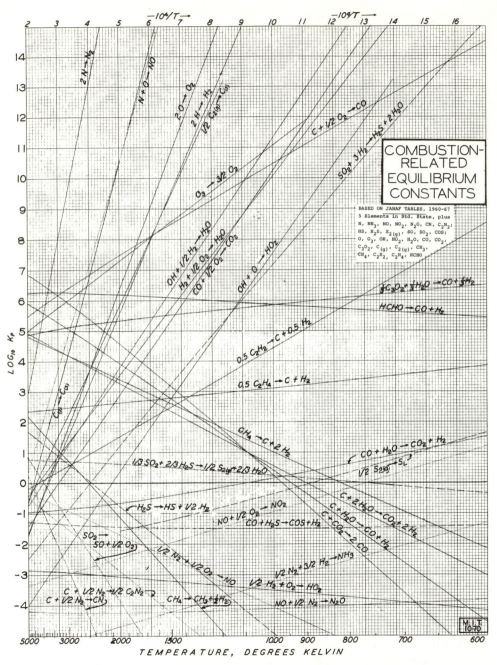

Fig. 10–6. Effect of temperature on the chemical equilibrium constant K_T. For the reaction $aA + bB \rightleftharpoons cC + dD$, $K_T = p_C^c p_D^d / p_A^a p_B^b$, with partial pressures expressed in atmosphere. H_2O is vapor; carbon, unless otherwise noted, is β graphite. Reproduced by permission from Professor H. C. Hottel, of MIT, who prepared the chart.

For temperatures close to 298°K and for systems in which ΔG_R^0 is not a strong function of temperature, we can estimate K_T by modifying Eq. (10–23),

$$\ln K_T \approx -\frac{\Delta G_R^0}{RT}, \tag{10–31}$$

where K_T is the equilibrium constant at a temperature T which is close to 298°K.

Example 10–8. In the presence of platinum or vanadium pentoxide catalysts, sulfur dioxide combines with oxygen to form sulfur trioxide:

$$SO_{2\,(g)} + \tfrac{1}{2}O_{2\,(g)} \rightarrow SO_{3\,(g)}.$$

Analysis of the contents of an actual equilibrium reaction chamber at 560°C and 1 atm reveals that

$$K_{833°K} = 20\ (atm)^{-1}.$$

i) Compute $\Delta G_R^{0\,(833)}$.

ii) How do the above values compare to ΔG_R^0 and K?

iii) Is Eq. (10–28) applicable to this reaction system?

Answer

i) Using Eq. (10–23) we find that

$$\Delta G_R^{0\,(833)} = -RT \ln \frac{20\ (atm)^{-1}}{1\ (atm)^{-1}} = -1.987 \times 833 \times \ln 20 = -4961\ cal/g\text{-mol}.$$

ii) When we insert data from Table 10–1 in Eq. (10–14), we find that

$$\Delta G_R^0 = \Delta G_F^0(SO_3) - \Delta G_F^0(SO_2) = -88,520 - (-71,790) = -16,730\ cal/g\text{-mol}.$$

Therefore

$$\ln K = -\frac{\Delta G_R^0}{R(298)} = \frac{16,730}{1.987 \times 298} = 28.26.$$

At STP, ΔG_R^0 and K differ considerably from the 560°C values. The smaller K at the higher temperature is in keeping with an exothermic ΔH_R^0.

iii) To test the applicability of Eq. (10–28), we first compute ΔH_R^0, using data from Table 2–5:

$$\Delta H_R^0 = -94,450 + 70,960 = -23,490\ cal/g\text{-mol}.$$

Then we rearrange Eq. (10–28) to evaluate K_T:

$$\ln K_T = -\frac{\Delta H_R^0}{RT} + \frac{\Delta H_R^0}{R(298)} + \ln K$$

$$= +\frac{23,490}{1.987(833)} - \frac{23,490}{1.987(298)} + 28.26 = +3.00,$$

or

$$K_T = 20.1.$$

This is a very good agreement with Eq. (10–28) and indicates that $\Delta H_R^0 \approx \Delta H_R^0{}^{(833)}$. This temperature span is too large to enable us to use Eq. (10–31).

Exercise 10–9. Evaluate K_{833} in Example 10–8 in units of mm Hg. Does this new value of K affect the equilibrium composition at 560°C?

10–12
Equilibrium mixtures of real gases

The equilibrium constant for reactions of real gases is given by Eq. (10–11),*

$$K = \frac{(f_{Ce})^c(f_{De})^d}{(f_{Ae})^a(f_{Be})^b},$$

where f_{Ce} is the fugacity of component C in the equilibrium mixture of real gases, etc. There is no completely satisfactory general method for evaluating the fugacities of mixtures. The usual compromise procedure is to assume that the real gases form an ideal mixture and use Eqs. (9–28) and (8–13a),

$$f_{Ce} = f_{C\mathbb{P}} \times y_C = (\gamma_{C\mathbb{P}}\mathbb{P})y_C,$$

where $\gamma_{C\mathbb{P}} = f/\mathbb{P}$ = the fugacity coefficient of pure C at the total pressure $\mathbb{P}$ of the equilibrium mixture. Therefore

$$K = \frac{(\gamma_C)^c(\gamma_D)^d}{(\gamma_A)^a(\gamma_B)^b} \times \frac{\mathbb{P}^c\mathbb{P}^d}{\mathbb{P}^a\mathbb{P}^b} \times \frac{(y_C)^c(y_D)^d}{(y_A)^a(y_B)^b} \tag{10–32}$$

or

$$K = K_\gamma \times K_y \times \mathbb{P}^{c+d-a-b}, \tag{10–33}$$

where K_γ and K_y stand for the fugacity coefficient and mole fraction ratios, respectively. Both K_γ and K_y vary with the total pressure, whereas the true equilibrium constant K is independent of pressure, at constant temperature.

The following example, taken from Weber and Meissner† and based on data by Larson,‡ illustrates the advantages of using fugacities to compute K, as well as the limitations of Eq. (9–28) for computing mixture fugacities.

Example 10–9. The following equilibrium compositions at various pressures were measured in an ammonia reactor at 500°C.

---

* The standard state to which K (for real gases) is referred is pure reactants and products in an ideal-gas state at 1 atm and 298°K.

† Weber and Meissner, *ibid.*

‡ A. T. Larson, *J. Am. Chem. Soc.* **46**, 367 (1924).

P, atm	y, mole fraction		
	H_2	N_2	NH_3
10	0.7409	0.2470	0.0121
100	0.6705	0.2235	0.1061
1000	0.3190	0.1063	0.5747

Using the above data, calculate K for the reaction,

$$N_{2(g)} + 3H_{2(g)} \rightarrow 2NH_{3(g)}.$$

Answer

$$K = \frac{(f_{NH_3})^2}{(f_{N_2})(f_{H_2})^3}. \tag{a}$$

If the gases are assumed to be ideal, then (a) becomes

$$K_p = \frac{(p_{NH_3})^2}{(p_{N_2})(p_{H_2})^3} = \frac{(y_{NH_3})^2}{(y_{N_2})(y_{H_2})^3 \mathbb{P}^2}. \tag{b}$$

If the gases are assumed real, but forming ideal mixtures, then we can use fugacity coefficients as in Eq. (10–33), and

$$K_c = \frac{(\gamma_{NH_3})^2}{(\gamma_{N_2})(\gamma_{H_2})^3} \times \frac{(y_{NH_3})^2}{(y_{N_2})(y_{H_2})^3} \times \mathbb{P}^{-2}. \tag{c}$$

We find γ or f/p from Fig. 8–1, using appropriate reduced temperatures and pressures. Here is a comparison of K_p and K_c as computed from equations (b) and (c):

atm	$K_p \times 10^6$	$K_c \times 10^6$
10	14	14
100	17	14
1000	96	22

In view of the smaller variation in K_c than K_p, the assumption of an ideal *solution* of real gases is seen to represent an improvement over simply assuming ideal-gas behavior. The variation in the value of K_c at 1000 atm indicates that Eq. (9–28) is not valid at this high pressure.

10–13
Homogeneous reactions in liquid or solid phases

From a free-energy point of view, all chemical-reaction processes are gas- or vapor-phase reactions. This is so because all reactions can be considered as occurring between the *equilibrium vapors* of the components of the reaction, irrespective of the actual state of aggregation of the reactants, the equilibrium vapors having the same free energy as the liquid or solid from which they emanate. We used this idea before when we were determining the free energy of mixing of liquid solutions (Fig. 9–4).

Consider the general nongaseous reaction process in which the pure reactants and pure products can form a *single homogeneous solution* at the equilibrium state:

$$r \text{ Reactants (liquid)} \rightarrow n \text{ Products (liquid)}.$$

We can compute the ΔG_R^0 for this process over the path shown in Fig. 10–7:

$\Delta G_R^0 = \Delta G$ of expanding pure reactant *vapors* from their *vapor pressure* at 298°K (and 1 atm total pressure) to their *partial pressure* in the reaction equilibrium solution,

plus the

ΔG of compressing the product vapors from their *partial pressures* in the equilibrium solution to the *vapor pressures* of pure products at 298°K (and 1 atm total pressure).

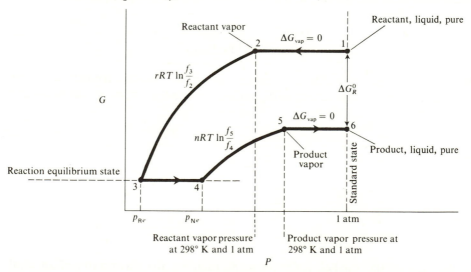

Fig. 10–7. The computation path for the ΔG_R^0 for a nongaseous homogeneous reaction at 298°K and 1 atm.

Since the effect of a total pressure of 1 atm on the vapor pressure of liquids is negligible (see the discussion of the Poynting effect and Eq. 8–18), the vapor pressure we use to find ΔG_R^0 is the normal equilibrium vapor pressure at 298°K:

$$\Delta G_R^0 = \sum_r rRT \ln \frac{f_{Re}}{f_{PR}} + \sum_n nRT \ln \frac{f_{PN}}{f_{Ne}}, \qquad (10\text{–}34)$$

where f_{Re} and f_{Ne} are the fugacities of reactant and product components in the equilibrium mixture, and f_{PR} and f_{PN} are the fugacities of pure reactant and product *vapors* at their *equilibrium vapor pressures*, P_R and P_N, at 298°K.

If we now recall the definition of activity, $a_i \equiv f_i/f_0$, then we may write Eq. (10–34) as

$$\Delta G_R^0 = -RT \ln \frac{\Pi_n(a_{\text{products } e})^n}{\Pi(a_{\text{reactants } e})^r}, \qquad (10\text{–}35)$$

where the activities are those of the components in the equilibrium solution.

When we can treat the reaction vapors as ideal gases, Eq. (10–34) becomes

$$\Delta G_R^0 = \sum_{\text{Reactants}} rRT \ln \frac{p_{Re}}{P_R} + \sum_{\text{Products}} nRT \ln \frac{P_N}{p_{Ne}}.$$

If in addition the equilibrium reaction mixture is an ideal liquid solution, then from Raoult's law (Eq. 10–9), we have

$$\frac{p_{Ie}}{P_I} = x_{Ie}.$$

Therefore

$$\Delta G_R^0 = -RT \frac{\Pi_n(x_{Ne})^n}{\Pi_r(x_{Re})^r}, \qquad (10\text{–}36)$$

where x_{Ie} is the mole fraction of component I in the equilibrium *liquid* solution.

From Eqs. (10–35) and (10–36), it is evident that the equilibrium constant, K, can be expressed in terms of activities and mole fractions, as well as in terms of partial pressures and fugacities. The most general expression for K is that derived from Eq. (10–35):

$$K = \frac{\Pi_n(a_{\text{products } e})^n}{\Pi_r(a_{\text{reactants } e})^r}. \qquad (10\text{–}37)$$

10–14
Heterogeneous reactions

Chemical reactions frequently occur between materials which are in different phases and/or states of aggregation. Such reactions involving different phases are called HETEROGENEOUS REACTIONS, in contrast to *homogeneous reactions*, which involve materials that are all in the same phase. An example of a heterogeneous reaction is the reaction of *aqueous* hydrochloric acid solution with *solid* zinc to form *gaseous* hydrogen.

Further examples are the reactions considered in Example 10–3 (b, c, and d). As with the nongaseous reaction processes considered in Section 10–13, we can treat these reactions, from a free-energy point of view, as occurring between the equilibrium vapors of the reaction components. We can therefore compute the ΔG_R^0 via a computation path analogous to that in Fig. 10–7.

For a reaction process that starts with pure reactants and ends with pure product,

$$aA \text{ (liquid)} + bB \text{ (solid)} \rightarrow cC \text{ (gas)}, \tag{10–38}$$

each at 298°K and 1 atm, we calculate as follows:

$$\Delta G_R^0 = \int_1^{p_{Ae}} (V\,dP)_A + \int_1^{p_{Be}} (V\,dP)_B + \int_{p_{Ce}}^1 (V\,dP)_C$$

$$= +aRT\ln\frac{f_{Ae}}{f_{0A}} + bRT\ln\frac{f_{Be}}{f_{0B}} - cRT\ln\frac{f_{Ce}}{f_{0C}}. \tag{10–39}$$
$$\text{Liquid}\qquad\qquad\text{Solid}\qquad\qquad\text{Gas}$$

Now if the equilibrium mixture contains each of the components as *separate* and *pure* phases, then the ratio of the equilibrium fugacities of pure liquid A and pure solid B to their fugacities at the standard state, f_{Ae}/f_{0A} and f_{Be}/f_{0B}, will be *unity* (except at extremely high pressures; see Examples 8–2 and 8–3).

Let's state this another way: A heterogeneous reaction process may proceed (at a phase boundary) *without the formation of solutions*. Therefore the incompressible components present in the equilibrium state are pure separate phases which have the same fugacity (concentration) in the equilibrium state as they do in the standard state (except at extremely high pressures). Their equilibrium *activity* is therefore *unity*.

Equation (10–39) therefore reduces to

$$\Delta G_R^0 = -RT\ln\left(\frac{f_{Ce}}{f_{0C}}\right)^c, \tag{10–40}$$
$$\text{Gas}$$

and since $\Delta G_R^0 = -RT\ln K$, then

$$K = \left(\frac{f_{Ce}}{f_{0C}}\right)^c = (a_{Ce})^c,$$

or, if C is ideal, then

$$K = (p_{Ce})^c.$$

In the following example, the equilibrium constant uniquely determines the equilibrium pressure of the gas.

Example 10–10. Lime is made by heating limestone:

$$CaCO_{3\,(s)} \rightarrow CaO_{(s)} + CO_{2\,(g)}.$$

At 1000°K, the equilibrium pressure of CO_2 is 0.0483 atm. Calculate $\Delta G_R^{0\,(1000)}$ and K_{1000}.

Answer: We assume that $CaCO_3$ and CaO are separate solid phases and not solid solutions, and also that CO_2 is in an ideal-gas state:

$$\Delta G_R^{0\,(1000)} = -RT \ln \frac{(a_{CaO\,e})(a_{CO_2\,e})}{a_{CaCO_3\,e}}.$$

But the equilibrium activities of the separate solid phases are:

$$a_{CaO\,e} = 1 = a_{CaCO_3\,e}$$

and of CO_2 is

$$a_{CO_2\,e} = \frac{p_{CO_2\,e}}{1} = 0.0483.$$

Therefore

$$K_{1000} = 0.0483,$$

$$\Delta G_R^{0\,(1000)} = -1.987 \times 1000 \times \ln\,(0.0483) = +2610 \text{ cal/g-mol}.$$

Exercise 10–10

i) If, in the preceding example, the partial pressure of CO_2 were to be maintained at 0.01 atm by pumping off CO_2, what would be the final composition in a 1000°K reactor initially supplied with pure $CaCO_3$?

ii) What are the $\Delta G_R^{0\,(1180)}$ and K_{1180} for the above reaction? At 1180°K, the equilibrium pressure of CO_2 is 1 atm.

10–15
Simultaneous reactions

To be perfectly general, the reaction between aA and bB should not stop at cC and dD, as in Eq. (10–4), but should continue along alternative reaction routes and with further reactions between C and D or A and C, etc., to form a host of additional products in confused abundance. The equilibrium mixture resulting from the reaction of A and B will then contain *all the possible products* that can result from *any combination* of the reaction components. The concentration of any component of the equilibrium mixture will adjust itself to suit the equilibrium constants of *all* the possible reactions in which that component participates. Equilibrium compositions are therefore computed by solving simultaneously all the independent equilibrium equations of the system. The number of equations equals the least number that includes all reactants present to an appreciable extent and accounts for the formation of all possible products. Information on the initial concentration of reactants is also required for the computation.

Example 10–11. An important industrial process for the production of hydrogen and "synthesis gas" (CO + H_2) is the catalytic reaction between steam and methane at high temperatures (600°C) to produce CO and H_2. (The thermodynamic analysis of the reaction is based on that presented by Hougen *et al.**) The principal reaction at 600°C is

$$CH_{4\,(g)} + H_2O_{(g)} \rightleftarrows CO_{(g)} + 3H_{2\,(g)}; \qquad K_1 = \frac{(a_{H_2})^3(a_{CO})}{(a_{CH_4})(a_{H_2O})} = 0.574. \qquad (1)$$

But (1) may be considered as a combination of

$$CH_{4\,(g)} \rightleftarrows C_{(s)} + 2H_{2\,(g)}; \qquad K_2 = \frac{(a_{H_2})^2}{a_{CH_4}} = 2.13 \qquad (2)$$

and

$$C_{(s)} + H_2O_{(g)} \rightleftarrows CO_{(g)} + H_{2\,(g)}; \qquad K_3 = \frac{(a_{CO})(a_{H_2})}{a_{H_2O}} = 0.269. \qquad (3)$$

In addition, the following are possible reactions involving the components of (1) and (2) at 600°C:

$$CO_{(g)} + H_2O_{(g)} \rightleftarrows CO_{2\,(g)} + H_{2\,(g)}; \qquad K_4 = 2.21, \qquad (4)$$

$$2CO_{(g)} \rightleftarrows C_{(s)} + CO_{2\,(g)}; \qquad K_5 = 8.14, \qquad (5)$$

$$CO_{2\,(g)} \rightleftarrows CO_{(g)} + \tfrac{1}{2}O_{2\,(g)}; \qquad K_6 = 4.9 \times 10^{-13}, \qquad (6)$$

$$H_2O_{(g)} \rightleftarrows H_{2\,(g)} + \tfrac{1}{2}O_{2\,(g)}; \qquad K_7 = 1.2 \times 10^{-12}, \qquad (7)$$

$$2CH_{4\,(g)} \rightleftarrows C_2H_{6\,(g)} + H_{2\,(g)}; \qquad K_8 = 5.5 \times 10^{-5}, \qquad (8)$$

$$CO_{(g)} + 2H_{2\,(g)} \rightleftarrows CH_3OH_{(g)}; \qquad K_9 = 6 \times 10^{-13}. \qquad (9)$$

It is apparent from the magnitude of the equilibrium constants for reactions (6) through (9) that no appreciable amounts of ethane (C_2H_6), methyl alcohol (CH_3OH), or O_2 will be present at equilibrium at 600°C. Also reactions (2) and (3) tell us that high concentrations of steam prevent the formation of C at equilibrium. This occurs because a large a_{H_2O} (in K_3) forces reaction (3) to the right, using up C and increasing a_{H_2}, but a large a_{H_2} tends to move reaction (2) to the left, also using up any free C. Thus the formation of C is inhibited by concentrations of steam and CH_4 which keep the K_3 ratio less than 0.269, and the K_2 ratio greater than 2.13. Thus only reactions (1) and (4) are needed to compute equilibrium compositions.

---

* O. A. Hougen, K. M. Watson, and R. A. Ragatz, *ibid.*, page 1046.

We shall now calculate the composition of the equilibrium mixture at 600°C and 1 atm, given that the reactor is charged with 1 g-mol of methane and 5 g-mol of steam and is held at 600°C and 1 atm.

Assume: Gases in ideal-gas state; no carbon present at equilibrium, because of excess steam charged.

Feed composition: 1 g-mol CH_4;   5 g-mol H_2O.

Let

$$x = \text{g-mol } CH_4 \text{ reacted by reaction (1)},$$

$$y = \text{g-mol } CO \quad \text{reacted by reaction (4)}.$$

Therefore, at equilibrium,

$$1 - x = \text{g-mol of } CH_4 \text{ in reaction chamber}$$
$$5 - x - y = \text{g-mol of } H_2O \text{ in reaction chamber}$$
$$x - y = \text{g-mol of } CO \text{ in reaction chamber}$$
$$3x + y = \text{g-mol of } H_2 \text{ in reaction chamber}$$
$$y = \text{g-mol of } CO_2 \text{ in reaction chamber}$$

$$6 + 2x = \text{Total moles in reaction chamber}$$

Since the gases are ideal and the total pressure is 1 atm, the activities of the gases in the equilibrium mixture are equivalent to their mole fractions (see Eq. 9–32). Therefore

$$K_1 = \frac{\left(\dfrac{3x + y}{6 + 2x}\right)^3 \left(\dfrac{x - y}{6 + 2x}\right)}{\left(\dfrac{1 - x}{6 + 2x}\right)\left(\dfrac{5 - x - y}{6 + 2x}\right)} = 0.574,$$

$$K_4 = \frac{y(3x + y)}{(x - y)(5 - x - y)} = 2.21.$$

The equations are solved by plotting computed values of y against assumed values of x for K_1 and K_4, and noting the intersection of the curves obtained.

Thus $x = 0.9124$ and $y = 0.633$, and the equilibrium composition is as follows.

			mole %
CH_4	$1 - x$	= 0.0876	1.1
H_2O	$5 - x - y$	= 3.4546	44.2
CO	$x - y$	= 0.2794	3.6
H_2	$3x + y$	= 3.3702	43.0
CO_2	y	= 0.6330	8.1
Total	$6 + 2x$	= 7.8248	100.0

To check whether the assumption of no carbon is valid (for reaction 2), we write

$$\frac{(a_{H_2})^2}{a_{CH_4}} = \frac{(0.430)^2}{0.011} = 16.5 > (K_2 = 2.13).$$

For reaction (3),

$$\frac{(a_{CO})(a_{H_2})}{a_{H_2O}} = \frac{0.035 \times 0.430}{0.442} = 0.034 < (K_3 = 0.269).$$

For reaction (5),

$$\frac{a_{CO_2}}{(a_{CO})^2} = \frac{0.081}{(0.036)^2} = 66 > (K_5 = 8.14).$$

The values of the activity ratios are such as to eliminate carbon from the equilibrium state.

It is interesting—as well as important—to note that, in spite of the unfavorable equilibrium constant, methanol is produced industrially by reaction (9). The keys to a successful industrial process are the development of special catalysts which accelerate reaction (9) over other possible reactions and the employment of extremely high-pressure reactors which increase the equilibrium yield of methanol. Thus, when one is considering the yields of a given reaction, reaction rates (kinetics) as well as thermodynamic considerations must be taken into account. If the reaction rate is very slow, equilibrium conditions may never be attained, in spite of highly favorable K-values.

10–16
Electrochemical processes

In the preceding subsections the chemical work available through composition change was released by means of isothermal turbines or expansion engines attached to a van't Hoff equilibrium box. Unfortunately such devices are *gedanken apparate* only, and have never been put to practical use because semipermeable membranes having sufficiently high selectivity and permeation rates have not yet been developed.

The practical method of obtaining the chemical work of a process involving a composition change (the change can be a concentration change or a chemical reaction) is to carry out the process in an electrochemical or galvanic cell. This may be done for any process of composition change which proceeds by an ionic mechanism. The chemical work is then obtained in the form of electrical work done on the cell's surroundings. If the cell operates at constant T and P, then the maximum chemical work is a form of W_{other} and is equal to $-\Delta G_{T,P}$ of the reaction process (Eq. 3–25).

For reversible operation at conditions of constant temperature and pressure,

$$-dG_{T,P} = dW_{other}$$
$$= -\mathbf{V} \, dq = \mathbf{V} z \mathscr{F} \, dn, \qquad (10\text{–}41)$$

where $\mathbf{V}$ = voltage through which the charge, q, is moved
    = electromotive force of the cell for reversible operation,

   n = number of gram-moles of material reacted
    = number of gram-moles of ions reacting in the cell,

   $\mathbf{z}$ = valence of the reacting ions, or the valence change of the ions during reaction, and

   $\mathscr{F}$ = Faraday's constant = 96,487 coulombs per gram-equivalent of reacting ions = electrical charge needed to discharge 1 g-mol of univalent material at an electrode = 23,050 cal/volt.

If the reactants and products in the cell are in a standard state (pure, with pressure or activity equal to unity, and temperature at 298°K), it follows that the change in free energy per mole for the cell process will be the standard free energy of reaction ΔG_R^0, and that the reversible cell voltage will be a characteristic *standard voltage*, $\mathbf{V}^0$. Therefore, from Eq. (10–41), we obtain

$$\Delta G_R^0 = -\mathbf{V}^0\mathbf{z}\mathscr{F}. \tag{10–42}$$

The free energy of reaction at 298°K for reactants and products at conditions (activities) other than standard (unity) is given by Eq. (10–22). It therefore follows that the cell voltage at nonstandard activities is

$$\Delta \bar{G}_R = -\mathbf{V}\mathbf{z}\mathscr{F} = \Delta G_R^0 + RT \ln \frac{\Pi_n \, (a_{\text{products}})^n}{\Pi_r \, (a_{\text{reactants}})^r}.$$

When we substitute Eq. (10–42) in the above, we obtain

$$\mathbf{V} = \mathbf{V}^0 - \frac{RT}{\mathbf{z}\mathscr{F}} \ln \frac{\Pi_n(a_{\text{products}})^n}{\Pi_r(a_{\text{reactants}})^r}. \tag{10–43}$$

The activities on the right-hand side of Eq. (10–43) are the *initial* activities of the reactants and the *final* activity of the products, i.e., the activities in the actual cell, not the activities of the equilibrium mixture. If these are unity, then $\mathbf{V} = \mathbf{V}^0$. If they are equilibrium values, then $\mathbf{V} = 0$. Equation (10–43) is called the NERNST EQUATION.

Figure 10–8 shows an electrochemical cell, called the *Daniell cell*. The overall chemical changes occurring in this cell are given by the equation

$$Zn + CuSO_4 \rightarrow Cu + ZnSO_4.$$

This cell is representative of many electrochemical cells. It contains two electrodes (strips of Zn and Cu metal) immersed in solutions of electrolytes (in this case saturated solutions of $ZnSO_4$ and $CuSO_4$, respectively). A porous barrier separates the two electrolytes and retards their mixing with each other.

By convention, reactions in cells are written, and cells are drawn, so that the direction of spontaneous change is from left to right. In the Daniell cell, metallic zinc goes into solution and zinc sulfate crystals are deposited, while copper sulfate

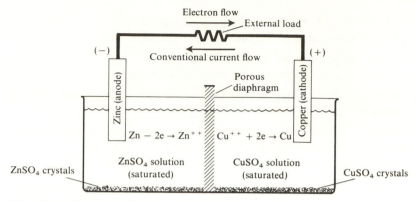

Fig. 10–8. The Daniell cell.

crystals dissolve and metallic copper deposits on the copper electrode. The electrode at which a cell reactant loses electrons is called the *anode*, and that at which electrons are gained by a cell reactant is called the *cathode*. The cell voltage is considered positive when the reaction proceeds in the spontaneous direction.

The overall process may be considered as comprising two electrode processes, called *half-cell processes*:

$$Zn \rightarrow Zn^{++} \text{ (ion)} + 2 \text{ electrons,}$$

$$Cu^{++} \text{ (ion)} + 2 \text{ electrons} \rightarrow Cu.$$

The zinc electrode (anode) tends to develop an excess of electrons, and the copper electrode (cathode) tends to have a deficiency of them. Thus electrons flow from the zinc to the copper electrode if an external circuit connecting the two is provided. This direction of electron flow is opposite to the direction that is conventionally assigned to the flow of electric current. The electromotive force, V, of the cell is the potential difference between the electrodes at zero current.

The cell may be represented as

$$Zn \mid ZnSO_4(\text{saturated}) \parallel CuSO_4(\text{saturated}) \mid Cu,$$

where the single vertical line represents a phase boundary and the double vertical lines represent a junction between two liquid phases, which divides the cell into two half cells.

10–17
Standard electrode voltages

We can construct tables of standard electrode voltages or standard half-cell voltages by using cells having hydrogen gas at 1 atm as one electrode and the material of interest—with ions at *unit activity*—as the other electrode. The electromotive force (emf) of such a cell at 298°K is, by convention, recorded as the *standard electrode* or *half-cell voltage* of the non-hydrogen electrode, the half-cell voltage of the hydrogen being taken as zero. The hydrogen electrode usually consists of a strip of porous platinum (platinum black) over which H_2 gas at 1 atm is bubbled. Table 10–3 presents selected standard electrode voltages. The standard voltage of a cell consisting of a pair of these electrodes is the sum of their potentials with the sign of the *cathode* reversed. The lower of any two listed electrodes always functions as the cathode in a cell in which the reaction proceeds in its normal direction.

Table 10–3 Electrode potentials at 25°C*

Reaction	Volts
$Li = Li^+ + e$	2.960
$K = K^+ + e$	2.924
$Na = Na^+ + e$	2.716
$\frac{1}{2}H_2 + OH^- = H_2O + e$	0.830
$Zn = Zn^{++} + 2e$	0.762
$Cr = Cr^{++} + 2e$	0.56
$Fe = Fe^{++} + 2e$	0.44
$Cd = Cd^{++} + 2e$	0.401
$Ni = Ni^{++} + 2e$	0.23
$Sn = Sn^{++} + 2e$	0.14
$\frac{1}{2}H_2 = H^+ + e$	0.000
Normal calomel electrode	-0.281
0.1N calomel electrode	-0.333
$Cu = Cu^{++} + 2e$	-0.344
$Fe^{++} = Fe^{3+} + e$	-0.748
$Ag = Ag^+ + e$	-0.798
$Sn^{++} = Sn^{4+} + 2e$	-1.256
$Cl^- = \frac{1}{2}Cl_2 + e$	-1.358

* *International Critical Tables*, Vol. 6, page 332.

Example 10–12. Standard zinc and copper (Cu^{++}) electrodes form a cell which operates at 25°C. (Note that this is not the Daniell cell, which uses saturated solutions of zinc and copper as electrolytes to surround the electrodes, but a related cell in which the electrode solutions have unit ion activity.)

a) Write the half-cell reactions and the overall reaction.

b) Calculate the standard emf of the cell.

c) Calculate ΔG_R^0 for the cell reaction.

d) Calculate the K for the cell reaction.

Answer

(a) and (b): Since Cu is listed below Zn in Table 10–3, we know that it functions as cathode (donor of electrons to the reactants). Therefore the half-cell reactions are:

$$\text{Cathode:}\quad Cu^{++} + 2e \rightarrow Cu \qquad\qquad = 0.344$$

$$\text{Anode:}\quad Zn \rightarrow Zn^{++} + 2e \qquad\qquad = 0.762$$

$$\text{Overall:}\quad Cu^{++} + Zn \rightarrow Cu + Zn^{++} = 1.106 \text{ volt}$$

c) $\Delta G_R^0 = -Vz\mathscr{F} = -(1.106)(2)(96{,}487) \text{ volt-coul} = -51{,}055 \text{ g-cal.}$

d) $\ln K = -\Delta G_R^0/RT = 51{,}055/1.987 \times 298 = 86.2, \qquad K = 2.7 \times 10^{37}.$

Note that

$$K = \frac{(a_{Cu\,e})(a_{Zn^{++}\,e})}{(a_{Zn\,e})(a_{Cu^{++}\,e})} = \frac{a_{Zn^{++}\,e}}{a_{Cu^{++}\,e}},$$

because the copper and zinc present as pure metals in the equilibrium mixture have unit activities (Section 10–14).

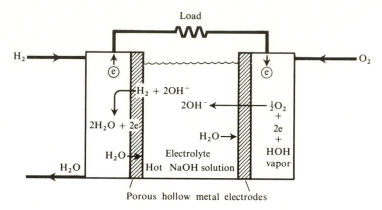

Fig. 10–9. A hydrogen-oxygen fuel cell.

10–18
Fuel cells

Fuel cells are electrolytic cells used for power production. They are designed so that cell reactants may be replenished continuously, as they are consumed. A common type of fuel cell derives its power by carrying out "combustion" reactions electrolytically. An example is the hydrogen–oxygen fuel cell shown schematically in Fig. 10–9. The half-cell reactions at the fuel (hydrogen) electrode and at the oxygen electrode are, respectively,

$$H_2 + 2OH^- \rightarrow 2H_2O + 2 \text{ electrons,}$$
$$\tfrac{1}{2}O_2 + H_2O + 2 \text{ electrons} \rightarrow 2OH^-,$$

so that the overall cell reaction is

$$H_2 + \tfrac{1}{2}O_2 \rightarrow H_2O.$$

Exercise 10–11

i) Designate the cathode and anode electrodes in the hydrogen–oxygen fuel cell.

ii) Compute the maximum work and voltage which this cell can deliver at 25°C when fed with H_2 and O_2 at 1 atm and when product is removed as liquid at 25°C and 1 atm. The half-cell voltage of both electrodes is that for the formation or neutralization of OH^- ions.

iii) How do these factors change if the cathode is supplied with air instead of pure O_2?

iv) As in (iii), but with reactants each supplied at 100 atm.

v) Compare the heat given off during a standard combustion process (that is, ΔH_R^0) with that transferred to the surroundings under the conditions given in part (ii), in order to maintain the cell at 25°C.

Problems

1. Calculate the change in standard free energy and the equilibrium constants at 25°C, 1 atm for the reactions:

a) $Na_2CO_{3\ (s)} + 2HCl_{(g)} \rightarrow 2NaCl_{(s)} + CO_{2\ (g)} + H_2O_{(\ell)}$.
b) $Fe_2O_{3\ (s)} + CO_{(g)} \rightarrow CO_{2\ (g)} + 2FeO_{(s)}$.
c) $3C_2H_{2\ (g)} \rightarrow C_6H_{6\ (\ell)}$.
d) $HgO_{(s)} + 2HCl \rightarrow H_2O_{(\ell)} + HgCl_{2\ (s)}$.

2. For the reaction $N_{2\ (g)} + 3H_{2\ (g)} = 2NH_{3\ (g)}$ at 723°K and 10 atm, the equilibrium composition of the gas phase is 74.1 mole % H_2, 24.7 mole % N_2, and 1.2 mole % NH_3. What is the standard free-energy change for this reaction at 723°K?

3. Using Fig. 10–6 and assuming ideal-gas behavior, calculate the equilibrium composition of the reaction mixture for the reaction

$$CO_{(g)} + H_2O_{(g)} \rightarrow CO_{2\ (g)} + H_{2\ (g)}$$

under the following given conditions:

a) Feed with 1 mol $CO_{(g)}$ and 1 mol H_2O at 1000°K and 1 atm.
b) Feed with 3 mol $CO_{(g)}$, 1 mol CO at 1000°K and 1 atm.
c) Feed with 1 mol $CO_{(g)}$, 2 mol $H_2O_{(g)}$, and 2 mol $CO_{2\ (g)}$ at 1000°K and 1 atm.
d) Same as (a), except at 20 atm.
e) Same as (a), except including 3 mol $N_{2\ (g)}$ in the feed.

You may neglect the side reaction $CO_2 = CO + \frac{1}{2}O_2$ at high temperature.

4. Consider the dissociation reaction $O_3 = \frac{3}{2}O_2$ at 1 atm and constant temperature. The volume of the equilibrium mixture is 25% greater than that of the pure ozone. Assuming ideal-gas behavior, determine

a) The degree of dissociation,
b) The equilibrium constant,
c) The dissociation temperature (from Fig. 10–6).

5. Given the reaction $N_2O_4 = 2NO_2$.

a) At 55°C and 1 atm pressure the average molecular weight of the partially dissociated gas is 61.2. Calculate the degree of dissociation, α.

b) Nitrogen is introduced into the mixture to bring the pressure up to 3 atm. What is the new value of α after the equilibrium is re-established?

c) Calculate the standard free-energy change of the reaction.

d) How many moles of N_2O_4 must be added to a 10-liter vessel at 55°C in order that the concentration of NO_2 will be 0.1 mol per liter?

6. The equilibrium constants of the reaction $C + 2H_2 = CH_4$ have the following values.

Kelvin temperature	1473	1573	1648	1673	1723
Equilibrium constant	2.44	1.54	1.09	0.89	0.75

a) Determine the average enthalpy of formation graphically.

b) Calculate ΔH_R^{0T} at 1500°K.

c) By extrapolation, obtain $K_{1000°K}$ and $\Delta G_R^0 \, (1000)$.

7. a) Using the data in Fig. 10–6, determine the enthalpy of dissociation of $H_2O_{(g)}$ at 5000°K according to the chemical equation

$$H_2O_{(g)} \rightarrow H_2 \,_{(g)} + \tfrac{1}{2}O_2 \,_{(g)}.$$

b) At 1 atm pressure, what is the temperature when one-half of $H_2O_{(g)}$ is dissociated?

c) Same as (b), except for 20% dissociation at 2 atm.

8. For the reaction $Fe_2O_3 \,_{(s)} + 3CO_{(g)} \rightarrow 2Fe_{(s)} + 3CO_2 \,_{(g)}$, the following values of K_p are known (see Eq. 10–12):

T, °C	100	250	1000
K_p	1100	100	0.0721

At 1120°C for the reaction $2CO_2 \,_{(g)} = 2CO_{(g)} + O_2 \,_{(g)}$, the equilibrium constant $K_p = 1.4 \times 10^{-12}$ atm. At what equilibrium partial pressure would O_2 have to be supplied to a vessel containing 1 mol of solid Fe_2O_3 at 1120°C in order to prevent the formation of Fe?

9. Using C_P data from Appendix I and data from Table 10–1, calculate the standard free-energy change and the equilibrium constant at 2000°K and 1 atm for the reaction

$$SO_2 + \tfrac{1}{2}O_2 = SO_3.$$

10. Evaluate the equilibrium constant of the reaction $2HI_{(g)} = H_2 \,_{(g)} + I_2 \,_{(g)}$ at 127°C. Assume that gases obey ideal-gas laws. The following data are available.

$$\Delta H_{\text{Reaction}}^{400°K} = 2.70 \text{ kcal/g-mol } I_2.$$

Absolute S (at 127°C and 1 atm), cal/mol·°K:

$H_2 \,_{(g)}$	33.23
$I_2 \,_{(g)}$	64.89
$HI_{(g)}$	51.38

11. a) Use C_P data from Appendix I and data from Table 10–1 to compute the equilibrium constant of the reaction

$$H_2 \,_{(g)} + \tfrac{1}{2}O_2 \,_{(g)} = H_2O_{(g)} \qquad \text{at } 727°C.$$

b) Now find the equilibrium constants for the combustion processes

$$C_{(s)} + CO_{2\ (g)} = 2CO_{(g)}, \qquad CO_{2\ (g)} + H_{2\ (g)} = CO_{(g)} + H_2O,$$

at 727°C (from Fig. 10–6) and calculate the standard free-energy change at 727°C for the reaction

$$C_{(s)} + \tfrac{1}{2}O_2 = CO.$$

Use the results of part (a) in your computation. Assume ideal-gas behaviors.

12. One gram-mole of solid $NiCl_2$ is placed in an evacuated vessel which has a volume of 100 l. When equilibrium is attained, the pressure in the vessel at 1000°K is found to be 6.6×10^{-3} atm. Is there any solid Ni formed in the vessel at equilibrium? If so, how much? The following data are given:

	$-(G^0_{1000°K} - H^0_{298°K})/T,$ cal/g-mole·°K	$\Delta H^0_{298°K},$ kcal/g-mol
$NiCl_{2\ (s)}$	32.69*	−72.97**
$Cl_{2\ (g)}$	55.43†	0
$Ni_{(s)}$	10.77*	0
$Ni_{(g)}$	46.45†	102.8‡

   * Reference state is pure solid at 1 atm pressure.
   † Reference state is pure ideal gas at 1 atm pressure.
 ** Enthalpy of formation from $Ni_{(s)}$ and $Cl_{2\ (g)}$.
   ‡ Enthalpy of sublimation.

The only chemical reaction we need to consider is the decomposition of $NiCl_2$. State and justify any additional assumptions involved in your analysis.

13. The change in free energy for the reaction $CO + 2H_2 \rightarrow CH_3OH$ at 1 atm at 323°C is 11.1 kcal/g-mol. For a mixture of 1 mol CO_2 and 2 mol H_2, find the composition at equilibrium for each of the following conditions.

a) The mixture forms an ideal solution at 1 atm.
b) The mixture forms an ideal solution at 500 atm.
c) The mixture behaves as an ideal gas at 500 atm.

14. One mole of a dimer gas $A_{2\ (g)}$ is to be expanded isothermally and reversibly in an ideal piston-cylinder device from 10 atm to 1 atm. The $A_{2\ (g)}$ partially decomposes into $A_{(g)}$ and is in equilibrium with $A_{(g)}$ at all times. Prove that the change in free energy at constant temperature T for the isothermal expansion is

$$\Delta G_T = 2RT \ln\left(\frac{\sqrt{K_p + 4} - \sqrt{K_p}}{\sqrt{K_p + 40} - \sqrt{K_p}}\right),$$

where K_p is the equilibrium constant for the reaction, $A_{2\ (g)} = 2A_{(g)}$ in atmospheres.

15. Given the gaseous reaction: $aA + bB \rightarrow cC + dD$.

a) What is the significance of a positive standard Gibbs free energy of reaction

$$(\Delta G^{298°K}_{Reaction} > 0)?$$

b) The pure reactants, each at a pressure P_1 and temperature T_1, are sent in equimolal proportions through an isothermal reactor from which an equilibrium mixture of reactants and products at total pressure P_1 issues. What is the ΔG of the process?

c) Suppose that the exit mixture (whose composition is known) is not at equilibrium. Show how you would compute the ΔG of the process.

16. The valuable acrylic polymers are made from the deadly gas HCN. HCN can be made in a flow reactor. Suppose that such a reactor is fed with an equimolal mixture of CH_4 and NH_3 at 25°C and discharges an equilibrium mixture of reactants and products at 25°C.

a) How does the ΔG for this process compare with ΔG_R?

b) How much heat must be added or removed from the reactor to keep the process isothermal?

The equilibrium constant at 25°C is 0.55×10^{-32}. The enthalpy of reaction at 25°C is 60 kcal.

17. Consider the combustion reaction $CO + \frac{1}{2}O_2 = CO_2$ as occurring in an adiabatic reactor at 1 atm. Both the CO and the O_2 enter the reactor at 25°C. Use the heat capacity–temperature relation in Appendix I to determine:

a) the degree of reaction,

b) the final equilibrium composition,

c) the adiabatic flame temperature (final equilibrium temperature).

18. Suppose that we wish to find the composition of hydrogen fluoride in the gas phase, using the following experimental data on HF which were recorded in an equilibrium gas phase.

T, °C	Total pressure, atm	Density, g/liter
25	1.4	3.800
50	2.8	5.450
50	1.4	1.399
100	5.8	3.978

Standard ΔG^0 of formation of HF $= -64.7$ kcal/g-mol.

Can the data be used to choose between the following equilibrium reactions:

$$\text{(i) } 4HF \to H_4F_4 \quad \text{or} \quad \text{(ii) } 6HF \to H_6F_6?$$

Do the data agree with either of these reactions? If so, what is the ΔS^0_{298} for the reaction? If not, what other reactions might explain the data?

19. Determine the equilibrium composition of the reaction mixture for the following two reactions at 8 atm, given that the reactor feed is 2 mol of M and 1 mol of N. Assume:

$$\text{(1) } M + N = P + Q, \quad K_{p1} = 2.67$$
$$\text{(2) } M + P = 2R, \quad K_{p2} = 3.20$$

20. Hydrogen gas is burned isothermally in a moist air mixture at 25°C and 1 atm. Given the following information, determine the composition of the equilibrium mixture.

a) The equilibrium constant for the reaction $H_2 + \frac{1}{2}O_2 = H_2O_{(g)}$ at 25°C and 1 atm is 1.2×10^{40}.

b) The initial composition of the moist air mixture is:

Species	H_2	N_2	O_2	$H_2O_{(liq)}$
Mole %	5	73	20	2

c) The vapor pressure for $H_2O_{(liq)}$ at 25°C and 1 atm is 0.0313 atm.

21. The main reactions in the pyrolysis of propane are the following:

$$C_3H_8 \rightarrow C_3H_6 + H_2, \qquad C_3H_8 \rightarrow C_2H_4 + CH_4.$$

Assume that only these two reactions occur, and that equilibrium is reached with respect to both reactions. The following data are available.

	$-(G^0_{1500°K} - H^0_{298°K})/T,*$ cal/g-mol · °K	$H^0_{298°K},$ kcal/g-mol
H_2	35.59	0
CH_4	52.84	-15.99
C_2H_4	63.94	14.52
C_3H_6	81.43	8.47
C_3H_8	85.86	-19.48

Enthalpy of formation (for CH_4, C_2H_4, C_3H_6, C_3H_8)

* The reference state is pure ideal gas at 1 atm.

Calculate the composition of the reaction mixture obtained by heating propane at 1500°K and 10 atm.

22. Calculate the change in standard free energy and the equilibrium constant for the following electrochemical cell reactions:

a) $Sn + Pb^{++} = Pb + Sn^{++}$

b) $Fe^{++} + Hg^{++} = Fe^{3+} + \frac{1}{2}Hg_2^{++}$

Given the following data:

$$V^0_{Sn,Sn^{++}} = 0.140 \text{ V}, \qquad V^0_{Pb,Pb^{++}} = 0.126 \text{ V}$$

$$V^0_{Fe^{++},Fe^{3+}} = -0.771 \text{ V}, \qquad V^0_{Hg_2^{++},Hg^{++}} = -0.910 \text{ V}$$

23. Compute the maximum work and voltage obtained from the following:

a) A hydrogen–oxygen fuel cell which is fed with H_2 and O_2 at 25°C and 10 atm. The reaction product is removed as liquid at 25°C and 10 atm.

b) As in (a), but the cathode is supplied with compressed air at 10 atm instead of pure oxygen.

WILLIAM THOMSON, LORD KELVIN,
1824–1907

Section 11
IRREVERSIBLE THERMODYNAMICS

Classical thermodynamics is preoccupied with reversible processes and systems in equilibrium states. When it treats irreversible or nonequilibrium processes, it does so through the equilibrium terminal states of these processes by substituting reversible processes for the actual processes which connect the terminal states. The conclusions it draws are generally those that depend only on the terminal states.

The reason that classical thermodynamics is preoccupied with equilibrium states lies in the difficulty of specifying the properties of nonequilibrium systems. Our discussions of statistical thermodynamics (Section 5) showed that the temperature (or β), the chemical potential (or α), and the thermodynamic properties are either parameters or functions of a statistical distribution function (Eq. 5–14). When a system is changing rapidly, a stable distribution does not exist and no single distribution function is capable of describing the entire system. Therefore the distribution parameters and the properties dependent on the distribution function have no meaning, or at best can be defined over small regions only.

11–1
The steady state

There are some *nonequilibrium* irreversible processes for which we can specify intensive thermodynamic properties, and even spatial gradients of these properties. These processes are characterized by properties which *do not vary with time*, but which do *vary from point to point* in the system. Such processes are said to be in a STEADY STATE. A simple example of a steady-state process is the conduction of heat along a copper bar connecting two heat reservoirs which are at different constant temperatures (Fig. 11–1). The bar reaches a steady state characterized by a stationary temperature profile like that shown in Fig. 11–1. The intensive properties of the bar are *time invariant at any point*, but do change from point to point in the horizontal direction. Heat is being conducted at a constant rate from the hot to the cold end of the bar. The process is irreversible because heat is being trans-

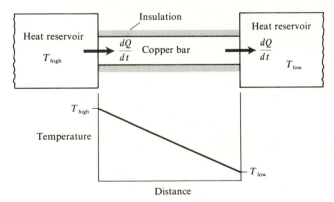

Fig. 11–1. Steady-state heat flow: a nonequilibrium irreversible process. Thermodynamic properties of the bar do not vary with time, but vary from point to point in the system.

ferred over a finite temperature difference and the net entropy of the bar and of the heat reservoirs (i.e., the universe) increases.

Exercise 11–1. Write an expression for the change in entropy of the universe which takes place during t seconds of operation of the system in Fig. 11–1.

11–2
Coupled flows

Steady-state processes may involve the simultaneous flow of two or more quantities which interact with each other. For example, if a voltage difference is impressed across the ends of the copper bar in Fig. 11–1, both electric current and heat will flow through the bar simultaneously. It is found that the flow of current alters the flow of heat in a manner that is quite *independent of the resistance heating* of the bar. In addition, the flow of heat through the bar affects the flow of electric current. Such interdependent flows are called COUPLED FLOWS.

Coupled flows of heat and electricity occur in thermocouples, thermoelectric generators, and in thermoelectric refrigerators. Coupled flows of mass and heat occur in thermal diffusion columns and in catalytic chemical reactors. There are coupled flows of mass and electricity involved in electrochemical cells. If we replaced the copper bar in Fig. 11–1 with a long tube containing a homogeneous mixture of gases, the tube would eventually reach a steady state in which both the temperature and composition would establish fixed gradients along the length of the tube. The flow of heat through the tube causes a simultaneous, coupled migration of matter through the tube.

It is to such *steady-state* irreversible processes involving *coupled flows* that thermodynamics is presently being extended. The applications to date are relatively limited. They are interesting both for the light they throw on certain phenomena and for the insights they provide into the broader nature of thermodynamics. This so-called irreversible thermodynamics or steady-state thermodynamics yields useful information on coupling phenomena only. It does not provide information about the rate at which individual flows occur.

11–3
Supplementary postulates*

To apply thermodynamics to steady-state coupled-flow processes, three postulates must be added to the basic laws of equilibrium thermodynamics:

---

* The postulates presented reflect the Onsager [1] and "Belgian school" [2] approach. A rather different approach (which is in some ways a lesser departure from equilibrium analysis) has been suggested by Tykodi [3].

1) L. Onsager, *Phys. Rev.* **37**, 405 (1931); **38**, 2265 (1931).
2) S. de Groot, *Thermodynamics of Irreversible Processes*, Amsterdam: North-Holland Publishing Co., 1951. I. Prigogine, *Introduction to the Thermodynamics of Irreversible Processes*, Springfield, Ill.: Thomas, 1955.
3) R. J. Tykodi, *Thermodynamics of Steady States*, New York: Macmillan, 1967.

Postulate 1. The rate of flow of matter or energy can be described by linear equations of the form:

Flow rate ∝ driving force

or

$$J_i = L_i \mathbb{F}_i, \tag{11–1}$$

where J_i is called the *ith flux*, L_i is a proportionality constant, and $\mathbb{F}_i$ is the *force* or potential associated with the flow of *i*.

Linear relations for various kinds of flow are well known. Some examples are as follows.

a) *Fourier's law of heat flow:*

$$\frac{dQ}{A\,dt} = k\,\frac{dT}{dx}, \tag{11–1a}$$

where J_i is $dQ/A\,dt$, the heat flux; $\mathbb{F}_i$ is dT/dx, the temperature gradient; and L_i is k, the thermal conductivity.

b) *Ohm's law of electrical flow:*

$$I = \left(\frac{1}{\mathbb{R}}\right) \Delta \mathbf{V}. \tag{11–1b}$$

c) *Fick's law of diffusion:*

$$\frac{dn}{A\,dt} = D\,\frac{dc}{dx}. \tag{11–1c}$$

When a number of flows occur simultaneously, it is *assumed* that the *driving forces* primarily associated with each type of flow also *contribute* (or "*couple*") *to all other flows* in a *linear fashion*. Thus, for simultaneous coupled flows of materials *i, j,* and *k*, Eq. (11–1) becomes:

$$\boxed{\begin{aligned}
J_i &= L_{ii}\,\mathbb{F}_i + L_{ij}\,\mathbb{F}_j + L_{ik}\,\mathbb{F}_k \\
J_j &= L_{ji}\,\mathbb{F}_i + L_{jj}\,\mathbb{F}_j + L_{jk}\,\mathbb{F}_k \\
J_k &= L_{ki}\,\mathbb{F}_i + L_{kj}\,\mathbb{F}_j + L_{kk}\,\mathbb{F}_k
\end{aligned}}$$

all of which may be abbreviated to:

$$\boxed{J_i = \sum_j L_{ij}\,\mathbb{F}_j.} \tag{11–2}$$

Note that the summation varies *j* while *i* remains constant for each J_i flux. Equation (11–2) represents a set of *i* simultaneous linear equations (or an *i*th-order matrix).

The L_{ij} coefficients ($i \neq j$) are called the COUPLING COEFFICIENTS. They are the constants which relate the j driving forces ($\mathbb{F}_j$) to the i flows (J_i). The L_{ii}, L_{jj}, or L_{kk} are called DIRECT or CONJUGATE COEFFICIENTS. They relate the i driving forces to the i flows. Equation (11–2) is the general statement of the first postulate, which we can now restate as:

1. An equation of the form of (11–2) generally exists for a steady-state system involving multiple flows.

Equation (11–2) usually holds true at *small* departures from equilibrium. However, its validity is *not* a necessity, but must be established experimentally in the system under study. Therefore the analysis described in this section is only applicable to systems for which equations having the form of (11–2) are known.

Postulate 2. Irreversible processes *dissipate energy* per unit volume (or lose available energy) at a rate given by

$$\boxed{\frac{T}{V} \frac{dS_{\text{irreversible}}}{dt} = \sum J_i \mathbb{F}_i \, .}$$

(11–3)

To discuss this postulate, we must introduce the concept of entropy generation. We can illustrate the generation of entropy by considering the steady-state copper conductor between the heat reservoirs of Fig. 11–1. At steady state, the high-temperature reservoir loses heat at a constant rate, $-dQ/dt$, which is equal in magnitude to the rate of heat gain, $+dQ/dt$, by the low-temperature reservoir. The rate of entropy change for the reservoirs and bar is the *net rate* of entropy change of the *reservoirs*:

$$\frac{dS_{\text{irreversible}}}{dt} = -\frac{dQ}{dt} \cdot \frac{1}{T_{\text{high}}} + \frac{dQ}{dt} \cdot \frac{1}{T_{\text{low}}} = \frac{dQ}{dt} \left(\frac{1}{T_{\text{low}}} - \frac{1}{T_{\text{high}}} \right) > 0. \quad (11\text{–}4)$$

If one considers entropy as a quantity that flows through the conducting bar, then more entropy leaves the right end of the bar than enters the left end ($\dot{Q}/T_{\text{low}} > \dot{Q}/T_{\text{high}}$). The difference in entropy can be thought of as being *generated* in the conductor.

All irreversible steady-state flows involve a $dS_{\text{irreversible}}/dt$ which is positive, and is called the *entropy generation rate*.

The term $T(dS_{\text{irreversible}}/dt)$ is the rate of dissipation of *energy* by the irreversible flows. An energy rate may be expressed in terms of *generalized forces times generalized rates of displacements*, as in classical mechanics. Hence the rate of dissipation of energy per unit volume is given by

$$\frac{T}{V} \dot{S}_{\text{irreversible}} = \sum_i J_i \mathbb{F}_i, \quad (11\text{–}3)$$

where J_i is, as before, the generalized displacement or irreversible flux, $\mathbb{F}_i$ is the conjugate generalized force, and $\dot{S}_{\text{irreversible}} \equiv dS_{\text{irreversible}}/dt$.

Equation (11–3) *tells us how to select conjugate forces and fluxes* for use in Eq. (11–2). The sum of the product of conjugate fluxes and forces must equal the total rate of energy dissipation per unit volume. The thermodynamic forces that meet the Eq. (11–3) criterion do not always resemble conventional forces.

Postulate 3. *Reciprocal coupling coefficients are equal,* that is, the coefficient relating the *i*th force to the *j*th flux is the same as that relating the *j*th force to the *i*th flux:

$$\boxed{L_{ij} = L_{ji},} \qquad i \neq j. \tag{11–5}$$

This postulate is called the ONSAGER RECIPROCAL RELATION, after the man who first presented a formal proof of it.

Equation (11–5) is derived by recourse to statistical mechanics and to a fundamental principle called *microscopic reversibility* (which states that there are no cyclical routes to equilibrium at the microscopic level, and that any equilibrium process and its reverse occur at the same rate). A simpler proof of Eq. (11–5) for first-order chemical reactions, using only microscopic reversibility, is presented by Denbigh.*

We can also derive the reciprocal relation by regarding the energy dissipation and dissipation rate as continuous functions of the intensive driving forces which promote fluxes. Thus, if the energy dissipated per unit volume of system is

$$\frac{TS_{\text{irreversible}}}{V} = \theta(\mu_1, \mu_2, \dots),$$

where θ is a continuous function of intensive properties $\mu_1, \mu_2, \dots$, etc., then it follows from Eq. (6–8a) that

$$d\theta = \left(\frac{\partial \theta}{\partial \mu_A}\right) d\mu_A + \left(\frac{\partial \theta}{\partial \mu_B}\right) d\mu_B + \cdots + \cdots$$

If the dissipation occurs in time dt, then

$$\dot{\theta} = \left(\frac{\partial \dot{\theta}}{\partial \mu_A}\right) d\mu_A + \left(\frac{\partial \dot{\theta}}{\partial \mu_B}\right) d\mu_B + \cdots \tag{11–6}$$

But from Eq. (11–3), we know that

$$\dot{\theta} = J_A \mathbb{F}_A + J_B \mathbb{F}_B + \cdots$$

where

$$J_A = \frac{\partial \dot{\theta}}{\partial \mu_A}, \qquad J_B = \frac{\partial \dot{\theta}}{\partial \mu_B}, \qquad \mathbb{F}_A = d\mu_A, \qquad \mathbb{F}_B = d\mu_B. \tag{11–7}$$

---

* K. Denbigh, *The Thermodynamics of the Steady State*, London: Methuen, 1951.

If Eq. (11–6) is continuous, then it must meet the Euler criterion (Eq. 6–14), which requires that

$$\frac{\partial J_A}{\partial \mu_B} = \frac{\partial J_B}{\partial \mu_A}.$$ (11–8)

If we now apply Eq. (11–2) to the flux of materials A and B, we obtain

$$\begin{aligned} J_A &= L_{AA}\mathbb{F}_A + L_{AB}\mathbb{F}_B, \\ J_B &= L_{BA}\mathbb{F}_A + L_{BB}\mathbb{F}_B. \end{aligned}$$ (11–9)

Since each of the fluxes is a total derivative with respect to time, we may write Eq. (11–9) as

$$J_A = \left(\frac{\partial J_A}{\partial \mu_A}\right) d\mu_A + \left(\frac{\partial J_A}{\partial \mu_B}\right) d\mu_B,$$ (11–10)

$$J_B = \left(\frac{\partial J_B}{\partial \mu_A}\right) d\mu_A + \left(\frac{\partial J_B}{\partial \mu_B}\right) d\mu_B.$$

Comparing Eq. (11–9) with Eqs. (11–10) and (11–8), we get

$$L_{BA} = L_{AB}.$$ (11–5)

11–4
Applications: the thermocouple

Thermoelectric phenomena were among the first steady-state processes subjected to thermodynamic analysis. Lord Kelvin (William Thomson) was the first to attempt the analysis.

Consider a differential slice of a bar which simultaneously conducts heat and electricity (Fig. 11–2). The rate of generation of entropy by the *heat flow* ($\dot{S}_{heat}$) is

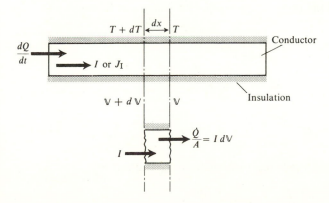

Fig. 11–2. A bar which simultaneously conducts heat and electricity.

given by Eq. (11–4). If $T_{\text{high}} = T_{\text{low}} + dT$, Eq. (11–4) becomes

$$\frac{\dot{S}_{\text{heat}}}{A} = \frac{dQ}{A\, dt} \left(\frac{1}{T} - \frac{1}{T + dT} \right)$$

$$= -\frac{dQ}{A\, dt} \times d\left(\frac{1}{T}\right) = J_Q \frac{dT}{T^2}, \qquad (11\text{–}11)$$

where J_Q is $dQ/A\, dt$, the rate of heat flow per unit area per second. Dividing through by dx, we obtain

$$\frac{\dot{S}_{\text{heat}}}{V} = J_Q \frac{dT/dx}{T^2}. \qquad (11\text{–}12)$$

The rate of generation of entropy by the *electricity flow* is found as follows.

The rate at which heat is generated in the differential volume of bar by the electric current is:

$$\dot{Q}_{\text{electric}}/V = I\, d\mathbf{V}/dx, \qquad (11\text{–}13)$$

where $d\mathbf{V}/dx$ is the voltage gradient and I is the electric current in coulombs/$\text{cm}^2 \cdot \text{sec}$.

The rate of production of entropy per unit volume is the electric-heat rate (Eq. 11–13) divided by the temperature of the slice, or

$$\frac{\dot{S}_{\text{electricity}}}{V} = \frac{\dot{Q}_{\text{electricity}}}{VT} = \frac{I(d\mathbf{V}/dx)}{T} = J_I \frac{d\mathbf{V}/dx}{T}. \qquad (11\text{–}14)$$

And the total entropy rate is

$$\frac{\dot{S}_{\text{irreversible}}}{V} = \frac{\dot{S}_{\text{heat}}}{V} + \frac{\dot{S}_{\text{electricity}}}{V}.$$

We may now insert Eqs. (11–12) and (11–13) in Eq. (11–3) to obtain the total energy-dissipation rate and the form of the conjugate fluxes and forces:

$$\frac{T\dot{S}_{\text{irreversible}}}{V} = \frac{T}{V}(\dot{S}_{\text{heat}} + \dot{S}_{\text{electricity}})$$

$$= T\left(J_Q \frac{dT/dx}{T^2} + J_I \frac{d\mathbf{V}/dx}{T} \right)$$

$$= J_Q \left(\frac{dT/dx}{T} \right) + J_I(d\mathbf{V}/dx) = \sum_i J_i \mathbb{F}_i. \qquad (11\text{–}15)$$

According to Eq. (11–12), the forces conjugate to J_Q and J_I are

$$\frac{dT/dx}{T} \quad \text{and} \quad \frac{d\mathbf{V}}{dx},$$

respectively. Therefore, using Eq. (11–2), we can express the fluxes as

$$J_I = I = L_{II}\, dV/dx + L_{IQ}\, \frac{dT/dx}{T}, \tag{11–16}$$

$$J_Q = L_{QI}\, dV/dx + L_{QQ}\, \frac{dT/dx}{T}. \tag{11–17}$$

The coupling coefficients, L_{IQ} and L_{QI}, are determined by setting the forces or fluxes equal to zero. Thus, when $I = 0$, from Eq. (11–16),

$$\left(\frac{dV}{dT}\right)_{I=0} = -\frac{L_{IQ}}{L_{II}}\frac{1}{T}, \tag{11–18}$$

and when $dT/dx = 0$,

$$\boxed{\left(\frac{J_Q}{J_I}\right)_T = \frac{L_{QI}}{L_{II}} \equiv Q^\dagger.} \tag{11–19}$$

The left-hand ratio is a measure of the heat flow *induced* per unit of whatever else is flowing in the system at constant temperature. It is called the HEAT OF TRANSPORT or the HEAT OF COUPLING, $Q^\dagger$.

If we substitute Eq. (11–5) and Eq. (11–18) in Eq. (11–19), we obtain

$$\left(\frac{J_Q}{J_I}\right)_T \frac{1}{T} = \frac{Q^\dagger}{T} = -\left(\frac{dV}{dT}\right)_{I=0} = S^\dagger, \tag{11–20}$$

where $Q^\dagger/T$ represents an *entropy flux*, $S^\dagger$, which is called the ENTROPY OF COUPLING. The values of $Q^\dagger$ and $S^\dagger$ are characteristics of the conducting material.

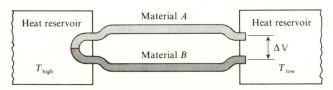

Fig. 11–3. A thermocouple.

A thermocouple consists of two dissimilar materials joined as in Fig. 11–3. If the junctions are held at different temperatures, a voltage is generated which is related to temperature in a nearly linear fashion. The voltage may be found by integrating over the entire circuit in Fig. 11–3, using the last two terms in Eq. (11–20). Thus:

$$\Delta V_{AB} = \int_{T_{low}}^{T_{high}} S_A^\dagger\, dT + \int_{T_{high}}^{T_{low}} S_B^\dagger\, dT = \int_{T_{low}}^{T_{high}} (S_A^\dagger - S_B^\dagger)\, dT. \tag{11–21}$$

In thermoelectric parlance, $\Delta V_{AB}/\Delta T$ is called the *thermoelectric power*.

11–5
Peltier effect

When an electric current passes through a thermocouple junction, if the coupled heat (J_Q) dragged along by the electricity flux has a *different magnitude* on each side of the junction, a *heat effect* must occur. This heat effect is in addition to the usual resistance heating $(I \Delta V)$ of the junction. It is called the *Peltier effect*.

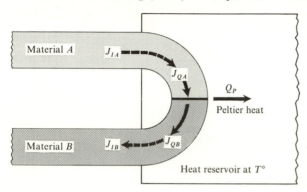

Fig. 11–4. Peltier effect at a thermocouple junction. The coupled heat flux, J_Q, has different magnitudes in materials A and B; therefore a heat effect occurs at the isothermal junction of A and B when electric current passes through the junction.

Figure 11–4 is an enlarged view of a thermocouple junction in a heat reservoir. The *isothermal* heat flux in conductor A is, from Eq. (11–20),

$$J_{QA} = (TS_A^\dagger)J_{IA}, \tag{11–22}$$

whereas the corresponding heat flux in conductor B is

$$J_{QB} = (TS_B^\dagger)J_{IB}. \tag{11–23}$$

J_{IA} equals J_{IB} and is simply the flow of electric current, I, through the junction. (The circuit in Fig. 11–3 is now closed.) *If the junction temperature is to remain constant, the difference in coupled heat flux entering and leaving the junction must be added to or removed from the junction.* Hence the amount of heat transferred to the reservoir is, from Eqs. (11–22) and (11–23),

$$Q_P = IT(S_A^\dagger - S_B^\dagger). \tag{11–24}$$

This phenomenon, which is readily observed, was first reported in 1834 by Jean C. A. Peltier. The term $T(S_A^\dagger - S_B^\dagger)$ is called the *Peltier coefficient*.

Thermoelectric refrigerators and heat pumps are devices whose operation depends on the Peltier effect.

11–6
Thomson effect

The procedures of irreversible thermodynamics can be used to demonstrate still another phenomenon of the coupled flow of heat and electricity: the *Thomson effect*.

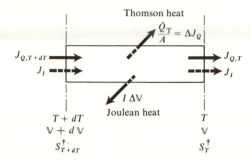

Fig. 11–5. The Thomson effect: a homogeneous conductor through which heat and electricity flow simultaneously.

If heat and electricity flow simultaneously through a homogeneous conductor (Fig. 11–5), there is a heat effect over and above the Joulean ($I \, \Delta V$) heat if the heat of coupling ($Q^\dagger$) or the entropy of coupling ($S^\dagger$) differ at T and $T + dT$. The magnitude of this heat effect is, from Fig. 11–5 and Eq. (11–20) or (11–22),

$$\dot{Q}_{\text{Thomson}} = J_{Q,T+dT} - J_{Q,T} = I[(T + dT)S^\dagger_{T+dT} - TS^\dagger_T]. \qquad (11\text{–}25)$$

Now if

$$S^\dagger_{T+dT} = S^\dagger_T + \left(\frac{\partial S^\dagger}{\partial T}\right) dT, \qquad (11\text{–}26)$$

then, substituting Eq. (11–26) in (11–25) and neglecting $(dT)^2$, we obtain

$$\frac{\dot{Q}_T}{A} = I\left[T\left(\frac{\partial S^\dagger}{\partial T}\right) dT + S^\dagger \, dT \right].$$

The total heat effect is the sum of the Thomson and Joulean heat, or

$$\frac{\dot{Q}}{A} = I \, dV + \frac{\dot{Q}_T}{A} = I\left[dV + T\left(\frac{\partial S^\dagger}{\partial T}\right) dT + S^\dagger \, dT \right]. \qquad (11\text{–}27)$$

Recalling Eq. (11–16), we may express dV as

$$dV = \frac{I \, dx}{L_{II}} - \frac{L_{IQ}}{L_{II}} \frac{dT}{T}.$$

Since Eq. (11–16) reduces to Ohm's law ($I = \Delta \mathbf{V}/\mathbb{R}$) when $dT = 0$, therefore $L_{II} = dx/d\mathbb{R}$. Also, from Eqs. (11–19) and (11–20), we obtain

$$\frac{L_{IQ}}{L_{II}} = TS^{\dagger}.$$

Therefore

$$d\mathbf{V} = I \, d\mathbb{R} - S^{\dagger} \, dT,$$

which, when substituted in Eq. (11–27), yields

$$\dot{Q} = I \left[I \, d\mathbb{R} + T \left(\frac{\partial S^{\dagger}}{\partial T} \right) dT \right]. \tag{11–28}$$

The first term in Eq. (11–28) is the *Joulean* or *resistance heating*. The second term,

$$I \left[T \left(\frac{\partial S^{\dagger}}{\partial T} \right) \right] dT,$$

is the *Thomson heat*, which is often written as

$$-I(\sigma)dT,$$

where

$$\sigma \equiv -T \frac{\partial S^{\dagger}}{\partial T} \tag{11–29}$$

is called the *Thomson coefficient*.

For a thermocouple made of materials A and B, the net Thomson coefficient is

$$\boxed{\sigma_A - \sigma_B = -T \frac{d(S_A^{\dagger} - S_B^{\dagger})}{dT}.} \tag{11–30}$$

But Eq. (11–21) shows that

$$\boxed{\frac{d\mathbf{V}_{AB}}{dT} = S_A^{\dagger} - S_B^{\dagger}.} \tag{11–31}$$

Therefore we may write Eq. (11–30) as

$$\boxed{\sigma_A - \sigma_B = -T \frac{d^2 \mathbf{V}_{AB}}{dT^2}.} \tag{11–32}$$

Equations (11–31) and (11–32) are the fundamental equations of thermoelectricity.

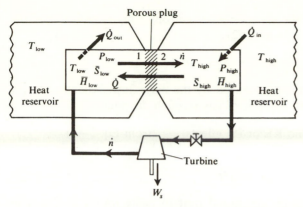

Fig. 11–6. Thermomolecular flow, an example of a coupled flow process. Two containers of gas maintained at different temperatures, are connected by a porous plug. The flow of heat from the hot to the cold side of the plug induces a flow of matter in the opposite direction, against the pressure gradient.

11–7
Thermal transpiration

Another example of a coupled-flow process which yields to the above method of analysis is the system shown in Fig. 11–6, in which two containers of a gas, held at *different temperatures*, are connected by means of a porous plug. At steady state, it is found that the pressure in the high-temperature container is higher than the pressure in the low-temperature container, and that heat leaves the high-temperature reservoir and enters the low-temperature reservoir. Kinetic theory can be used to demonstrate that the steady-state temperatures and pressures are related as follows:

$$\frac{P_{\text{high}}}{P_{\text{low}}} = \left(\frac{T_{\text{high}}}{T_{\text{low}}}\right)^{1/2}. \qquad (11\text{–}33)$$

The phenomenon is called the *Knudsen effect*.

If the gas at P_{high} is allowed to flow slowly through an external circuit and expanded back to P_{low}, as in Fig. 11–6, then an equal quantity of gas will flow through the porous plug from the *low-* to the *high*-pressure chamber. The flow of gas through the porous plug *against* the pressure gradient is induced by the flow of heat through the plug, down a temperature gradient. This is an example of the coupling of mass and heat flows.

Considering the porous plug in Fig. 11–6, the entropy generated by the heat flux through the plug is the same as in the previous heat conduction example (Eq. 11–11):

$$\dot{S}_{\text{heat}} = \dot{Q}\,\frac{\Delta T}{T^2} \qquad \text{(when } T_{\text{high}} - T_{\text{low}} = \Delta T \text{ is small)}.$$

The *mass flow*, $\dot{n}$, through the plug generates entropy simply by virtue of the transfer of matter from state (1) to state (2):

$$\dot{S}_{\text{mass}} = \dot{n}(\bar{S}_2 - \bar{S}_1) = \dot{n}\,\Delta\bar{S}.$$

An additional reverse heat flow, induced by the mass flow, becomes evident when we apply the First Law to the flow of gas through the plug:

$$\dot{Q} = W_s + \Delta H \qquad \text{(no KE or PE effects)} \qquad \text{(from Eq. 2–45)}$$

and since $W_s = 0$, then

$$\dot{Q} = \Delta H.$$

The heat reservoirs must exchange additional heat to make up for this gain in enthalpy, at a rate $-\dot{n}\,\Delta\bar{H}$. Therefore the total rate of generation of entropy is

$$\dot{S}_{\text{irreversible}} = \frac{\dot{Q}\,\Delta T}{T^2} - \dot{n}\,\frac{\Delta\bar{H}}{T} + \dot{n}\,\Delta\bar{S}, \tag{11–34}$$

and substituting Eqs. (3–22c) and (3–29), and replacing Δ by d, we obtain

$$\dot{S}_{\text{irreversible}} = \dot{Q}\,\frac{dT}{T^2} - \dot{n}\,\frac{d\bar{G}_T}{T} = \dot{Q}\,\frac{dT}{T^2} - \dot{n}\,\frac{\bar{V}\,dP}{T}. \tag{11–35}$$

To obtain thermodynamic forces, we express Eq. (11–35) in the form of the energy-dissipation rate (Eq. 11–3):

$$\frac{T\dot{S}_{\text{irreversible}}}{V_{\text{plug}}} = \frac{\dot{Q}}{A}\left(\frac{dT/dx}{T}\right) - \frac{\dot{n}}{A}\left(\bar{V}\,\frac{dP}{dx}\right).$$

Then from Eq. (11–2), we obtain

$$\frac{\dot{Q}}{A} = J_Q = L_{QQ}\,\frac{dT/dx}{T} - L_{Qn}\,\bar{V}\,\frac{dP}{dx}, \tag{11–36}$$

$$\frac{\dot{n}}{A} = J_n = L_{nQ}\,\frac{dT/dx}{T} - L_{nn}\,\bar{V}\,\frac{dP}{dx}. \tag{11–37}$$

To find the coupling coefficients, we set first $\dot{n}$ and then ΔT equal to zero:

$$\dot{n} = 0: \qquad \frac{L_{nQ}}{L_{nn}} = T\bar{V}\left(\frac{dP}{dT}\right)_{\dot{n}=0}, \tag{11–38}$$

$$dT = 0: \qquad \frac{\dot{Q}}{\dot{n}} = \frac{L_{Qn}}{L_{nn}} = Q^\dagger. \tag{11–39}$$

Equation (11–39) is the effusion analog of Eq. (11–19), and gives the *heat of coupling for thermal diffusion.*

Combining Eqs. (11–38) and (11–39), using Eq. (11–5), we obtain

$$\bar{V}\left(\frac{\Delta P}{\Delta T}\right)_{\dot{n}=0} = \frac{Q^\dagger}{T} = S^\dagger, \qquad (11\text{–}40)$$

where $S^\dagger = Q^\dagger/T$ is the *entropy of coupling* (see Eq. 11–20).

Exercise 11–2. Using Eqs. (11–40) and (11–33), prove that, for an ideal gas,

$$S^\dagger = \frac{R}{2}.$$

11–8
Other transport processes

For common transport phenomena, the thermodynamics forces which can be used in Eqs. (11–2) and (11–3) are:

Transport process	$\mathbb{F}_i$ (force)
Heat flow	$(dT/dx)/T$
Electric flow	$d\mathbb{V}/dx$
Mass flow, effusion	$\bar{V}\, dP/dx$
Chemical reaction (first order)	$d\mu_i/dx$

Units of both $\mathbb{F}_i$ and J_i must be chosen to correspond to a rate of energy dissipation. The selection of proper forces for chemical reactions other than first-order ones is an open question, since such reaction rates are not linear functions of concentrations of reactants.

APPENDIXES

APPENDIX I Molar Heat Capacities of Gases at Zero Pressure*

$$C_p^0 = a + bT + cT^2 + dT^3; \quad T = °K$$

		a	$b \times 10^2$	$c \times 10^5$	$d \times 10^9$	Temperature Range, °K	Error Max. %	Error Avg. %
Paraffinic Hydrocarbons								
Methane	CH_4	4.750	1.200	0.3030	−2.630	273–1500	1.33	0.57
Ethane	C_2H_6	1.648	4.124	−1.530	1.740	273–1500	0.83	0.28
Propane	C_3H_8	−0.966	7.279	−3.755	7.580	273–1500	0.40	0.12
n–Butane	C_4H_{10}	0.945	8.873	−4.380	8.360	273–1500	0.54	0.24
i–Butane	C_4H_{10}	−1.890	9.936	−5.495	11.92	273–1500	0.25	0.13
n–Pentane	C_5H_{12}	1.618	10.85	−5.365	10.10	273–1500	0.56	0.21
n–Hexane	C_6H_{14}	1.657	13.19	−6.844	13.78	273–1500	0.72	0.20
Monoolefinic Hydrocarbons								
Ethylene	C_2H_4	0.944	3.735	−1.993	4.220	273–1500	0.54	0.13
Propylene	C_3H_6	0.753	5.691	−2.910	5.880	273–1500	0.73	0.17
1–Butene	C_4H_8	−0.240	8.650	−5.110	12.07	273–1500	0.25	0.18
i–Butene	C_4H_8	1.650	7.702	−3.981	8.020	273–1500	0.11	0.06
cis-2–Butene	C_4H_8	−1.778	8.078	−4.074	7.890	273–1500	0.78	0.14
trans-2–Butene	C_4H_8	2.340	7.220	−3.403	6.070	273–1500	0.54	0.12
Cycloparaffinic Hydrocarbons								
Cyclopentane	C_5H_{10}	−12.957	13.087	−7.447	16.41	273–1500	1.00	0.25
Methylcyclopentane	C_6H_{12}	−12.114	15.380	−8.915	20.03	273–1500	0.86	0.23
Cyclohexane	C_6H_{12}	−15.935	16.454	−9.203	19.27	273–1500	1.57	0.37
Methylcyclohexane	C_7H_{14}	−15.070	18.972	−10.989	24.09	273–1500	0.92	0.22
Aromatic Hydrocarbons								
Benzene	C_6H_6	−8.650	11.578	−7.540	18.54	273–1500	0.34	0.20
Toluene	C_7H_8	−8.213	13.357	−8.230	19.20	273–1500	0.29	0.18
Ethylbenzene	C_8H_{10}	−8.398	15.935	−10.003	23.95	273–1500	0.34	0.19
Styrene	C_8H_8	−5.968	14.354	−9.150	22.03	273–1500	0.37	0.23
Cumene	C_9H_{12}	−9.452	18.686	−11.869	28.80	273–1500	0.36	0.17

Acetylenes and Diolefins

Name	Formula					Range (°K)		
Acetylene	C_2H_2	5.21	2.2008	-1.559	4.349	273-1500	1.46	0.59
Methylacetylene	C_3H_4	4.21	4.073	-2.192	4.713	273-1500	0.36	0.13
Dimethylacetylene	C_4H_6	3.54	5.838	-2.760	4.974	273-1500	0.70	0.16
Propadiene	C_3H_4	2.43	4.693	-2.781	6.484	273-1500	0.37	0.19
1,3-Butadiene	C_4H_6	-1.29	8.350	-5.582	14.24	273-1500	0.91	0.47
Isoprene	C_5H_8	-0.44	10.418	-6.762	16.93	273-1500	0.99	0.43

Combustion Gases (Low Range)

Name	Formula					Range (°K)		
Nitrogen	N_2	6.903	-0.03753	0.1930	-0.6861	273-1800	0.59	0.34
Oxygen	O_2	6.085	0.3631	-0.1709	0.3133	273-1800	1.19	0.28
Air		6.713	0.04697	0.1147	-0.4696	273-1800	0.72	0.33
Hydrogen	H_2	6.952	-0.04576	0.09563	-0.2079	273-1800	1.01	0.26
Carbon monoxide	CO	6.726	0.04001	0.1283	-0.5307	273-1800	0.89	0.37
Carbon dioxide	CO_2	5.316	1.4285	-0.8362	1.784	273-1800	0.67	0.22
Water vapor	H_2O	7.700	0.04594	0.2521	-0.8587	273-1800	0.53	0.24

Combustion Gases (High Range)

Name	Formula					Range (°K)		
Nitrogen	N_2	6.529	0.1488	-0.02271	—	273-3800	2.05	0.72
Oxygen	O_2	6.732	0.1505	-0.01791	—	273-3800	3.24	1.20
Air		6.557	0.1477	-0.02148	—	273-3800	1.64	0.70
Hydrogen	H_2	6.424	0.1039	-0.007804	—	273-3800	2.14	0.79
Carbon dioxide	CO_2	See footnote † for special equation.					2.65	0.54
Carbon monoxide	CO	6.480	0.1566	-0.02387	—	273-3800	1.86	1.01
Water vapor	H_2O	6.970	0.3464	-0.04833	—	273-3800	2.03	0.66

Sulfur Compounds

Name	Formula					Range (°K)		
Sulfur	S_2	6.499	0.5298	-0.3888	0.9520	273-1800	0.99	0.38
Sulfur dioxide	SO_2	6.157	1.384	-0.9103	2.057	273-1800	0.45	0.24
Sulfur trioxide	SO_3	3.918	3.483	-2.675	7.744	273-1300	0.29	0.13
Hydrogen sulfide	H_2S	7.070	0.3128	0.1364	-0.7867	273-1800	0.74	0.37
Carbon disulfide	CS_2	7.390	1.489	-1.096	2.760	273-1800	0.76	0.47
Carbonyl sulfide	COS	6.222	1.536	-1.058	2.560	273-1800	0.94	0.49

*Selected from K. A. Kobe and associates, "Thermochemistry for the Petrochemical Industry," *Petroleum Refiner*, Jan. 1949 through Nov. 1954. Reprinted with permission. In the original article, constants are also given for T in degrees centigrade, degrees Fahrenheit, and degrees Rankine.

† Equation for CO_2, 273 to 3800 °K: $c_p° = 18.036 - 0.00004474T - 158.08/\sqrt{T}$.

Table 1. Saturated steam: temperature table

T, °F	P, psia	$\overline{V}_{liq}$	$\Delta\overline{V}_{vap}$	$\overline{V}_{gas}$	$\overline{H}_{liq}$	$\Delta\overline{H}_{vap}$	$\overline{H}_{gas}$	$\overline{S}_{liq}$	$\Delta\overline{S}_{vap}$	$\overline{S}_{gas}$	T, °F
		Specific volume, ft³/lb			Enthalpy, Btu/lb			Entropy, Btu/lb · °R			
32.0	0.08859	0.016022	3304.7	3304.7	0.0179	1075.5	1075.5	0.0000	2.1873	2.1873	32.0
34.0	0.09600	0.016021	3061.9	3061.9	1.996	1074.4	1076.4	0.0041	2.1762	2.1802	34.0
36.0	0.10395	0.016020	2839.0	2839.0	4.008	1073.2	1077.2	0.0081	2.1651	2.1732	36.0
38.0	0.11249	0.016019	2634.1	2634.2	6.018	1072.1	1078.1	0.0122	2.1541	2.1663	38.0
40.0	1.12163	0.016019	2445.8	2445.8	8.027	1071.0	1079.0	0.0162	2.1432	2.1594	40.0
42.0	0.13143	0.016019	2272.4	2272.4	10.035	1069.8	1079.9	0.0202	2.1325	2.1527	42.0
44.0	0.14192	0.016019	2112.8	2112.8	12.041	1068.7	1080.7	0.0242	2.1217	2.1459	44.0
46.0	0.15314	0.016020	1965.7	1965.7	14.047	1067.6	1081.6	0.0282	2.1111	2.1393	46.0
48.0	0.16514	0.016021	1830.0	1830.0	16.051	1066.4	1082.5	0.0321	2.1006	2.1327	48.0
50.0	0.17796	0.016023	1704.8	1704.8	18.054	1065.3	1083.4	0.0361	2.0901	2.1262	50.0
52.0	0.19165	0.016024	1589.2	1589.2	20.057	1064.2	1084.2	0.0400	2.0798	2.1197	52.0
54.0	0.20625	0.016026	1482.4	1482.4	22.058	1063.1	1085.1	0.0439	2.0695	2.1134	54.0
56.0	0.22183	0.016028	1383.6	1383.6	24.059	1061.9	1086.0	0.0478	2.0593	2.1070	56.0
58.0	0.23843	0.016031	1292.2	1292.2	26.060	1060.8	1086.9	0.0516	2.0491	2.1008	58.0
60.0	0.25611	0.016033	1207.6	1207.6	28.060	1059.7	1087.7	0.0555	2.0391	2.0946	60.0
62.0	0.27494	0.016036	1129.2	1129.2	30.059	1058.5	1088.6	0.0593	2.0291	2.0885	62.0
64.0	0.29497	0.016039	1056.5	1056.5	32.058	1057.4	1089.5	0.0632	2.0192	2.0824	64.0
66.0	0.31626	0.016043	989.0	989.1	34.056	1056.3	1090.4	0.0670	2.0094	2.0764	66.0
68.0	0.33889	0.016046	926.5	926.5	36.054	1055.2	1091.2	0.0708	1.9996	2.0704	68.0
70.0	0.36292	0.016050	868.3	868.4	38.052	1054.0	1092.1	0.0745	1.9900	2.0645	70.0
72.0	0.38844	0.016054	814.3	814.3	40.049	1052.9	1093.0	0.0783	1.9804	2.0587	72.0
74.0	0.41550	0.016058	764.1	764.1	42.046	1051.8	1093.8	0.0821	1.9708	2.0529	74.0
76.0	0.44420	0.016063	717.4	717.4	44.043	1050.7	1094.7	0.0858	1.9614	2.0472	76.0
78.0	0.47461	0.016067	673.8	673.9	46.040	1049.5	1095.6	0.0895	1.9520	2.0415	78.0
80.0	0.50683	0.016072	633.3	633.3	48.037	1048.4	1096.4	0.0932	1.9426	2.0359	80.0
82.0	0.54093	0.016077	595.5	595.5	50.033	1047.3	1097.3	0.0969	1.9334	2.0303	82.0
84.0	0.57702	0.016082	560.3	560.3	52.029	1046.1	1098.2	0.1006	1.9242	2.0248	84.0
86.0	0.61518	0.016087	227.5	527.5	54.026	1045.0	1099.0	0.1043	1.9151	2.0193	86.0
88.0	0.65551	0.016093	496.8	496.8	56.022	1043.9	1099.9	0.1079	1.9060	2.0139	88.0
90.0	0.69813	0.016099	468.1	468.1	58.018	1042.7	1100.8	0.1115	1.8970	2.0086	90.0
92.0	0.74313	0.016105	441.3	441.3	60.014	1041.6	1101.6	0.1152	1.8881	2.0033	92.0
94.0	0.79062	0.016111	416.3	416.3	62.010	1040.5	1102.5	0.1188	1.8792	1.9980	94.0
96.0	0.84072	0.016117	392.8	392.9	64.006	1039.3	1103.3	0.1224	1.8704	1.9928	96.0
98.0	0.89356	0.016123	370.9	370.9	66.003	1038.2	1104.2	0.1260	1.8617	1.9876	98.0
100.0	0.94924	0.016130	350.4	350.4	67.999	1037.1	1105.1	0.1295	1.8530	1.9825	100.0
102.0	1.00789	0.016137	331.1	331.1	69.995	1035.9	1105.9	0.1331	1.8444	1.9775	102.0
104.0	1.06965	0.016144	313.1	313.1	71.992	1034.8	1106.8	0.1366	1.8358	1.9725	104.0
106.0	1.1347	0.016151	296.16	296.18	73.99	1033.6	1107.6	0.1402	1.8273	1.9675	106.0
108.0	1.2030	0.016158	280.28	280.30	75.98	1032.5	1108.5	0.1437	1.8188	1.9626	108.0
110.0	1.2750	0.016165	265.37	265.39	77.98	1031.4	1109.3	0.1472	1.8105	1.9577	110.0
112.0	1.3505	0.016173	251.37	251.38	79.98	1030.2	1110.2	0.1507	1.8021	1.9528	112.0
114.0	1.4299	0.016180	238.21	238.22	81.97	1029.1	1111.0	0.1542	1.7938	1.9480	114.0
116.0	1.5133	0.016188	225.84	225.85	83.97	1027.9	1111.9	0.1577	1.7856	1.9433	116.0
118.0	1.6009	0.016196	214.20	214.21	85.97	1026.8	1112.7	0.1611	1.7774	1.9386	118.0
120.0	1.6927	0.016204	203.25	203.26	87.97	1025.6	1113.6	0.1646	1.7693	1.9339	120.0
122.0	1.7891	0.016213	192.94	192.95	89.96	1024.5	1114.4	0.1680	1.7613	1.9293	122.0
124.0	1.8901	0.016221	183.23	183.24	91.96	1023.3	1115.3	0.1715	1.7533	1.9247	124.0
126.0	1.9959	0.016229	174.08	174.09	93.96	1022.2	1116.1	0.1749	1.7453	1.9202	126.0
128.0	2.1068	0.016238	165.45	165.47	95.96	1021.0	1117.0	0.1783	1.7374	1.9157	128.0
130.0	2.2230	0.016247	157.32	157.33	97.96	1019.8	1117.8	0.1817	1.7295	1.9112	130.0
132.0	2.3445	0.016256	149.64	149.66	99.95	1018.7	1118.6	0.1851	1.7217	1.9068	132.0
134.0	2.4717	0.016265	142.40	142.41	101.95	1017.5	1119.5	0.1884	1.7140	1.9024	134.0
136.0	2.6047	0.016274	135.55	135.57	103.95	1016.4	1120.3	0.1918	1.7063	1.8980	136.0
138.0	2.7438	0.016284	129.09	129.11	105.95	1015 2	1121.1	0.1951	1.6986	1.8937	138.0
140.0	2.8892	0.016293	122.98	123.00	107.95	1014.0	1122.0	0.1985	1.6910	1.8895	140.0
142.0	3.0411	0.016303	117.21	117.22	109.95	1012.9	1122.8	0.2018	1.6534	1.8852	142.0
144.0	3.1997	0.016312	111.74	111.76	111.95	1011.7	1123.6	0.2051	1.6759	1.8810	144.0
146.0	3.3653	0.016322	106.58	106.59	113.95	1010.5	1124.5	0.2084	1.6684	1.8769	146.0
148.0	3.5381	0.016332	101.68	101.70	115.95	1009.3	1125.3	0.2117	1.6610	1.8727	148.0
150.0	3.7184	0.016343	97.05	97.07	117.95	1008.2	1126.1	0.2150	1.6536	1.8686	150.0
152.0	3.9065	0.016353	92.66	92.68	119.95	1007.0	1126.9	0.2183	1.6463	1.8646	152.0
154.0	4.1025	0.016363	88.50	88.52	121.95	1005.8	1127.7	0.2216	1.6390	1.8606	154.0
156.0	4.3068	0.016374	84.56	84.57	123.95	1004.6	1128.6	0.2248	1.6318	1.8566	156.0
158.0	4.5197	0.016384	80.82	80.83	125.96	1003.4	1129.4	0.2281	1.6245	1.8526	158.0
160.0	4.7414	0.016395	77.27	77.29	127.96	1002.2	1130.2	0.2313	1.6174	1.8487	160.0
162.0	4.9722	0.016406	73.90	73.92	129.96	1001.0	1131.0	0.2345	1.6103	1.8448	162.0
164.0	5.2124	0.016417	70.70	70.72	131.96	999.8	1131.8	0.2377	1.6032	1.8409	164.0
166.0	5.4623	0.016428	67.67	67.68	133.97	998.6	1132.6	0.2409	1.5961	1.8371	166.0
168.0	5.7223	0.016440	64.78	64.80	135.97	997.4	1133.4	0.2441	1.5892	1.8333	168.0
170.0	5.9926	0.016451	62.04	62.06	137.97	996.2	1134.2	0.2473	1.5822	1.8295	170.0
172.0	6.2736	0.016463	59.43	59.45	139.98	995.0	1135.0	0.2505	1.5753	1.8258	172.0
174.0	6.5656	0.016474	56.95	56.97	141.98	993.8	1135.8	0.2537	1.5684	1.8221	174.0
176.0	6.8690	0.016486	54.59	54.61	143.99	992.6	1136.6	0.2568	1.5616	1.8184	176.0
178.0	7.1840	0.016498	52.35	52.36	145.99	991.4	1137.4	0.2600	1.5548	1.8147	178.0

Table 1. Saturated steam: temperature table (*continued*)

T, °F	P, psia	Specific volume, ft³/lb			Enthalpy, Btu/lb			Entropy, Btu/lb · °R			T, °F
		$\overline{V}_{liq}$	$\Delta\overline{V}_{vap}$	$\overline{V}_{gas}$	$\overline{H}_{liq}$	$\Delta\overline{H}_{vap}$	$\overline{H}_{gas}$	$\overline{S}_{liq}$	$\Delta\overline{S}_{vap}$	$\overline{S}_{gas}$	
180.0	7.5110	0.016510	50.21	50.22	148.00	990.2	1138.2	0.2631	1.5480	1.8111	180.0
182.0	7.850	0.016522	48.172	18.189	150.01	989.0	1139.0	0.2662	1.5413	1.8075	182.0
184.0	8.203	0.016534	46.232	46.249	152.01	987.8	1139.8	0.2694	1.5346	1.8040	184.0
186.0	8.568	0.016547	44.383	44.400	154.02	986.5	1140.5	0.2725	1.5279	1.8004	186.0
188.0	8.947	0.016559	42.621	42.638	156.03	985.3	1141.3	0.2756	1.5213	1.7969	188.0
190.0	9.340	0.016572	40.941	40.957	158.04	984.1	1142.1	0.2787	1.5148	1.7934	190.0
192.0	9.747	0.016585	39.337	39.354	160.05	982.8	1142.9	0.2818	1.5082	1.7900	192.0
194.0	10.168	0.016598	37.808	37.824	162.05	981.6	1143.7	0.2848	1.5017	1.7865	194.0
196.0	10.605	0.016611	36.348	36.364	164.06	980.4	1144.4	0.2879	1.4952	1.7831	196.0
198.0	11.058	0.016624	34.954	34.970	166.08	979.1	1145.2	0.2910	1.4888	1.7798	198.0
200.0	11.526	0.016637	33.622	33.639	168.09	977.9	1146.0	0.2940	1.4824	1.7764	200.0
204.0	12.512	0.016664	31.135	31.151	172.11	975.4	1147.5	0.3001	1.4697	1.7698	204.0
208.0	13.568	0.016691	28.862	28.878	176.14	972.8	1149.0	0.3061	1.4571	1.7632	208.0
212.0	14.696	0.016719	26.782	26.799	180.17	970.3	1150.5	0.3121	1.4447	1.7568	212.0
216.0	15.901	0.016747	24.878	24.894	184.20	967.8	1152.0	0.3181	1.4323	1.7505	216.0
220.0	17.186	0.016775	23.131	23.148	188.23	965.2	1153.4	0.3241	1.4201	1.7442	220.0
224.0	18.556	0.016805	21.529	21.545	192.27	962.6	1154.9	0.3300	1.4081	1.7380	224.0
228.0	20.015	0.016834	20.056	20.073	196.31	960.0	1156.3	0.3359	1.3961	1.7320	228.0
232.0	21.567	0.016864	18.701	18.718	200.35	957.4	1157.8	0.3417	1.3842	1.7260	232.0
236.0	23.216	0.016895	17.454	17.471	204.40	954.8	1159.2	0.3476	1.3725	1.7201	236.0
240.0	24.968	0.016926	16.304	16.321	208.45	952.1	1160.6	0.3533	1.3609	1.7142	240.0
244.0	26.826	0.016958	15.243	15.260	212.50	949.5	1162.0	0.3591	1.3494	1.7085	244.0
248.0	28.796	0.016990	14.264	14.281	216.56	946.8	1163.4	0.3649	1.3379	1.7028	248.0
252.0	30.883	0.017022	13.358	13.375	220.62	944.1	1164.7	0.3706	1.3266	1.6972	252.0
256.0	33.091	0.017055	12.520	12.538	224.69	941.4	1166.1	0.3763	1.3154	1.6917	256.0
260.0	35.427	0.017089	11.745	11.762	228.76	938.6	1167.4	0.3819	1.3043	1.6862	260.0
264.0	37.894	0.017123	11.025	11.042	232.83	935.9	1168.7	0.3876	1.2933	1.6808	264.0
268.0	40.500	0.017157	10.358	10.375	236.91	933.1	1170.0	0.3932	1.2823	1.6755	268.0
272.0	43.249	0.017193	9.738	9.755	240.99	930.3	1171.3	0.3987	1.2715	1.6702	272.0
276.0	46.147	0.017228	9.162	9.180	245.08	927.5	1172.5	0.4043	1.2607	1.6650	276.0
280.0	49.200	0.017264	8.627	8.644	249.17	924.6	1173.8	0.4098	1.2501	1.6599	280.0
284.0	52.414	0.01730	8.1280	8.1453	253.3	921.7	1175.0	0.4154	1.2395	1.6548	284.0
288.0	55.795	0.01734	7.6634	7.6807	257.4	918.8	1176.2	0.4208	1.2290	1.6498	288.0
292.0	59.350	0.01738	7.2301	7.2475	261.5	915.9	1177.4	0.4263	1.2186	1.6449	292.0
296.0	63.084	0.01741	6.8259	6.8433	265.6	913.0	1178.6	0.4317	1.2082	1.6400	296.0
300.0	67.005	0.01745	6.4483	6.4658	269.7	910.0	1179.7	0.4372	1.1979	1.6351	300.0
304.0	71.119	0.01749	6.0955	6.1130	273.8	907.0	1180.9	0.4426	1.1877	1.6303	304.0
308.0	75.433	0.01753	5.7655	5.7830	278.0	904.0	1182.0	0.4479	1.1776	1.6256	308.0
312.0	79.953	0.01757	5.4566	5.4742	282.1	901.0	1183.1	0.4533	1.1676	1.6209	312.0
316.0	84.688	0.01761	5.1673	5.1849	286.3	897.9	1184.1	0.4586	1.1576	1.6162	316.0
320.0	89.643	0.01766	4.8961	4.9138	290.4	894.8	1185.2	0.4640	1.1477	1.6116	320.0
324.0	94.826	0.01770	4.6418	4.6595	294.6	891.6	1186.2	0.4692	1.1378	1.6071	324.0
328.0	100.245	0.01774	4.4030	4.4208	298.7	888.5	1187.2	0.4745	1.1280	1.6025	328.0
332.0	105.907	0.01779	4.1788	4.1966	302.9	885.3	1188.2	0.4798	1.1183	1.5981	332.0
336.0	111.820	0.01783	3.9681	3.9859	307.1	882.1	1189.1	0.4850	1.1086	1.5936	336.0
340.0	117.992	0.01787	3.7699	3.7878	311.3	878.8	1190.1	0.4902	1.0990	1.5892	340.0
344.0	124.430	0.01792	3.5834	3.6013	315.5	875.5	1191.0	0.4954	1.0894	1.5849	344.0
348.0	131.142	0.01797	3.4078	3.4258	319.7	872.2	1191.1	0.5006	1.0799	1.5806	348.0
352.0	138.138	0.01801	3.2423	3.2603	323.9	868.9	1192.7	0.5058	1.0705	1.5763	352.0
356.0	145.424	0.01806	3.0863	3.1044	328.1	865.5	1193.6	0.5110	1.0611	1.5721	356.0
360.0	153.010	0.01811	2.9392	2.9573	332.3	862.1	1194.4	0.5161	1.0517	1.5678	360.0
364.0	160.903	0.01816	2.8002	2.8184	336.5	858.6	1195.2	0.5212	1.0424	1.5637	364.0
368.0	169.113	0.01821	2.6691	2.6873	340.8	855.1	1195.9	0.5263	1.0332	1.5595	368.0
372.0	177.648	0.01826	2.5451	2.5633	345.0	851.6	1196.7	0.5314	1.0240	1.5554	372.0
376.0	186.517	0.01831	2.4279	2.4462	349.3	848.1	1197.4	0.5365	1.0148	1.5513	376.0
380.0	195.729	0.01836	2.3170	2.3353	353.6	844.5	1198.0	0.5416	1.0057	1.5473	380.0
384.0	205.294	0.01842	2.2120	2.2304	357.9	840.8	1198.7	0.5466	0.9966	1.5432	384.0
388.0	215.220	0.01847	2.1126	2.1311	362.2	837.2	1199.3	0.5516	0.9876	1.5392	388.0
392.0	225.516	0.01853	2.0184	2.0369	366.5	833.4	1199.9	0.5567	0.9786	1.5352	392.0
396.0	236.193	0.01858	1.9291	1.9477	370.8	829.7	1200.4	0.5617	0.9696	1.5313	396.0
400.0	247.259	0.01864	1.8444	1.8630	375.1	825.9	1201.0	0.5667	0.9607	1.5274	400.0
404.0	258.725	0.01870	1.7640	1.7827	379.4	822.0	1201.5	0.5717	0.9518	1.5234	404.0
408.0	270.600	0.01875	1.6877	1.7064	383.8	818.2	1201.9	0.5766	0.9429	1.5195	408.0
412.0	282.894	0.01881	1.6152	1.6340	388.1	814.2	1202.4	0.5816	0.9341	1.5157	412.0
416.0	295.617	0.01887	1.5463	1.5651	392.5	810.2	1202.8	0.5866	0.9253	1.5118	416.0
420.0	308.780	0.01894	1.4808	1.4997	396.9	806.2	1203.1	0.5915	0.9165	1.5080	420.0
424.0	322.391	0.01900	1.4184	1.4374	401.3	802.2	1203.5	0.5964	0.9077	1.5042	424.0
428.0	336.463	0.01906	1.3591	1.3782	405.7	798.0	1203.7	0.6014	0.8990	1.5004	428.0
432.0	351.00	0.01913	1.30266	1.32179	410.1	793.9	1204.0	0.6063	0.8903	1.4966	432.0
436.0	366.03	0.01919	1.24887	1.26806	414.6	789.7	1204.2	0.6112	0.8816	1.4928	436.0
440.0	381.54	0.01926	1.19761	1.21687	419.0	785.4	1204.4	0.6161	0.8729	1.4890	440.0
444.0	397.56	0.01933	1.14874	1.16806	423.5	781.1	1204.6	0.6210	0.8643	1.4853	444.0
448.0	414.09	0.01940	1.10212	1.12152	428.0	776.7	1204.7	0.6259	0.8557	1.4815	448.0
452.0	431.14	0.01947	1.05764	1.07711	432.5	772.3	1204.8	0.6308	0.8471	1.4778	452.0
456.0	448.73	0.01954	1.01518	1.03472	437.0	767.8	1204.8	0.6356	0.8385	1.4741	456.0

Table 1. Saturated steam: temperature table (*continued*)

T, °F	P, psia	$\overline{V}_{liq}$	$\Delta\overline{V}_{vap}$	$\overline{V}_{gas}$	$\overline{H}_{liq}$	$\Delta\overline{H}_{vap}$	$\overline{H}_{gas}$	$\overline{S}_{liq}$	$\Delta\overline{S}_{vap}$	$\overline{S}_{gas}$	T, °F
		Specific volume, ft³/lb			Enthalpy, Btu/lb			Entropy, Btu/lb · °R			
460.0	466.87	0.01961	0.97463	0.99424	441.5	763.2	1204.8	0.6405	0.8299	1.4704	460.0
464.0	485.56	0.01969	0.93588	0.95557	446.1	758.6	1204.7	0.6454	0.8213	1.4667	464.0
468.0	504.83	0.01976	0.89885	0.91862	450.7	754.0	1204.6	0.6502	0.8127	1.4629	468.0
472.0	524.67	0.01984	0.86345	0.88329	455.2	749.3	1204.5	0.6551	0.8042	1.4592	472.0
476.0	545.11	0.01992	0.82958	0.84950	459.9	744.5	1204.3	0.6599	0.7956	1.4555	476.0
480.0	566.15	0.02000	0.79716	0.81717	464.5	739.6	1204.1	0.6648	0.7871	1.4518	480.0
484.0	587.81	0.02009	0.76613	0.78622	469.1	734.7	1203.8	0.6696	0.7785	1.4481	484.0
488.0	610.10	0.02017	0.73641	0.75658	473.8	729.7	1203.5	0.6745	0.7700	1.4444	488.0
492.0	633.03	0.02026	0.70794	0.72820	478.5	724.6	1203.1	0.6793	0.7614	1.4407	492.0
496.0	656.61	0.02034	0.68065	0.70100	483.2	719.5	1202.7	0.6842	0.7528	1.4370	496.0
500.0	680.86	0.02043	0.65448	0.67492	487.9	714.3	1202.2	0.6890	0.7443	1.4333	500.0
504.0	705.78	0.02053	0.62938	0.64991	492.7	709.0	1201.7	0.6939	0.7357	1.4296	504.0
508.0	731.40	0.02062	0.60530	0.62592	497.5	703.7	1201.1	0.6987	0.7271	1.4258	508.0
512.0	757.72	0.02072	0.58218	0.60289	502.3	698.2	1200.5	0.7036	0.7185	1.4221	512.0
516.0	784.76	0.02081	0.55997	0.58079	507.1	692.7	1199.8	0.7085	0.7099	1.4183	516.0
520.0	812.53	0.02091	0.53864	0.55956	512.0	687.0	1199.0	0.7133	0.7013	1.4146	520.0
524.0	841.04	0.02102	0.51814	0.53916	516.9	681.3	1198.2	0.7182	0.6926	1.4108	524.0
528.0	870.31	0.02112	0.49843	0.51955	521.8	675.5	1197.3	0.7231	0.6839	1.4070	528.0
532.0	900.34	0.02123	0.47947	0.50070	526.8	669.6	1196.4	0.7280	0.6752	1.4032	532.0
536.0	931.17	0.02134	0.46123	0.48257	531.7	663.6	1195.4	0.7329	0.6665	1.3993	536.0
540.0	962.79	0.02146	0.44367	0.46513	536.8	657.5	1194.3	0.7378	0.6577	1.3954	540.0
544.0	995.22	0.02157	0.42677	0.44834	541.8	651.3	1193.1	0.7427	0.6489	1.3915	544.0
548.0	1028.49	0.02169	0.41048	0.43217	546.9	645.0	1191.9	0.7476	0.6400	1.3876	548.0
552.0	1062.59	0.02182	0.39479	0.41660	552.0	638.5	1190.6	0.7525	0.6311	1.3837	552.0
556.0	1097.55	0.02194	0.37966	0.40160	557.2	632.0	1189.2	0.7575	0.6222	1.3797	556.0
560.0	1133.38	0.02207	0.36507	0.38714	562.4	625.3	1187.7	0.7625	0.6132	1.3757	560.0
564.0	1170.10	0.02221	0.35099	0.37320	567.6	618.5	1186.1	0.7674	0.6041	1.3716	564.0
568.0	1207.72	0.02235	0.33741	0.35975	572.9	611.5	1184.5	0.7725	0.5950	1.3675	568.0
572.0	1246.26	0.02249	0.32429	0.34678	578.3	604.5	1182.7	0.7775	0.5859	1.3634	572.0
576.0	1285.74	0.02264	0.31162	0.33426	583.7	597.2	1180.9	0.7825	0.5766	1.3592	576.0
580.0	1326.17	0.02279	0.29937	0.32216	589.1	589.9	1179.0	0.7876	0.5673	1.3550	580.0
584.0	1367.7	0.02295	0.28753	0.31048	594.6	582.4	1176.9	0.7927	0.5580	1.3507	584.0
588.0	1410.0	0.02311	0.27608	0.29919	600.1	574.7	1174.8	0.7978	0.5485	1.3464	588.0
592.0	1453.3	0.02328	0.26499	0.28827	605.7	566.8	1172.6	0.8030	0.5390	1.3420	592.0
596.0	1497.8	0.02345	0.25425	0.27770	611.4	558.8	1170.2	0.8082	0.5293	1.3375	596.0
600.0	1543.2	0.02364	0.24384	0.26747	617.1	550.6	1167.7	0.8134	0.5196	1.3330	600.0
604.0	1589.7	0.02382	0.23374	0.25757	622.9	542.2	1165.1	0.8187	0.5097	1.3284	604.0
608.0	1637.3	0.02402	0.22394	0.24796	628.8	533.6	1162.4	0.8240	0.4997	1.3238	608.0
612.0	1686.1	0.02422	0.21442	0.23865	634.8	524.7	1159.5	0.8294	0.4896	1.3190	612.0
616.6	1735.9	0.02444	0.20516	0.22960	640.8	515.6	1156.4	0.8348	0.4794	1.3141	616.6
620.0	1786.9	0.02466	0.19615	0.22081	646.9	506.3	1153.2	0.8403	0.4689	1.3092	620.0
624.0	1839.0	0.02489	0.18737	0.21226	653.1	496.6	1149.8	0.8458	0.4583	1.3041	624.0
628.0	1892.4	0.02514	0.17880	0.20394	659.5	486.7	1146.1	0.8514	0.4474	1.2988	628.0
632.0	1947.0	0.02539	0.17044	0.19583	665.9	476.4	1142.2	0.8571	0.4364	1.2934	632.0
636.0	2002.8	0.02566	0.16226	0.18792	672.4	465.7	1138.1	0.8628	0.4251	1.2879	636.0
640.0	2059.9	0.02595	0.15427	0.18021	679.1	454.6	1133.7	0.8686	0.4134	1.2821	640.0
644.0	2118.3	0.02625	0.14644	0.17269	685.9	443.1	1129.0	0.8746	0.4015	1.2761	644.0
648.0	2178.1	0.02657	0.13876	0.16534	692.9	431.1	1124.0	0.8806	0.3893	1.2699	648.0
652.0	2239.2	0.02691	0.13124	0.15816	700.0	418.7	1118.7	0.8868	0.3767	1.2634	652.0
656.0	2301.7	0.02728	0.12387	0.15115	707.4	405.7	1113.1	0.8931	0.3637	1.2567	656.0
660.0	2365.7	0.02768	0.11663	0.14431	714.9	392.1	1107.0	0.8995	0.3502	1.2498	660.0
664.0	2431.1	0.02811	0.10947	0.13757	722.9	377.7	1100.6	0.9064	0.3361	1.2425	664.0
668.0	2498.1	0.02858	0.10229	0.13087	731.5	362.1	1093.5	0.9137	0.3210	1.2347	668.0
672.0	2566.6	0.02911	0.09514	0.12424	740.2	345.7	1085.9	0.9212	0.3054	1.2266	672.0
676.0	2636.8	0.02970	0.08799	0.11769	749.2	328.5	1077.6	0.9287	0.2892	1.2179	676.0
680.0	2708.6	0.03037	0.08080	0.11117	758.5	310.1	1068.5	0.9365	0.2720	1.2086	680.0
684.0	2782.1	0.03114	0.07349	0.10463	768.2	290.2	1058.4	0.9447	0.2537	1.1984	684.0
688.0	2857.4	0.03204	0.06595	0.09799	778.8	268.2	1047.0	0.9535	0.2337	1.1872	688.0
692.0	2934.5	0.03313	0.05797	0.09110	790.5	243.1	1033.6	0.9634	0.2110	1.1744	692.0
696.0	3013.4	0.03455	0.04916	0.08371	804.4	212.8	1017.2	0.9749	0.1841	1.1591	696.0
700.0	3094.3	0.03662	0.03857	0.07519	822.4	172.7	995.2	0.9901	0.1490	1.1390	700.0
702.0	3135.5	0.03824	0.03173	0.06997	835.0	144.7	979.7	1.0006	0.1246	1.1252	702.0
704.0	3177.2	0.04108	0.02192	0.06300	854.2	102.0	956.2	1.0169	0.0876	1.1046	704.0
705.0	3198.3	0.04427	0.01304	0.05730	873.0	61.4	934.4	1.0329	0.0527	1.0856	705.0
705.47*	3208.2	0.05078	0.00000	0.05078	906.0	0.0	906.0	1.0612	0.0000	1.0612	705.47*

*Critical temperature

Table 2. Saturated steam: pressure table

P, psia	T, °F	Specific volume, ft³/lb			Enthalpy, Btu/lb			Entropy, Btu/lb · °R			P, psia
		$\overline{V}_{liq}$	$\Delta\overline{V}_{vap}$	$\overline{V}_{gas}$	$\overline{H}_{liq}$	$\Delta\overline{H}_{vap}$	$\overline{H}_{gas}$	$\overline{S}_{liq}$	$\Delta\overline{S}_{vap}$	$\overline{S}_{gas}$	
0.08865	32.018	0.016022	3302.4	3302.4	0.0003	1075.5	1075.5	0.0000	2.1872	2.1872	0.08865
0.25	59.323	0.016032	1235.5	1235.5	27.382	1060.1	1087.4	0.0542	2.0425	2.0967	0.25
0.50	79.586	0.016071	641.5	641.5	47.623	1048.6	1096.3	0.0925	1.9446	2.0370	0.50
1.0	101.74	0.016136	333.59	333.60	69.73	1036.1	1105.8	0.1326	1.8455	1.9781	1.0
5.0	162.24	0.016407	73.515	73.532	130.20	1000.9	1131.1	0.2349	1.6094	1.8443	5.0
10.0	193.21	0.016592	38.404	38.420	161.26	982.1	1143.3	0.2836	1.5043	1.7879	10.0
14.696	212.00	0.016719	26.782	26.799	180.17	970.3	1150.5	0.3121	1.4447	1.7568	14.696
15.0	213.03	0.016726	26.274	26.290	181.21	969.7	1150.9	0.3137	1.4415	1.7552	15.0
20.0	227.96	0.016834	20.070	20.087	196.27	960.1	1156.3	0.3358	1.3962	1.7320	20.0
30.0	250.34	0.017009	13.7266	13.7436	218.9	945.2	1164.1	0.3682	1.3313	1.6995	30.0
40.0	267.25	0.017151	10.4794	10.4965	236.1	933.6	1169.8	0.3921	1.2844	1.6765	40.0
50.0	281.02	0.017274	8.4967	8.5140	250.2	923.9	1174.1	0.4112	1.2474	1.6586	50.0
60.0	292.71	0.017383	7.1562	7.1736	262.2	915.4	1177.6	0.4273	1.2167	1.6440	60.0
70.0	302.93	0.017482	6.1875	6.2050	272.7	907.8	1180.6	0.4411	1.1905	1.6316	70.0
80.0	312.04	0.017573	5.4536	5.4711	282.1	900.9	1183.1	0.4534	1.1675	1.6208	80.0
90.0	320.28	0.017659	4.8779	4.8953	290.7	894.6	1185.3	0.4643	1.1470	1.6113	90.0
100.0	327.82	0.017740	4.4133	4.4310	298.5	888.6	1187.2	0.4743	1.1284	1.6027	100.0
110.3	334.79	0.01782	4.0306	4.0484	305.8	883.1	1188.9	0.4834	1.1115	1.5950	110.0
120.0	341.27	0.01789	3.7097	3.7275	312.6	877.8	1190.4	0.4919	1.0960	1.5879	120.0
130.0	347.33	0.01796	3.4364	3.4544	319.0	872.8	1191.7	0.4998	1.0815	1.5813	130.0
140.0	353.04	0.01803	3.2010	3.2190	325.0	868.0	1193.0	0.5071	1.0681	1.5752	140.0
150.0	358.43	0.01809	2.9958	3.0139	330.6	863.4	1194.1	0.5141	1.0554	1.5695	150.0
160.0	363.55	0.01815	2.8155	2.8336	336.1	859.0	1195.1	0.5206	1.0435	1.5641	160.0
170.0	368.42	0.01821	2.6556	2.6738	341.2	854.8	1196.0	0.5269	1.0322	1.5591	170.0
180.0	373.08	0.01827	2.5129	2.5312	346.2	850.7	1196.9	0.5328	1.0215	1.5543	180.0
190.0	377.53	0.01833	2.3847	2.4030	350.9	846.7	1197.6	0.5384	1.0113	1.5498	190.0
200.0	381.80	0.01839	2.2689	2.2873	355.5	842.8	1198.3	0.5438	1.0016	1.5454	200.0
210.0	385.91	0.01844	2.16373	2.18217	359.9	839.1	1199.0	0.5490	0.9923	1.5413	210.0
220.0	389.88	0.01850	2.06779	2.08629	364.2	835.4	1199.6	0.5540	0.9834	1.5374	220.0
230.0	393.70	0.01855	1.97991	1.99846	368.3	831.8	1200.1	0.5588	0.9748	1.5336	230.0
240.0	397.39	0.01860	1.89909	1.91769	372.3	828.4	1200.6	0.5634	0.9665	1.5299	240.0
250.0	400.97	0.01865	1.82452	1.84317	376.1	825.0	1201.1	0.5679	0.9585	1.5264	250.0
260.0	404.44	0.01870	1.75548	1.77418	379.9	821.6	1201.5	0.5722	0.9508	1.5230	260.0
270.0	407.80	0.01875	1.69137	1.71013	383.6	818.3	1201.9	0.5764	0.9433	1.5197	270.0
280.0	411.07	0.01880	1.63169	1.65049	387.1	815.1	1202.3	0.5805	0.9361	1.5166	280.0
290.0	414.25	0.01885	1.57597	1.59482	390.6	812.0	1202.6	0.5844	0.9291	1.5135	290.0
300.0	417.35	0.01889	1.52384	1.54274	394.0	808.9	1202.9	0.5882	0.9223	1.5105	300.0
350.0	431.73	0.01912	1.30642	1.32554	409.8	794.2	1204.0	0.6059	0.8909	1.4968	350.0
400.0	444.60	0.01934	1.14162	1.16095	424.2	780.4	1204.6	0.6217	0.8630	1.4847	400.0
450.0	456.28	0.01954	1.01224	1.03179	437.3	767.5	1204.8	0.6360	0.8378	1.4738	450.0
500.0	467.01	0.01975	0.90787	0.92762	449.5	755.1	1204.7	0.6490	0.8148	1.4639	500.0
550.0	476.94	0.01994	0.82183	0.84177	460.9	743.3	1204.3	0.6611	0.7936	1.4547	550.0
600.0	486.20	0.02013	0.74962	0.76975	471.7	732.0	1203.7	0.6723	0.7738	1.4461	600.0
650.0	494.89	0.02032	0.68811	0.70843	481.9	720.9	1202.8	0.6828	0.7552	1.4381	650.0
700.0	503.08	0.02050	0.63505	0.65556	491.6	710.2	1201.8	0.6928	0.7377	1.4304	700.0
750.0	510.84	0.02069	0.58880	0.60949	500.9	699.8	1200.7	0.7022	0.7210	1.4232	750.0
800.0	518.21	0.02087	0.54809	0.56896	509.8	689.6	1199.4	0.7111	0.7051	1.4163	800.0
850.0	525.24	0.02105	0.51197	0.53302	518.4	679.5	1198.0	0.7197	0.6899	1.4096	850.0
900.0	531.95	0.02123	0.47968	0.50091	526.7	669.7	1196.4	0.7279	0.6753	1.4032	900.0
950.0	538.39	0.02141	0.45064	0.47205	534.7	660.0	1194.7	0.7358	0.6612	1.3970	950.0
1000.0	544.58	0.02159	0.42436	0.44596	542.6	650.4	1192.9	0.7434	0.6476	1.3910	1000.0
1050.0	550.53	0.02177	0.40047	0.42224	550.1	640.9	1191.0	0.7507	0.6344	1.3851	1050.0
1100.0	556.28	0.02195	0.37863	0.40058	557.5	631.5	1189.1	0.7578	0.6216	1.3794	1100.0
1150.0	561.82	0.02214	0.35859	0.38073	564.8	622.2	1187.0	0.7647	0.6091	1.3738	1150.0
1200.0	567.19	0.02232	0.34013	0.36245	571.9	613.0	1184.8	0.7714	0.5969	1.3683	1200.0
1250.0	572.38	0.02250	0.32306	0.34556	578.8	603.8	1182.6	0.7780	0.5850	1.3630	1250.0
1300.0	577.42	0.02269	0.30722	0.32991	585.6	594.6	1180.2	0.7843	0.5733	1.3577	1300.0
1350.0	582.32	0.02288	0.29250	0.31537	592.3	585.4	1177.8	0.7906	0.5620	1.3525	1350.0
1400.0	587.07	0.02307	0.27871	0.30178	598.8	576.5	1175.3	0.7966	0.5507	1.3474	1400.0
1450.0	591.70	0.02327	0.26584	0.28911	605.3	567.4	1172.8	0.8026	0.5397	1.3423	1450.0
1500.0	596.20	0.02346	0.25372	0.27719	611.7	558.4	1170.1	0.8085	0.5288	1.3373	1500.0
1550.0	600.59	0.02366	0.24235	0.26601	618.0	549.4	1167.4	0.8142	0.5182	1.3324	1550.0
1600.0	604.87	0.02387	0.23159	0.25545	624.2	540.3	1164.5	0.8199	0.5076	1.3274	1600.0
1650.0	609.05	0.02407	0.22143	0.24551	630.4	531.3	1161.6	0.8254	0.4971	1.3225	1650.0
1700.0	613.13	0.02428	0.21178	0.23607	636.5	522.2	1158.6	0.8309	0.4867	1.3176	1700.0
1750.0	617.12	0.02450	0.20263	0.22713	642.5	513.1	1155.6	0.8363	0.4765	1.3128	1750.0
1800.0	621.02	0.02472	0.19390	0.21861	648.5	503.8	1152.3	0.8417	0.4662	1.3079	1800.0
1850.0	624.83	0.02495	0.18558	0.21052	654.5	494.6	1149.0	0.8470	0.4561	1.3030	1850.0
1900.0	628.56	0.02517	0.17761	0.20278	660.4	485.2	1145.6	0.8522	0.4459	1.2981	1900.0
1950.0	632.22	0.02541	0.16999	0.19540	666.3	475.8	1142.0	0.8574	0.4358	1.2931	1950.0
2000.0	635.80	0.02565	0.16266	0.18831	672.1	466.2	1138.3	0.8625	0.4256	1.2881	2000.0
2100.0	642.76	0.02615	0.14885	0.17501	683.8	446.7	1130.5	0.8727	0.4053	1.2780	2100.0
2200.0	649.45	0.02669	0.13603	0.16272	695.5	426.7	1122.2	0.8828	0.3848	1.2676	2200.0
2300.0	655.89	0.02727	0.12406	0.15133	707.2	406.0	1113.2	0.8929	0.3640	1.2569	2300.0
2400.0	662.11	0.02790	0.11287	0.14076	719.0	384.8	1103.7	0.9031	0.3430	1.2460	2400.0
2500.0	668.11	0.02859	0.10209	0.13068	731.7	361.6	1093.3	0.9139	0.3206	1.2345	2500.0
2600.0	673.91	0.02938	0.09172	0.12110	744.5	337.6	1082.0	0.9247	0.2977	1.2225	2600.0
2700.0	679.53	0.03029	0.08165	0.11194	757.3	312.3	1069.7	0.9356	0.2741	1.2097	2700.0
2800.0	684.96	0.03134	0.07171	0.10305	770.7	285.1	1055.8	0.9468	0.2491	1.1958	2800.0
2900.0	690.22	0.03262	0.06158	0.09420	785.1	254.7	1039.8	0.9588	0.2215	1.1803	2900.0
3000.0	695.33	0.03428	0.05073	0.08500	801.8	218.4	1020.3	0.9728	0.1891	1.1619	3000.0
3100.0	700.28	0.03681	0.03771	0.07452	824.0	169.3	993.3	0.9914	0.1460	1.1373	3100.0
3200.0	705.08	0.04472	0.01191	0.05663	875.5	56.1	931.6	1.0351	0.0482	1.0832	3200.0
3208.2*	705.47	0.05078	0.00000	0.05078	906.0	0.0	906.0	1.0612	0.0000	1.0612	3208.2*

*Critical pressure

Table 3. Superheated steam

P, psia (sat. temp.)		Sat. water	Sat. steam	200	250	300	350	400	450	500	600	700	800	900	1000	1100	1200
1 (101.74)	Sh			98.26	148.26	198.26	248.26	298.26	348.26	398.26	498.26	598.26	698.26	798.26	898.26	998.26	1098.26
	v	0.01614	333.6	392.5	422.4	452.3	482.1	511.9	541.7	571.5	631.1	690.7	750.2	809.8	869.4	929.1	988.7
	h	69.73	1105.8	1150.2	1172.9	1195.7	1218.7	1241.8	1265.1	1288.6	1336.1	1384.5	1431.0	1480.8	1531.4	1583.0	1635.4
	s	0.1326	1.9781	2.0509	2.0841	2.1152	2.1445	2.1722	2.1985	2.2237	2.2708	2.3144	2.3512	2.3892	2.4251	2.4592	2.4918
5 (162.24)	Sh			37.76	87.76	137.76	187.76	237.76	287.76	337.76	437.76	537.76	637.76	737.76	837.76	937.76	1037.76
	v	0.01641	73.53	78.14	84.21	90.24	96.25	102.24	108.23	114.21	126.15	138.08	150.01	161.94	173.86	185.78	197.70
	h	130.20	1131.1	1148.6	1171.7	1194.8	1218.0	1241.3	1264.7	1288.2	1335.9	1384.3	1433.6	1483.7	1534.7	1586.7	1639.6
	s	0.2349	1.8443	1.8716	1.9054	1.9369	1.9664	1.9943	2.0208	2.0460	2.0932	2.1369	2.1776	2.2159	2.2521	2.2866	2.3194
10 (193.21)	Sh			6.79	56.79	106.79	156.79	206.79	256.79	306.79	406.79	506.79	606.79	706.79	806.79	906.79	1006.79
	v	0.01659	38.42	38.84	41.93	44.98	48.02	51.03	54.04	57.04	63.03	69.00	74.98	80.94	86.91	92.87	98.84
	h	161.26	1143.3	1146.6	1170.2	1193.7	1217.1	1240.6	1264.1	1287.8	1335.5	1384.0	1433.4	1483.5	1534.6	1586.6	1639.5
	s	0.2836	1.7879	1.7928	1.8273	1.8593	1.8892	1.9173	1.9439	1.9692	2.0166	2.0603	2.1011	2.1394	2.1757	2.2101	2.2430
14.696 * (212.00)	Sh				38.00	88.00	138.00	188.00	238.00	288.00	388.00	488.00	588.00	688.00	788.00	888.00	988.00
	v	0.0167	26.828		28.44	30.52	32.61	34.65	36.73	38.75	42.83	46.91	50.97	55.03	59.09	63.19	67.25
	h	180.07	1150.4		1169.2	1192.0	1215.4	1238.9	1262.1	1285.4	1333.0	1381.4	1430.5	1480.4	1531.1	1582.7	1635.1
	s	0.3120	1.7566		1.7838	1.8148	1.8446	1.8727	1.8989	1.9238	1.9709	2.0145	2.0551	2.0932	2.1292	2.1634	2.1960
15 (213.03)	Sh				36.97	86.97	136.97	186.97	236.97	286.97	386.97	486.97	586.97	686.97	786.97	886.97	986.97
	v	0.01673	26.290		27.837	29.899	31.939	33.963	35.977	37.985	41.986	45.978	49.964	53.946	57.926	61.905	65.882
	h	181.21	1150.9		1168.7	1192.5	1216.2	1239.9	1263.6	1287.3	1335.2	1383.8	1433.2	1483.4	1534.5	1586.5	1639.4
	s	0.3137	1.7552		1.7809	1.8134	1.8437	1.8720	1.8988	1.9242	1.9717	2.0155	2.0563	2.0946	2.1309	2.1653	2.1982
20 (227.96)	Sh				22.04	72.04	122.04	172.04	222.04	272.04	372.04	472.04	572.04	672.04	772.04	872.04	972.04
	v	0.01683	20.087		20.788	22.356	23.900	25.428	26.946	28.457	31.466	34.465	37.458	40.447	43.435	46.420	49.405
	h	196.27	1156.3		1167.1	1191.4	1215.4	1239.2	1263.0	1286.9	1334.9	1383.5	1432.9	1483.2	1534.3	1586.3	1639.3
	s	0.3358	1.7320		1.7475	1.7805	1.8111	1.8397	1.8666	1.8921	1.9397	1.9836	2.0244	2.0628	2.0991	2.1336	2.1665
25 (240.07)	Sh				9.93	59.93	109.93	159.93	209.93	259.93	359.93	459.93	559.93	659.93	759.93	859.93	959.93
	v	0.01693	16.301		16.558	17.829	19.076	20.307	21.527	22.740	25.153	27.557	29.954	32.348	34.740	37.130	39.518
	h	208.52	1160.6		1165.6	1190.2	1214.5	1238.5	1262.5	1286.4	1334.6	1383.3	1432.7	1483.0	1534.2	1586.2	1639.2
	s	0.3535	1.7141		1.7212	1.7547	1.7856	1.8145	1.8415	1.8672	1.9149	1.9588	1.9997	2.0381	2.0744	2.1089	2.1418
30 (250.34)	Sh					49.66	99.66	149.66	199.66	249.66	349.66	449.66	549.66	649.66	749.66	849.66	949.66
	v	0.01701	13.744			14.810	15.859	16.892	17.914	18.929	20.945	22.951	24.952	26.949	28.943	30.936	32.927
	h	218.93	1164.1			1189.0	1213.6	1237.8	1261.9	1286.0	1334.2	1383.0	1432.5	1482.8	1534.0	1586.1	·1639.0
	s	0.3682	1.6995			1.7334	1.7647	1.7937	1.8210	1.8467	1.8946	1.9386	1.9795	2.0179	2.0543	2.0888	2.1217
35 (259.29)	Sh					40.71	90.71	140.71	190.71	240.71	340.71	440.71	540.71	640.71	740.71	840.71	940.71
	v	0.01708	11.896			12.654	13.562	14.453	15.334	16.207	17.939	19.662	21.379	23.092	24.803	26.512	28.220
	h	228.03	1167.1			1187.8	1212.7	1237.1	1261.3	1285.5	1333.9	1382.8	1432.3	1482.7	1533.9	1586.0	1638.9
	s	0.3809	1.6872			1.7152	1.7468	1.7761	1.8035	1.8294	1.8774	1.9214	1.9624	2.0009	2.0372	2.0717	2.1046
40 (267.25)	Sh					32.75	82.75	132.75	182.75	232.75	332.75	432.75	532.75	632.75	732.75	832.75	932.75
	v	0.01715	10.497			11.036	11.838	12.624	13.398	14.165	15.685	17.195	18.699	20.199	21.697	23.194	24.689
	h	236.14	1169.8			1186.6	1211.7	1236.4	1260.8	1285.0	1333.6	1382.5	1432.1	1482.5	1533.7	1585.8	1638.8
	s	0.3921	1.6765			1.6992	1.7312	1.7608	1.7883	1.8143	1.8624	1.9065	1.9476	1.9860	2.0224	2.0569	2.0899
45 * * (274.43)	Sh					25.57	75.57	125.57	175.57	225.57	325.57	425.57	525.57	625.57	725.57	825.57	925.57
	v	0.01722	9.403			9.782	10.503	11.206	11.897	12.584	13.939	15.284	16.623	17.959	19.292	20.623	21.954
	h	243.47	1172.1			1185.4	1210.8	1235.7	1260.2	1284.6	1333.3	1382.3	1432.0	1482.4	1533.6	1585.7	1638.8
	s	0.4021	1.6671			1.6949	1.7174	1.7472	1.7749	1.8010	1.8492	1.8934	1.9345	1.9730	2.0094	2.0439	2.0769
50 (281.02)	Sh					18.98	68.98	118.98	168.98	218.98	318.98	418.98	518.98	618.98	718.98	818.98	918.98
	v	0.1727	8.514			8.769	9.424	10.062	10.688	11.306	12.529	13.741	14.947	16.150	17.350	18.549	19.746
	h	250.21	1174.1			1184.1	1209.9	1234.9	1259.6	1284.1	1332.9	1382.0	1431.7	1482.2	1533.4	1585.6	1638.6
	s	0.4112	1.6586			1.6720	1.7048	1.7349	1.7628	1.7890	1.8374	1.8816	1.9227	1.9613	1.9977	2.0322	2.0652
55 * * (287.07)	Sh					12.93	62.93	112.93	162.93	212.93	312.93	412.93	512.93	612.93	712.93	812.93	912.93
	v	0.01733	7.787			7.947	8.550	9.134	9.706	10.270	11.385	12.489	13.587	14.682	15.775	16.865	17.954
	h	256.42	1176.0			1182.9	1208.9	1234.3	1259.1	1283.6	1332.6	1381.8	1431.6	1482.0	1533.3	1585.5	1638.5
	s	0.4196	1.6510			1.6602	1.6934	1.7238	1.7518	1.7781	1.8267	1.8710	1.9123	1.9507	1.9871	2.0217	2.0546
60 (292.71)	Sh					7.29	57.29	107.29	157.29	207.29	307.29	407.29	507.29	607.29	707.29	807.29	907.29
	v	0.1738	7.174			7.257	7.815	8.354	8.881	9.400	10.425	11.438	12.446	13.450	14.452	15.452	16.450
	h	262.21	1177.6			1181.6	1208.0	1233.5	1258.5	1283.2	1332.3	1381.5	1431.3	1481.8	1533.2	1585.3	1638.4
	s	0.4273	1.6440			1.6492	1.6793	1.7134	1.7417	1.7681	1.8168	1.8612	1.9024	1.9410	1.9774	2.0120	2.0450
65 (297.98)	Sh					2.02	52.02	102.02	152.02	202.02	302.02	402.02	502.02	602.02	702.02	802.02	902.02
	v	0.01743	6.653			6.675	7.195	7.697	8.186	8.667	9.615	10.552	11.484	12.412	13.337	14.261	15.183
	h	267.63	1179.1			1180.3	1207.0	1232.7	1257.9	1282.7	1331.9	1381.3	1431.1	1481.6	1533.0	1585.2	1638.3
	s	0.4344	1.6375			1.6390	1.6731	1.7040	1.7324	1.7590	1.8077	1.8522	1.8935	1.9321	1.9685	2.0031	2.0361
70 (302.93)	Sh						47.07	97.07	147.07	197.07	297.07	397.07	497.07	597.07	697.07	797.07	897.07
	v	0.01748	6.205				6.664	7.133	7.590	8.039	8.922	9.793	10.659	11.522	12.382	13.240	14.097
	h	272.74	1180.6				1206.0	1232.0	1257.3	1282.2	1331.6	1381.0	1430.9	1481.5	1532.9	1585.1	1638.2
	s	0.4411	1.6316				1.6640	1.6951	1.7237	1.7504	1.7993	1.8439	1.8852	1.9238	1.9603	1.9949	2.0279
75 (307.61)	Sh						42.39	92.39	142.39	192.39	292.39	392.39	492.39	592.39	692.39	792.39	892.39
	v	0.01753	5.814				6.204	6.645	7.074	7.494	8.320	9.135	9.945	10.750	11.553	12.355	13.155
	h	277.56	1181.9				1205.0	1231.2	1256.7	1281.7	1331.3	1380.7	1430.7	1481.3	1532.7	1585.0	1638.1
	s	0.4474	1.6260				1.6554	1.6868	1.7156	1.7424	1.7915	1.8361	1.8774	1.9161	1.9526	1.9872	2.0202

Sh = superheat, °F
v = specific volume, cu ft per lb

h = enthalpy, Btu per lb
s = entropy, Btu per °R per lb

*Values from STEAM TABLES, Properties of Saturated and Superheated Steam
 Published by COMBUSTION ENGINEERING, INC., Copyright 1940
**Values interpolated from ASME STEAM TABLES

Table 3. Superheated steam (*continued*)

P, psia (sat. temp.)		Sat. water	Sat. steam	T, °F 350	400	450	500	550	600	700	800	900	1000	1100	1200	1300	1400
80 (312.04)	Sh			37.96	87.96	137.96	187.96	237.96	287.96	387.96	487.96	587.96	687.96	787.96	887.96	987.96	1087.96
	v	0.01757	5.471	5.801	6.218	6.622	7.018	7.408	7.794	8.560	9.319	10.075	10.829	11.581	12.331	13.081	13.829
	h	282.15	1183.1	1204.0	1230.5	1256.1	1281.3	1306.2	1330.9	1380.5	1430.5	1481.1	1532.6	1584.9	1638.0	1692.0	1746.8
	s	0.4534	1.6208	1.6473	1.6790	1.7080	1.7349	1.7602	1.7842	1.8289	1.8702	1.9089	1.9454	1.9800	2.0131	2.0446	2.0750
85 (316.26)	Sh			33.74	83.74	133.74	183.74	233.74	283.74	383.74	483.74	583.74	683.74	783.74	883.74	983.74	1083.74
	v	0.01762	5.167	5.445	5.840	6.223	6.597	6.966	7.330	8.052	8.768	9.480	10.190	10.898	11.604	12.310	13.014
	h	286.52	1184.2	1203.0	1229.7	1255.5	1280.8	1305.8	1330.6	1380.2	1430.3	1481.0	1532.4	1584.7	1637.9	1691.9	1746.8
	s	0.4590	1.6159	1.6396	1.6716	1.7008	1.7279	1.7532	1.7772	1.8220	1.8634	1.9021	1.9386	1.9733	2.0063	2.0379	2.0682
90 (320.28)	Sh			29.72	79.72	129.72	179.72	229.72	279.72	379.72	479.72	579.72	679.72	779.72	879.72	979.72	1079.72
	v	0.01766	4.895	5.128	5.505	5.869	6.223	6.572	6.917	7.600	8.277	8.950	9.621	10.290	10.958	11.625	12.290
	h	290.69	1185.3	1202.0	1228.9	1254.9	1280.3	1305.4	1330.2	1380.0	1430.1	1480.8	1532.3	1584.6	1637.8	1691.8	1746.7
	s	0.4643	1.6113	1.6323	1.6646	1.6940	1.7212	1.7467	1.7707	1.8156	1.8570	1.8957	1.9323	1.9669	2.0000	2.0316	2.0619
95 (324.13)	Sh			25.87	75.87	125.87	175.87	225.87	275.87	375.87	475.87	575.87	675.87	775.87	875.87	975.87	1075.87
	v	0.01770	4.651	4.845	5.205	5.551	5.889	6.221	6.548	7.196	7.838	8.477	9.113	9.747	10.380	11.012	11.643
	h	294.70	1186.2	1200.9	1228.1	1254.3	1279.8	1305.0	1329.9	1379.7	1429.9	1480.6	1532.1	1584.5	1637.7	1691.7	1746.6
	s	0.4694	1.6069	1.6253	1.6580	1.6876	1.7149	1.7404	1.7645	1.8094	1.8509	1.8897	1.9262	1.9609	1.9940	2.0256	2.0559
100 (327.82)	Sh			22.18	72.18	122.18	172.18	222.18	272.18	372.18	472.18	572.18	672.18	772.18	872.18	972.18	1072.18
	v	0.01774	4.431	4.590	4.935	5.266	5.588	5.904	6.216	6.833	7.443	8.050	8.655	9.258	9.860	10.460	11.060
	h	298.54	1187.2	1199.9	1227.4	1253.7	1279.3	1304.6	1329.6	1379.5	1429.7	1480.4	1532.0	1584.4	1637.6	1691.6	1746.5
	s	0.4743	1.6027	1.6187	1.6516	1.6814	1.7088	1.7344	1.7586	1.8036	1.8451	1.8839	1.9205	1.9552	1.9883	2.0199	2.0502
105 (331.37)	Sh			18.63	68.63	118.63	168.63	218.63	268.63	368.63	468.63	568.63	668.63	768.63	868.63	968.63	1068.63
	v	0.01778	4.231	4.359	4.690	5.007	5.315	5.617	5.915	6.504	7.086	7.665	8.241	8.816	9.389	9.961	10.532
	h	302.24	1188.0	1198.8	1226.6	1253.1	1278.8	1304.2	1329.2	1379.2	1429.4	1480.3	1531.8	1584.2	1637.5	1691.5	1746.4
	s	0.4790	1.5988	1.6122	1.6455	1.6755	1.7031	1.7288	1.7530	1.7981	1.8396	1.8785	1.9151	1.9498	1.9828	2.0145	2.0448
110 (334.79)	Sh			15.21	65.21	115.21	165.21	215.21	265.21	365.21	465.21	565.21	665.21	765.21	865.21	965.21	1065.21
	v	0.01782	4.048	4.149	4.468	4.772	5.068	5.357	5.642	6.205	6.761	7.314	7.865	8.413	8.961	9.507	10.053
	h	305.80	1188.9	1197.7	1225.8	1252.5	1278.3	1303.8	1328.9	1379.0	1429.2	1480.1	1531.7	1584.1	1637.4	1691.4	1746.4
	s	0.4834	1.5950	1.6061	1.6396	1.6698	1.6975	1.7233	1.7476	1.7928	1.8344	1.8732	1.9099	1.9446	1.9777	2.0093	2.0397
115 (338.08)	Sh			11.92	61.92	111.92	161.92	211.92	261.92	361.92	461.92	561.92	661.92	761.92	861.92	961.92	1061.92
	v	0.01785	3.881	3.957	4.265	4.558	4.841	5.119	5.392	5.932	6.465	6.994	7.521	8.046	8.570	9.093	9.615
	h	309.25	1189.6	1196.7	1225.0	1251.8	1277.9	1303.3	1328.6	1378.7	1429.0	1479.9	1531.6	1584.0	1637.2	1691.4	1746.3
	s	0.4877	1.5913	1.6001	1.6340	1.6644	1.6922	1.7181	1.7425	1.7877	1.8294	1.8682	1.9049	1.9396	1.9727	2.0044	2.0347
120 (341.27)	Sh			8.73	58.73	108.73	158.73	208.73	258.73	358.73	458.73	558.73	658.73	758.73	858.73	958.73	1058.73
	v	0.01789	3.7275	3.7815	4.0786	4.3610	4.6341	4.9009	5.1637	5.6813	6.1928	6.7006	7.2060	7.7096	8.2119	8.7130	9.2134
	h	312.58	1190.4	1195.6	1224.1	1251.2	1277.4	1302.9	1328.2	1378.4	1428.8	1479.8	1531.4	1583.9	1637.1	1691.3	1746.2
	s	0.4919	1.5879	1.5943	1.6286	1.6592	1.6872	1.7132	1.7376	1.7829	1.8246	1.8635	1.9001	1.9349	1.9680	1.9996	2.0300
130 (347.33)	Sh			2.67	52.67	102.67	152.67	202.67	252.67	352.67	452.67	552.67	652.67	752.67	852.67	952.67	1052.67
	v	0.01796	3.4544	3.4699	3.7489	4.0129	4.2652	4.5151	4.7589	5.2384	5.7118	6.1814	6.6486	7.1140	7.5781	8.0411	8.5033
	h	318.95	1191.7	1193.4	1222.5	1249.9	1276.4	1302.1	1327.5	1377.9	1428.4	1479.4	1531.1	1583.6	1636.9	1691.1	1746.1
	s	0.4998	1.5813	1.5833	1.6182	1.6493	1.6775	1.7037	1.7283	1.7737	1.8155	1.8545	1.8911	1.9259	1.9591	1.9907	2.0211
140 (353.04)	Sh				46.96	96.96	146.96	196.96	246.96	346.96	446.96	546.96	646.96	746.96	846.96	946.96	1046.96
	v	0.01803	3.2190		3.4661	3.7143	3.9526	4.1844	4.4119	4.8588	5.2995	5.7364	6.1709	6.6036	7.0349	7.4652	7.8946
	h	324.96	1193.0		1220.8	1248.7	1275.3	1301.3	1326.8	1377.4	1428.0	1479.1	1530.8	1583.4	1636.7	1690.9	1745.9
	s	0.5071	1.5752		1.6085	1.6400	1.6686	1.6949	1.7196	1.7652	1.8071	1.8461	1.8828	1.9176	1.9508	1.9825	2.0123
150 (358.43)	Sh				41.57	91.57	141.57	191.57	241.57	341.57	441.57	541.57	641.57	741.57	841.57	941.57	1041.57
	v	0.01809	3.0139		3.2208	3.4555	3.6799	3.8978	4.1112	4.5298	4.9421	5.3507	5.7568	6.1612	6.5642	6.9661	7.3671
	h	330.65	1194.1		1219.1	1247.4	1274.3	1300.5	1326.1	1376.9	1427.6	1478.7	1530.5	1583.1	1636.5	1690.7	1745.7
	s	0.5141	1.5695		1.5993	1.6313	1.6602	1.6867	1.7115	1.7573	1.7992	1.8383	1.8751	1.9099	1.9431	1.9748	2.0052
160 (363.55)	Sh				36.45	86.45	136.45	186.45	236.45	336.45	436.45	536.45	636.45	736.45	836.45	936.45	1036.45
	v	0.01815	2.8336		3.0060	3.2288	3.4413	3.6469	3.8480	4.2420	4.6295	5.0132	5.3945	5.7741	6.1522	6.5293	6.9055
	h	336.07	1195.1		1217.4	1246.0	1273.3	1299.6	1325.4	1376.4	1427.2	1478.4	1530.3	1582.9	1636.3	1690.5	1745.6
	s	0.5206	1.5641		1.5906	1.6231	1.6522	1.6790	1.7039	1.7499	1.7919	1.8310	1.8678	1.9027	1.9359	1.9676	1.9980
170 (368.42)	Sh				31.58	81.58	131.58	181.58	231.58	331.58	431.58	531.58	631.58	731.58	831.58	931.58	1031.58
	v	0.01821	2.6738		2.8162	3.0288	3.2306	3.4255	3.6158	3.9879	4.3536	4.7155	5.0749	5.4325	5.7888	6.1440	6.4983
	h	341.24	1196.0		1215.6	1244.7	1272.2	1298.8	1324.7	1375.8	1426.8	1478.0	1530.0	1582.6	1636.1	1690.4	1745.4
	s	0.5269	1.5591		1.5823	1.6152	1.6447	1.6717	1.6968	1.7428	1.7850	1.8241	1.8610	1.8959	1.9291	1.9608	1.9913
180 (373.08)	Sh				26.92	76.92	126.92	176.92	226.92	326.92	426.92	526.92	626.92	726.92	826.92	926.92	1026.92
	v	0.01827	2.5312		2.6474	2.8508	3.0433	3.2286	3.4093	3.7621	4.1084	4.4508	4.7907	5.1289	5.4657	5.8014	6.1363
	h	346.19	1196.9		1213.8	1243.4	1271.2	1297.9	1324.0	1375.3	1426.3	1477.7	1529.7	1582.4	1635.9	1690.2	1745.3
	s	0.5328	1.5543		1.5743	1.6078	1.6376	1.6647	1.6900	1.7362	1.7784	1.8176	1.8545	1.8894	1.9227	1.9545	1.9849
190 (377.53)	Sh				22.47	72.47	122.47	172.47	222.47	322.47	422.47	522.47	622.47	722.47	822.47	922.47	1022.47
	v	0.01833	2.4030		2.4961	2.6915	2.8756	3.0525	3.2246	3.5601	3.8889	4.2140	4.5365	4.8572	5.1766	5.4949	5.8124
	h	350.94	1197.6		1212.0	1242.0	1270.1	1297.1	1323.3	1374.8	1425.9	1477.4	1529.4	1582.1	1635.7	1690.0	1745.1
	s	0.5384	1.5498		1.5667	1.6006	1.6307	1.6581	1.6835	1.7299	1.7722	1.8115	1.8484	1.8834	1.9166	1.9484	1.9789
200 (381.80)	Sh				18.20	68.20	118.20	168.20	218.20	318.20	418.20	518.20	618.20	718.20	818.20	918.20	1018.20
	v	0.01839	2.2873		2.3598	2.5480	2.7247	2.8939	3.0583	3.3783	3.6915	4.0008	4.3077	4.6128	4.9165	5.2191	5.5209
	h	355.51	1198.3		1210.1	1240.6	1269.0	1296.2	1322.6	1374.3	1425.5	1477.0	1529.1	1581.9	1635.4	1689.8	1745.0
	s	0.5438	1.5454		1.5593	1.5938	1.6242	1.6518	1.6773	1.7239	1.7663	1.8057	1.8426	1.8776	1.9109	1.9427	1.9732

Sh = superheat, °F
v = specific volume, cu ft per lb

h = enthalpy, Btu per lb
s = entropy, Btu per °R per lb

Table 3. Superheated steam (*continued*)

P, psia (sat. temp.)		Sat. water	Sat. steam	400	450	500	550	600	700	800	900	1000	1100	1200	1300	1400	1500
210 (385.91)	Sh			14.09	64.09	114.09	164.09	214.09	314.09	414.09	514.09	614.09	714.09	814.09	914.09	1014.09	1114.09
	v	0.01844	2.1822	2.2364	2.4181	2.5880	2.7504	2.9078	3.2137	3.5128	3.8080	4.1007	4.3915	4.6811	4.9695	5.2571	5.5440
	h	359.91	1199.0	1208.02	1239.2	1268.0	1295.3	1321.9	1373.7	1425.1	1476.7	1528.8	1581.6	1635.2	1689.6	1744.8	1800.8
	s	0.5490	1.5413	1.5522	1.5872	1.6180	1.6458	1.6715	1.7182	1.7607	1.8001	1.8371	1.8721	1.9054	1.9372	1.9677	1.9970
220 (389.88)	Sh			10.12	60.12	110.12	160.12	210.12	310.12	410.12	510.12	610.12	710.12	810.12	910.12	1010.12	1110.12
	v	0.01850	2.0863	2.1240	2.2999	2.4638	2.6199	2.7710	3.0642	3.3504	3.6327	3.9125	4.1905	4.4671	4.7426	5.0173	5.2913
	h	364.17	1199.6	1206.3	1237.8	1266.9	1294.5	1321.2	1373.2	1424.7	1476.3	1528.5	1581.4	1635.0	1689.4	1744.7	1800.6
	s	0.5540	1.5374	1.5453	1.5808	1.6120	1.6400	1.6658	1.7128	1.7553	1.7948	1.8318	1.8668	1.9002	1.9320	1.9625	1.9919
230 (393.70)	Sh			6.30	56.30	106.30	156.30	206.30	306.30	406.30	506.30	606.30	706.30	806.30	906.30	1006.30	1106.30
	v	0.01855	1.9985	2.0212	2.1919	2.3503	2.5008	2.6461	2.9276	3.2020	3.4726	3.7406	4.0068	4.2717	4.5355	4.7984	5.0606
	h	368.28	1200.1	1204.4	1236.3	1265.7	1293.6	1320.4	1372.7	1424.2	1476.0	1528.2	1581.1	1634.8	1689.3	1744.5	1800.5
	s	0.5588	1.5336	1.5385	1.5747	1.6062	1.6344	1.6604	1.7075	1.7502	1.7897	1.8268	1.8618	1.8952	1.9270	1.9576	1.9869
240 (397.39)	Sh			2.61	52.61	102.61	152.61	202.61	302.61	402.61	502.61	602.61	702.61	802.61	902.61	1002.61	1102.61
	v	0.01860	1.9177	1.9268	2.0928	2.2462	2.3915	2.5316	2.8024	3.0661	3.3259	3.5831	3.8385	4.0926	4.3456	4.5977	4.8492
	h	372.27	1200.6	1202.4	1234.9	1264.6	1292.7	1319.7	1372.1	1423.8	1475.6	1527.9	1580.9	1634.6	1689.1	1744.3	1800.4
	s	0.5634	1.5299	1.5320	1.5687	1.6006	1.6291	1.6552	1.7025	1.7452	1.7848	1.8219	1.8570	1.8904	1.9223	1.9528	1.9822
250 (400.97)	Sh				49.03	99.03	149.03	199.03	299.03	399.03	499.03	599.03	699.03	799.03	899.03	999.03	1099.03
	v	0.01865	1.8432		2.0016	2.1504	2.2909	2.4262	2.6872	2.9410	3.1909	3.4382	3.6837	3.9278	4.1709	4.4131	4.6546
	h	376.14	1201.1		1233.4	1263.5	1291.8	1319.0	1371.6	1423.4	1475.3	1527.6	1580.6	1634.4	1688.9	1744.2	1800.2
	s	0.5679	1.5264		1.5629	1.5951	1.6239	1.6502	1.6976	1.7405	1.7801	1.8173	1.8524	1.8858	1.9177	1.9482	1.9776
260 (404.44)	Sh				45.56	95.56	145.56	195.56	295.56	395.56	495.56	595.56	695.56	795.56	895.56	995.56	1095.56
	v	0.01870	1.7742		1.9173	2.0619	2.1981	2.3289	2.5808	2.8256	3.0663	3.3044	3.5408	3.7758	4.0097	4.2427	4.4750
	h	379.90	1201.5		1231.9	1262.4	1290.9	1318.2	1371.1	1423.0	1474.9	1527.3	1580.4	1634.2	1688.7	1744.0	1800.1
	s	0.5722	1.5230		1.5573	1.5899	1.6189	1.6453	1.6930	1.7359	1.7756	1.8128	1.8480	1.8814	1.9133	1.9439	1.9732
270 (407.80)	Sh				42.20	92.20	142.20	192.20	292.20	392.20	492.20	592.20	692.20	792.20	892.20	992.20	1092.20
	v	0.01875	1.7101		1.8391	1.9799	2.1121	2.2388	2.4824	2.7186	2.9509	3.1806	3.4084	3.6349	3.8603	4.0849	4.3087
	h	383.56	1201.9		1230.4	1261.2	1290.0	1317.5	1370.5	1422.6	1474.6	1527.1	1580.1	1634.0	1688.5	1743.9	1800.0
	s	0.5764	1.5197		1.5518	1.5848	1.6140	1.6406	1.6885	1.7315	1.7713	1.8085	1.8437	1.8771	1.9090	1.9396	1.9690
280 (411.07)	Sh				38.93	88.93	138.93	188.93	288.93	388.93	488.93	588.93	688.93	788.93	888.93	988.93	1088.93
	v	0.01880	1.6505		1.7665	1.9037	2.0322	2.1551	2.3909	2.6194	2.8437	3.0655	3.2855	3.5042	3.7217	3.9384	4.1543
	h	387.12	1202.3		1228.8	1260.0	1289.1	1316.8	1370.0	1422.1	1474.2	1526.8	1579.9	1633.8	1688.4	1743.7	1799.8
	s	0.5805	1.5166		1.5464	1.5798	1.6093	1.6361	1.6841	1.7273	1.7671	1.8043	1.8395	1.8730	1.9050	1.9356	1.9649
290 (414.25)	Sh				35.75	85.75	135.75	185.75	285.75	385.75	485.75	585.75	685.75	785.75	885.75	985.75	1085.75
	v	0.01885	1.5948		1.6988	1.8327	1.9578	2.0772	2.3058	2.5269	2.7440	2.9585	3.1711	3.3824	3.5926	3.8019	4.0106
	h	390.60	1202.6		1227.3	1258.9	1288.1	1316.0	1369.5	1421.7	1473.9	1526.5	1579.6	1633.5	1688.2	1743.6	1799.7
	s	0.5844	1.5135		1.5412	1.5750	1.6048	1.6317	1.6799	1.7232	1.7630	1.8003	1.8356	1.8690	1.9010	1.9316	1.9610
300 (417.35)	Sh				32.65	82.65	132.65	182.65	282.65	382.65	482.65	582.65	682.65	782.65	882.65	982.65	1082.65
	v	0.01889	1.5427		1.6356	1.7665	1.8883	2.0044	2.2263	2.4407	2.6509	2.8585	3.0643	3.2689	3.4721	3.6746	3.8764
	h	393.99	1202.9		1225.7	1257.7	1287.2	1315.2	1368.9	1421.3	1473.6	1526.2	1579.4	1633.3	1688.0	1743.4	1799.6
	s	0.5882	1.5105		1.5361	1.5703	1.6003	1.6274	1.6758	1.7192	1.7591	1.7964	1.8317	1.8652	1.8972	1.9278	1.9572
310 (420.36)	Sh				29.64	79.64	129.64	179.64	279.64	379.64	479.64	579.64	679.64	779.64	879.64	979.64	1079.64
	v	0.01894	1.4939		1.5763	1.7044	1.8233	1.9363	2.1520	2.3600	2.5638	2.7650	2.9644	3.1625	3.3594	3.5555	3.7509
	h	397.30	1203.2		1224.1	1256.5	1286.3	1314.5	1368.4	1420.9	1473.2	1525.9	1579.2	1633.1	1687.8	1743.3	1799.4
	s	0.5920	1.5076		1.5311	1.5657	1.5960	1.6233	1.6719	1.7153	1.7553	1.7927	1.8280	1.8615	1.8935	1.9241	1.9536
320 (423.31)	Sh				26.69	76.69	126.69	176.69	276.69	376.69	476.69	576.69	676.69	776.69	876.69	976.69	1076.69
	v	0.01899	1.4480		1.5207	1.6462	1.7623	1.8725	2.0823	2.2843	2.4821	2.6774	2.8708	3.0628	3.2538	3.4438	3.6332
	h	400.53	1203.4		1222.5	1255.2	1285.3	1313.7	1367.8	1420.5	1472.9	1525.6	1578.9	1632.9	1687.6	1743.1	1799.3
	s	0.5956	1.5048		1.5261	1.5612	1.5918	1.6192	1.6680	1.7116	1.7516	1.7890	1.8243	1.8579	1.8899	1.9206	1.9500
330 (426.18)	Sh				23.82	73.82	123.82	173.82	273.82	373.82	473.82	573.82	673.82	773.82	873.82	973.82	1073.82
	v	0.01903	1.4048		1.4684	1.5915	1.7050	1.8125	2.0168	2.2132	2.4054	2.5950	2.7828	2.9692	3.1545	3.3389	3.5227
	h	403.70	1203.6		1220.9	1254.0	1284.4	1313.0	1367.3	1420.0	1472.5	1525.3	1578.7	1632.7	1687.5	1742.9	1799.2
	s	0.5991	1.5021		1.5213	1.5568	1.5876	1.6153	1.6643	1.7079	1.7480	1.7855	1.8208	1.8544	1.8864	1.9171	1.9466
340 (428.99)	Sh				21.01	71.01	121.01	171.01	271.01	371.01	471.01	571.01	671.01	771.01	871.01	971.01	1071.01
	v	0.01908	1.3640		1.4191	1.5399	1.6511	1.7561	1.9552	2.1463	2.3333	2.5175	2.7000	2.8811	3.0611	3.2402	3.4186
	h	406.80	1203.8		1219.2	1252.8	1283.4	1312.2	1366.7	1419.6	1472.2	1525.0	1578.4	1632.5	1687.3	1742.8	1799.0
	s	0.6026	1.4994		1.5165	1.5525	1.5836	1.6114	1.6606	1.7044	1.7445	1.7820	1.8174	1.8510	1.8831	1.9138	1.9432
350 (431.73)	Sh				18.27	68.27	118.27	168.27	268.27	368.27	468.27	568.27	668.27	768.27	868.27	968.27	1068.27
	v	0.01912	1.3255		1.3725	1.4913	1.6002	1.7028	1.8970	2.0832	2.2652	2.4445	2.6219	2.7980	2.9730	3.1471	3.3205
	h	409.83	1204.0		1217.5	1251.5	1282.4	1311.4	1366.2	1419.2	1471.8	1524.7	1578.2	1632.3	1687.1	1742.6	1798.9
	s	0.6059	1.4968		1.5119	1.5483	1.5797	1.6077	1.6571	1.7009	1.7411	1.7787	1.8141	1.8477	1.8798	1.9105	1.9400
360 (434.41)	Sh				15.59	65.59	115.59	165.59	265.59	365.59	465.59	565.59	665.59	765.59	865.59	965.59	1065.59
	v	0.01917	1.2891		1.3285	1.4454	1.5521	1.6525	1.8421	2.0237	2.2009	2.3755	2.5482	2.7196	2.8898	3.0592	3.2279
	h	412.81	1204.1		1215.8	1250.3	1281.5	1310.6	1365.6	1418.7	1471.5	1524.4	1577.9	1632.1	1686.9	1742.5	1798.8
	s	0.6092	1.4943		1.5073	1.5441	1.5758	1.6040	1.6536	1.6976	1.7379	1.7754	1.8109	1.8445	1.8766	1.9073	1.9368
380 (439.61)	Sh				10.39	60.39	110.39	160.39	260.39	360.39	460.39	560.39	660.39	760.39	860.39	960.39	1060.39
	v	0.01925	1.2218		1.2472	1.3606	1.4635	1.5598	1.7410	1.9139	2.0825	2.2484	2.4124	2.5750	2.7366	2.8973	3.0572
	h	418.59	1204.4		1212.4	1247.7	1279.5	1309.0	1364.5	1417.9	1470.8	1523.8	1577.4	1631.6	1686.5	1742.2	1798.5
	s	0.6156	1.4894		1.4982	1.5360	1.5683	1.5969	1.6470	1.6911	1.7315	1.7692	1.8047	1.8384	1.8705	1.9012	1.9307

Sh = superheat, °F
v = specific volume, cu ft per lb

h = enthalpy, Btu per lb
s = entropy, Btu per °R per lb

Table 3. Superheated steam (*continued*)

P, psia (sat. temp.)		Sat. water	Sat. steam	*T*, °F 450	500	550	600	650	700	800	900	1000	1100	1200	1300	1400	1500
400 (444.60)	Sh			5.40	55.40	105.40	155.40	205.40	255.40	355.40	455.40	555.40	655.40	755.40	855.40	955.40	1055.40
	v	0.01934	1.1610	1.1738	1.2841	1.3836	1.4763	1.5646	1.6499	1.8151	1.9759	2.1339	2.2901	2.4450	2.5987	2.7515	2.9037
	h	424.17	1204.6	1208.8	1245.1	1277.5	1307.4	1335.9	1363.4	1417.0	1470.1	1523.3	1576.9	1631.2	1686.2	1741.9	1798.2
	s	0.6217	1.4847	1.4894	1.5282	1.5611	1.5901	1.6163	1.6406	1.6850	1.7255	1.7632	1.7988	1.8325	1.8647	1.8955	1.9250
420 (449.40)	Sh			.60	50.60	100.60	150.60	200.60	250.60	350.60	450.60	550.60	650.60	750.60	850.60	950.60	1050.60
	v	0.01942	1.1057	1.1071	1.2148	1.3113	1.4007	1.4856	1.5676	1.7258	1.8795	2.0304	2.1795	2.3273	2.4739	2.6196	2.7647
	h	429.56	1204.7	1205.2	1242.4	1275.4	1305.8	1334.5	1362.3	1416.2	1469.4	1522.7	1576.4	1630.8	1685.8	1741.6	1798.0
	s	0.6276	1.4802	1.4808	1.5206	1.5542	1.5835	1.6100	1.6345	1.6791	1.7197	1.7575	1.7932	1.8269	1.8591	1.8899	1.9195
440 (454.03)	Sh				45.97	95.97	145.97	195.97	245.97	345.97	445.97	545.97	645.97	745.97	845.97	945.97	1045.97
	v	0.01950	1.0554		1.1517	1.2454	1.3319	1.4138	1.4926	1.6445	1.7918	1.9363	2.0790	2.2203	2.3605	2.4998	2.6384
	h	434.77	1204.8		1239.7	1273.4	1304.2	1333.2	1361.1	1415.3	1468.7	1522.1	1575.9	1630.4	1685.5	1741.2	1797.7
	s	0.6332	1.4759		1.5132	1.5474	1.5772	1.6040	1.6286	1.6734	1.7142	1.7521	1.7878	1.8216	1.8538	1.8847	1.9143
460 (458.50)	Sh				41.50	91.50	141.50	191.50	241.50	341.50	441.50	541.50	641.50	741.50	841.50	941.50	1041.50
	v	0.01959	1.0092		1.0939	1.1852	1.2691	1.3482	1.4242	1.5703	1.7117	1.8504	1.9872	2.1226	2.2569	2.3903	2.5230
	h	439.83	1204.8		1236.9	1271.3	1302.5	1331.8	1360.0	1414.4	1468.0	1521.5	1575.4	1629.9	1685.1	1740.9	1797.4
	s	0.6387	1.4718		1.5060	1.5409	1.5711	1.5982	1.6230	1.6680	1.7089	1.7469	1.7826	1.8165	1.8488	1.8797	1.9093
480 (462.82)	Sh				37.18	87.18	137.18	187.18	237.18	337.18	437.18	537.18	637.18	737.18	837.18	937.18	1037.18
	v	0.01967	0.9668		1.0409	1.1300	1.2115	1.2881	1.3615	1.5023	1.6384	1.7716	1.9030	2.0330	2.1619	2.2900	2.4173
	h	444.75	1204.8		1234.1	1269.1	1300.8	1330.5	1358.8	1413.6	1467.3	1520.9	1574.9	1629.5	1684.7	1740.6	1797.2
	s	0.6439	1.4677		1.4990	1.5346	1.5652	1.5925	1.6176	1.6628	1.7038	1.7419	1.7777	1.8116	1.8439	1.8748	1.9045
500 (467.01)	Sh				32.99	82.99	132.99	182.99	232.99	332.99	432.99	532.99	632.99	732.99	832.99	932.99	1032.99
	v	0.01975	0.9276		0.9919	1.0791	1.1584	1.2327	1.3037	1.4397	1.5708	1.6992	1.8256	1.9507	2.0746	2.1977	2.3200
	h	449.52	1204.7		1231.2	1267.0	1299.1	1329.1	1357.7	1412.7	1466.6	1520.3	1574.4	1629.1	1684.4	1740.3	1796.9
	s	0.6490	1.4639		1.4921	1.5284	1.5595	1.5871	1.6123	1.6578	1.6990	1.7371	1.7730	1.8069	1.8393	1.8702	1.8998
520 (471.07)	Sh				28.93	78.93	128.93	178.93	228.93	328.93	428.93	528.93	628.93	728.93	828.93	928.93	1028.93
	v	0.01982	0.8914		0.9466	1.0321	1.1094	1.1816	1.2504	1.3819	1.5085	1.6323	1.7542	1.8746	1.9940	2.1125	2.2302
	h	454.18	1204.5		1228.3	1264.8	1297.4	1327.7	1356.5	1411.8	1465.9	1519.7	1573.9	1628.7	1684.0	1740.0	1796.7
	s	0.6540	1.4601		1.4853	1.5223	1.5539	1.5818	1.6072	1.6530	1.6943	1.7325	1.7684	1.8024	1.8348	1.8657	1.8954
540 (475.01)	Sh				24.99	74.99	124.99	174.99	224.99	324.99	424.99	524.99	624.99	724.99	824.99	924.99	1024.99
	v	0.01990	0.8577		0.9045	0.9884	1.0640	1.1342	1.2010	1.3284	1.4508	1.5704	1.6880	1.8042	1.9193	2.0336	2.1471
	h	458.71	1204.4		1225.3	1262.5	1295.7	1326.3	1355.3	1410.9	1465.1	1519.1	1573.4	1628.2	1683.6	1739.7	1796.4
	s	0.6587	1.4565		1.4786	1.5164	1.5485	1.5767	1.6023	1.6483	1.6897	1.7280	1.7640	1.7981	1.8305	1.8615	1.8911
560 (478.84)	Sh				21.16	71.16	121.16	171.16	221.16	321.16	421.16	521.16	621.16	721.16	821.16	921.16	1021.16
	v	0.01998	0.8264		0.8653	0.9479	1.0217	1.0902	1.1552	1.2787	1.3972	1.5129	1.6266	1.7388	1.8500	1.9603	2.0699
	h	463.14	1204.2		1222.2	1260.3	1293.9	1324.9	1354.2	1410.0	1464.4	1518.6	1572.9	1627.8	1683.3	1739.4	1796.1
	s	0.6634	1.4529		1.4720	1.5106	1.5431	1.5717	1.5975	1.6438	1.6853	1.7237	1.7598	1.7939	1.8263	1.8573	1.8870
580 (482.57)	Sh				17.43	67.43	117.43	167.43	217.43	317.43	417.43	517.43	617.43	717.43	817.43	917.43	1017.43
	v	0.02006	0.7971		0.8287	0.9100	0.9824	1.0492	1.1125	1.2324	1.3473	1.4593	1.5693	1.6780	1.7855	1.8921	1.9980
	h	467.47	1203.9		1219.1	1258.0	1292.1	1323.4	1353.0	1409.2	1463.7	1518.0	1572.4	1627.4	1682.9	1739.1	1795.9
	s	0.6679	1.4495		1.4654	1.5049	1.5380	1.5668	1.5929	1.6394	1.6811	1.7196	1.7556	1.7898	1.8223	1.8533	1.8831
600 (486.20)	Sh				13.80	63.80	113.80	163.80	213.80	313.80	413.80	513.80	613.80	713.80	813.80	913.80	1013.80
	v	0.02013	0.7697		0.7944	0.8746	0.9456	1.0109	1.0726	1.1892	1.3008	1.4093	1.5160	1.6211	1.7252	1.8284	1.9309
	h	471.70	1203.7		1215.9	1255.6	1290.3	1322.0	1351.8	1408.3	1463.0	1517.4	1571.9	1627.0	1682.6	1738.8	1795.6
	s	0.6723	1.4461		1.4590	1.4993	1.5329	1.5621	1.5884	1.6351	1.6769	1.7155	1.7517	1.7859	1.8184	1.8494	1.8792
650 (494.89)	Sh				5.11	55.11	105.11	155.11	205.11	305.11	405.11	505.11	605.11	705.11	805.11	905.11	1005.11
	v	0.02032	0.7084		0.7173	0.7954	0.8634	0.9254	0.9835	1.0979	1.1969	1.2979	1.3969	1.4944	1.5909	1.6864	1.7813
	h	481.89	1202.8		1207.6	1249.6	1285.7	1318.3	1348.7	1406.0	1461.2	1515.9	1570.7	1625.9	1681.6	1738.0	1794.9
	s	0.6828	1.4381		1.4430	1.4858	1.5207	1.5507	1.5775	1.6249	1.6671	1.7059	1.7422	1.7765	1.8092	1.8403	1.8701
700 (503.08)	Sh					46.92	96.92	146.92	196.92	296.92	396.92	496.92	596.92	696.92	796.92	896.92	996.92
	v	0.02050	0.6556			0.7271	0.7928	0.8520	0.9072	1.0102	1.1078	1.2023	1.2948	1.3858	1.4757	1.5647	1.6530
	h	491.60	1201.8			1243.4	1281.0	1314.6	1345.6	1403.7	1459.4	1514.4	1569.4	1624.8	1680.7	1737.2	1794.3
	s	0.6928	1.4304			1.4726	1.5090	1.5399	1.5673	1.6154	1.6580	1.6970	1.7335	1.7679	1.8006	1.8318	1.8617
750 (510.84)	Sh					39.16	89.16	139.16	189.16	289.16	389.16	489.16	589.16	689.16	789.16	889.16	989.16
	v	0.02069	0.6095			0.6676	0.7313	0.7882	0.8409	0.9386	1.0306	1.1195	1.2063	1.2916	1.3759	1.4592	1.5419
	h	500.89	1200.7			1236.9	1276.1	1310.7	1342.5	1401.5	1457.6	1512.9	1568.2	1623.8	1679.8	1736.4	1793.6
	s	0.7022	1.4232			1.4598	1.4977	1.5296	1.5577	1.6065	1.6494	1.6886	1.7252	1.7598	1.7926	1.8239	1.8538
800 (518.21)	Sh					31.79	81.79	131.79	181.79	281.79	381.79	481.79	581.79	681.79	781.79	881.79	981.79
	v	0.02087	0.5690			0.6151	0.6774	0.7323	0.7828	0.8759	0.9631	1.0470	1.1289	1.2093	1.2885	1.3669	1.4446
	h	509.81	1199.4			1230.1	1271.1	1306.8	1339.3	1399.1	1455.8	1511.4	1566.9	1622.7	1678.9	1735.7	1792.9
	s	0.7111	1.4163			1.4472	1.4869	1.5198	1.5484	1.5980	1.6413	1.6807	1.7175	1.7522	1.7851	1.8164	1.8464
850 (525.24)	Sh					24.76	74.76	124.76	174.76	274.76	374.76	474.76	574.76	674.76	774.76	874.76	974.76
	v	0.02105	0.5530			0.5683	0.6296	0.6829	0.7315	0.8205	0.9034	0.9830	1.0606	1.1366	1.2115	1.2855	1.3588
	h	518.40	1198.0			1223.0	1265.9	1302.8	1336.0	1396.8	1454.0	1510.0	1565.7	1621.6	1678.0	1734.9	1792.3
	s	0.7197	1.4096			1.4347	1.4763	1.5102	1.5396	1.5899	1.6336	1.6733	1.7102	1.7450	1.7780	1.8094	1.8395
900 (531.95)	Sh					18.05	68.05	118.05	168.05	268.05	368.05	468.05	568.05	668.05	768.05	868.05	968.05
	v	0.02123	0.5009			0.5263	0.5869	0.6388	0.6858	0.7713	0.8504	0.9262	0.9998	1.0720	1.1430	1.2131	1.2825
	h	526.70	1196.4			1215.5	1260.6	1298.6	1332.7	1394.4	1452.2	1508.5	1564.4	1620.6	1677.1	1734.1	1791.6
	s	0.7279	1.4032			1.4223	1.4659	1.5010	1.5311	1.5822	1.6263	1.6662	1.7033	1.7382	1.7713	1.8028	1.8329

Sh = superheat, °F
v = specific volume, cu ft per lb
h = enthalpy, Btu per lb
s = entropy, Btu per °R per lb

Table 3. Superheated steam (*continued*)

P, psia (sat. temp.)		Sat. water	Sat. steam	550	600	650	700	750	800	850	900	1000	1100	1200	1300	1400	1500
									T, °F								
950 (538.39)	Sh			11.61	61.61	111.61	161.61	211.61	261.61	311.61	361.61	461.61	561.61	661.61	761.61	861.61	961.61
	v	0.02141	0.4721	0.4883	0.5485	0.5993	0.6449	0.6871	0.7272	0.7656	0.8030	0.8753	0.9455	1.0142	1.0817	1.1484	1.2143
	h	534.74	1194.7	1207.6	1255.1	1294.4	1329.3	1361.5	1392.0	1421.5	1450.3	1507.0	1563.2	1619.5	1676.2	1733.3	1791.0
	s	0.7358	1.3970	1.4098	1.4557	1.4921	1.5228	1.5500	1.5748	1.5977	1.6193	1.6595	1.6967	1.7317	1.7649	1.7965	1.8267
1000 (544.58)	Sh			5.42	55.42	105.42	155.42	205.42	255.42	305.42	355.42	455.42	555.42	655.42	755.42	855.42	955.42
	v	0.02159	0.4460	0.4535	0.5137	0.5636	0.6080	0.6489	0.6875	0.7245	0.7603	0.8295	0.8966	0.9622	1.0266	1.0901	1.1529
	h	542.55	1192.9	1199.3	1249.3	1290.1	1325.9	1358.7	1389.6	1419.4	1448.5	1505.4	1561.9	1618.4	1675.3	1732.5	1790.3
	s	0.7434	1.3910	1.3973	1.4457	1.4833	1.5149	1.5426	1.5677	1.5908	1.6126	1.6530	1.6905	1.7256	1.7589	1.7905	1.8207
1050 (550.53)	Sh				49.47	99.47	149.47	199.47	249.47	299.47	349.47	449.47	549.47	649.47	749.47	849.47	949.47
	v	0.02177	0.4222		0.4821	0.5312	0.5745	0.6142	0.6515	0.6872	0.7216	0.7881	0.8524	0.9151	0.9767	1.0373	1.0973
	h	550.15	1191.0		1243.4	1285.7	1322.4	1355.8	1387.2	1417.3	1446.6	1503.9	1560.7	1617.4	1674.4	1731.8	1789.6
	s	0.7507	1.3851		1.4358	1.4748	1.5072	1.5354	1.5608	1.5842	1.6062	1.6469	1.6845	1.7197	1.7531	1.7848	1.8151
1100 (556.28)	Sh				43.72	93.72	143.72	193.72	243.72	293.72	343.72	443.72	543.72	643.72	743.72	843.72	943.72
	v	0.02195	0.4006		0.4531	0.5017	0.5440	0.5826	0.6188	0.6533	0.6865	0.7505	0.8121	0.8723	0.9313	0.9894	1.0468
	h	557.55	1189.1		1237.3	1281.2	1318.8	1352.9	1384.7	1415.2	1444.7	1502.4	1559.4	1616.3	1673.5	1731.0	1789.0
	s	0.7578	1.3794		1.4259	1.4664	1.4996	1.5284	1.5542	1.5779	1.6000	1.6410	1.6787	1.7141	1.7475	1.7793	1.8097
1150 (561.82)	Sh				39.18	89.18	139.18	189.18	239.18	289.18	339.18	439.18	539.18	639.18	739.18	839.18	939.18
	v	0.02214	0.3807		0.4263	0.4746	0.5162	0.5538	0.5889	0.6223	0.6544	0.7161	0.7754	0.8332	0.8899	0.9456	1.0007
	h	564.78	1187.0		1230.9	1276.6	1315.2	1349.9	1382.2	1413.0	1442.8	1500.9	1558.1	1615.2	1672.6	1730.2	1788.3
	s	0.7647	1.3738		1.4160	1.4582	1.4923	1.5216	1.5478	1.5717	1.5941	1.6353	1.6732	1.7087	1.7422	1.7741	1.8045
1200 (567.19)	Sh				32.81	82.81	132.81	182.81	232.81	282.81	332.81	432.81	532.81	632.81	732.81	832.81	932.81
	v	0.02232	0.3624		0.4016	0.4497	0.4905	0.5273	0.5615	0.5939	0.6250	0.6845	0.7418	0.7974	0.8519	0.9055	0.9584
	h	571.85	1184.8		1224.2	1271.8	1311.5	1346.9	1379.7	1410.8	1440.9	1499.4	1556.9	1614.2	1671.6	1729.4	1787.6
	s	0.7714	1.3683		1.4061	1.4501	1.4851	1.5150	1.5415	1.5658	1.5883	1.6298	1.6679	1.7035	1.7371	1.7691	1.7996
1300 (577.42)	Sh				22.58	72.58	122.58	172.58	222.58	272.58	322.58	422.58	522.58	622.58	722.58	822.58	922.58
	v	0.02269	0.3299		0.3570	0.4052	0.4451	0.4804	0.5129	0.5436	0.5729	0.6287	0.6822	0.7341	0.7847	0.8345	0.8836
	h	585.58	1180.2		1209.9	1261.9	1303.9	1340.8	1374.6	1406.4	1437.1	1496.3	1554.3	1612.0	1669.8	1727.9	1786.3
	s	0.7843	1.3577		1.3860	1.4340	1.4711	1.5022	1.5296	1.5544	1.5773	1.6194	1.6578	1.6937	1.7275	1.7596	1.7902
1400 (587.07)	Sh				12.93	62.93	112.93	162.93	212.93	262.93	312.93	412.93	512.93	612.93	712.93	812.93	912.93
	v	0.02307	0.3018		0.3176	0.3667	0.4059	0.4400	0.4712	0.5004	0.5282	0.5809	0.6311	0.6798	0.7272	0.7737	0.8195
	h	598.83	1175.3		1194.1	1251.4	1296.1	1334.5	1369.3	1402.0	1433.2	1493.2	1551.8	1609.9	1668.0	1726.3	1785.0
	s	0.7966	1.3474		1.3652	1.4181	1.4575	1.4900	1.5182	1.5436	1.5670	1.6096	1.6484	1.6845	1.7185	1.7508	1.7815
1500 (596.20)	Sh				3.80	53.80	103.80	153.80	203.80	253.80	303.80	403.80	503.80	603.80	703.80	803.80	903.80
	v	0.02346	0.2772		0.2820	0.3328	0.3717	0.4049	0.4350	0.4629	0.4894	0.5394	0.5869	0.6327	0.6773	0.7210	0.7639
	h	611.68	1170.1		1176.3	1240.2	1287.9	1328.0	1364.0	1397.4	1429.2	1490.1	1549.2	1607.7	1666.2	1724.8	1783.7
	s	0.8085	1.3373		1.3431	1.4022	1.4443	1.4782	1.5073	1.5333	1.5572	1.6004	1.6395	1.6759	1.7101	1.7425	1.7734
1600 (604.87)	Sh					45.13	95.13	145.13	195.13	245.13	295.13	395.13	495.13	595.13	695.13	795.13	895.13
	v	0.02387	0.2555			0.3026	0.3415	0.3741	0.4032	0.4301	0.4555	0.5031	0.5482	0.5915	0.6336	0.6748	0.7153
	h	624.20	1164.5			1228.3	1279.4	1321.4	1358.5	1392.8	1425.2	1486.9	1546.6	1605.6	1664.3	1723.2	1782.3
	s	0.8199	1.3274			1.3861	1.4312	1.4667	1.4968	1.5235	1.5478	1.5916	1.6312	1.6678	1.7022	1.7347	1.7657
1700 (613.13)	Sh					36.87	86.87	136.87	186.87	236.87	286.87	386.87	486.87	586.87	686.87	786.87	886.87
	v	0.02428	0.2361			0.2754	0.3147	0.3468	0.3751	0.4011	0.4255	0.4711	0.5140	0.5552	0.5951	0.6341	0.6724
	h	636.45	1158.6			1215.3	1270.5	1314.5	1352.9	1388.1	1421.2	1483.8	1544.0	1603.4	1662.5	1721.7	1781.0
	s	0.8309	1.3176			1.3697	1.4183	1.4555	1.4867	1.5140	1.5388	1.5833	1.6232	1.6601	1.6947	1.7274	1.7585
1800 (621.02)	Sh					28.98	78.98	128.98	178.98	228.98	278.98	378.98	478.98	578.98	678.98	778.98	878.98
	v	0.02472	0.2186			0.2505	0.2906	0.3223	0.3500	0.3752	0.3988	0.4426	0.4836	0.5229	0.5609	0.5980	0.6343
	h	648.49	1152.3			1201.2	1261.1	1307.4	1347.2	1383.3	1417.1	1480.6	1541.4	1601.2	1660.7	1720.1	1779.7
	s	0.8417	1.3079			1.3526	1.4054	1.4446	1.4768	1.5049	1.5302	1.5753	1.6156	1.6528	1.6876	1.7204	1.7516
1900 (628.56)	Sh					21.44	71.44	121.44	171.44	221.44	271.44	371.44	471.44	571.44	671.44	771.44	871.44
	v	0.02517	0.2028			0.2274	0.2687	0.3004	0.3275	0.3521	0.3749	0.4171	0.4565	0.4940	0.5303	0.5656	0.6002
	h	660.36	1145.6			1185.7	1251.3	1300.2	1341.4	1378.4	1412.9	1477.4	1538.8	1599.1	1658.8	1718.6	1778.4
	s	0.8522	1.2981			1.3346	1.3925	1.4338	1.4672	1.4960	1.5219	1.5677	1.6084	1.6458	1.6808	1.7138	1.7451
2000 (635.80)	Sh					14.20	64.20	114.20	164.20	214.20	264.20	364.20	464.20	564.20	664.20	764.20	864.20
	v	0.02565	0.1883			0.2056	0.2488	0.2805	0.3072	0.3312	0.3534	0.3942	0.4320	0.4680	0.5027	0.5365	0.5695
	h	672.11	1138.3			1168.3	1240.9	1292.6	1335.4	1373.5	1408.7	1474.1	1536.2	1596.9	1657.0	1717.0	1777.1
	s	0.8625	1.2881			1.3154	1.3794	1.4231	1.4578	1.4874	1.5138	1.5603	1.6014	1.6391	1.6743	1.7075	1.7389
2100 (642.76)	Sh					7.24	57.24	107.24	157.24	207.24	257.24	357.24	457.24	557.24	657.24	757.24	857.24
	v	0.02615	0.1750			0.1847	0.2304	0.2624	0.2888	0.3123	0.3339	0.3734	0.4099	0.4445	0.4778	0.5101	0.5418
	h	683.79	1130.5			1148.5	1229.8	1284.9	1329.3	1368.4	1404.4	1470.9	1533.6	1594.7	1655.2	1715.4	1775.7
	s	0.8727	1.2780			1.2942	1.3661	1.4125	1.4486	1.4790	1.5060	1.5532	1.5948	1.6327	1.6681	1.7014	1.7330
2200 (649.45)	Sh					.55	50.55	100.55	150.55	200.55	250.55	350.55	450.55	550.55	650.55	750.55	850.55
	v	0.02669	0.1627			0.1636	0.2134	0.2458	0.2720	0.2950	0.3161	0.3545	0.3897	0.4231	0.4551	0.4862	0.5165
	h	695.46	1122.2			1123.9	1218.0	1276.8	1323.1	1363.3	1400.0	1467.6	1530.9	1592.5	1653.3	1713.9	1774.4
	s	0.8828	1.2676			1.2691	1.3523	1.4020	1.4395	1.4708	1.4984	1.5463	1.5883	1.6266	1.6622	1.6956	1.7273
2300 (655.89)	Sh						44.11	94.11	144.11	194.11	244.11	344.11	444.11	544.11	644.11	744.11	844.11
	v	0.02727	0.1513				0.1975	0.2305	0.2566	0.2793	0.2999	0.3372	0.3714	0.4035	0.4344	0.4643	0.4935
	h	707.18	1113.2				1205.3	1268.4	1316.7	1358.1	1395.7	1464.2	1528.3	1590.3	1651.5	1712.3	1773.1
	s	0.8929	1.2569				1.3381	1.3914	1.4305	1.4628	1.4910	1.5397	1.5821	1.6207	1.6565	1.6901	1.7219

Sh = superheat, °F
v = specific volume, cu ft per lb

h = enthalpy, Btu per lb
s = entropy, Btu per °R per lb

Table 3. Superheated steam (*continued*)

P, psia (sat. temp.)		Sat. water	Sat. steam	700	750	800	850	900	950	1000	1050	1100	1150	1200	1300	1400	1500
									T, °F								
2400 (662.11)	Sh			37.89	87.89	137.89	187.89	237.89	287.89	337.89	387.89	437.89	487.89	537.89	637.89	737.89	837.89
	v	0.02790	0.1408	0.1824	0.2164	0.2424	0.2648	0.2850	0.3037	0.3214	0.3382	0.3545	0.3703	0.3856	0.4155	0.4443	0.4724
	h	718.95	1103.7	1191.6	1259.7	1310.1	1352.8	1391.2	1426.9	1460.9	1493.7	1525.6	1557.0	1588.1	1649.6	1710.8	1771.8
	s	0.9031	1.2460	1.3232	1.3808	1.4217	1.4549	1.4837	1.5095	1.5332	1.5553	1.5761	1.5959	1.6149	1.6509	1.6847	1.7167
2500 (668.11)	Sh			31.89	81.89	131.89	181.89	231.89	281.89	331.89	381.89	431.89	481.89	531.89	631.89	731.89	831.89
	v	0.02859	0.1307	0.1681	0.2032	0.2293	0.2514	0.2712	0.2896	0.3068	0.3232	0.3390	0.3543	0.3692	0.3980	0.4259	0.4529
	h	731.71	1093.3	1176.7	1250.6	1303.4	1347.4	1386.7	1423.1	1457.5	1490.7	1522.9	1554.6	1585.9	1647.8	1709.2	1770.4
	s	0.9139	1.2345	1.3076	1.3701	1.4129	1.4472	1.4766	1.5029	1.5269	1.5492	1.5703	1.5903	1.6094	1.6456	1.6796	1.7116
2600 (673.91)	Sh			26.09	76.09	126.09	176.09	226.09	276.09	326.09	376.09	426.09	476.09	526.09	626.09	726.09	826.09
	v	0.02938	0.1211	0.1544	0.1909	0.2171	0.2390	0.2585	0.2765	0.2933	0.3093	0.3247	0.3395	0.3540	0.3819	0.4088	0.4350
	h	744.47	1082.0	1160.2	1241.1	1296.5	1341.9	1382.1	1419.2	1454.1	1487.7	1520.2	1552.2	1583.7	1646.0	1707.7	1769.1
	s	0.9247	1.2225	1.2908	1.3592	1.4042	1.4395	1.4696	1.4964	1.5208	1.5434	1.5646	1.5848	1.6040	1.6405	1.6746	1.7068
2700 (679.53)	Sh			20.47	70.47	120.47	170.47	220.47	270.47	320.47	370.47	420.47	470.47	520.47	620.47	720.47	820.47
	v	0.03029	0.1119	0.1411	0.1794	0.2058	0.2275	0.2468	0.2644	0.2809	0.2965	0.3114	0.3259	0.3399	0.3670	0.3931	0.4184
	h	757.34	1069.7	1142.0	1231.1	1289.5	1336.3	1377.5	1415.2	1450.7	1484.6	1517.5	1549.8	1581.5	1644.1	1706.1	1767.8
	s	0.9356	1.2097	1.2727	1.3481	1.3954	1.4319	1.4628	1.4900	1.5148	1.5376	1.5591	1.5794	1.5988	1.6355	1.6697	1.7021
2800 (684.96)	Sh			15.04	65.04	115.04	165.04	215.04	265.04	315.04	365.04	415.04	465.04	515.04	615.04	715.04	815.04
	v	0.03134	0.1030	0.1278	0.1685	0.1952	0.2168	0.2358	0.2531	0.2693	0.2845	0.2991	0.3132	0.3268	0.3532	0.3785	0.4030
	h	770.69	1055.8	1121.2	1220.6	1282.2	1330.7	1372.8	1411.2	1447.2	1481.6	1514.8	1547.3	1579.3	1642.2	1704.5	1766.5
	s	0.9468	1.1958	1.2527	1.3368	1.3867	1.4245	1.4561	1.4838	1.5089	1.5321	1.5537	1.5742	1.5938	1.6306	1.6651	1.6975
2900 (690.22)	Sh			9.78	59.78	109.78	159.78	209.78	259.78	309.78	359.78	409.78	459.78	509.78	609.78	709.78	809.78
	v	0.03262	0.0942	0.1138	0.1581	0.1853	0.2068	0.2256	0.2427	0.2585	0.2734	0.2877	0.3014	0.3147	0.3403	0.3649	0.3887
	h	785.13	1039.8	1095.3	1209.6	1274.7	1324.9	1368.0	1407.2	1443.7	1478.5	1512.1	1544.9	1577.0	1640.4	1703.0	1765.2
	s	0.9588	1.1803	1.2283	1.3251	1.3780	1.4171	1.4494	1.4777	1.5032	1.5266	1.5485	1.5692	1.5889	1.6259	1.6605	1.6931
3000 (695.33)	Sh			4.67	54.67	104.67	154.67	204.67	254.67	304.67	354.67	404.67	454.67	504.67	604.67	704.67	804.67
	v	0.03428	0.0850	0.0982	0.1483	0.1759	0.1975	0.2161	0.2329	0.2484	0.2630	0.2770	0.2904	0.3033	0.3282	0.3522	0.3753
	h	801.84	1020.3	1060.5	1197.9	1267.0	1319.0	1363.2	1403.1	1440.2	1475.4	1509.4	1542.4	1574.8	1638.5	1701.4	1763.8
	s	0.9728	1.1619	1.1966	1.3131	1.3692	1.4097	1.4429	1.4717	1.4976	1.5213	1.5434	1.5642	1.5841	1.6214	1.6561	1.6888
3100 (700.28)	Sh				49.72	99.72	149.72	199.72	249.72	299.72	349.72	399.72	449.72	499.72	599.72	699.72	799.72
	v	0.03681	0.0745		0.1389	0.1671	0.1887	0.2071	0.2237	0.2390	0.2533	0.2670	0.2800	0.2927	0.3170	0.3403	0.3628
	h	823.97	993.3		1185.4	1259.1	1313.0	1358.4	1399.0	1436.7	1472.3	1506.6	1539.9	1572.6	1636.7	1699.8	1762.5
	s	0.9914	1.1373		1.3007	1.3604	1.4024	1.4364	1.4658	1.4920	1.5161	1.5384	1.5594	1.5794	1.6169	1.6518	1.6846
3200 (705.08)	Sh				44.92	94.92	144.92	194.92	244.92	294.92	344.92	394.92	444.92	494.92	594.92	694.92	794.92
	v	0.04472	0.0566		0.1300	0.1588	0.1804	0.1987	0.2151	0.2301	0.2442	0.2576	0.2704	0.2827	0.3065	0.3291	0.3510
	h	875.54	931.6		1172.3	1250.9	1306.9	1353.4	1394.9	1433.1	1469.2	1503.8	1537.4	1570.3	1634.8	1698.3	1761.2
	s	1.0351	1.0832		1.2877	1.3515	1.3951	1.4300	1.4600	1.4866	1.5110	1.5335	1.5547	1.5749	1.6126	1.6477	1.6806
3300	Sh																
	v				0.1213	0.1510	0.1727	0.1908	0.2070	0.2218	0.2357	0.2488	0.2613	0.2734	0.2966	0.3187	0.3400
	h				1158.2	1242.5	1300.7	1348.4	1390.7	1429.5	1466.1	1501.0	1534.9	1568.1	1632.9	1696.7	1759.9
	s				1.2742	1.3425	1.3879	1.4237	1.4542	1.4813	1.5059	1.5287	1.5501	1.5704	1.6084	1.6436	1.6767
3400	Sh																
	v				0.1129	0.1435	0.1653	0.1834	0.1994	0.2140	0.2276	0.2405	0.2528	0.2646	0.2872	0.3088	0.3296
	h				1143.2	1233.7	1294.3	1343.4	1386.4	1425.9	1462.9	1498.3	1532.4	1565.8	1631.1	1695.1	1758.5
	s				1.2600	1.3334	1.3807	1.4174	1.4486	1.4761	1.5010	1.5240	1.5456	1.5660	1.6042	1.6396	1.6728
3500	Sh																
	v				0.1048	0.1364	0.1583	0.1764	0.1922	0.2066	0.2200	0.2326	0.2447	0.2563	0.2784	0.2995	0.3198
	h				1127.1	1224.6	1287.8	1338.2	1382.2	1422.2	1459.7	1495.5	1529.9	1563.6	1629.2	1693.6	1757.2
	s				1.2450	1.3242	1.3734	1.4112	1.4430	1.4709	1.4962	1.5194	1.5412	1.5618	1.6002	1.6358	1.6691
3600	Sh																
	v				0.0966	0.1296	0.1517	0.1697	0.1854	0.1996	0.2128	0.2252	0.2371	0.2485	0.2702	0.2908	0.3106
	h				1108.6	1215.3	1281.2	1333.0	1377.9	1418.6	1456.5	1492.6	1527.4	1561.3	1627.3	1692.0	1755.9
	s				1.2281	1.3148	1.3662	1.4050	1.4374	1.4658	1.4914	1.5149	1.5369	1.5576	1.5962	1.6320	1.6654
3800	Sh																
	v				0.0799	0.1169	0.1395	0.1574	0.1729	0.1868	0.1996	0.2116	0.2231	0.2340	0.2549	0.2746	0.2936
	h				1064.2	1195.5	1267.6	1322.4	1369.1	1411.2	1450.1	1487.0	1522.4	1556.8	1623.6	1688.9	1753.2
	s				1.1888	1.2955	1.3517	1.3928	1.4265	1.4558	1.4821	1.5061	1.5284	1.5495	1.5886	1.6247	1.6584
4000	Sh																
	v				0.0631	0.1052	0.1284	0.1463	0.1616	0.1752	0.1877	0.1994	0.2105	0.2210	0.2411	0.2601	0.2783
	h				1007.4	1174.3	1253.4	1311.6	1360.2	1403.6	1443.6	1481.3	1517.3	1552.2	1619.8	1685.7	1750.6
	s				1.1396	1.2754	1.3371	1.3807	1.4158	1.4461	1.4730	1.4976	1.5203	1.5417	1.5812	1.6177	1.6516
4200	Sh																
	v				0.0498	0.0945	0.1183	0.1362	0.1513	0.1647	0.1769	0.1883	0.1991	0.2093	0.2287	0.2470	0.2645
	h				950.1	1151.6	1238.6	1300.4	1351.2	1396.0	1437.1	1475.5	1512.2	1547.6	1616.1	1682.6	1748.0
	s				1.0905	1.2544	1.3223	1.3686	1.4053	1.4366	1.4642	1.4893	1.5124	1.5341	1.5742	1.6109	1.6452
4400	Sh																
	v				0.0421	0.0846	0.1090	0.1270	0.1420	0.1552	0.1671	0.1782	0.1887	0.1986	0.2174	0.2351	0.2519
	h				909.5	1127.3	1223.3	1289.0	1342.0	1388.3	1430.4	1469.7	1507.1	1543.0	1612.3	1679.4	1745.3
	s				1.0556	1.2325	1.3073	1.3566	1.3949	1.4272	1.4556	1.4812	1.5048	1.5268	1.5673	1.6044	1.6389

Sh = superheat, °F
v = specific volume, cu ft per lb

h = enthalpy, Btu per lb
s = entropy, Btu per °R per lb

APPENDIX III

A. Legendre transforms

Although the defined thermodynamic properties H, A, and G were established rather arbitrarily, they are related to the fundamental properties in an intriguing mathematical way. We shall now show that these defined properties are in fact mathematical transforms of the fundamental property U.

Suppose that $y = y(x)$ is represented by the curve

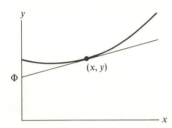

Now the tangent to the curve at point (x, y) is described by the equation

$$y = [M]x + \Phi, \tag{a}$$

where $[M]$ is the slope of the tangent at x, y (that is, dy/dx), and Φ is its intercept on the y-axis. The value of y at any x may be found if the corresponding $[M]$ and Φ values are given. In other words, $y = y(x)$ can also be represented by Eq. (a), which is the family of lines tangent to the function.

Equation (a) can obviously be rearranged to

$$\Phi = y - [M]x. \tag{b}$$

The intercept Φ is now called the *Legendre transform of y.*

Now let us apply this transformation-of-variables idea to the accumulated internal energy of a pure homogeneous system which is free of internal constraints and which does expansion work only. (Such a system is sometimes called a *simple system.*) For this system,

$$U = U(S, V),$$

which, we know from Section 6, is a function represented by a surface in a U-S-V coordinate framework. At any constant volume, the surface intersects the constant-volume plane in a curve given by the function

$$U = U_V(S). \tag{d}$$

It follows from Eqs. (a) and (b) that the Legendre transform of Eq. (d) is

$$\Phi_V = U - [M]_V S. \tag{e}$$

428

But $[M]$ is the slope of the $U(S)$ function in a constant V plane or

$$[M]_V = \left(\frac{\partial U}{\partial S}\right)_V = T. \qquad \text{(From Eq. 6–10)}$$

Therefore Eq. (e) becomes

$$\Phi_V = U - TS \equiv A. \qquad \text{(From Eq. 3–34)}$$

Similarly, if we consider the variation of U along a constant-S plane,

$$U = U_S(V),$$

and from Eqs. (a) and (b), we obtain

$$\Phi_S = U - [M]_S V. \qquad \text{(f)}$$

But

$$[M]_S = \left(\frac{\partial U}{\partial V}\right)_S = -P. \qquad \text{(From Eq. 6–11)}$$

Therefore Eq. (f) becomes

$$\Phi_S = U + PV \equiv H. \qquad \text{(From Eq. 2–16)}$$

Lastly, if we have a function of two variables,

$$y = y(x, z),$$

the surface formed by this function can be represented by the family of planes tangent to the surface. The equation of these planes is

$$y = \Phi + [M]x + [N]z, \qquad \text{(g)}$$

where Φ is now the intercept of a plane with the z-axis and $[M]$ and $[N]$ are the slopes of the plane in the x- and z-directions. That is,

$$[M] = \left(\frac{\partial y}{\partial x}\right)_z; \qquad [N] = \left(\frac{\partial y}{\partial z}\right)_x. \qquad \text{(h)}$$

It follows that

$$\Phi = y - [M]x - [N]z \qquad \text{(i)}$$

is the Legendre transform of $y(x, z)$.

Applying Eqs. (g), (h), and (i) to (c), we find that the Legendre transform of $U(S, V)$ is

$$\Phi = U - \left(\frac{\partial U}{\partial S}\right)_V S - \left(\frac{\partial U}{\partial V}\right)_S V$$

$$= U - TS + PV \equiv G.$$

$$\text{(From Eqs. 3–22, 6–10, 6–11)}$$

To summarize: The defined properties of a simple system are actually the Legendre transforms of the internal energy:

$$\Phi = U - TS + PV \equiv G,$$

$$\Phi_S = U \qquad + PV \equiv H,$$

$$\Phi_V = U - TS \qquad \equiv A.$$

B. Euler's theorem for homogeneous functions

The derivation of Eq. (9–66) from Eq. (9–53) is more elegantly achieved by using Euler's theorem for homogeneous functions.

A function $f(x, y, z)$ is said to be *homogeneous of degree m* if the function is multiplied by λ^m whenever x, y, z are multiplied by λ. Thus

$$f(\lambda x, \lambda y, \lambda z) = \lambda^m f(x, y, z). \tag{a}$$

For example, the internal-energy function

$$U = U(S, V, n_i) \tag{b}$$

is a homogeneous function of the first degree, because U becomes $\lambda^1 U$ when S, V, and n_i are increased by a factor of λ. In fact, the characteristic of homogeneity of the first degree with respect to mole numbers is the mathematical definition of what we call an *extensive* property.

Now Euler's theorem states that

$$x\frac{\partial f}{\partial x} + y\frac{\partial f}{\partial y} + z\frac{\partial f}{\partial z} = mf(x, y, z) \tag{c}$$

for a homogeneous function of degree m.

If we apply Eq. (c) to Eq. (b), remembering that $m = 1$, we obtain

$$S\left(\frac{\partial U}{\partial S}\right)_{V, n_i} + V\left(\frac{\partial U}{\partial V}\right)_{S, n_i} + \sum n_i \left(\frac{\partial U}{\partial n_i}\right)_{S, V, n_i} = U, \tag{d}$$

and from Eqs. (6–10), (6–11), and (9–63), we see that Eq. (d) is

$$TS - PV + \sum n_i \mu_i = U. \tag{9–66}$$

Q.E.D.

Proof of the Euler theorem can be found in J. W. Mellor, *Higher Mathematics for Students of Physics and Chemistry*, New York: Dover, 1955, page 75, and in other standard textbooks of advanced mathematics.

INDEX

Activity, 294, 301, 303, 328, 385, 386, 388
  of liquids, 302
  of solids, 302
Activity coefficient, 328, 341, 380
Adiabatic process, 35, 45
  chemical, 100
  reversible, 47
Air conditioners, 178
Air liquefaction, 181
Allowed energy, 189
Amagat's law, 309, 310
Arbeit function, 150, 204
Automobile (gasoline) engine, 168
Available work energy, 157
Avogadro's number, 203, 218
Azeotropes, 342, 343

Benedict-Webb-Rubin equation, 260, 294
Berthelot equation of state, 259
Bias-free distribution, 194, 195
Boltzmann constant, 203, 213
Boltzmann factor, 202
Bose-Einstein statistics, 219
Bosons, 219
Boyle temperature, 256
Brayton cycle, 174
Bubble point, 314

Carathéodory, C., 131
Carnot, Lazare, 18, 115
  Sadi, 115
Carnot cycle, 118, 121, 136
Carnot engine, 121
Chemical equilibrium, 356, 357
Chemical potential, 218, 323, 337, 338, 341, 345, 347, 413
Chemical-reaction process, 357, 359
Chemical work, 322, 336
Clapeyron equation, 242, 288
Clapeyron-Clausius equation, 243

Clausius, Rudolf, 113, 128
Clausius inequality, 158, 211
Clausius' theorem, 129
Closed system, 24
Closed-cycle engine, 168
Coefficient of compressibility, 246
  of performance, 123
  of thermal expansion, 246
Combustion, external, 168
  internal, 168
Compressibility, coefficient of volumetric, 246
Compressibility factor, 256, 262, 324
Compressible flow, 72
Compression ratio, 170
Constant-pressure process, 45
Constant-temperature process, 46
Continuity equation, 65
Conversion factors, 21
COP, 123
Corresponding states, law of, 261
Coupled flows, 401
Coupling coefficients, 403
Coupling mass, heat flows, 411
Critical point table, 258, 262
Critical temperature table, 58, 262
Cycles, power, 167
  refrigeration, 167, 176, 177
Cyclical process, 56

Dalton's law, 309, 310
Daniell cell, 389, 390
Degeneracy, 192
Degrees of freedom, 191, 347
Departure functions, 253, 274
  generalized equation of state, 277
Diesel, Rudolf, 167, 170
Diesel engine, 170
Dieterici equation of state, 259

Direct, conjugate coefficients, 403
Distribution function, 195

Efficiency, 19, 123
    and temperature, 124
    thermal, maximum, 121, 123
Electrochemical cell, 388
Endothermic reaction, 76
Energy, accumulated, 8–11
    allowed, 189
    available work, 157
    conservation, 27
    external, 8
    internal, 8, 9, 10, 30, 189
    kinetic, 8
    potential, 9
    transitory, 11
Energy states, 192
Energy units, conversion factors, 21
    closed-cycle, 168
    four-stroke, 169
    open-cycle, 168
    steam, 171
    2T, 123
Ensemble, 222
Ensemble average, 222, 223
Enthalpy, 33
    of atomization, 95
    of chemical reaction, 78, 90
    of combustion; standard, table, 86, 87–89
    of formation; standard, table, 79, 80–85
    of formation of atoms, table, 96
    partial molar, nonideal binary solution, 334
    of reaction, 78
    of solution, 95
    of solution; standard, table, 80–85
Enthalpy departure functions, 272
Entropy, 114, 128, 210–212
    absolute, 141, 367
    absolute and molecular structure, 142
    absolute, table, 364, 365, 368
    of coupling, 407, 412
    and integrating factors, 131
    of mixing, 314, 315, 317
    of phase change, 141
    of vaporization, 141
Entropy departure, 272

Entropy increase, irreversible process, 137
Equal *a priori* probability, 225
Equation of state, 250, 256–288
Equilibrium, 151, 154
    mass and thermal, 217
Equilibrium chemical reaction, 356, 357
Equilibrium constant, chemical, 362, 373, 376, 380, 383, 386
Equilibrium constant, versus temperature chart, 378
Equilibrium criteria, 156
Equilibrium state, 16, 152
Ergodic, quasi-ergodic hypothesis, 222
Euler criterion for integrability, 241
Euler's theorem, homogeneous functions, 430
Exact and inexact differentials, 50, 131
Exothermic reaction, 76
Expansion coefficients, 246
Expectation value, 192, 194, 208
Extensive properties, 16, 330, 430

Fermi-Dirac distributions, 219
Fermions, 219
Fick's law, 402
First law of thermodynamics, 2, 7, 27
First and second laws, combined, 237
Flame temperature, 100
Flow process, 63
    non-steady, 74
Flow work, 64
Fourier's law, 402
Free energy
    of formation, table, 364, 365
    Gibbs, 143, 218, 356
    Helmholtz, 150, 204
    of mixing, 317, 319
    partial molar, 219
    of reaction, standard, 357, 359, 367
Free expansion, 61, 137
Friction, fluid, 73
Fuel cell, 392
Fugacity, 294, 303, 360–362
    effect of pressure on, 299
    of liquids and solids, 298
    real gas solutions, 326
Fugacity coefficient, 296, 297, 303
Fugacity ratio, 296, 297

Gas chromatograph, 350
Gaussian distribution, 215
Gedanken apparat, 117, 357
Generalized equation of state, 261, 295
   improved, 263
Gibbs, J. Willard, 143, 307, 398
Gibbs equation, 237
Gibbs free energy; function, 143
Gibbs-Duhem equation, 339
Gram-force, 9

Heat, 11, 12, 21, 32
   of coupling, 407, 412
   of transport, 407
Heat capacity (*see* Specific heat)
Heat engine, 114, 116
Heat pump, 122, 178
Heat reservoirs, 115
Helmholtz free energy; function, 150, 204
Helmholtz function, of mixing, 317
Henry's law, 326, 341
Heterogeneous reactions, 384
Heterogeneous system, 345
Homogeneity of the first degree, 430
Homogeneous functions, Euler's theorem, 430
Homogeneous reactions, liquid solid, 382

Ideal gas, 42, 190
Ideal-gas law constant, 43
Ideal solutions, 312
   real materials, 324, 325
Indeterminate multiplicrs, Lagrange's method, 196
Indicator diagram, 170
Information, energy exchange, 75
Information theory, 195, 223, 226
Intensive properties, 16, 330
Inversion temperature, Joule-Thomson, 182
Irreversible processes, 61, 116, 138, 400
Irreversible thermodynamics, 3, 400
Isentropic process, 132
Isolated system, 24, 192
Isothermal expansion, 19

Jacobians, 244–251, 269, 271
Jet engine, 70

Joule, James Prescott, 7, 22
Joule cycle, 174
Joule-Thomson coefficient, 182, 252

K factors, 327
Kelvin (William Thomson), 399, 405
Kinetic energy, 8
Knudsen effect, 411
Kopp's law, 288

Lagrangian multiplier, 196
Legendre transforms, 428
Lewis, G. N., 293, 294
Liquids and solids, thermodynamic properties of, 283
Lost work, 158

Macroscopic, microscopic, 3, 188
Mass unit, 9
Maximum work, 19
Maxwell relations, 239–241, 244, 245
Maxwell's distribution, 213
Minimum work, 20
   of separation, 323
Mixing
   entropy, 314–317
   free energy, 317–319
Mixing processes, 331, 356
Mollier diagram, 283, 284
Multiphase systems, 345

Nernst, Walther, 141
Nernst equation, 389
Nonideal gas, 56, 57
Nuclear energy, 75

Ohm's law, 402
Onsager reciprocal relation, 404
Open system, 24, 63, 215, 336
Otto cycle, air-standard, 168

Partial molar properties, 329, 332, 334, 335
Partition function, 199, 206
   multiparticle, 207
   single-particle, 204, 206, 207
Peltier effect, 408
Phase rule, 346
Phase space, 222

Pipeline flow, 67
Planck's constant, 189–191, 213
Pound-force, 9
Pound-mass, 9
Power, 21, 167
Poynting effect, 149, 299
Pressure, 203
    expected, 208
    partial, 309
Probability, 192, 204
Property, 56
Pseudo-critical properties, 324

Quality, steam, 109
Quantum mechanics, 189, 205
Quantum number, 189
    rotational, 191
    vibrational, 190
Quantum state, 190
    multiparticle, 191
Quasi-static processes, 76

Ramjet, 175
Rankine cycle, 171
Raoult's law, 312, 326, 341
Real gases, equilibrium constant, 380
Redlich-Kwong equation of state, 259
Reduced properties, 261, 262
Refrigerant, 176
Refrigeration
    absorption, 179
    steam jet, 181
    tons capacity, 176
    cycle, 121, 176
Refrigerator, 122
Reversibility, 18, 115
Reversible compression, 20
Reversible expansion, 119
Reversible processes, 116
Reversible work, 18, 20, 158, 209
    ideal-gas system, 44
Rumford, Count (Benjamin Thompson), 22

Saturated liquid, 57
Saturated vapor, 57
Schrödinger equation, 189, 221
Second law of thermodynamics, 2, 114
Separation, minimum work of, 323

Shaft work, 63
Sign convention; work, heat, 28
Simple system, 424
Simultaneous reactions, 385
Slug, unit, 9
Solutions
    ideal, 308, 312, 325
    real liquids, 326
Specific heat
    constant P, 38, 245
    constant V, 38, 245
    ratio of, 245
    table, 37, 245, 416
Spontaneous process, 115
State principle, 29, 237
State property, 30
Statistical mechanics, 205
Statistical thermodynamics, 3, 188
Steady-flow processes, 65, 66
Steady state, 400
Steam, 56, 414
    dry, 173
    wet, 173
Steam engines, 171
Steam quality, 109, 173
Steam table, 283, 414
Stirling's approximation, 225, 315
Stoichiometric, 98
Supercharging, 183
Supercritical temperature, 58
Superheated vapor, 58
Superheaters, 172
Systems: open; closed; isolated, 23, 24, 192, 215, 218

Temperature
    absolute, 5
    thermodynamic scale of, 126
    units and conversions, 5
Temperature-entropy diagram, 136
Thermochemistry, 76
Thermocouple, 403
Thermodynamic properties, 29, 50
Thermodynamic state, 29
Thermodynamic surfaces, 50, 239
Thermoelectric power, 407
Third law of thermodynamics, 141
Thompson, Benjamin, 22

Thomson coefficient; heat, 410
Thomson effect, 409
Throttling, adiabatic, 73
Throttling calorimeter, 288
Time-average property, 222
Tons, refrigeration capacity, 176
Total differential, 239, 329
Transpiration, thermal, 411
Triple point, 5
Trouton's law, 141, 288
Turbine, adiabatic reversible, 69

Uncertainty, 195
Uncertainty principle, Heisenberg, 194
Universal gas constant, 42, 43

van der Waals, 255
van der Waals equation of state, 258, 294
van Laar equation, 342
van't Hoff equation, 355, 377

Vaporization equilibrium constant, K factor, 327
Variance, 231, 347
Venturi meter, 68
Virial equations of state, 261
Voltage, standard electrode, 390, 391
Volume, partial molar, 332

Work, 12
  available, 157
  chemical, 322
  elastic, 12, 13
  electrical, 14
  lost, 158, 159, 211
  other, 143, 322, 323
  shaft, 63
  surface, 15
Work functions (see Free energy), 142, 150
Work functions and equilibrium, 154

Zeroth law, 3

ANSWERS TO SELECTED EXERCISES

Section 1

Exercise 1–1. Add 11.1% to $\frac{1}{2}$ of (°F − 32). Adding 10% produces an error of 1% which may be acceptable for rapid estimation.

Section 2

Exercise 2–1. $W_{\text{earth}} = 20{,}000$ ft-lb$_f$; $W_{\text{moon}} = 3{,}333$ ft-lb$_f$.

Exercise 2–2.

$$W_{1-b-2} = P_1 \int_1^b dV + \int_b^2 P(0) = P_1(V_2 - V_1),$$

$$W_{1-c-2} = K \int_1^2 (-P)\frac{dP}{P^2} = K \ln \frac{P_1}{P_2}.$$

Exercise 2–3. Maximum work output in moving from state (1) to (2) = reversible work = (−)minimum work needed to move from (2) back to (1).
Reversible process: Infinitesimal rate; no heat, noise, accelerations, oscillations or other energy-dissipating or frictional effects.
Irreversible process: Finite rate; energy is dissipated in friction, deceleration, noise, heat etc.

Exercise 2–4. $216{,}000$ ft-lb$_f$ = 277.5 Btu = 69,930 cal = 292,760 joules = 0.0813 kW-hr = 3.25×10^{-9} g-rest mass.

Exercise 2–5.
i) $W = 9.632$ ft-lb$_f$ (work done against atmosphere).
ii) 16.2 psia; 1.5 psia.
iii) 10.62 ft-lb$_f$; 0.98 ft-lb$_f$.
iv) Reduce piston area. Cylinder pressure will be higher.

Exercise 2–6. A 67.2 ft-lb$_f$; B 67.5 ft-lb$_f$; C 68.2 ft-lb$_f$.

Exercise 2–7. Fig. 2–8(c).

Exercise 2–8. If the machines in Fig. 2–9 are always out of balance, they will operate. Are they then out of balance?

It does not suffice to say that the machines do not operate because they would then violate the First Law. The First Law is one of the rules we laid down in establishing the game of Thermodynamics, and the machines did *not* participate in making up the game.

The Escher frontispiece is another matter, understood by students of perception and perspective. The artist has intentionally violated the rules of perspective.

Exercise 2–9. Neglecting air friction, loss of fluid from splash, and assuming that ΔE_P is converted into heat, we get:

System	W	Q, cal	ΔU, cal	ΔE_P, cal
a) Ball only	0	−1382	0	−1382
b) Wine only	0	+1382	1382	0
c) Wine and ball	0	0	1382	−1382

Exercise 2–10. Energy dimensions: force × length.

Exercise 2–11. $Q_V < Q_P$.

Exercise 2–13.

$$\Delta U = U_C - (U_A + U_B) = Q - W = -30{,}000 - P(V_C - (V_A + V_B)) \le -29{,}000 \text{ cal}$$
$$U_C \ll (U_A + U_B)$$
$$H_C \ll (U_A + U_B)$$
$$|Q_V| < |Q_P|.$$

Exercise 2–15.

$$1 \, \frac{\text{Btu}}{\text{lb-mol} \cdot {}^\circ\text{F}} = 1 \, \frac{\text{cal}}{\text{g-mol} \cdot {}^\circ\text{K}}.$$

Exercise 2–16.

$$C_P \text{ (liquid or solid)} \approx C_V \text{ (liquid or solid)}$$

because

$$C_P = \left(\frac{\partial \overline{H}}{\partial T}\right)_P = \left(\frac{\partial \overline{U}}{\partial T}\right)_P + P\left(\frac{\partial V}{\partial T}\right)_P$$

and $(\partial V / \partial T)_P$ is very small for liquids and solids as compared to gases; and

$$\left(\frac{\partial \overline{U}}{\partial T}\right)_P \approx \left(\frac{\partial \overline{U}}{\partial T}\right)_V.$$

Exercise 2–17. $\Delta \overline{U}_{\text{vap}} = 970.3 - 71.7 = 898.6 \text{ Btu/lb}_m.$

Exercise 2–18. $\Delta \overline{H}_{\text{fusion}} < \Delta U_{\text{fusion}}; (\Delta U_{\text{fusion}} = U_{\text{liquid}} - U_{\text{solid}}).$

Exercise 2–19.

$$\rho_i = \frac{P}{RT} \cdot (\text{MW}_i); \qquad \overline{V}_{70,\text{/atm}} = \frac{RT}{P}.$$

Exercise 2–20.

$$C_P = \tfrac{7}{2} \cdot 1.987 \text{ cal/g-mol } {}^\circ\text{K} = 6.954 \text{ cal/g-mol } {}^\circ\text{K}.$$

The values in Table 2–3 are near 7.

Exercise 2–21. $P_2 = 3.79 \text{ atm}; W = 519{,}000 \text{ ft-lb}_f.$

Exercise 2–22. $d(y/x).$

Exercise 2–24. $292.7 \, {}^\circ\text{F}; \Delta U = 536.4 \text{ Btu}.$

Exercise 2–26.

a) Choosing $P_1 = 20$ psia, $T_1 = 400 \, {}^\circ\text{F}$, $\overline{V}_1 = 25.43 \text{ ft}^3$, we have

$$\overline{U}_1 = 1145.1 \text{ Btu} = \overline{U}_2 \qquad (\Delta U = 0)$$
$$\overline{V}_2 = 51.04$$
$$T_2 = 402 \, {}^\circ\text{F} \qquad (\Delta T \ne 0 \text{ for a real gas}).$$

b) $Q = 645 \text{ Btu}, \Delta U = 0, W = 645 \text{ Btu}.$

Exercise 2–29. $u_1 = 422.6\sqrt{P_1 - P_2}.$

Section 3

Exercise 3–2.
a) Process 1–2 cannot continue "indefinitely."
b) $(W/Q)_{1-2} = 1; (W/Q)_{2-3} = P_2(\Delta V)/\Delta U + P_2(\Delta V) < 1.$

Exercise 3–4.
b) $W_{cycle} < 0$.
c) COP may be greater or smaller than unity. A large COP is desirable.
d) $Q_{out}/W_{cycle} > 1$.
e) Heat pump will deliver 3000 watts *plus* the heat energy absorbed from the house exterior, to the interior. The resistance heater can deliver only 3000 W of heat energy. Thus the operating cost of the heat pump is less than that of the resistance heater. However, the initial cost of the heat pump is very much greater (by a factor of > 10) than that of the resistance heater. Therefore decision as to which to use will depend on capital available, amortization rate, etc.

Exercise 3–7. $\Delta S = 0.067$ Btu/°R; $T_2 = 368.1$ °F; $P_2 = 122.8$ psia.

Exercise 3–8.
a) $dG_T = V \, dP - dW_{other}$.
b) Isothermal $dW_{reversible} = P \, dV = -V \, dP = -dG_T$ (ideal gas).
c) Equation (3–25) is general. Equation (b) above applies only to an isothermal expansion-work system using an ideal gas.

Exercise 3–9. Equation (2–46).

Exercise 3–10. $G = A + PV$; $dA_{T,V} = 0$, for systems that do $P \, dV$ work only.

Exercise 3–11.
a) $\Delta S_{universe} = [(-79.8 \times 20)/273] + 1 \times 100 \ln 289/273 = -0.08$ cal/°K.
b) No.
c) Yes. Heat is flowing "uphill."
d) No!

Exercise 3–12.
a) $\Delta S_{universe} = (1/T)(\overset{\text{system}}{\Delta H_{vap}} - \overset{\text{surr.}}{\Delta H_{vap}}) = 0$. T and P are constant.
b) $\Delta G_{T,P} = 0 = \Delta H_{vap} - T\Delta S_{vap} = \Delta H_{vap} - T(\Delta H_{vap}/T)$.

Exercise 3–14. -12.5 Btu; $+12.5$ °F.

Exercise 3–16. $dS_{adiabatic} = d$ (lost work) ≥ 0.

Section 4

Exercise 4–1.
i) Yes.
ii) No!

iii) $\eta_{Carnot} - \eta_{Otto} = \left(1 - \dfrac{T_6}{T_4}\right) - \left(1 - \dfrac{T_5 - T_6}{T_4 - T_3}\right) = \left(\dfrac{V_2}{V_3}\right)^{1-\gamma} \times \left(1 - \dfrac{T_3}{T_4}\right) > 0.$

Exercise 4–2. Increasing compression ratio increases efficiency.

$$\eta = \frac{Q_{3-4} + Q_{5-6}}{Q_{3-4}} = 1 - \left(\frac{T_5 - T_6}{T_4 - T_3}\right), \tag{4-1}$$

$$\eta = 1 - \left(\frac{T_5 - T_6}{T_4 - T_3}\right) = 1 - \left(\frac{T_5 - T_2}{T_4 - T_3}\right).$$

But

$$\frac{T_2}{T_3} = \left(\frac{V_2}{V_3}\right)^{(1-\gamma)}; \qquad \frac{T_4}{T_5} = \left(\frac{V_4}{V_5}\right)^{(1-\gamma)} = \left(\frac{V_2}{V_3}\right)^{(\gamma-1)}.$$

Therefore

$$\eta = 1 - T_4(V_2/V_3)^{1-\gamma} - T_3(V_2/V_3)^{1-\gamma}/(T_4 - T_3)$$

or

$$\eta = 1 - r^{(1-\gamma)} \tag{4-3}$$

Exercise 4–4.

$$\eta_a = 0.204; \qquad \eta_b = 0.269.$$

Superheating improves the thermal efficiency of the cycle.

Exercise 4–5. Controls: temperature sensor in cold box which opens expansion valve as temperature rises and closes it as temperature drops; low pressure sensor in cooling coil or compressor inlet which actuates compressor when cooling coil pressure rises above certain limit and shuts down compressor when coil pressure drops below a certain limit; high-pressure sensor at compressor discharge which shuts down compressor if discharge pressure rises beyond safety limits of condenser.

Point s is compressed saturated vapor about to start condensing.

Exercise 4–6.
ii) For a cycle,

$$Q_{\text{cycle}} = W_{\text{cycle}} = Q_L - Q_H + Q_R - Q'_{\text{out}} = -W_s.$$

iii) No. The heat fall from T_R to T_H causes an entropy gain that is greater than the entropy loss inherent in the heat pumped from T_L to T_H (or provides more than the available work energy needed to pump heat from T_L to T_H.)

Exercise 4–7. 35°F; 4.2%.

Exercise 4–8. $-\Delta H = W_S$. This is in contrast to the throttling processes where $\Delta H = 0$. Does not work for an ideal gas because T is constant when H is constant. Thus $(\partial T/\partial P)_H = 0$ for ideal gas.

Section 5

Exercise 5–1. Allowed energy magnitudes and velocities are independent of temperature. We shall see that the *probability* of a particle having a given allowed velocity or energy does depend on temperature.

Exercise 5–2. See Table 5–1.

Exercise 5–3. $\ln \mathscr{P}_l = -\psi$. Therefore $\mathscr{P}_l = $ constant. But $\sum_{l=1}^{w} \mathscr{P}_l = 1 = w\mathscr{P}_l$.

Exercise 5–4. If $\mathscr{E}_l$ is constant, then (5–13) or (5–14) states that

$$\mathscr{P}_l = \text{constant} = \frac{1}{w}$$

$$S = -k \sum_{l=1}^{w} \mathscr{P}_l \ln \mathscr{P}_l = -k \sum_{1}^{w} \frac{1}{w} \ln \frac{1}{w} = k \ln w.$$

Degeneracy of energy state $\mathscr{E}_l = w$.

Exercise 5–5. The constraint $\sum_l \mathscr{P}_l n_l = \langle N \rangle$ gives rise to the additional Lagrangian multiplies α, in the maximization procedure.

Exercise 5–6.

$$S = -k \sum \mathscr{P}_l \ln \mathscr{P}_l = -k \sum \mathscr{P}_l(-\psi - \beta \mathscr{E}_l) = k(\psi + \beta U).$$

Exercise 5–7.

$$A \equiv U - TS = U - \frac{S}{k\beta} = U - \frac{k\psi - k\beta U}{k\beta} = \frac{-\psi}{\beta}.$$

Exercise 5–9. $w_6 = 7.$

Exercise 5–10.

i) $+0.41k.$

ii) $\Delta S = 0.$ Reversible work input.

iii) There are 10 states with $\mathscr{P}_l$ ranging from $\frac{1}{27}$ to $\frac{6}{27}$.

$$S_{iii} = -k \sum_{l=1}^{10} \mathscr{P}_l \ln \mathscr{P}_l.$$

Exercise 5–11. Note that the 1500 °K distribution is broader.

Exercise 5–13.

$$\frac{\partial U}{\partial T} = \frac{\partial U}{\partial \beta} \frac{\partial \beta}{\partial T} = \left(\frac{\partial^2 \Omega}{\partial \beta^2} \right) V k \beta^2.$$

Section 6

Exercise 6–1.

$$\left(\frac{\partial T}{\partial L} \right)_S = \left(\frac{\partial \tau}{\partial S} \right)_L.$$

Exercise 6–2.

$$\frac{[T, S]}{[L, S]} = - \frac{[\tau, L]}{[L, S]} \qquad \text{analog of (6–24a)}$$

Analogs of the Maxwell relations are generated by dividing (6–24a′) by each of the four Jacobean combinations of T and S with τ and L; that is, $[L, S]$, $[\tau, S]$, $[T, L]$, and $[T, \tau]$. Hence

$$\left(\frac{\partial T}{\partial L} \right)_S = \left(\frac{\partial \tau}{\partial S} \right)_L; \qquad \left(\frac{\partial T}{\partial \tau} \right)_S = - \left(\frac{\partial L}{\partial S} \right)_\tau$$

$$\left(\frac{\partial S}{\partial L} \right)_T = - \left(\frac{\partial \tau}{\partial T} \right)_L; \qquad \left(\frac{\partial S}{\partial \tau} \right)_T = \left(\frac{\partial L}{\partial T} \right)_\tau$$

Exercise 6–5. $(\partial T / \partial P)_S = \bar{V} \beta_T T / C_P.$ $N_2 = 0.078$, $H_2O = 1.37 \times 10^{-3}$ °K/atm, Pb = 1.6×10^{-3} °K/atm.

Problem 6–3. From Eq. (6–17),

$$\frac{dP}{dT} = \left(\frac{\Delta H}{\Delta V} \right)_{\text{fusion}} \cdot \frac{1}{T},$$

$$T = -0.17 \,°C.$$

Problem 6–4.

$$\left(\frac{\partial T}{\partial P} \right)_H = \frac{T}{C_P} \left[\left(\frac{\partial V}{\partial T} \right)_P - \frac{V}{T} \right].$$

For an ideal gas,

$$\left(\frac{\partial V}{\partial T} \right)_P = \frac{V}{T},$$

therefore

$$\left(\frac{\partial T}{\partial P} \right)_H = 0 \qquad \text{(ideal gas).}$$

Section 7

Exercise 7–2.

$$P = \frac{RT}{(\overline{V} - b)} - \frac{a}{\overline{V}^2},$$ (1)

$$\left(\frac{\partial P_c}{\partial \overline{V}_c}\right)_{T_c} = -\frac{RT}{(\overline{V} - b)^2} + \frac{2a}{\overline{V}^3} = 0,$$ (2)

$$\left(\frac{\partial^2 P_c}{\partial \overline{V}}\right)_{T_c} = +\frac{2RT}{(\overline{V} - b)^3} - \frac{6a}{\overline{V}^4} = 0.$$ (3)

Combine (1), (2), and (3) to obtain (7–4) and (7–5).

Exercise 7–3. Let

$$e^x = 1 + x + \frac{x^2}{2!} + \frac{x^3}{3!} + \cdots,$$

where $x = -a/RT\overline{V}$, to obtain Eq. (7–10).

Exercise 7–4.

i) Reduced properties are dimensionless. Critical properties have normal units of temperature, volume and pressure.

ii)

	CH_4	O_2	H_2O_{vap}	
T	229 °K	186 °K	776	$T_r = 1.2$
P	50.4 atm	55.2	240	$P_r = 1.1$
V	4.67 ft³/lb-mol	343	3.32	
iii) P_r	0.0218	0.0199	0.0045	$P = 1$ atm
T_r	1.96	2.4	0.576	$T = 212$ °F

Exercise 7–5. 9.4 atm, 138 psia, by iteration.

Exercise 7–6. 2052 cal/g-mol.

Exercise 7–7.

$$\Delta \overline{G} = RT_0 \ln \frac{P}{P_0}$$

Exercise 7–9.

$$V^* - V \approx 20 \text{ cm}^3/\text{g-mol} \quad (0.08\% \text{ error}).$$
$$H^* - H = 110 \text{ cal/g-mol}.$$
$$S^* - S = 0.0048 \text{ cal/°K g-mol}.$$

Exercise 7–10. -170 cal/g-mol CH_4.

Exercise 7–11. $+3290$ cal/g-mol.

Exercise 7–12. 3215 cal/g-mol (read from intersections of isotherm with saturation curve).

Exercise 7–13.

i) Trouton's rule prediction of ΔH_{vap} is 5% high.

ii) 785 cal/g (Trouton's). 790.3 (Steam table).

iii) Predicted: 0.212 cal/g · °K.

Exercise 7–14. 96.6%, 467°F.

Exercise 7–15. Evaporator pressure $= 23$ psia $= P_4$. Condenser temperature $= 120°F$. Liquid : vapor at (4) $= 1.5$.

Section 8

Exercise 8–1.

b) $\Delta G_T = \int_{\bar{V}_2}^{\bar{V}_1} \left[-\frac{RT}{\bar{V}} - \frac{2}{\bar{V}^2} \left(B_0 RT - A_0 - \frac{C_0}{T^2} \right) - \frac{3}{\bar{V}^3} (bRT - a) - \frac{6a\alpha}{\bar{V}^6} \right.$

$\left. - \frac{3c(1 + \gamma/\bar{V}^2)}{\bar{V}^3 T^2} e^{-\gamma/\bar{V}^2} - \frac{2\gamma c e^{-\gamma/\bar{V}^2}}{\bar{V}^5 T^2} - \frac{2\gamma c}{\bar{V}^5 T^2} e^{-\gamma/\bar{V}^2} \left(1 - \frac{\gamma}{\bar{V}^2} \right) \right] d\bar{V}.$

Exercise 8–2.

$$\Delta G_T = RT \ln \frac{f_2}{P_1} = 1.987 \times 323 \ln 57.26 = 2595 \text{ cal/g-mol},$$

$$\Delta G_T = \Delta H - T\Delta S = -1034 + 323 (11.21) = 2590 \text{ cal/g-mol}.$$

Agreement within 0.2% is surprisingly good.

Exercise 8–3. $f_{ice} = 0.0026$ atm; $f_{H_2O(100)} = 1$ atm; $f_{H_2O(300)} = 6.9$ atm.

Exercise 8–4. No. Solids and liquids normally occur at total pressures considerably above the vapor pressures corresponding to their temperatures. Their fugacites, however, must be the same as that of their equilibrium vapors.

Exercise 8–5. (i) 0.065 atm, (ii) 98 atm, (iii) 98 atm.

Exercise 8–6. 1.0; 0.8.

Exercise 8–7. 1.000, 1.0007, 1.077.

Exercise 8–8. 117 cal/g-mol.

Exercise 8–9. Yes. $(\partial \ln P/\partial T)_P = 0$.

Section 9

Exercise 9–1. Molecular weight = 30.3; total weight = 586 lb.

Exercise 9–2. 5/5; 252; 5.51k; 4.95k.

Exercise 9–3. -820 cal/g-mol.

Exercise 9–4.

$$\Delta \bar{G} \text{ (1 atm)} = -493 \text{ cal/g-mol};$$
$$\Delta \bar{G} \text{ (1000 atm)} = -489 \text{ cal/g-mol};$$
$$W = -\Delta G.$$

Exercise 9–6. G, S, A, composition.

$$\Delta S = -1.14 \text{ Btu/lb-mol K}°; \qquad \Delta \bar{G} = \Delta \bar{A} = 606 \text{ Btu/lb-mol}.$$

Exercise 9–7. 35.2 atm. Use Eqs. (9–26), (9–27), Fig. 7–3, and Eq. (7.13). P is found in one iteration starting with $z = 1$. $P = 35.2$ atm.

Exercise 9–8. $K = P_i/\mathbb{P}$. K will vary with composition only to the extent that P_i varies, which in turn depends on the extent that the boiling point of the solution changes with solution composition.

Exercise 9–9. (a) is true for $\hat{S}_i$, $\hat{G}_i$, and $\hat{A}_i$. Both (a) and (b) are true for $\hat{U}_i$ and $\hat{H}_i$.

Exercise 9–10. 0.00; 0.003 ft³/lb on basis of absolute volumes. -0.003; 0.00 if pure shot and balls are chosen as the reference state.

Exercise 9–11. $\hat{V}_{heptane} = 0.00$. $\hat{V}_i$ depends on reference state. In plotting, datum at $n_w = 10.24$ appears to be in error. $\hat{V}_{H_2O}$ approaches zero as $n > 10$.

Exercise 9–12. As n_{H_2O} increases, we are effectively adding water to water so that there is no thermal effect.

Exercise 9–14.

$\widehat{V}_a = -2.03$ cm^3/g-mol; $= -0.07$, at 25 mole-% and 75 mole-% alcohol.

$\widehat{V}_w = -0.47$ cm^3/g-mol; $= -2.40$, at 25 mole-% and 75 mole-% alcohol.

Exercise 9–13. $\widehat{H}_{NaOH} = -10{,}200$ cal/g-mol; $\widehat{H}_{H_2O} = 0.0$.

Exercise 9–17.

ii)
$$S\,dT - V\,dP + \sum_i n_i\,d\mu_i - X_i\,dF_i = 0.$$

Exercise 9–18. If $p_1 = P_1 x_1$, then

$$x_1 \frac{d \ln p_1}{dx_1} = 1 \quad \text{and} \quad d \ln p_2 = d \ln x_2.$$

Therefore,

$$\ln p_2 = \ln x_2 + \ln K.$$

Exercise 9–20. $F = 2 + 2 - 2 = 2$. Pressure and composition remain independent of temperature when only one (liquid) phase is present. As soon as vapor appears F drops from 3 to 2. Therefore, if both temperature and pressure are controlled, *composition* of the liquid phase and the vapor phase will depend on the temperature. Composition of the liquid phase is found by solving for $x_{(benzene)}$ in the equation $x_B P_B + (1 - x_B)P_T = \mathbb{P}$, where P_B and P_T are the vapor pressures of the pure components at the system temperatures. Vapor compositions are found by using Eqs. (9–5) and (9–9).

Exercise 9–21. $F_{(initial)} = 11$; (a) 10, (b) 8, (c) 7, (d) 5, (e) 4, (f) 3.

Section 10

Exercise 10–1. $\Delta G_R^0 > 0$. Process (Pure $B \to$ Pure D) requires a work input. Therefore it does not occur spontaneously.

Exercise 10–2. $\Delta G = aRT \ln p_{Ae} \neq \Delta G_R^0$.

Exercise 10–3. (Pressure units) $^{c+d-a-b}$. No. We can only take logarithms of pure positive numbers. $1/K_{initial}$.

Exercise 10–4. -8329 cal/g-mol; 1.26×10^6.

Exercise 10–5. Total moles in equilibrium reaction chamber is $(1.5 - 0.5x)$. Microscopically smaller amount of CO_2 will be formed.

Exercise 10–6.

$$\Delta S_R^{0T} = \Delta S_R^0 + \sum_{reactants} n_j \int_T^{298} C_{P_j} \frac{dT}{T} + \sum_{products} n_i \int_{298}^T C_{P_i} \frac{dT}{T}$$

Exercise 10–7.

i) $\Delta G_{NO_2} = 2715$ cal/g-mol.

ii) $\Delta G_R = \Delta G_{NO_2} + \Delta G_{N_2O_4} = 2715 - 62.7 = 2652$.

iii) $\Delta G_R = \Delta G_R^0 + RT \ln \frac{3}{2}(3)^{1/2} = -60{,}886$ cal.

Exercise 10–8. From slope $\Delta H_R^0 = -67{,}600$. From table $\Delta H_R^0 = -67{,}636$.

Exercise 10–9. 0.0263 (mm Hg)$^{-1}$. No.

Exercise 10–10. (i) Pure CaO. (ii) $\Delta G_R^{0(1180)} = 0$, $K = 1$.

Exercise 10–11. (i) O_2 cathode; H_2 anode, (ii) V^0 1.23 volts. $\Delta G_R^0 = -56.690 = -W_{max}$.

iii)
$$\mathbf{V} = \mathbf{V}^0 + \frac{RT}{\mathbf{z}\mathscr{F}} \ln (0.21)^{1/2} = 1.22 \text{ volts,}$$

$$\Delta G_R = -\mathbf{V}\mathbf{z}\mathscr{F} = -56,200 \text{ cal/g-mol.}$$

iv)
$$\mathbf{V} = \mathbf{V}^0 + \frac{RT}{\mathbf{z}\mathscr{F}} \ln 1000 = 2.119 \text{ volts,}$$

$$\Delta G = -97,800 \text{ cal.}$$

v) Heat exchanged during reversible cell operation is
$$T \Delta S_R^0 = \Delta G_R^0 - \Delta H_R^0 = +11,627 \text{ cal,}$$
whereas
$$\Delta H_R^0 = -68,317 \text{ cal/g-mol.}$$

Section 11

Exercise 11–1.
$$S_{universe} = Q \left(\frac{1}{T_C} - \frac{1}{T_H} \right).$$

Exercise 11–2.
$$S^* = \bar{V}\frac{dP}{dT} = \frac{\bar{V}P}{2T} = \frac{R}{2}.$$